FOOD SECURITY IN NUTRIENT-STRESSED ENVIRONMENTS: EXPLOITING PLANTS' GENETIC CAPABILITIES

Food Security in Nutrient-Stressed Environments: Exploiting Plants' Genetic Capabilities

Edited by

J.J. ADU-GYAMFI

JAPAN
Official Development Assistance

Japan International Research Center
for Agricultural Sciences (JIRCAS)
1-2 Ohwashi, Tsukuba, Ibaraki 305-8686, Japan

International Crops Research Institute
for the Semi-Arid Tropics (ICRISAT)
Patancheru 502 324, Andhra Pradesh, India

Partly reprinted from *Plant and Soil*, Volume 245, No. 1 (2002).

KLUWER ACADEMIC PUBLISHERS
DORDRECHT / BOSTON / LONDON

Library of Congress Cataloging-in-Publication Data

A C.I.P. Catalogue record for this book is available from the library of Congress.

ISBN 1-4020-0603-9

Published by Kluwer Academic Publishers,
P.O. Box 17, 3300 AA Dordrecht, The Netherlands

Sold and distributed in North, Central and South America
by Kluwer Academic Publishers,
101 Philip Drive, Norwell, MA 02061, U.S.A.

In all other countries, sold and distributed
by Kluwer Academic Publishers,
P.O. Box 322, 3300 AH Dordrecht, The Netherlands

Printed on acid-free paper

Printed in the Netherlands

Contents

FOOD SECURITY IN NUTRIENT-STRESSED ENVIRONMENTS:
EXPLOITING PLANTS' GENETIC CAPABILITIES

Summary and Recommendations of an International Workshop held on 27–30 September 1999, ICRISAT, Patancheru, India

J.J. Adu-Gyamfi (Ed.), Food security in nutrient-stressed environments: exploiting plants' genetic capabilities: 1–3, 2002.

Welcome address

L. D. Swindale
Director General, International Crops Research Institute for the Semi-Arid Tropics (ICRISAT), Patancheru 502 324, Andhra Pradesh, India

This International Workshop on 'Food Security in Nutrient-Stressed Environments: Exploiting Plants' Genetic Capabilities' is the culmination of 15 years' work in partnership between the International Crops Research Institute for the Semi-Arid Tropics (ICRISAT) the Government of Japan (GoJ) and various other associates.

In addition to its core contributions to ICRISAT – which have been considerable – the GoJ, through the Japan International Research Center for Agricultural Sciences (JIRCAS) initiated a Special Project with ICRISAT, in November 1984 entitled 'Development of Sustainable Cultivation of Upland Crops in the Semi-Arid Tropics (SAT)'.

The Project was conducted in three phases (I 1984–1989, II 1989–1994, and III 1995–1999). During that time a number of Japanese scientists came to work at ICRISAT. After they returned to Japan, I can assure you, they remained in very close touch with our Institute, both through their work, and through the Friends of ICRISAT Group. During its 15-year period the Project attempted to address the issues of food security and poverty alleviation by exploiting how genetic options can best be combined with management options to improve nutrient availability in SAT cropping systems, with the basic understanding that nutrients are limited in the soils of the region and fertilizer additions are not easy to acquire for various reasons.

In July 1984, I signed the MoU between the Tropical Agriculture Research Center, (TARC), now Japan International Research Center for Agricultural Sciences, Ministry of Agriculture, Forestry and Fisheries, Japan and ICRISAT. Today, I am very privileged to be here to witness the successful completion of the Project and hear about the results. Later in 1984, a team of Japanese scientists, Drs Ae, and Arihara (whom I am delighted to see here today), and Dr Okada joined ICRISAT to begin Phase I of the Project to study the method of soil management and pulses cultivation in the semi-arid tropics with emphasis on the adaptation of pulses to moisture and nutrient, especially phosphorus (P), stress.

We had known for a long time that pigeonpea was unresponsive to P and could use it in ways that other crops could not. We had tried to understand the mechanisms prior to this Project, and had made some progress. But the Project clarified that root exudates had a role to play and specifically for pigeonpea. That was the central thrust of Phase I – it was much debated – and led to experiments designed to test and prove the hypothesis wrong. So seldom do we follow Karl Popper's axiom that experiments should be designed to prove the hypothesis wrong – we are only too happy to prove ourselves right – but attempting to prove that we are wrong helps the discipline to advance.

Phase I proved there was a mechanism involving root exudates contributing to the availability of P in pigeonpea – this led to consideration of whether additional means to increase utilization of exudates could take place. Results are summarized in the ICRISAT publication, *Phosphorus Nutrition of Grain Legumes in the Semi-Arid Tropics*.

Phase I led to an FAO-ICRISAT Expert Consultancy Workshop on 'Genetic Manipulation of Crop Plants to Enhance Integrated Nutrient Management in cropping systems–1. Phosphorus,' 15–18 March 1994. The workshop attempted to establish a global consortium of researchers focusing their efforts on improving the ability of crop plants to acquire P, particularly through sources from which it is only sparingly available.

I attended a conference in Japan in early 1998 that brought together people from all over the world who were working on similar tropics. This conference on 'New Concepts of Plant Nutrient Acquisition' was held in Tsukuba, Japan, from 22 to 27 March 1998 to summarize the newly developing concepts on mechanisms of plants to acquire soil nutrients, and to apply them to practical agriculture. I also had the opportunity to make a contribution to a preconference symposium on identifying appropriate agricultural technology for Africa and possible research areas for collaboration with the Government of Japan. I would like to quote from my conclusions. "There is evidence of a small but growing number of successes in West Africa with conventional crop improvement research. Input-responsive and relatively stable cultivars of the major cereals, sorghum, maize, and millet, the main grain legumes, cowpea and groundnut, and the major root crop, cassava been produced by research institutes, released by governments, and are slowly reaching the hands of the farmers. But one can not avoid the conclusion that some inputs must be obtained and made available to farmers". More than 20 years ago FAO assembled the results of hundreds of fertilizer trials carried out in West Africa by the Freedom from Hunger Fertilizer Program that I led for several years. The results showed consistent improvements in yield from use of fertilizer – with the farmers' varieties of crops and landraces. Today with improved cultivars the responses are clearly greater. The obvious input is P, and numerous studies have concluded that P is needed by West African soils and that it can be used profitably by a range of crops in a range of situations. P is the nutrient most available in the region, most countries have deposits of rock phosphates – Togo gets its main foreign exchange by exporting rock phosphate to Europe. Deposits exist in Niger and Senegal where phosphoric acid is now produced to support agronomic productivity in Senegal and India. If N is the most important nutrient and it can not be obtained readily from fertilizers then agriculture based on legumes must be the appropriate response in the region. Legumes in rotation and intercropped can provide food, fodder, fuel, protection from the winds, and cash. They also provide N to subsequent crops.

Phase II scientists worked on roots themselves rather than exudates. The objectives were:

- To establish methods for quantifying soil–root interactions in various cropping systems.
- To study root activities, functions and N uptake by component crops.
- To study rooting behavior of pigeonpea in relation to waterlogging and N availability.
- To examine soil solutions to estimate the soil's capacity for N supply.
- To prepare the N balance sheet in pigeonpea-based cropping systems.
- To develop a working model of the rooting profile for pigeonpea in cropping systems.
- To compare the different management options of N fertilizer application for the enhancement of N utilization.

Considerable work was done. Project staff produced the JIRCAS publication *Roots and Nitrogen in Cropping Systems of the Semi-Arid Tropics*. Hopefully this will contribute to our knowledge and opportunities for genetic improvement.

In Phase III, the Project scope widened to include genetic manipulation. Scientists are aware that no improved cultivar has a reasonable chance of success unless the constraint of soil fertility is addressed through improvement of soil fertility or adaptation of crops. You can not get something for nothing – and no genetic manipulation will help you do it.

Work has continued and achievements are to be reported here. The scientists have been conscientious and well-focused. This reflects on individuals, the organization of the program, and the obvious attention to detail by Japanese scientists who try to make good progress. We learned a lot, but it is a difficult task. However, progress has been made and field-management practices to increase productivity of sorghum and pigeonpea through generation of crop/water/nutrient response curves for optimum nutrient and water management developed. These response curves could be used to model the yield of cultivars in target nutrient and drought-stressed environments of probable productivity and responses and also to learn more about gaps in research.

A lot of basic research still needs to be done on P in plants. We know about the role of P and its relationship to energy transfer. But I don't think we know nearly enough about P partitioning in plants, particularly how that partitioning changes over the growing season as the crop develops.

We have learned from soil scientists about forms of P in soil, their relative solubilities and diffusion of P in soil solutions but there is still a lot to be learned, and the interactions at the root surface between P and plants need to be elucidated. None, or very little of this is going to be done at ICRISAT. That is not our role, but these are things that should be pursued in agricultural research centers and universities, and I hope in 10–15 years we can have another conference like this where that information can be contributed to strategic and applied research of the sort we do here.

The CGIAR over the last 10 years has been working to establish system-wide programs. The Group agrees that centers can not be scientifically autonomous. There is much to be gained from research across centers and a system-wide approach. Several centers work in the same agro-ecological zones, ICRISAT is not the only center in the SAT. System-wide initiatives involve intercenter cooperation and cooperation with governments of the region. But it is very difficult to make such initiatives work properly, it involves institutional cooperation and some institutions have their own infrastructures, and transaction costs can be very high. Only a few system-wide programs are reasonably successful. But the scientists have forged partnerships in other ways. Scientists cooperate because they are interested in each others' work and see opportunities to learn. These partnerships are set up by scientists themselves each contributing to the effort, and interchange is by results. Such partnerships are easy to set up, have low transaction costs, and can be highly effective, like the ones that exist on virtually every study of major viruses in plants and animals. We invite scientists with basic knowledge of the role of phosphorus in plants to share with those working on genetic manipulation and agronomic practices. We want to encourage scientist-to-scientist partnerships.

I shall conclude by expressing my thanks to the Organizing Committee of this conference. I hope you have good meetings and discussions – I will be interested in the results.

J.J. Adu-Gyamfi (Ed.), Food security in nutrient-stressed environments: exploiting plants' genetic capabilities: 5–8, 2002.

Government of Japan Project activities at ICRISAT 1984–1999

J. J. Adu-Gyamfi[1], O. Ito[2], J. Arihara[3], H. Nakano[4] & C. Johansen[1]
[1]*International Crops Research Institute for the Semi-Arid Tropics (ICRISAT) PMB 3491, Kano, Nigeria*
[2]*Japan International Research Center for Agricultural Sciences (JIRCAS), 1-2 Ohwashi, Tsukuba, Ibaraki 305-0851, Japan*
[3]*National Agriculture Research Center (NARC), 3-1-1 Kannondai, Tsukuba, Ibaraki 305-8666, Japan*
[4]*Hokkaido National Agricultural Experimental Station (HNAES), Shinsei, Memuro, Kasaigun, Hokkaido 052-0071, Japan*

Introduction

> The little we already know has been very important. BUT what we don't yet understand may prove to be our greatest resource for the future PROVIDED that research to deepen our understanding is sustained – L T Evans.

The Alfisols and Vertisols of the semi-arid tropics (SAT) have the potential to support regional food needs, but are limited by low availability of nutrients, especially phosphorus (P) and nitrogen (N). The situation is further exacerbated by low and unreliable rainfall, and limited groundwater resources. Despite these problems that make food production an unpredictable enterprise, some 850 million (one sixth of humanity) inhabitants of these regions are directly involved in agriculture for their living.

Some say they should not even be there, but if they were to leave their homes, where should they go? Achieving food security – ensuring that sufficient food is available, that supplies are relatively stable, and that those in need can obtain it – has long been the goal of global development investors (donors). Realizing that declining soil fertility in the SAT is the fundamental agronomic cause of declining food production, earlier projects that aimed to improve soil conditions by increasing fertilizer use to achieve sustainable yields were very successful in the better-endowed regions of the Asian SAT. However, in the less-endowed SAT in Africa and Asia, such approaches have not been successful because the required inputs are either not available to farmers in the right place at the right time, or are unaffordable, and in many case a risky investment. This in turn has led to low rates of adoption of improved crop cultivars because yield gains from them have generally proven transient.

Given the above constraints, one of the possible solutions to enhancing productivity in harsh SAT environments is to improve the adaptation of crop plants to the existing soil conditions by selecting and breeding crops that require only moderate external inputs to approach their genetic potential. This involves improving the nutrient acquisition abilities of existing crops, and/or optimizing nutrient cycling via the use of locally available plant nutrient sources.

The Government of Japan (GoJ), realizing the importance of food security in the SAT, has made substantial contributions to the Consultative Group on International Agricultural Research (CGIAR) to help improve the living conditions and enhance the household income of smallholder resource-poor farmers in the SAT. In addition to its core contributions to the CGIAR, the GoJ, through the Japan International Research Center for Agricultural Sciences (JIRCAS) initiated a special project with ICRISAT, in November 1984 entitled 'Development of Sustainable Cultivation of Upland Crops in the Semi-Arid Tropics'. The Project was conducted in three phases (I 1984–1989, II 1989–1994, and III 1994–1999). The Japanese scientists assigned to the Project collaborated with ICRISAT scientists to conduct basic and strategic research to improve and sustain productivity in nutrient-and drought-stressed environments. During its 15-year period, the Project attempted to address the issues of food security and poverty alleviation by exploiting how genetic options can best be combined with management options to improve nutrient availability in SAT cropping systems to thereby enhance the productivity of degraded soils.

Achievements

Realizing that P is a major limiting factor limiting crop productivity in the SAT, during Phase I scientists concentrated on the management options to enhance P availability in cropping systems. The finding that pigeonpea can increase the pool of P available to cropping systems by accessing iron-bound P (Fe-P) via its unique root exudates, provided the first clue to explaining why pigeonpea is better able than other crop species to absorb P from Fe-bound P, the most prevalent form of unavailable P in Alfisols. Phase I scientists further identified the mechanisms by which this occurred, particularly the role of organic and phenolic acids of P-solubilizing root exudates. The research identified a phenolic compound 'piscidic acid' (*p*-hydroxybenzyl tartaric acid) that could chelate Fe^{3+}, thereby releasing P from Fe-bound P in the soil. These findings, apart from having particular relevance to low-P Alfisols in which Fe-P is the dominant form of soil P, suggested scope for possible enhancement of P efficiency through genetic manipulation of the release of root exudates. Other related research pertaining to the inhibition of root respiration of pigeonpea by waterlogging, developing appropriate diagnostic tests for P, and fertilizer-P management strategy helped accelerate the adoption of pigeonpea in cropping systems as a soil fertility replenishment crop. The results are summarized in an ICRISAT publication *Phosphorus Nutrition of Grain Legumes in the Semi-Arid Tropics*. The findings were also published in refereed journals (Ae et al., 1990, 1993; Otani and Ae, 1996; Otani et al., 1996). Phase I led to an FAO-ICRISAT Expert Consultancy Workshop on 'Genetic Manipulation of Crop Plants to Enhance Integrated Nutrient Management in Cropping Systems–1 Phosphorus' held at ICRISAT 15–18 March 1994. This workshop attempted to establish a global consortium of researchers focusing their efforts on improving the ability of crop plants to acquire P, particularly through sources from which it is only sparingly available. During this workshop, research areas with prospects for impact on agriculture in the medium term were identified and appropriate outputs and activities proposed. A further workshop on 'New Concepts of Plant Nutrient Acquisition' was held in Tsukuba, Japan, to summarize the newly developing concepts on the mechanisms by which plants acquire soil nutrients, and to apply these concepts to practical agriculture.

Despite the important role of roots in nutrient and water acquisition, they have been given scant attention in field experiments due to their hidden belowground nature and to logistical difficulties in data collection. During Phase II in 1989–1994, the Project scientists devoted efforts to establishing an overall picture of the profile distribution of water, nitrogen (N), root systems, and root activities and function of ICRISAT mandate crops. They established a relationship between rooting behavior and N flow in pigeonpea-based cropping systems with the development of a simple root model using field data on water and nutrient flow, distribution of root systems, N uptake, N_2 fixation, biomass, and yield. The findings of Phase II are unique in that they increased understanding of roots and nutrient dynamics in cropping systems, focusing on intercropping, a common practice in the SAT. These studies are summarized in a JIRCAS publication *Roots and Nitrogen in Cropping Systems of the Semi-Arid Tropics*. Several journal articles (Adu-Gyamfi et al., 1997a,b; Fujita et al., 1995, Gayatri Devi et al., 1996; Ito et al., 1996a–c; Katayama et al., 1995, 1998, 1999; Matsunaga et al., 1994; Rao et al., 1993, 1997; Tobita et al., 1994) were published as a result of the research conducted during Phase II. In addition to the journal articles, more that 15 conference papers were published.

With the promising information of possible low-P and low-N mechanisms in pigeonpea, sorghum, and pearl millet evidenced by the root exudates, and deep-rooting ability of these crops, scientists in the current Phase III (1995–1999) concentrated on providing a linkage between nutrient physiology and genetic manipulation of crop nutrient use efficiency. The scientists realized that in low-fertility, adverse-rainfall regions of the SAT, no improved cultivar has a reasonable chance of achieving substantial and sustainable yields in farmers' fields unless the critical constraint of soil fertility is addressed through the adaptation of crops to existing soil conditions. Attention was therefore concentrated on identifying and exploiting physio-genetic systems that increase extraction and utilization of nutrients by crops. Work focused mainly on:

- Understanding the candidate mechanisms of plant adaptation by quantifying the nature and extent of genetic variation in the target production system.
- Identifying physiological traits suitable for genetic manipulation of crops that are more stable, and production in nutrient-stressed environments.

These efforts have resulted in a better understanding of the physiological characteristics required for future ideotypes of pigeonpea and sorghum developed for nutrient-stressed environments. Furthermore, the research has helped to strengthen the knowledge base for initiating breeding programs using nutrient-efficient germplasm accessions as parents in crosses. This sets the stage for eventual gene transfer to enhance food security in nutrient-stressed environments. During the past 5 years, scientists established genetic variability in the P-solubilizing activity of pigeonpea root exudates and have quantified the amount of organic and phenolic compounds in these root exudates that could be used as indicators for selection of P-efficient genotypes. This Phase has helped to bridge ICRISAT research on natural resource management and genetic enhancement, particularly by establishing links between nutrient physiology and genetic improvement of crops.

Field management practices to increase productivity of sorghum and pigeonpea through generation of crop/water/nutrient response curves for optimum nutrient and water management have been developed. These response curves could be used to model the yield of cultivars in target nutrient and drought-stressed environments.

Capacity building

The Project has supplemented ICRISAT programs in important areas of basic research, not otherwise possible to address, particularly in view of declining core resources. In addition, collaborative ties between ICRISAT and Japanese Universities particularly those of Hiroshima, Hokkaido, and Tokyo were strengthened. The Project also contributed to training Research Fellows – K Okada, S Tobita, J J Adu-Gyamfi and T Nakamura (currently working in JIRCAS) M Yoshida, and S Ishikawa; Visiting Scientists and Scholars – T Matsumoto, T Goto, H Kokubun, M Moriyama, N Tanaka, K Maeda, Y Ohwaki. A Kubota, M Kondo, J Kashiwagi and A Yamamoto, to mention but a few, and improved the creativity and job competence of support staff. A total budget of approximately US$ 4.5m was provided by the Japanese Government during the 15-year period. Besides paying staff salaries, the Project also invested in capital equipment including an ^{15}N spectrophotometer, spectrophotometer, neutron probe, isotopic analyzer, Kjeldhal digestion and distillation units, ultra-centrifuges, photocopiers, computers, printers, camera, refrigerators, and ovens.

There is now a wealth of accessible knowledge that remains to be blended with the realities of smallholder farming communities in the SAT. ICRISAT must build on these achievements and exploit the comparative advantages in the continuing link between nutrient physiology and genetic manipulation for nutrient efficiency in crops.

The potential contribution of physiological research to genetic improvement of crop plants has long been expounded, but the traditional separation of physiological and breeding research tends to persist. It is hoped this book will stimulate interest in this fascinating dimension of improving the adaptation of crop plants to nutrient-stressed environments through an integrated approach involving plant breeders, crop physiologists, statisticians, and crop modelers.

Acknowledgements

We thank all who have worked on the Project and our collaborators in ICRISAT and other advanced research institutions for their dedication and hard work, all who have contributed to the organization and running of this meeting, and all the participants for their attendance and diligence in preparing and participating in discussions. We acknowledge the help of Sue Hainsworth in the preparation of material prior to the meeting, the production of the commemorative publication in English and Japanese, and for editorial work on the full proceedings.

References

Adu-Gyamfi J J, Ito O, Yoneyama T, Gayatri Devi and Katayama K 1997a Timing of N fertilization on N_2 fixation, N recovery and soil profile nitrate dynamics on sorghum/pigeonpea intercrops on Alfisols of the semi-arid tropics. Nutr. Cycl. Agrosyst. 48, 197–208.

Adu-Gyamfi J J, Ito O, Yoneyama, T, Gayatri Devi and Katayama K 1997b Nitrogen management and biological nitrogen fixation in sorghum/pigeonpea intercropping on Alfisols of the semi-arid tropics. Soil Sci. Plant Nutr. 43, 1061–1066.

Ae N, Arihara J, Okada K, Yoshihara T and Johansen C 1990 Phosphorus uptake by pigeonpea and its role in cropping systems of the Indian subcontinent. Science 284, 477–480.

Ae N, Arihara J, Okada K, Yoshihara T, Otani T and Johansen C 1993 The role of piscidic and secreted by pigeonpea roots grown in an Alfisol with low P fertility. *In* Genetics Aspects of Plant Mineral Nutrition. Eds. P J Randall, E Delhaize, R A Richards and R Munns. pp 279–288. Kluwer Academic Publishers, Dordrecht, The Netherlands.

Fujita K, Chaudhary M I, Adu-Gyamfi J J and Yoshizawa D 1995 Dinitrogen fixation and growth responses to phosphorus and aluminium application in pigeonpea (*Cajanus cajan* L). Soil Sci. Plant Nutr. 41, 729–735.

Gayatri Devi, Ito O, Matsunaga R, Tobita S, Rao T P, Vidyalakshmi N and Lee K K 1996 Simulating root system development of short-duration pigeonpea. Exp. Agric. 32, 67–78.

Ito O, Johansen C, Adu-Gyamfi J J, Katayama K, Kumar Rao J V D K and Rego T J (Eds.) 1996a Roots and Nitrogen in Cropping Systems of The Semi-Arid Tropics. JIRCAS (Japan International Research Centre for Agricultural Sciences) International Agriculture Series No. 3. 691 pp.

Ito O, Matsunaga R, Katayama K, Tobita, S, Adu-Gyamfi J J, Kashiwagi J, Rao T P and Gayatri Devi 1996b Dynamics of roots and nitrogen in cropping systems of the semi-arid tropics. JIRCAS J. 3, 33–47.

Ito O, Matsunaga R, and Rao T P 1996c Diurnal changes in nonstructural carbohydrates in pigeonpea leaves. Soil Sci. Plant Nutr. 42, 419–424.

Johansen C Lee K K and Sahrawat K L (Eds.) 1991 Phosphorus Nutrition of Grain Legumes in the Semi-Arid Tropics. ICRISAT (International Crops Research Institute for the Semi-Arid Tropics) Patancheru, Andhra Pradesh, India. 264 pp.

Katayama K, Ito O, Matsunaga R, Adu-Gyamfi J J, Rao T P and Anders M M 1995 Nitrogen balance and root behavior in four pigeonpea-based intercropping systems. Fertil. Res. 42, 315–319.

Katayama K, Ito O and Rao TP 1998 Seeding characteristics and retention of current photosynthates in leaves in relation to initial growth in pigeonpea (*Cajanus cajan* L. Millsp.) and cowpea (*Vigna sinensis* Endl.) Soil Sci. Plant Nutr. 44 (3), 477–480.

Katayama K, Ito O, Adu-Gyamfi J J, Rao T P, Dacanay E V and Yoneyama T 1999 Effects of NPK fertilizer combinations on yield and nitrogen balance in sorghum or pigeonpea on a Vertisol in the semi-arid tropics. Soil Sci. Plant Nutr. 45, 143–150.

Matsunaga R, Ito O, Tobita S, Rao T P and Johansen C 1994 Response of short-duration pigeonpea to nitrogen application after short-term waterlogging on a Vertisol. Field Crops Res. 38, 167–174.

Otani T and Ae N 1996 Phosphorus (P) uptake mechanisms of crops grown in soils of low P status I. Screening of crops for efficient P uptake, Soil Sci. Plant Nutr. 42, 155–163.

Otani T, Ae N and Tanaka H 1996 Phosphorus (P) uptake mechanisms of crops grown in soils of low P status: II Significance of organic acids in root exudates of pigeonpea . Soil Sci. Plant Nutr. 42, 553–560.

Rao, T P, Ito O and Matsunaga R 1993 Differences in uptake kinetics of ammonium and nitrate in cereals and legumes. Plant Soil 154, 67–72.

Rao T P, Ito O, Matsunaga R and Yoneyama T 1997 Kinetics of ^{15}N-labelled nitrate uptake by maize (*Zea mays* L) root segments. Soil Sci. Plant Nutr. 43, 491–498.

Tobita S, Ito O, Matsunaga R, Rao T P, Rego T J , Johansen C and Yoneyama T 1994 Field evaluation of nitrogen fixation and use of nitrogen fertilizer by sorghum/pigeonpea intercropping on an Alfisol in the Indian semi-arid tropics. Biol. Fertil. Soils. 17, 241–248.

J.J. Adu-Gyamfi (Ed.), *Food security in nutrient-stressed environments: exploiting plants' genetic capabilities*: 9–23, 2002.

Physiological traits for crop yield improvement in low N and P environments

Thomas R. Sinclair[1] & Vincent Vadez
USDA- Agricultural Research Service and Agronomy Department, P.O. Box 110965, University of Florida, Gainesville, Florida 32611-0965, USA. [1]*Corresponding author**

Received 10 May 2000. Accepted in revised form 25 September 2000

Key words: legume, nitrogen, nutrient uptake, phosphorus, yield limits

Abstract

Nitrogen and phosphorus are recognized as essential elements in crop production, but the full extent of the requirement for these elements in the physiological processes leading to crop growth seems not to be always fully appreciated. Virtually all the biochemical compounds in plants that support development and growth contain N and/or P. Deficiencies in either element lead to a lost ability for plant growth such that there is a quantitative relationship between crop yield and accumulation by plants of each of these elements. Few options appear to exist to greatly diminish the requirement for either element in crop growth and the formation of seed yield. Consequently, crop yields cannot be increased without increased acquisition of N and P by plants. If the soil environment does not offer these elements, then crop yield will necessarily be restricted. While little opportunity exists to increase N recovery under low nutrient environments, several options can be investigated for increasing P accumulation by the crop. Ultimately, however, the rigid limitation on yields of inadequate N means that without external supplies of N for the cropping system, biological fixation of N must be enhanced to increase N input. In particular, it appears that considerable research needs to be focused on whole-plant processes in legumes that lead to enhanced symbiotic N fixation. A critical aspect of increased legume production will be improved management of P to allow legumes to achieve high N fixation rates and yields.

Introduction

Nitrogen and phosphorus have long been regarded as essential elements that must be provided to crops to obtain good growth. However, the constraining nature of the 'essential' status of these elements on the physiology of crop growth is sometimes overlooked, particularly when issues relative to increasing crop yields are examined. In this paper, we examine a series of questions that help to define the role of N and P in crop growth. What is the physiological role of each of these elements? Can the need for N and P be decreased in the plant? Can N and P allocation to the grain be increased to improve yields? What is the quantitative yield limitation imposed by N and P? Two additional questions related to the physiology of crop plants are discussed in regards to increasing the acquisition of N and P. Can crop N and P uptake from the soil be increased by altering physiological traits of the plant? What is the role of legumes in improving N levels in low-nutrient environments? In the discussion of each of these questions, the intention is to give some perspectives about the opportunities for physiological improvement of crops that might result in increased yields in low-nutrient environments.

What are the essential physiological roles of N and P in crop growth?

Nitrogen and phosphorus are both integral components of virtually all the biochemical compounds that make plant life possible, as discussed below. There

* FAX No: 352-392-6139; Tel.: 352-392-6180.
E-mail: trsincl@gnv.ifas.ufl.edu

This paper previously appeared in *Plant and Soil* **245**, 2002, pp. 1–15.

is no conceivable alternative to these two elements in constructing the biochemical machinery of plants. It is absolutely clear that both N and P are essential elements in their structural, biochemical and physiological roles contributing to crop growth.

Furthermore, the dependence of crop growth on N and P is a quantitative one. Virtually all of the compounds that participate in metabolism, cell growth and tissue development contain N and/or P. Therefore, diminished levels of N and P have many ramifications that result in quantitative decreases in the rate of growth, and ultimately yield. Commonly, decreases in either N or P, which is likely under low-nutrient conditions, leads directly to decreases in the levels of the critical compounds required to sustain high rates of growth. It should be no surprise, therefore, that there is frequently a close correlation between crop yield and the amount of N and P that the plants have accumulated.

Nitrogen

Nitrogen is an essential component of amino acids and nucleic acids. Consequently, without N it is not possible to synthesize the necessary proteins, enzymes, DNA and RNA required in virtually all plant cells for their initial development, sustained growth and functioning to support other tissues of the plant. Deficiencies in reduced N necessarily results in less biochemical machinery to catalyze plant metabolism and to generate new cells. Consequently, nitrogen deficiencies result in decreased crop leaf area, photosynthetic assimilation and seed growth (Sinclair, 1990).

The biochemical process in the plant that has quantitatively the greatest demand for N is photosynthesis. In fact, RuBisCO, which is the enzyme that initially captures CO_2, accounts for over half the protein in leaves of C_3 plants. In C_4 species, RuBisCO may account for a quarter of the protein in the leaves and additional enzymes of the organic acid cycle for CO_2 capture add to the dependence of photosynthesis on N content in the leaves. Therefore, declines in N content in leaves of all crop species are invariably associated with declines in leaf photosynthetic rates. A close correlation between maximum leaf CO_2 assimilation rate and N per unit leaf area has now be documented in a number of crop species including maize (*Zea mays* L.) and sorghum (*Sorghum bicolor* L.) (Muchow and Sinclair, 1994), sunflower (*Helianthus annuus* L.) (Gimenez et al., 1994), rice (Peng et al., 1995), potato (*Solanum tuberosum* L.) (Vos

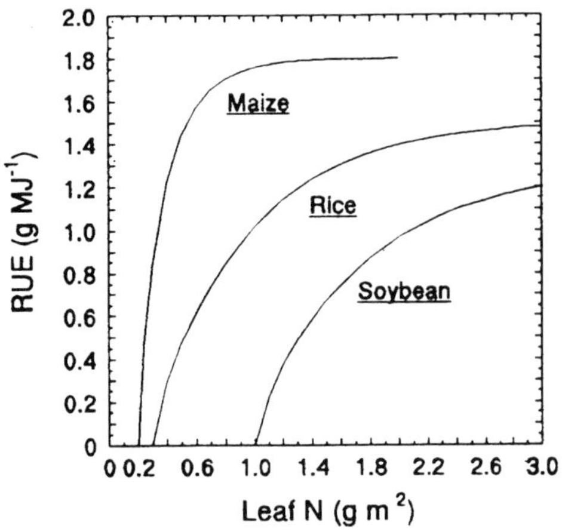

Figure 1. Theoretical relationship between radiation use efficiency (RUE) and average canopy leaf nitrogen content per unit leaf area for maize, rice and soybean (Sinclair and Horie, 1989).

and Van Der Putten, 1998 and sugarcane (*Saccharum* spp.) (Meinzer and Zhu, 1998). Indeed, much of the variation in leaf photosynthetic capacity for different cultivars, age of leaves and growth conditions can be attributed directly to differences in leaf N content (Sinclair and Horie, 1989).

A recent report by Meinzer and Zhu (1998) indicated that an additional consequence of the loss of activity in the photosynthetic enzymes is a decrease in apparent quantum efficiency. They found in sugarcane that the quantum yield of CO_2 assimilation is decreased linearly with decreasing N content. They attributed the decrease in quantum yield to a decline in the activity of the enzymes of the C_3 cycle. There is also the possibility that N deficiency in the leaves may result in increased photoinhibition of photosynthesis (Henley et al., 1991).

Crop radiation use efficiency (RUE, ratio of accumulated crop mass divided by cumulative intercepted solar radiation) is directly dependent on maximum leaf photosynthesis rate (Hammer and Wright, 1994; Sinclair and Horie, 1989). Because photosynthetic activity is dependent on leaf N content, it follows that RUE is also closely linked to leaf N content. A number of experimental studies have demonstrated such a linkage (Sinclair and Muchow, 1999). Theoretical studies (Hammer and Wright, 1994; Sinclair and Horie, 1989) have also developed functional relationships between RUE and leaf N content for different crop species, as illustrated in Figure 1.

The requirement for N in the construction of enzymes and nucleic acids for new cells results in a particularly large influence on new cell development under N deficiency. One of the obvious manifestations of this limitation is in the inhibition of leaf area development under low N conditions. For example, Muchow (1988) reported substantial decreases under deficit N conditions in the area of individual leaves of maize and sorghum plants even though the number of leaves was virtually unchanged. Similarly, Gimenez et al. (1994) found large decreases in sunflower leaf area with low nitrogen fertility, whereas the CO_2 assimilation capability was decreased only to a small extent. In potato, Vos and Van Der Putten (1998) reported decreases in the area of individual leaves with decreasing N fertility, but leaf photosynthetic rates were not different among N treatments. Consistent with this last report, the inability to generate new cells for expanding leaf area under N deficit is commonly more serious in limiting crop growth than the potential decrease in photosynthetic activity per unit leaf area.

Phosphorus

Phosphorus is an essential component of cell structures, mainly as nucleic acids and phospholipids. It is especially critical in establishing the enzymatic machinery in energy storage and transfer, which in many cases involves membrane processes. Not surprisingly, P deficiency results in a loss in cell integrity (Ratnayake et al., 1978). The bonding properties of P also make it crucial for metabolic processes that are nucleotide-based, e.g. ADP, NAD and NADP, because of its unique energy-transfer properties. A general consequence of P deficiency is a decrease in the energy charge of cells [(ATP + 0.5ADP)/(ATP+ADP+AMP)] (Duff et al., 1989; Sa and Israel, 1991).

Inorganic P (Pi) is essential because it is used as a 'currency' in the generation of high-energy phosphate compounds such as ATP from ADP, which is especially critical in respiration. According to Bieleski (1973), P found in DNA/RNA/lipid-P/nucleotide-P would be found in typical proportions of 0.2/2/1.5/1 and Pi usually represents about 2/3 of total plant P. With the development of P deficiency, it is primarily Pi that decreases in plant tissue, prior to a decrease in adenylate-P (Rufty et al., 1993).

The concentration of Pi in plant cell cytoplasm of P-sufficient plants is commonly in the range of 5–10 mM (Bieleski, 1973), which is assumed to be necessary to sustain optimal activity (Foyer and Spencer, 1986). Rebeille et al. (1984) reported that mitochondrial respiration of sycamore cells (*Acer pseudoplatanus* L.) was not affected at a Pi concentration as low as 3 mm. Maintenance of a stable cytoplasmic Pi concentration appears to be essential for many enzyme reactions (Schachtman et al., 1998) to sustain optimal activity. This requirement, however, is met by using P stored in the cell vacuole to buffer cytoplasmic P concentration (Lee and Ratcliffe, 1993; Mimura et al., 1990).

Phosphorus distribution within plant cells is regulated so as to maintain the cytoplasm P concentration at a fairly constant level and vacuole P concentration varies to accommodate surpluses or deficiencies that may develop in cells (Lauer et al., 1989; Lee et al., 1990). Concentrations of P in the vacuole can vary from near zero under P-deficit conditions to up to 25 mm under non-limiting conditions. Consequently, physiological activity of a tissue has been observed to be fairly stable over a wide range in bulk tissue P concentration as long as vacuoles are able to buffer changes in P so that the cytoplasm P concentration remains at adequate levels. This buffering capacity makes it difficult sometimes to resolve the sequence of physiological effects of P deficiency, unless cytoplasm Pi concentration is determined apart from bulk tissue P.

One manifestation of the ability to stabilize cytoplasmic P levels is that photosynthesis rates are stable over a wide range of decreasing bulk leaf P. For example, in maize, several studies have shown a relative insensitivity of leaf photosynthesis to decreasing P (Khamis et al., 1990; Usuda and Shimogawara, 1991). Starch levels have actually been observed to increase in P deficit leaves (Fredeen et al., 1989, 1990; Lauer et al., 1989). Mollier and Pellerin (1999), who observed in maize that leaf area development was the first and most dramatic influence of P deprivation, found that RUE decreased only at a late stage in the low-P treatment of their experiment.

Developing cells do not have mature vacuoles, and as a result, do not have a large stored reserve of P to buffer growth under P-deficient conditions. Therefore, developing cells that require P for synthesis and maintenance of membrane integrity are especially vulnerable when the P supply is low. Leaf area development, as an example, decreases drastically under P deficiency in sugar beet (*Beta vulgaris* L.) (Rao and Terry, 1989) and in soybean (Rao et al., 1993). Fredeen et al. (1989) also reported for soybean that the first and most dramatic response to decreased P was decreased expansion of individual leaves.

In contrast to the sensitivity of leaf development to P deficiency, root development under low P is sustained (Anghinoni and Barber, 1980; Cakmak et al., 1994) or even stimulated (Bates and Lynch, 1996; Rufty et al., 1993). As a consequence, the root–shoot ratio is increased under low P conditions relative to high P conditions (Freeden et al., 1989; Yan et al., 1996). This response in root growth has been hypothesized to result from the fact that shoot development is readily decreased by P deficiency so that additional carbon is allocated to the roots (Mollier and Pellerin, 1999). This hypothesis is consistent with observed increases in root carbohydrates levels (Cakmak et al., 1994) under P-deficient conditions.

Inorganic P is directly involved in enzymatic processes that require phosphorylation of carbon compounds such as during glycolysis (Duff et al., 1994) or during the dark phase of photosynthesis. Photosynthetic metabolism under P deficiency can be dramatically slowed because of a decreased rate of RUBP regeneration and a slower activity of the Calvin cycle enzymes (Fredeen et al., 1989, 1990). Although the photosynthetic response commonly occurs later than the reduction of leaf area upon P deprivation, it is of importance for plants growing under constant P deficiency.

Studies have shown that one of the major consequences of inadequate P is a decrease in the energy metabolism of the plant. A central reason for this response is the decrease of an adequate Pi supply to the mitochondria for the phosphorylation of ADP in ATP. For example, Rufty et al. (1993) found that low P in soybean (*Glycine max* L. Merr.) resulted in decreased [ATP] in roots, and this was associated with depressed nitrate uptake and transport by the roots. Sa and Israel (1991) reported that low levels of P resulted in low levels of ATP and energy charge in the plant-cell fraction of soybean nodules. In their study, the negative influence of low P availability on energy metabolism resulted in greatly decreased nitrogen fixation activity.

Can the need for N and P in plant growth be decreased?

Nitrogen

The plot of RUE against leaf N content (Figure 1) leads to an interesting possibility for differences in growth potential among species under N-limited conditions. In situations where the N availability only allows low leaf N content, there is a clear distinction among species. C_4 species as represented by maize in Figure 1 have the potential to sustain high values of RUE, while in other species, RUE, and consequently growth, falls to very low levels. For example, consider the case in Figure 1 where the N availability allows crop leaf N content to be only about 0.5 g N m^{-2} leaf area. Such a low leaf N is insufficient to even sustain soybean growth and rice is relegated to a very low RUE of only about 0.5 g MJ^{-1}. In contrast, maize is predicted to be able to sustain a RUE of 1.5 g MJ^{-1} even with this assumed low leaf N content. In this example, C_4 photosynthesis confers about a three-fold advantage to maize over rice for RUE. However, as discussed later, a high RUE in itself is not sufficient for high grain yield because the overall benefit of the C_4 trait for mass accumulation still must be matched with N and P acquisition to construct seed mass. Therefore, incorporation of the C_4 photosynthesis pathway into C_3 species may not bring any major benefit under low-N environments.

Moreover, it should be noted in Figure 1 that the tremendous advantage of the C_4 trait under low leaf N is diminished considerably when leaves of each species are allowed to reach N levels that result in maximal RUE for each species, and where incorporating a C_4 pathway in rice would confer a maximum RUE increase of only about 20%. That is, under conditions that give high leaf N for each species, RUE for maize is still greatest among the species at 1.7 g MJ^{-1}, but RUE in rice is about 1.4 g MJ^{-1} and soybean is 1.2 g MJ^{-1}. Therefore, under high fertility conditions, C_4 photosynthesis would be expected to confer as a maximum only about a 20% advantage in RUE for rice even if a fully functional C_4 system could be incorporated into its leaves.

Since a decrease in leaf area development is commonly the most substantial limitation on yield under low N fertility, options to decrease the sensitivity of leaf area development to N deficiency would be quite beneficial. One solution would be to produce thinner leaves, which presumably would require less N. That is, fewer cell numbers of the thinner leaves would decrease the requirement for N. The tradeoff, however, is that thinner leaves would likely lead to lower photosynthetic activity per unit leaf area. Sinclair and Horie (1989) presented calculations on crop mass production for various assumed levels of leaf N per unit area. Their results showed that decreasing leaf N per unit area, i.e. assuming thinner leaves, allowed leaf area to be greater but this advantage was almost completely offset by the lower photosynthetic activity per unit leaf area.

Another possibility for sustaining high mass accumulation under low N is to concentrate the available N into the leaves at the top of the leaf canopy. The theoretical analyses of Sinclair and Shiraiwa (1993) and Hammer and Wright (1994) showed a slight advantage in RUE for a decreasing gradient in leaf N content from the top to bottom of the crop canopy as compared to a canopy with no gradient. The analysis of Sinclair and Shiraiwa (1993) indicated that a canopy with a N gradient was more advantageous under conditions of low N accumulation than when N accumulation is high. The potential benefit of attempting to incorporate a N-gradient trait for the crop canopy is likely to be minimum, however, because such gradients have been found to already exist in crops even when supplied with adequate N (Pearcy and Seemann, 1990; Shiraiwa and Sinclair, 1993).

An interesting aside is the possible interaction of N storage in leaves and the trend towards developing crops with erect leaf angles. Sinclair and Sheehy (1999) calculated that high-yielding crops required large amounts of N to be stored in the leaves before seed growth. They speculated that to retain the large leaf area necessary to store the N, it was necessary for all of the individual leaves to receive a minimum amount of light so that they could be sustained. Therefore, to allow leaves at the bottom of the canopy to receive a minimum amount of light, it was necessary for the leaves higher in the canopy to be displayed at an erect angle. Their conclusion was that the advantage of erect leaves in modern crop varieties was to a large extent based on the need to retain leaves and store N in the leaves. In low-nutrient environments, the amount of N to be stored is low so that there is little need for erect-leaf crops.

Indeed, in low-nutrient environments, it is quite likely that horizontal-leaf crops would be more desirable. As discussed previously, loss of leaf area is one of the main consequences of inadequate N and P. To maximize light interception when only a low leaf area is produced, it is advantageous to display leaves in a more horizontal orientation. Furthermore, weed competition can be particularly troublesome under low-nutrient conditions. Crops with more horizontal leaves are more likely to shade competing weeds and lessen the negative impact of these weeds. Hence, crops to be grown in low-nutrient environments may well require a plant architecture that is markedly different from those being produced for nutrient-rich environments.

Phosphorus

The central role of P in energy transfer processes seemingly allows for few alternatives to decrease the requirement for P. There is evidence for alternate enzymes in the glycolysis pathway that minimize the requirement for P in respiration. Lambers (1982) reviewed evidence for the existence of a cyanide-resistant non-phosphorylating pathway of respiration in roots, which could be used as an energy overflow. More recently, Rychter et al. (1992) showed that the activity of this pathway was enhanced in mitochondria extracted from P- deficient bean roots. Duff et al. (1989) offered crucial evidence of enhanced activity of Pi-independent enzymes in the glycolytic pathway under low-P conditions. While these enzymes are less efficient energetically in glycolysis than the P-dependent enzymes, they offer a pathway for continued glycolysis under low Pi conditions (Theodorou and Plaxton, 1993). However, these intracellular enzymes showed a dramatic increase only after induction by severe P deficiency and with a lag compared to the increase in Pi influx capacity (Lefebvre et al., 1990). As of yet, there is no evidence whether this alternate glycolysis pathway is readily induced in crop species and whether there is genotypic variation for this trait under low P conditions.

There appears to be little opportunity to overcome the P requirement in the development of cell components, although internal P requirements seem to vary between and within species (Föhse et al., 1988). It is also striking that *Imperata cylindrica*, a challenging weed in deforested tropical soils (Hairiah et al., 1993) is able to thrive in low P soil with a shoot P concentration as low as 0.05%, four times less than shoot P concentration in bean or soybean. There are also reports showing genotypic differences exist in plant P concentrations. For example, Vadez et al. (1999) reported such differences among 48 genotypes of common bean. Similar to N, however, a decreased concentration of P in the vegetative tissues may not overcome the obligate requirement for P in the developing seed.

Can the allocation of N and P to grain be improved to achieve increased grain yield?

Both N and P are highly mobile in the plant. As a result, N and P that have been used in the development of the vegetative tissues of the plant can be transferred to the developing seed. The success in providing the accumulated N and P to the seed is indicated by the harvest index for each of the elements. That is, the ratio of the amount of the element in the grain relative to the amount of the element in the total above-ground portions of the plant indicates the success with which the element has been transferred to the grain.

The values of nitrogen harvest index (NHI) are consistently found to be high in modern crops. It is not uncommon for NHI to be 0.75 or more. Teyker et al. (1989) selected maize populations based on seedling nitrate uptake, but NHI remained essentially unchanged at an average of 0.74 across five populations. In soybean, Crafts-Brandner (1992) reported NHI subjected to various P treatments to be stable in the range of 0.79–0.86. Rice NHI has also been reported to be stable across cultivars and environments, although the values are somewhat lower than achieved in other crops. The studies of Tirol-Padre et al. (1996) and Singh et al. (1998) with rice resulted in NHI values generally in the range of 0.62–0.73. Consequently, it may be possible to yet increase NHI of rice to some extent.

Phosphorus is generally present as Pi or in mobile compounds so that P is readily available for transfer to developing seeds. As a consequence, phosphorus harvest index (PHI) is commonly quite high. Crafts-Brandner (1992) reported PHI greater than 0.8 for soybean when grown under low P. Values of about 0.8 for PHI were reported for 12 genotypes of common bean studied by Yan et al. (1995b), similar to that found in the bean cultivar Calima (Snapp and Lynch, 1996). In cases when PHI is at 0.8 or greater, it seems unlikely that the fraction of P that is allocated to the grain can be increased substantially.

What is the quantitative yield limitation based on N and P accumulation?

Sinclair (1999) recently employed a phenomenological equation to estimate the maximum yield that might be expected based on accumulated N.

$$Y < N_{up} \cdot NHI/G_N, \quad (1)$$

where Y = mass of grain yield, N_{up} is the total N uptake by the crop, and G_N is the N fraction of the seed.

The basis of this equation is that Y can be no greater than N available to the grain ($N_{up} \cdot NHI$) to produce the required fraction of protein in the developing seed. Maximum crop yield can be calculated using Equation (1) as a function of the total N uptake by the crop.

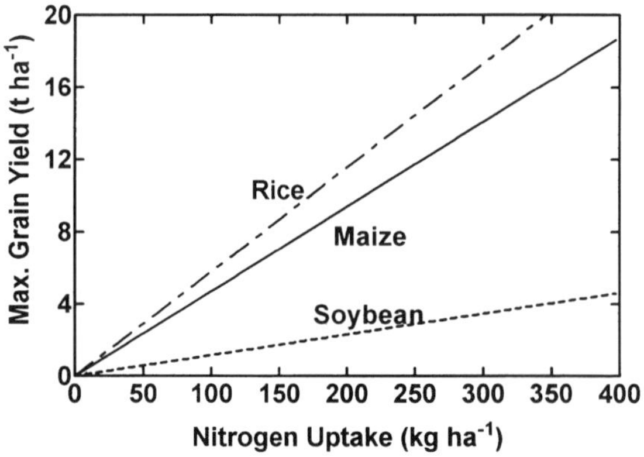

Figure 2. Calculated limits for grain yield of rice, maize and soybean at different levels of N accumulation by the crop. The nitrogen harvest index in each case was assumed to be 0.75 and the N content of the grain was assumed to be 13, 16 and 65 mg N g^{-1} for rice, maize and soybean, respectively.

As discussed previously, NHI is already high in many crops and it appears unlikely that this limit can be substantially improved.

One option to increase yield based on Equation (1) is to decrease G_N. A difficulty in decreasing G_N is that the food or feed value of the grain of many crop species is determined to a large extent by the value of G_N. Decreasing the protein content of rice or wheat, for example, will dramatically influence their nutritional value. Furthermore, the value of G_N tends to be stable. For example, Tirol-Padre et al. (1996) compared the grain N content of 180 rice genotypes of differing maturity dates and in two years of low fertility conditions. They found across all maturity dates a mean of about 10.5 mg N g^{-1}, and the CV within maturity group was less than 5%. Similarly, Singh et al. (1998) found in a comparison of 20 rice cultivars that grain N under low N fertility ranged from 8 to 10.5 mg N g^{-1}.

Equation (1) indicates that ultimately crop yield is directly limited by N_{up}, the amount of N accumulated through the growing season. Crop yield can be no greater than that allowed by N_{up} regardless of improvements in the accumulation of crop mass or in the biochemical pathways. This limitation can be shown graphically by plotting Y as a function of N_{up} as done in Figure 2. In these calculations, NHI was assumed to be 0.75 and G_N was assumed to be 65 mg N g^{-1} for soybean, 16 mg N g^{-1} for maize, and 13 mg N g^{-1} for rice.

It is clear in Figure 2 that in those nutrient-limited environments where N_{up} is low, the maximum yield is necessarily low. In those environments where a crop can accumulate no more than about 50 kg ha^{-1}, maximum yield is approximately 2 t ha^{-1}. Such levels of N_{up} and yield were essentially the limits of good agricultural practices (rotation, manure, etc.) before the advent of the use of chemical fertilizers (McCloud, 1998). Of course, a yield increase to 2 t ha^{-1} may be welcomed in many low-nutrient regions and the use of good, traditional management practices may offer sufficient nutrient to reach these yield levels. An important alternative indicated in Figure 2, however, is the use of a legume crop such as soybean. Soybean, which can obtain N from symbiotic fixation of atmospheric N_2, may be able to accumulate N well above 50 kg ha^{-1} and as a result achieve yields above the 2 t ha^{-1} limit in nutrient-limited environments without the addition of chemical N.

The approach presented in Equation (1) can also be used to define maximum crop yield based on the accumulation of P. To do this analysis for P, good information is needed on the level of P required to ensure unrestricted seed growth. These data, however, do not appear to be readily documented across a number of crop species. Data of P concentration in soybean were measured in eight soybean cultivars and found to vary from 5.7 to 7.1 mg P g^{-1} (Hanway and Weber, 1971a). Variation in seed P concentration was reported by Yan et al. (1995a) among 12 cultivars of common bean to be larger than soybean with a range of 2.7–5.0 mg P g^{-1}. Certainly, more of these types of data are required and an important research topic for the future might well be to understand the P levels that are required in seeds, as there seems to be larger variations in G_P than in G_N. Such data could lead to an analysis similar to that offered by Equation (1) of establishing maximum yield limits based on accumulated P (P_{up}).

Of course, some caution is needed in discussing independently the influence of N and P on the limits to crop yield. There is certainly interaction in the nutrient use by plants and the influence of this interaction on crop yield can be complex (Sinclair and Park, 1993). For example, Hanway and Weber (1971b) observed in soybean that seed N concentration was fairly constant, while seed P concentration varied to a much greater extent and was dependent on N supply. Therefore, if seed yield is closely dependent on N_{up} it might not be so closely dependent on P_{up}, which gives some leeway for seed production under limited P_{up}.

Can N and P uptake from the soil be increased in low-nutrient environments?

Since N and P accumulation are absolutely essential in obtaining maximum crop yields, it should not be surprising that crop plants have developed nutrient uptake systems that are very effective at recovering available N and P in the soil either as a result of natural selection or breeder selection. Uptake systems in the roots have high affinities for each element and are operational at very low concentrations (Schachtman et al., 1998). If the element is available at the root surfaces and there is a physiological deficiency for the element in the plant, roots have the capacity for uptake rates that are many fold greater than observed under normal field conditions.

Nitrogen

There are at least two uptake systems in plant roots for nitrate that allow rapid uptake of N. The Km for nitrate uptake in the low-concentration system is about 20 μM (Hole et al., 1990; Siddiqi et al., 1990) so roots are able to withdraw N from the soil solution to very low levels. There is, however, considerable evidence that N inflow rates per unit root length in field-grown crops are substantially less than the maximum inflow measured under controlled conditions (Feil, 1997). For example, the capacity for nitrate uptake by wheat roots has been found to be many fold greater than that required to support plant growth at all stages of plant development (Oscarson et al., 1995). This conclusion is further supported by the very high rates of N_{up} by field-grown rice that were observed for limited time periods when N supply in the soil was high (Cassman et al., 1998; Peng and Cassman, 1998).

The fact that N_{up} is not sustained at very high rates by crops under conditions of high N availability indicates a strong feedback system that substantially throttles N uptake capacity (Imsande and Touraine, 1994). The accumulation of nitrate in plants has been shown to have a direct and rapid feedback on the uptake system (RajRaman et al., 1995). In addition, the growth of the plant appears to feedback to regulate N content to a fairly constant level under adequate N supply. For example, Singh et al. (1998) found that there was no statistical difference in the accumulation of N among a number of rice genotypes of the same maturity classification.

One consequence of this tightly regulated feedback appears to be the common observation of a close relationship between %N and crop mass within a crop species when soil N availability is high (Lemaire and Gastal, 1997). Greenwood et al. (1990) proposed that a common relationship existed between %N and crop dry weight for those crops that have received N adequate to achieve maximum crop growth rates. This relationship defines a curve of the critical nitrogen concentration levels for each crop species, and observations below this curve identified conditions where crop growth was being limited by inadequate nutrient. While this approach does not lead to a physiological understanding of N use by the crop, it highlights the quantitative dependence of growth on accumulated nutrient. Sheehy et al. (1998), for example, used this approach to examine the limitation of N accumulation on rice growth.

Since the uptake capacity by roots for N is high, the total amount of available N in the soil defines the upper limit for N accumulation. Especially in low-nutrient environments, the challenge is to maximize the uptake of the available soil N. Nitrate is highly soluble and moves readily in the soil so that there is no need for an extraordinary root density to recover the N in the soil solution. Root length densities of only 1.0–1.5 cm cm^{-3} appear adequate to extract soil N. Greater root depth is of little benefit because the amount of N available at deep soil depths in nutrient-limited environments is usually quite small.

Increased N uptake from unfertilized soils is likely limited to maximizing the recovery of mineralized N from the soil. The simplest and most direct approach to obtaining more mineralized N is to extend the length of the growing season so that a growing crop is present over a longer time period to immediately take up the newly available N. Tirol-Padre et al. (1996) reported in a comparison of 180 rice cultivars that in a year of low nitrogen availability there was a high correlation ($r^2 = 0.81$) between grain yield and total N uptake. The variation in N uptake among the cultivars was, in turn, highly correlated with the duration of the season. Consequently, longer-season cultivars may be a useful alternative to increase N_{up} in low-N environments. The disadvantage of the long-season approach, however, is that the potential exposure of the crop to various stresses is increased.

Phosphorus

Similar to N, soluble P in the vicinity of roots is also readily recovered by plant roots. At least two uptake systems have been demonstrated for inorganic P, and the high-affinity system has a Km of 3 μM (Furihata et al., 1992; Schachtman et al., 1998). Consequently, P is recovered to low concentrations in the soil. This is consistent with the report of Crafts-Brandner (1992) that the amount of P accumulated by soybean was in direct proportion to the total amount that was available to the plants.

The low mobility of P in soils, however, is a serious limitation (Amijee et al., 1991). It is crucial that the P-absorbing surfaces in the soil are extensive and prolific to make contact with the available P. Clarkson (1985) concluded that P was only depleted in the rhizosphere to a distance of about 100–200 μM from the root surface. Lynch and Beebe (1995) concluded that it was essential to have “a vigorous, highly branched root system with many growing points” to maximize P uptake. Yan et al. (1995a) reported that those common bean genotypes that performed best on low P-available soils were those that produced the greatest root mass.

Probably more important than branched roots for P uptake is the presence of root hairs (Gahoonia and Nielsen, 1998). P depletion from the root surface at a distance that approximates the length of the root hairs (Clarkson, 1985) opens the possibility that longer root hairs would allow for an increase in P uptake (Itoh and Barber, 1983), while representing very little investment in carbon and P. Of course, a search for diversity in root hair length is likely to be an arduous and time-consuming effort. Such a search is made more difficult because root hair length varies with the P concentration to which the roots are exposed (Bates and Lynch, 1996).

Another approach to increase the P-absorbing surfaces is to take full advantage of the symbiosis formed between roots of some plant species and mycorrhiza. This topic has received considerable attention and has been reviewed by Smith and Gianinazzi-Pearson (1988). There appears to be a great challenge in managing mycorrhiza for annual crops such that the soil is adequately inoculated, plant infection and mycorrhiza development occur in a timely fashion, and a highly functional symbiosis develops. At this point, it appears that mycorrhiza may be of greatest benefit in cropping systems that include a perennial crop where a long time-period for the development of the symbiosis is possible.

In many cases, the essential requirement for increasing P uptake by a crop is a need to increase the availability of P in the soil. The concentration of the soluble fraction in the soil is usually low, seldom going above 5 μM (Bieleski, 1973; Clarkson and Grignon,

1991). The soluble fraction needs to be continuously replenished with P to avoid deficient conditions, which depends on an acceptable level of exchangeable P. An exchangeable P concentration of 15 ppm is considered as critical for bean production (Thung, 1991). The challenge, especially under inherently low P soils, is to sustain sufficient soluble and exchangeable P when P is present in a fixed form in the soil.

Indeed, much of the soil P is unavailable to plants because it can be fixed in the soil in both inorganic and organic forms. Hydroxide forms of Al, Fe or Ca account for much of the insoluble P in the soil. The solubility coefficients range from a low value for Ca–P to very low values for Al–P or Fe–P. The organic forms of P are present as live soil bacteria and as mostly inositol phosphate and phytate (Clarkson, 1985). There are options, however, to develop physiological traits of plants that will increase the transformation of P from its fixed form to forms that can be taken up by plant roots.

The availability of inorganic P is commonly enhanced under acidic conditions in the soil. Roots have the ability to release both organic acids and hydrogen ions from roots to acidify the surrounding soil so as to increase P availability (Lynch and Beebe, 1995). Differences have been identified among crop species in their ability to acidify soil and extract inorganic P. In an experiment on an unfertilized Alfisol soil with a high iron phosphate level, Ae et al. (1990) observed that pigeonpea thrived, while four other crop species died within 1 month after sowing. They found that pigeonpea was unique in that it acidified the soil by excreting piscidic acid, which solubilized $FePO_4$.

There also are indications that there may be genotypic differences within species for acidifying the soil and extracting P. Lynch and Beebe (1995) examined 364 genotypes of common bean and found substantial variation in yield when grown under P-deficient conditions, indicating a genetic potential for increasing P extraction. In comparing six contrasting common bean genotypes, Yan et al. (1996) did not find a difference in their ability to acquire P from aluminum phosphate, iron phosphate or inositol phosphate. One genotype, however, was superior in acidifying the rhizosphere and extracting P from CaHPO4.

Release of P from organic forms of P in the soil can be catalyzed by phosphatase enzymes. These enzymes either originate from exudates directly from plants or from microbes whose growth is stimulated by plant exudates. Goldstein et al. (1989) demonstrated increased protein secretion from tomato cells suspended in P-deficient media and that this secretion was associated with enhanced acid phosphatase activity. Regardless of the source of the phosphatases, differences in activity have been reported in comparisons of plant cultivars. Helal (1990) reported differences among cultivars of common bean in root phosphatase activity that were associated with differences in plant P uptake when the plants were grown on a solution containing organic P (inositol hexa phosphate). Similarly, Asmar et al. (1995) found differences in barley genotypes in their ability to induce soil phosphatase activity. The level of induced phosphatase activity was correlated with the hydrolysis of organic P compounds within the root hair zone. However, the exact role of phosphatase is not yet totally clear, as the major organic P compounds in soil, inositol P and phytate can be poor substrates for phosphatases (Clarkson, 1985). Furthermore, it is unlikely that plants would successfully compete for organic P with soil micro-organisms which permeate the soil. A possible role for phosphatases might be to recover Pi from organic P lost from the plant in the immediate vicinity of the roots.

Considering the fact that much of the P taken up by the plant is dependent on the release of P from unavailable forms, one approach to increase overall crop uptake is to extend the duration of the growth so that it has a longer time frame to accumulate P. Unlike N, there is no direct evidence that a long season allows time for more P to be released in the soil for root uptake. However, there may be a possibility that a longer season with delayed podfill would result in increased P uptake because roots remain functional for P uptake until late pod-filling (Snapp and Lynch, 1996). Also, Yan et al. (1995b) reported that some of the highest yielding common beans under low P are the type III and IV, which are climbing types. No details were given on time of harvest for individual genotypes, but climbing types commonly have a longer season than bushy types.

What is the opportunity for legumes in low-nutrient environments?

High yields are simply not possible without the accumulation of substantial amounts of N and P in the crop. One option for obtaining N, at least without relying on a soil source, is to take full advantage of those crop species that can symbiotically fix atmospheric N_2. In this situation, some of the carbon fixed by the plant is used to provide the energy required to 'fuel' the reduction of atmospheric N_2 for incorporation into organic molecules. While this process is energetically expensive, it might be very effective in low-N environments where assimilated carbon is abundant relative to N.

Grain legumes seem especially well suited to contribute to increased crop production in low-nutrient environments because they can be used both to produce a seed crop themselves and to generate high N-content plant residue. Since plant residues from grain legumes are sometimes considered poor sources of N because large amounts of the N is exported to the grain, it might be important to consider low harvest index, high N_2-fixing legumes or non-grain legumes. Rotations that include a legume as a green manure crop would be especially beneficial in increasing soil N fertility. Also, perennial legumes, which might be used in an agroforestry cropping system, offer an opportunity for long-term N_2 fixation as well as a long growth cycle to recover P from the soil.

Considerable attention has been given to increasing N_2 fixation activity of legumes. Much of the research, especially with the advent of molecular genetics, has focused on improved bacteria. Except in those locations where the appropriate bacteria to establish the symbiosis did not exist, little progress has been made in increasing legume N_2 fixation activity using improved bacteria. The performance of bacteria in isolation, however, does not translate well to plant performance under field conditions. The regulation of N_2 fixation activity tends to be a whole-plant phenomenon and unless the bacteria is ineffective at fixing N_2 there is little regulation at the bacteria level. Furthermore, it has proved to be very difficult to introduce alien bacteria into a soil environment where bacteria capable of forming a symbiosis already exist.

Attempts to increase N_2 fixation activity by altering the host plant have also achieved only limited success. In this case, much of the effort has focused on selecting hosts that result in greater nodulation. Again, nitrogen fixation activity is related to whole-plant activity and greater nodulation does not necessarily lead to greater N_2 fixation. Pracht et al. (1994) incorporated genes into soybean lines that produced hypernodulation on the roots but this resulted in no increase in either N_2 fixation activity or yield. A similar attempt has been made in the breeding of the 'RIZ' lines of beans (Kipe-Nolt and Giller, 1993) with no great success. On the other hand, a comparison of widely divergent lines of chickpea showed greater nodulation capacity was associated with greater N_2 fixation and grain yield (Rupela et al., 1997).

Another major limitation for enhancing N_2 fixation, especially in grain legumes such as common bean (Hardarson, 1993), is the great sensitivity of N_2 fixation to small amounts of residual soil nitrogen. Hardarson et al. (1993), however, identified genotypic variability among common bean for N_2 fixation with residual soil N.

One important success for increase N_2 fixation activity has occurred in alfalfa (*Medicago sativa* L.). Selection for several traits associated with increased N_2 fixation and plant growth resulted in the selection of parental genotypes and subsequently the development of a cultivar with increased N_2 fixation activity (Barnes et al., 1988). Since this concentrated and focused effort on selection for N_2 fixation activity has not been attempted in other legume crops, such an approach may offer a good option for increasing N_2 fixation productivity in low-N environments.

A constraint on the development of legumes for low-nutrient environments is the sensitivity of N_2 fixation to stress conditions. For example, N_2 fixation in several legumes was found to be very sensitive to soil drying (Sinclair and Serraj, 1995). Since many low-nutrient environments impose several stresses on the crop, intensive research is likely needed to incorporate a number of traits to achieve greater N_2 fixation and higher yields under these conditions. For example, Adu-Gyamfi et al. (1999) concluded that legumes for the low fertility regions of the semi-arid tropics "should have deep rooting systems to pump nutrients from the deep soil layers, moderately high nitrogen fixing ability, should be adapted to areas of poor rainfall, contribute a substantial amount of biomass to soil prior to harvest in the roots and through leaf fall, and have ability to grow under low P status soils".

Phosphorus may be a critical constraint of legumes under low-nutrient environments because there is a substantial need for P in the N_2 fixation process. One clear manifestation of this may be the common observation that application of P on a mixed grass-legume pasture will enhance the legume growth relative to the grass. The high requirement for P in legumes is consistent with the involvement of P in the high rates of energy transfer that must take place in the nodule. For example, Figure 3 illustrates the sensitivity of N_2 fixation in three common bean cultivars to varying levels of P.

Sa and Israel (1995) also found that N_2 fixation in soybean was very sensitive to low available P. They further indicated that P deficiency might be directly

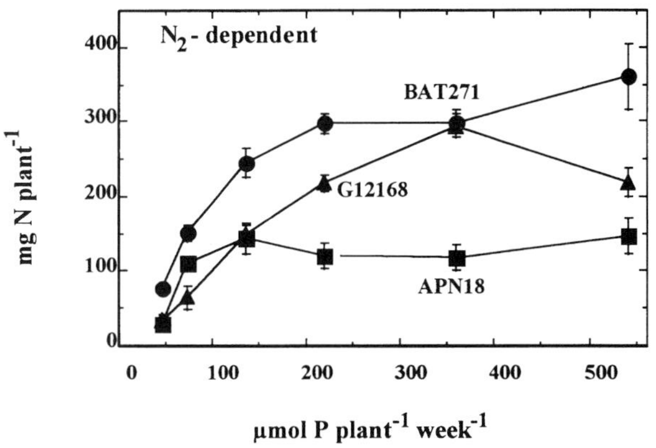

Figure 3. Nitrogen fixed in three common bean genotypes grown for 50 days in a N-free nutrient solution under various weekly supplies of P. Up to 200 μM P plant^{-1} plants absorbed all P supplied and could be compared on the basis of a similar P uptake. From Vadez et al. (submitted).

involved in the active loading of ureides, which are the compounds transported from soybean nodules, into the xylem for transport to the plant shoot. However, a contributing factor to the results of Sa and Israel (1995) was that the P deficiency severely decreased nodule mass so that the ability to produce ureide was drastically decreased. In any case, the high nodule ureide concentration observed under P deficiency might be explained by an increased specific nodule activity, which can be increased under P deficiency as a tolerance mechanism developed by certain bean cultivars (Vadez et al., 1997) or *Acacia mangium* (Ribet and Drevon, 1996; Sun et al., 1992).

Nodule P concentration is usually higher than in other plant tissues (Jakobsen, 1985; O'Hara et al., 1988; Robson et al., 1987). High P levels in nodules might be associated with the abundance of membranes, bacteroid, peri-bacteroid and other membranes, which are 20 times more abundant in infected than in non-infected cells of the nodule (Bergersen, 1994). Also, the energy charge is less affected in the bacteroid fraction of nodules than in the plant fraction of nodules (Sa and Israel, 1991). This might indicate that P is more tightly sequestered in infected than in non-infected cells of nodules. A consequence is that nodule P concentration is less affected by P deficiency (Adu-Gyamfi et al., 1989; Vadez et al., 1997). P deficiency seems to be readily accommodated by nodules of soybean (Ribet and Drevon, 1995) and common bean (Vadez et al., 1997) by decreasing the size of individual nodules.

A deficiency in soil P is not likely to be a hindrance to rhizobium development and the ability to initiate nodulation. Beck and Munns (1984) showed that rhizobium cells are able to take up P at very low solution concentrations (0.06–0.5 μM). In addition, the bacteria were able to accumulate excess P by forming polyphosphate granules that could sustain 3–4 subsequent generations of bacteria.

An important consideration for introducing legumes to a low-P environment, is the ability to sustain N_2 fixation activity when P levels are low. A key issue might be the distribution of P between the different organs of symbiotic plants under P deficiency. Those plants in which the allocation of P to nodules is a high priority are more likely to be able to sustain high N_2 fixation rates. Vadez et al. (1997, 1999) identified common bean cultivars that showed a capacity to fix 2–3 times more N at low, similar P-uptake levels (Figure 3). This was mainly explained by differences in the pattern of allocation of P in plants, with a higher allocation of P to nodules in tolerant genotypes. A similar result was found in superior fixing lines under P deficiency of mungbean (Gunawardena et al., 1992) where nearly 20% of total plant P was allocated to nodules. It has also been observed that a genotype in each species of *Acacia mangium* or pigeonpea is tolerant to P deficiency and appears to maintain a high level nodulation under low P conditions (Adu-Gyamfi et al., 1989; Sun et al., 1992). A comparison between common bean and cowpea receiving similar deficient P showed higher fixing capacity in cowpea than in common bean, which was correlated to a higher percentage of total plant P allocated to nodules in cowpea (Vadez and Gomez, pers. com.). Again, an advantage might be conferred to long-season cultivars, where competition for P between reproducing organs and nodules occur later in the cycle and allow for enhanced N_2 fixation. Indeed, Vadez et al. (1999) found in a screening of 48 genotypes receiving similar P that higher fixing lines were mostly long-season cultivars.

Conclusions

Nitrogen and phosphorus are essential components of virtually all the key molecules and tissues that make plant life possible. Deficiencies in either element result in quantitative decreases in the physiological activity of the crop. Leaf area development is one of the more sensitive responses of crop growth to N and P deficiencies. Photosynthesis rates are also decreased in response to severe limitations of N and P accumulation in the plant.

Some alterations in plant traits may be possible to ameliorate to a degree the large losses in plant activity that can occur with inadequate N and P. The N limitation in mass accumulation by C_3 crop species under low-N environments could be greatly decreased if the C_4 photosynthetic system could be introduced into C_3 species. This is a very complicated modification of the plant and certainly remains a speculative option. The P limitation for respiration can be minimized by increased utilization of enzymes that do not require P. This is an expensive alternative in terms of energy use in the plant, but it may be a good alternative under low-nutrient environments. Again, this approach is speculative and evidence of genotypic variation first needs to be gathered.

In the final analysis, however, increased grain yield requires increased N and P accumulation to synthesize the constituents of the developing seeds. Seeds have a clear requirement for both elements and it is not possible to increase crop yields without additional amounts of both elements to support seed growth. This perspective indicates that increased accumulation of N and P are obligate requirements to increase crop yield in low-nutrient environments.

Unfortunately, options for increasing nutrient uptake from the soil are somewhat limited. Roots have the ability to readily accumulate soluble N and P from the soil. Since soluble forms of N are highly mobile in the soil, there is no indication that altered root morphology would improve uptake. Phosphorus, on the other hand, is highly immobile and increased surface area in the rhizosphere for P uptake may well be useful for some crops. Specific modifications in length of root hairs or more extensive symbiosis with mycorrhiza might prove beneficial. Excretions from roots to stimulate the conversion of P from unavailable forms in the soil to more available forms offer an intriguing possibility.

One striking possibility to increase crop yields in low-nutrient environments is the increased use of legumes. Legumes offer the possibility of introducing N to the cropping system through symbiotic N_2 fixation. This fixed nitrogen can be used directly for seed production in grain legumes, or for soil enrichment by the incorporation into the soil of legume plant material. Legumes, however, still require P to sustain physiological activity and to grow tissues, including seeds. A major focus of research for low-nutrient environments may well be on the economy of P availability and accumulation by legumes. Approaches to enhance recovery of soil P by legumes will likely be a high priority to increase productivity of whole cropping systems in low-nutrient environments.

References

Adu-Gyamfi J J, Fujita K and Ogata S 1989 Phosphorus absorption and utilization efficiency of pigeon pea (*Cajanus Cajan* (L.) Millsp) in relation to dry matter production and dinitrogen fixation. Plant Soil 119, 315–324.

Adu-Gyamfi J J, Katayama K and Ito O 1999 Below-ground interactions and nitrogen management in cereal/legume intercropping in the semi-arid tropics. Jpn. J. Crop Sci. 67(extra issue 2), 233–236.

Ae N, Arihara J, Okada K, Yoshihara T and Johansen C 1990 Phosphorus uptake by pigeon pea and its role in cropping systems of the Indian subcontinent. Science 248, 477–480.

Amijee F, Barraclough P B and Tinker P B 1991 Modeling phosphorus uptake and utilization by plants. *In* Phosphorus Nutrition of Grain Legumes in the Semi-Arid Tropics. Eds. C Johansen, KK Lee and KL Sahrawat. pp. 63–76. ICRISAT, Patancheru, India.

Anghinoni I and Barber S A 1980 Phosphorus influx and growth characteristics of corn roots as influenced by phosphorus supply. Agron. J. 72, 685–688.

Asmar F, Gahoonia T S and Nielsen N E 1995 Barley genotypes differ in activity of soluble extracellular phosphatase and depletion of organic phosphorus in the rhizosphere soil. Plant Soil 172, 117–122.

Barnes D K, Sheaffer C C, Heichel G H, Smith D M and Peaden R N 1988 Registration of 'Nitro' alfalfa. Crop Sci. 28, 718.

Bates T R and Lynch J P 1996 Stimulation of root hair elongation in *Arabidopsis thaliana* by low phosphorus availability. Plant Cell Environ. 19, 529–438.

Beck D P and Munns D N 1984 Phosphate nutrition of *Rhizobium* spp. Appl. Environ. Microbiol. 47, 415–421.

Bergersen 1994 Distribution of O_2 within infected cells of soybean root nodules: A new simulation. Protoplasma 183, 49–60.

Bieleski R L 1973 Phosphate pools, phosphate transport and phosphate availability. Annu. Rev. Plant Physiol. 24, 225–252.

Cakmak I, Hengeler C and Marschner H 1994 Partitioning of shoot and root dry matter and carbohydrates in bean plants suffering from phosphorus, potassium and magnesium deficiency. J. Exp. Bot. 45, 1245–1250.

Cassman K G, Peng S, Olk D C, Ladha J K, Reichardt W, Dobermann A and Singh U 1998 Opportunities for increased nitrogen-use efficiency from improved resource management in irrigated rice systems. Field Crops Res. 56, 7–39.

Clarkson D T 1985 Factors affecting mineral nutrient acquisition by plants. Ann. Rev. Plant Physiol. 36, 77–115.

Clarkson D T and Grignon C 1991 The phosphate transport system and its regulation in roots. *In* Phosphorus Nutrition of Grain Legumes in the Semi-Arid Tropics. Eds. C Johansen, KK Lee and KL Sahrawat. pp. 49–59. ICRISAT, Patancheru, India.

Crafts-Brandner S J 1992 Significance of leaf phosphorus remobilization in yield production in soybean. Crop Sci. 32, 420–424.

Duff S M G, Moorhead G B G, Lefebvre D D and Plaxton W C 1989 Phosphate starvation inducible 'bypasses' of adenylate and phosphate dependent glycolytic enzymes in *Brassica nigra* suspension cells. Plant Physiol. 90, 1275–1278.

Duff S M G, Sarath G and Plaxton W 1994 The role of acid phosphatases in plant phosphorus metabolism. Physiol. Plant. 90, 791–800.

Feil B 1997 The inverse yield-protein relationship in cereals: Possibilities and limitations for genetically improving the grain protein yield. Trends Agron. 1, 103–119.

Föhse D, Claassen N and Jungk A 1988 Phosphorus efficiency of plants. I. External P requirement and P uptake efficiency of different plants species. Plant Soil 110, 101–109.

Foyer C and Spencer C 1986 The relationship between phosphate status and photosynthesis in leaves. Planta 167, 369–375.

Fredeen A L, Rao I M and Terry N 1989 Influence of phosphorous nutrition on growth and carbon partitioning in *Glycine max*. Plant Physiol. 89, 225–230.

Fredeen A L, Raab T K, Rao I M and Terry N 1990 Effects of phosphorus nutrition on photosynthesis in *Glycine max* (L.) Merr. Planta 181, 399–405.

Furihata T, Suzki M and Sakurai H 1992 Kinetic characterization of two phosphate uptake systems with different affinities in suspension-cultured *Catharanthus roseus* protoplasts. Plant Cell Physiol. 33, 1151–1157.

Gahoonia T S and Nielsen N E 1998 Direct evidence on participation of root hairs in phosphorus (^{32}P) uptake from soil. Plant Soil 198, 147–152.

Gimenez C, Connor D J and Rueda F 1994 Canopy development, photosynthesis and radiation-use efficiency in sunflower in response to nitrogen. Field Crops Res. 38, 15–27.

Goldstein A H, Mayfield S P, Danon A and Tibbot B K 1989 Phosphate starvation inducible metabolism in Lycopersicon esculentum. III. Changes in protein secretion under nutrient stress. Plant Physiol. 91, 175–192.

Greenwood D J, Lemaire G, Gosse G, Cruz P, Draycott A and Neeteson J J 1990 Decline in percentage N of C_3 and C_4 crops with increasing plant mass. Ann. Bot. 66, 425–436.

Gunawardena S F B N, Danso S K A and Zapata F 1992 Phosphorus requirement and nitrogen accumulation by three mungbean (*Vigna radiata* (L.) Welzk) cultivars. Plant Soil 147, 267–274.

Hairiah K, Van Noordwijk M and Setijono S 1993 Tolerance to acid soil conditions of the velvet bean *Mucana pruriens* var. *utilis* and *M. deeringiana*. II. Above-ground growth and control of *Imperata cylindrica*. Plant Soil 152, 175–185.

Hammer G L and Wright G C 1994 A theoretical analysis of nitrogen and radiation effects on radiation use efficiency in peanut. Aust. J. Agric. Res. 45, 575–589.

Hanway J J and Weber C R 1971a N, P and K percentages in soybean (*Glycine max* L. Merrill) plant parts. Agron. J. 63, 286–290.

Hanway J J and Weber C R 1971b Accumulation of N, P and K by soybean (*Glycine max* L. Merrill) plants. Agron. J. 63, 406–409.

Hardarson G 1993 Methods for enhancing symbiotic nitrogen fixation. Plant Soil 152, 1–17.

Hardarson G, Bliss F A, Cigales-Rivero M R, Henson R A, Kipe-Nolt J A, Longeri L, Manrique A, Pena-Cabriales J J, Pereira P A A, Sanabria C A and Tsai S M 1993 Genotypic variation in biological nitrogen fixation by common bean. Plant Soil 152, 59–70.

Helal H M 1990 Varietal differences in root phosphatase activity as related to the utilization of organic phosphates. Plant Soil 123, 161–163.

Henley W J, Levavasseur G, Franklin L A, Osmond C B and Ramus J 1991 Photoacclimation and photoinhibition in *Ulva rotundata* as influenced by nitrogen availability. Planta 184, 235–243.

Hole D J, Emran A M, Fares Y and Drew M C 1990 Induction of nitrate transport in maize roots, and kinetics of influx, measured with nitrogen-13. Plant Physiol. 93, 642–647.

Imsande J and Touraine B 1994 N demand and the regulation of nitrate uptake. Plant Physiol. 105, 3–7.

Itoh S and Barber S A 1983 Phosphorus uptake by six plant species as related to root hairs. Agron. J. 75, 457–461.

Jakobsen R 1985 The role of phosphorus in nitrogen fixation by young pea plants (*Pisum sativum*). Physiol. Plant. 64, 190–196.

Khamis S, Chaillou S and Lamaze T 1990 CO_2 assimilation and partitioning of carbon in maize plants deprived of orthophosphate. J. Exp. Bot. 41, 1619–1625.

Kipe-Nolt J A and Giller K G 1993 A field evaluation using the ^{15}N isotope dilution method of lines of *Phaseolus vulgaris* L. bred for increased nitrogen fixation. Plant Soil 152, 107–114.

Lambers H 1982 Cyanide-resistant respiration: A non-phosphorylating electron transport pathway acting as an energy overflow. Physiol. Plant. 55, 478–485.

Lauer M J, Blevins D G and Sierzputowska-Gracz H 1989 ^{31}P-nuclear magnetic resonance determination of phosphate compartmentation in leaves of reproductive soybeans (*Glycine max* L.) as affected by phosphate nutrition. Plant Physiol. 89, 1331–1336.

Lee R B, Ratcliffe R G and Southon T E 1990 ^{31}P NMR measurements of the cytoplasmic and vacuolar Pi content of mature maize roots: relationships with phosphorus status and phosphate fluxes. J. Exp. Bot. 41, 1063–1078.

Lee R B and Ratcliffe R G 1993 Subcellular distribution of inorganic phosphate, and levels of nucleoside triphosphate, in mature maize roots at low external phosphate concentrations: Measurements with ^{31}P-NMR. J. Exp. Bot. 44, 587–598.

Lefebvre D D, Duff S M G, Fife C A, Julien-Inalsingh C and Plaxton W C 1990 Response to phosphate deprivation in *Brassica nigra* suspension cells. Plant Physiol. 93, 504–511.

Lemaire G and Gastal F 1997 N uptake and distribution in plant canopies. *In* Diagnosis of the Nitrogen Status in Crops. Ed. G Lemaire. pp. 3–43. Springer, Berlin.

Lynch J P and Beebe S E 1995 Adaptation of beans (*Phaseolus vulgaris* L.) to low phosphorus availability. HortScience 30, 1165–1171.

McCloud D E 1998 Development of agricultural ecosystems. *In* Principles in Ecology in Plant Production. Eds. TR Sinclair and FP Gardner. pp. 49–61. CAB Intl., New York.

Meinzer F C and Zhu J 1998 Nitrogen stress reduces the efficiency of the C_4 CO_2 concentrating system and, therefore, the quantum yield, in *Saccharum* (sugarcane) species. J. Exp. Bot. 49, 1227–1234.

Mimura T, Dietz K -J, Kaiser W, Schramm M J, Kaiser G and Heber U 1990 Phosphate transport across biomembranes and cytosolic phosphate homeostasis in barley leaves. Planta 180, 139–146.

Mollier A and Pellerin S 1999 Maize root system growth and development as influenced by phosphorous deficiency. J. Exp. Bot. 50, 487–497.

Muchow R C 1988 Effect of nitrogen supply on the comparative productivity of maize and sorghum in a semi-arid tropical environment. I. Leaf growth and leaf nitrogen. Field Crops Res. 18, 1–16.

Muchow R C and Sinclair T R 1994 Nitrogen response of leaf photosynthesis and canopy radiation use efficiency in field-grown maize and sorghum. Crop Sci. 34, 721–727.

O'Hara G W, Boonkerd N and Dilworth M J 1988 Mineral constraints to nitrogen fixation. Plant Soil 108, 93–110.

Oscarson P, Lundborg T, Larrson M and Larsson C -M 1995 Genotypic differences in nitrate uptake and nitrogen utilization for spring wheat grown hydroponically. Crop Sci. 35, 1056–1062.

Pearcy R W and Seemann J R 1990 Photosynthetic induction state of leaves in a soybean canopy in relation to light regulation of ribulose-1,5-biosphosphate carboxylase and stomatal conductance. Plant Physiol. 94, 628–633.

Peng S and Cassman K G 1998 Upper threshholds (sic) of nitrogen uptake rates and associated nitrogen fertilizer efficiencies in irrigated rice. Agron. J. 90, 178–185.

Peng S, Cassman K G and Kropff M J 1995 Relationship between leaf photosynthesis and nitrogen content of field-grown rice in tropics. Crop Sci. 35, 1627–1630.

Pracht J E, Nickell C D, Harper J E and Bullock D G 1994 Agronomic evaluation of non-nodulating and hypernodulating mutants of soybean. Crop Sci. 34, 738–740.

RajRaman D, Spanswick R M and Walker L P 1995 The kinetics of nitrate uptake from flowing solutions by rice: Influence of pretreatment and light. Biores. Tech. 53, 125–132.

Rao M and Terry N 1989 Leaf phosphate status, photosynthesis and carbon partitioning in sugar beet. I. Changes in growth, gas exchange and Calvin Cycle enzymes. Plant Physiol. 90, 814–819.

Rao I M, Fredeen A L and Terry N 1993 Influence of phosphorus limitation on photosynthesis, carbon allocation and partitioning in sugar beet and soybean grown with a short photoperiod. Plant Physiol. Biochem. 31, 223–231.

Ratnayake M, Leonard R T and Menge J A 1978 Root exudation in relation to supply of phosphorus and its possible relevance to mycorrhizal formation. New Phytol. 81, 543–552.

Rebeille F, Bligny R and Douce R 1984 Is the cytosolic Pi concentration a limiting factor for plant cell respiration? Plant Physiol. 74, 355–359.

Ribet J and Drevon J J 1995 Increase in permeability to oxygen diffusion and in oxygen uptake of soybean nodules under limiting phosphorus nutrition. Physiol. Plant. 94, 298–304.

Ribet J and Drevon J -J 1996 The phosphorus requirement of N_2-fixing and urea-fed *Acacia mangium*. New Phytol. 132, 383–390.

Robson A D, O'Hara G W and Abbott L K 1987 Involvement of phosphorus in nitrogen fixation by subterranean clover (*Trifolium subterraneum* L.). Aust. J. Plant. Physiol. 8, 427–436.

Rufty T W Jr, Israel D W, Volk R J, Aiu J and Sa T 1993 Phosphate regulation of nitrate assimilation in soybean. J. Exp. Bot. 44, 879–891.

Rupela O P, Sharma L C and Danso S K A 1997 Evaluation of N_2 fixation by nodulation-variants of chickpea in India. *In* Improving Yield and Nitrogen Fixation of Grain Legumes in the Tropics and Sub-tropics of Asia. IAEA-TECDOC-1027 pp. 99–119. International Atomic Energy Agency. Vienna, Austria.

Rychter A M, Chauveau M, Bomsel J -L and Lance C 1992 The effect of phosphate deficiency on mitochondrial activity and adenylate levels in bean roots. Physiol. Plant. 84, 80–86.

Sa T -M and Israel D W 1991 Energy status and functioning of phosphorus-deficient soybean nodules. Plant Physiol. 97, 928–935.

Sa T -M and Israel D W 1995 Nitrogen assimilation in nitrogen-fixing soybean plants during phosphorus deficiency. Crop Sci. 35, 814–820.

Schachtman D P, Reid R J and Ayling S M 1998 Phosphorus uptake by plants: From soil to cell. Plant Physiol. 116, 447–453.

Sheehy J E, Dionora M J A, Mitchell P L, Peng S, Cassman K G, Lemaire G and Williams R L 1998 Critical nitrogen concentrations: implications for high-yielding rice (*Oryza sativa* L.) cultivars in the tropics. Field Crops Res. 59, 31–41.

Shiraiwa T and Sinclair T R 1993 Distribution of nitrogen among leaves in soybean canopies. Crop Sci. 33, 804–808.

Siddiqi M Y, Glass A D M, Ruth T J and Rufty T W Jr 1990 Studies of the uptake of nitrate in barley. I. Kinetics of 13NO3- influx. Plant Physiol. 93, 1426–1432.

Sinclair T R 1990 Nitrogen influence on the physiology of crop yield. *In* Theoretical Production Ecology: Reflections and Prospects. Eds. R Rabbinge, J Goudriaan, H van Keulen, F W T Penning de Vries and H H van Laar. pp. 41–55. Pudoc, Wageningen, The Netherlands.

Sinclair T R 1999 Options for sustaining and increasing the limiting yield-plateau of grain crops. Jpn. J. Crop Sci. 67(extra issue 2), 65–75.

Sinclair T R and Horie T 1989 Leaf nitrogen, photosynthesis, and crop radiation use efficiency: A review. Crop Sci. 29, 90–98.

Sinclair T R and Muchow R C 1999 Radiation use efficiency. Adv. Agron. 65, 215–265.

Sinclair T R and Park W I 1993 Inadequacy of Liebig limiting-factor paradigm for explaining varying crop yields. Agron. J. 85, 742–746.

Sinclair T R and Serraj R 1995 Legume nitrogen fixation and drought. Nature 378, 344.

Sinclair T R and Sheehy J E 1999 Erect leaves and photosynthesis in rice. Science 283, 1456–1457.

Sinclair T R and Shiraiwa T 1993 Soybean radiation-use efficiency as influenced by nonuniform specific leaf nitrogen distribution and diffuse radiation. Crop Sci. 33, 808–812.

Singh U, Ladha J K, Castillo E G, Punzalan G, Tirol-Padre A and Duqueza M 1998 Genotypic variation in nitrogen use efficiency in medium- and long-duration rice. Field Crops Res. 58, 35–53.

Smith S E and Gianinazzi-Pearson V 1988 Physiological interactions between symbionts in vesicular-arbuscular mycorrhizal plants. Ann. Rev. Plant. Physiol. 39, 221–244.

Snapp S S and Lynch J P 1996 Phosphorus distribution and remobilization in bean plants as influenced by phosphorus nutrition. Crop Sci. 36, 929–935.

Sun J S, Simpson R J and Sands R 1992 Nitrogenase activity of two genotypes of *Acacia mangium* as affected by phosphorus nutrition. Plant Soil 144, 51–58.

Teyker R H, Moll R H and Jackson W A 1989 Divergent selection among maize seedlings for nitrate uptake. Crop Sci. 29, 879–884.

Theodorou M E and Plaxton W C 1993 Metabolic adaptations of plant respiration to nutritional phosphate deprivation. Plant Physiol. 101, 339–344.

Thung M 1991 Bean agronomy in monoculture. *In* Common Beans: Research for Crop for Improvement. Eds. A Van Schoonhoven and O Voysest. pp. 737–835. CAB International/CIAT, Wallingford, UK.

Tirol-Padre A, Ladha J K, Singh U, Laureles E, Punzalan G and Akita S 1996 Grain yield performance of rice genotypes at sub-optimal levels of soil N as affected by N uptake and utilization efficiency. Field Crops Res. 46, 127–143.

Usuda H and Shimogawara K 1991 Phosphate deficiency in maize. I. Leaf phosphate status, growth, photosynthesis and carbon partitioning. Plant Cell Physiol. 32, 497–504.

Vadez V, Beck D P and Drevon J J 1997 Utilization of the acetylene reduction assay to screen for tolerance of symbiotic N_2-fixation to limiting P nutrition in common bean. Physiol. Plant. 99, 227–232.

Vadez V, Lasso J H, Beck D P and Drevon J J 1999 Variability of N_2-fixation in common bean (*Phaseolus vulgaris* L.) under P deficiency is related to P use efficiency – N_2-fixation tolerance to P deficiency. Euphytica 106, 231–242.

Vos J and Van Der Putten P E L 1998 Effect of nitrogen supply on leaf growth, leaf nitrogen economy and photosynthetic capacity of potato. Field Crops Res. 59, 63–72.

Yan X, Lynch J P and Beebe S E 1995a Genetic variation for phosphorus efficiency of common bean in contrasting soil types: I. Vegetative response. Crop Sci. 35, 1086–1093.

Yan X, Lynch J P and Beebe S E 1995b Genetic variation for phosphorus efficiency of common bean in contrasting soil types: I. Yield response. Crop Sci. 35, 1094–1099.

Yan X, Lynch J P and Beebe S E 1996 Utilization of phosphorus substrates by contrasting common bean genotypes. Crop Sci. 36, 936–941.

J.J. Adu-Gyamfi (Ed.), Food security in nutrient-stressed environments: exploiting plants' genetic capabilities: 25–36, 2002.

Prospects for using conventional techniques and molecular biological tools to enhance performance of 'orphan' crop plants on soils low in available phosphorus

C. T. Hash[1,4], R. E. Schaffert[2] & J. M. Peacock[3]
[1]*Genetic Resources and Enhancement Program (GREP), International Crops Research Institute for the Semi-Arid Tropics (ICRISAT), Patancheru 502 324, Andhra Pradesh, India.* [2]*Embrapa Milho e Sorgo (Embrapa/CNPMS), Sete Lagoas, Minas Gerais, Brazil.* [3]*International Center for Agricultural Research in Dry Areas (ICARDA), Arabian Peninsula Regional Program (APRP), PO Box 13979, Dubai, United Arab Emirates.* [4]*Corresponding author**

Key words: chickpea, groundnut, marker-assisted selection, pearl millet, phosphorus acquisition, pigeonpea, roots, sorghum

Abstract

Molecular biology, combined with Mendelian and quantitative genetics in quantitative trait locus (QTL) mapping and marker-assisted selection (MAS), provides powerful new tools to facilitate efficient genetic manipulation by plant breeders of complex traits such as drought tolerance and phosphorus (P) acquisition ability. This paper examines current opportunities for genetically manipulating the ability of crop plants to more efficiently acquire (i.e. access and take up) essential soil nutrients, using as examples P and several of the crops in the genetic improvement mandate of the International Crops Research Institute for the Semi-Arid Tropics (ICRISAT) – chickpea (*Cicer arietinum* L.), groundnut (*Arachis hypogaea* L.), pearl millet [*Pennisetum glaucum* (L.) R. Br.], pigeonpea [*Cajanus cajan* (L.) Millsp.], and sorghum [*Sorghum bicolor* (L.) Moench]. It is concluded that for at least some of these important, but often academically and economically orphaned tropical food, forage and feed-grain crops, the genetic variation and molecular tools that we will need already exist or can be expected to become available in the very near future. With appropriate, targeted research, these tools can permit empirical exploration of the potential for marker-facilitated mapping and manipulation of major genes that can contribute to enhanced ability of these crops to acquire P from sources with limited availability. With these tools, delivery of new versions of currently popular high-yielding, high quality, disease resistant crop cultivars, having genetically improved ability to acquire P currently in soils but unavailable for crop growth, could take as little as five to seven years. Sustainable use of such improved cultivars would require their utilization as components of integrated soil fertility management systems.

Introduction

Sustainable increases in global production of food and feed-grain crops – both legumes and cereals – will increasingly require integration of improved cultivars and cultural practices to economically achieve improvements in crop productivity and in stability of this productivity. This can be brought about in favorable environments in part by increasing the amounts of soil nutrients available to crops by application of mineral and organic fertilizers, and selection for short-statured crop genotypes that are more responsive to fertilization and less prone to lodging. However, high rates of fertilization in such environments can result in pollution of surface and sub-surface water supplies since much of the applied fertilizer nutrients are not taken up by the crop and instead leach into the subsoil or are carried away with soil particles in runoff water. In environments less favorable for agricultural production, moisture and soil toxicity constraints can interact very strongly with soil nutrient availability (Brück et al., 2000; Payne et al., 1995; Schaffert et al., 2000;

* E-mail: C.T.HASH@CGIAR.ORG

This paper previously appeared in *Plant and Soil* **245**, 2002, pp. 135–146.

Zaongo et al., 1994), sometimes making it difficult to obtain economic responses to individual fertilizers or other soil amendments (Subba Rao et al., 2000). Phosphorus (P) uptake by plants is reduced to near zero during water stress. The consequences of this on plant growth have been clearly outlined by Sinclair and Vadez (2000). The problem of drought stress is often complicated by soil fertility stress that impedes deep root development (e.g. Payne et al., 1995; Zaongo et al., 1994). This is a frequent occurrence and of particular concern across the tropics – from arid regions to sub-humid regions. Aluminium (Al) toxicity frequently occurs across the tropical acid soil savannas (Schaffert et al., 2000). A high level of aluminium saturation in the soil solution inhibits cell multiplication and division in growing roots, thereby greatly restricting root growth and consequently water and nutrient uptake. Many tropical soils, including those of the acid savannas in Africa and Latin America and of other regions, are low in available P and have a high potential for P fixation due to mineral compositions high in iron (Fe) and Al oxides (Weir, 1972, 1977). In these soils, P is one of the least available essential elements. In fact, although the total P present can be as high as in more productive soils having limited P fixation potential, P in these soils is often available only in micronutrient concentrations. These soil constraints, due to their nature, cannot be resolved completely by the simple application of irrigation water and other soil amendments. Enhancing crop growth under such conditions requires improving the ability of crop plants to acquire (i.e. gain access to and take up) nutrients from the soil, as there appears to be little opportunity to enhance efficiency of biomass production from a fixed amount of available nutrients (Sinclair and Vadez, 2000). It will be necessary to breed agronomically productive plant genotypes with the ability to increase the proportion of total soil nutrients available for crop growth. Fortunately, there appear to be exploitable levels of genetic variation in many crop species for root traits associated with superior ability to acquire nutrients that are growth limiting. Some plants with such root traits can perform close to their maximum capacity under conditions of low nutrient availability, and for this paper will be defined as being more nutrient efficient than plant genotypes requiring higher nutrient availability levels for near maximum productivity.

In both favorable and less favorable agricultural production environments, genetic improvements in the ability of crops to acquire soil nutrients – both naturally occurring and applied – will be critical to achieving sustainable increases in agricultural productivity. Genetic enhancement of root traits (Lynch, 1995), including root-induced changes in the chemistry of the rhizosphere that can enhance nutrient availability (Hinsinger, 1998), will be critical to improving crop acquisition of soil nutrients.

This paper will explore the potential to enhance the ability of crops to access sparingly available P sources – such as rock phosphate and that which is fixed by iron and aluminium oxides in heavily weathered tropical soils – using genetic variation already present in the germplasm collections of each of these crops and their wild relatives.

Importance of phosphorus in crop growth

The essential physiological roles of P in crop growth are reviewed in considerable detail by Sinclair and Vadez (2000), so only a brief summary is provided here. P is an integral component of the biochemical compounds that make life possible – playing essential structural, biochemical and physiological roles in crop growth. It is a component of nucleic acids and phospholipids, and plays an especially important role as the energy 'currency' of nucleotide-based metabolic pathways (that often involve membrane processes) in which energy storage and transfer occur. Stable cytoplasmic concentrations of inorganic P (Pi) often appear necessary for sustaining optimal levels of enzymatic activity. This requirement is normally met in plants by using vacuole-stored P as a buffer for cytoplasmic P concentrations. However, developing plant cells lack mature vacuoles, and hence a stored P reserve to permit growth under P-deficient conditions. Because of this, developing cells in growing meristems are particularly sensitive to P deficiency. Reduced rates and extents of leaf area expansion, and through this – and other processes – reduced rates of increase in individual plant photosynthetic capacity, can be the first and most dramatic responses of crops to P deficits. Of course, these responses have direct negative consequences for subsequent crop growth rates and biomass production via their effects on plant energy metabolism.

Options for increasing growth in low-phosphorus conditions

As the role of P in life processes appears to have been optimized during the course of evolution, there is likely to be little opportunity for reducing the amount of it that is required for plant growth. Therefore, in order to increase plant productivity, ways must be sought to increase the amount of P available for plant growth. Conceptually, this can occur either by moving stored P to the most rapidly growing plant tissues, or accessing additional P from the soil. Opportunities for enhancing the mobility of P between plant parts appear to be limited as it is generally present as Pi or in mobile compounds that are readily available for transfer between plant parts (Sinclair and Vadez, 2000), so that phosphate harvest index is commonly quite high. Indeed, phosphate harvest indices are sufficiently high [$\geq$ 0.6 in pearl millet (Wani et al., 1990) and $\geq$ 0.8 for some grain legumes] that the fraction of total crop plant P that is allocated to harvested portions cannot likely be increased substantially (see Inthapanya et al., 2000; Sinclair and Vadez, 2000). Apart from addition of P to the cropping system in the form of crop residues (Buerkert et al., 1998; Hafner et al., 1993; Ikpe et al., 1999), farmyard manure and/or livestock urine (Ikpe et al., 1999), rock phosphate and/or mineral fertilizers (Batiano et al., 2000; Brück et al., 2000; Ikpe et al., 1999; Inthapanya et al., 2000; Subbarao et al., 2000), it should be possible to improve the ability of plant roots to acquire P that is already present in soils but not currently accessible (Subbarao et al., 1997a,b). There appear to be several mechanisms by which plant roots could do this. These range from increasing the soil volume explored – effectively increasing root surface area by increasing root branching and/or length and/or enhancing symbiotic relationships with soil micro-organisms (Tinker, 1984) such as vesicular-arbuscular mycorrhizae (VAM) (Gavito and Miller, 1998; Kaeppler et al., 2000; Krishna et al., 1985; Li et al., 1991; Marschner and Dell, 1994; Smith et al., 1992) – to increasing the solubility of P that is normally sorbed to soil particles in the vicinity of root surfaces and hence unavailable for uptake by plants. Several of these mechanisms are dealt with in greater detail by others later in this publication.

How phosporus is acquired by plants – the role of root exudates

Soluble P in the rhizosphere is readily recovered by plant roots (Hinsinger, 1998). However, P commonly occurs at very low concentrations in the soil solution. Most P present in tropical soils is found in soil organic matter, as a component of sparingly soluble phosphate minerals, or sorbed to Fe and Al oxides (Hafner et al., 1993; Morris et al., 1992; Tiessen et al., 1993). Such forms of P, that are not soluble, are not directly available for uptake by roots.

Severe decreases in soil solution P concentration of the rhizosphere can shift the adsorption–desorption and dissolution–precipitation equilibria. However, because P sorption onto soil constituents is often poorly reversible, and most phosphate minerals in soils have low solubility, solution P concentrations must reach very low levels before significant equilibria shifts occur. Some plant species have very low solution P concentration requirements for near-maximum plant growth, as little as 0.1–0.5 μM for P-efficient grasses (Breeze et al., 1984; Föhse et al., 1988). However, even for such species, the rate of desorption of soil P or the rate of dissolution of P-bearing soil constituents must be larger that the fluxes of P uptake if P availability is not to limit growth. Surprisingly, several studies (for a summary see Hinsinger, 1998) indicate that P sometimes accumulates in the rhizosphere rather than being depleted. In *Lolium*, proton excretion can serve as the driving force for root-induced dissolution of rock phosphate that proceeds at a faster rate than P uptake. Modeling results indicate that with diffusion of protons away from roots and diffusion of dissolved P towards the roots, P can be expected to be depleted in the immediately vicinity of the roots and accumulate a short distance away. Small amounts of root growth into this P-accumulation zone could then result in substantially increased amounts of P available for uptake. Efficient uptake of the mineral nutrient from soil solution then occurs, resulting in large changes in ionic concentrations that again shift the equlibria involved in soil nutrient dynamics. Thus P uptake itself is not only the ultimate stage of the nutrient acquisition process, but can be regarded as the crucial driving force that enhances the proportion of soil P available for plant growth (Hinsinger, 1998).

As dynamics of the various forms of Pi are strongly pH-dependent, pH of the rhizosphere (to an even greater degree than that of the bulk soil solution) is a critical factor that can influence whether roots have access to P present in the soil. Root-mediated pH changes are the most documented chemical changes that have been observed to occur in the rhizosphere (Hinsinger, 1998). Root-induced acidification of the rhizosphere occurs when plant roots compensate for an excess uptake of cations over anions from the soil solution by releasing excess positive charge in the form of protons. The ability of some plants to utilize

P supplied in the form of phosphate rock is related to their capacity for proton excretion, and proton excretion is stimulated in some species under conditions of P-deficiency. Therefore, proton excretion by plant roots can be regarded as an adaptive strategy for P acquisition (Hinsinger, 1998).

Conversely, phosphate ions are strongly sorbed to Fe- and Al-oxides in acid soils (Ae et al., 1990; Weir, 1972, 1977). The charge of these P-sorbing minerals is pH dependent. Hydroxyl or bicarbonate ions excreted into the rhizosphere by plant roots can desorb some phosphate from these soil minerals, rendering P available for uptake. Thus both increases and decreases in rhizosphere pH induced by roots can enhance P nutritional status of plants, depending on the forms of Pi that are predominant in the soil. However, the consequences of root excretion of protons or hydroxyls is probably much greater than that indicated by rhizosphere pH changes as the critical factor influencing acquisition of P by plant roots is the actually flux of proton or hydroxyl equivalents excreted at the soil–root interface and not the resulting rhizosphere pH change (Hinsinger, 1998).

Exudation of organic acids from roots of P-deficient seedlings has also occasionally been reported to result in pH changes in the rhizosphere. However, the more important role of such organic acid exudation may be for the resulting organic anions to complex with Al or Fe, resulting in increased solubilization of P-Al and P-Fe compounds. Citrate, malate and oxalate anions, which can be exuded by roots of several acid soil tolerant grass and cereal genotypes, appear to be especially effective in forming complexes with Al and hence facilitate access to Al-bound P (Bolan et al., 1994; Jones, 1998; Pellet et al., 1995, 1996), while piscidic acid and its derivatives exuded from pigeon pea roots enhance the ability of this legume to access Fe-bound P (Ae et al., 1990).

Finally, there are several additional roles that root exudates can play in P acquisition. Phosphatase and phytase ectoenzymes can play important roles in catalytic hydrolysis of organic P and subsequent acquisition of soil P by plant roots. Such ectoenzymes can be produced by root-associated microbes that feed on root exudates, or can be excreted into the rhizosphere by the roots themselves in the absence of microbes. Stimulation by P-deficiency of direct excretion of phosphatase by plant roots has been demonstrated, so release of such enzymes can be considered another adaptive response to nutrient stress. Further, exudation of high-molecular-weight polysaccharides (commonly referred to as mucilage) can contribute to ion exchange capacity of root cell walls. In acid soils, Al can be detoxified by massive adsorption on mucilage, contributing to desorption of anions, such as phosphate, from soil particles (Hinsinger, 1998).

Thus, there is overwhelming evidence that associations of plant roots with VAM, and the exudates from plant roots, can contribute to significant increases in the supply of P to plant roots under conditions where only a limited portion of total soil-P is soluble. There are, therefore, two sets of target traits for assessment and possible utilization by plant breeders and physiologists. However, each set of traits is complex, and difficult to measure. Like other root-related traits, expression of phenotypic differences take place underground and out of sight of the breeder, so they will likely be expensive to assess and perhaps not possible to deal with in conventional plant breeding programs. Fortunately, there is a rapidly emerging set of tools suitable for dealing with traits such as these that are economically valuable but too complex, too difficult, too tedious and/or too expensive to accurately assess repeatedly on the large numbers of progenies required in an applied plant breeding program. These tools – quantitative trait locus (QTL) mapping and marker-assisted selection (MAS) – are the result of the melding together of Mendelian, molecular and quantitative genetics. They offer crop agronomists, breeders, nutritionists, physiologists, . . . , a new set of tools with which to effectively address complex target traits like enhanced drought tolerance or P-acquisition ability (Jones et al., 1997; Mohan et al., 1997; Prioul et al., 1997; Schneider et al., 1997). Conventional applied plant breeding approaches may also have potential (Bramel-Cox et al., 1991), but would require more cost-effective phenotypic screening procedures to permit repeated and accurate assessment of P-acquisition ability differences between large numbers of progenies.

QTL mapping and MAS

Molecular markers are rapidly being adopted by crop improvement researchers globally as an effective and appropriate tool for basic and applied studies addressing biological components in agricultural production systems (Jones et al., 1997; Mohan et al., 1997; Prioul et al., 1997). Molecular markers offer specific advantages in assessment of genetic diversity and in trait-specific crop improvement (Edwards, 1992;

Kochert, n.d.; Paterson et al., 1991). Use of markers in applied breeding programs can range from facilitating appropriate choice of parents for crosses, to mapping/tagging gene blocks associated with economically important traits (often termed QTLs) (Doerge et al., 1997). Gene tagging and QTL mapping in turn permit marker-assisted selection (MAS) in backcross (Frisch et al., 1999a,b; Ribaut et al., 1997), pedigree (Mohan et al., 1997), and population improvement programs. Plant numbers required to obtain the desired segregants can be readily predicted (Hash, 2000; Hash et al., 2000; Sedcole, 1977; Stam and Zeven, 1981), allowing well-structured plant breeding programs to be designed. MAS appears to be especially useful for crop traits that are otherwise difficult or impossible to deal with by conventional means (Moreau et al., 1998, 1999). The near-isogenic products of a marker-assisted backcrossing programme in turn provide genetic tools for crop physiologists and crop protection scientists to use in improving understanding the mechanisms of tolerance to various abiotic stresses such as extremes of temperature and water and/or nutrient availability (Jones et al., 1997; Prioul et al., 1997), in addition to mechanisms of resistance to biotic production constraints such as diseases, insect pests, nematodes, and parasitic weeds like *Striga*. QTL mapping of yield and quality components, and the components of other physiologically or biochemically complex pathways, can provide crop breeders with a better understanding of the basis for genetic correlations between economically important traits (linkage and/or pleiotropic relationships between gene blocks controlling associated traits; e.g. flowering time and biomass; inflorescence size and inflorescence number). This can facilitate more efficient incremental improvement of specific individual target traits like P-acquisition ability. Further, specific genomic regions associated with QTLs of large effect for one target trait can be identified having minimal effects on otherwise normally correlated traits, permitting an improvement in the first trait that need not be accompanied by counterbalancing reductions in others. Finally, these molecular marker tools can also be used in ways that allow more effective discovery and exploitation of the evolutionary relationships between organisms, through comparative genomics (Devos and Gale, 1997, 2000; Devos et al., 2000; Gale and Devos, 1998).

Opportunities

We would now like to switch gears, and look at opportunities to use these new genetic tools in ICRISAT mandate crops and related species, to identify and manipulate gene blocks contributing to within-species differences in P-acquisition ability, building on the recent examples of Wissuwa et al. (1998) in rice (*Oryza sativa* L.) and Kaeppler et al. (2000) in maize (*Zea mays* L.). In order for such an approach to be successful, we first require heritable genetic variation for the trait(s) of interest, effective screening procedures for efficient detection of these genetic differences (at least once, and under conditions that ultimately are relevant to farmers' fields in the breeding programme's target environment), adequate levels of marker polymorphism, and potential parents of mapping populations that differ in the both the trait(s) of interest (in economically important levels) and in marker genotype (at least in the vicinity of gene blocks contributing to the traits of interest. Therefore each of these is examined in turn.

Availability of genetic variability

There are considerable numbers of publications reporting genetic variability for various traits demonstrated to have, or with circumstantial evidence for, association with enhanced P-acquisition ability (Ae et al., 1990; Graham and Eissenstat, 1994; Kaeppler et al., 2000; Lynch, 1995; Mucchal et al., 1996; Pellet et al., 1995, 1996; Raghothama, 1999; Ryan et al., 1995; Subbarao et al., 1997a,b; Wani et al., 1990; Wissuwa et al., 1998). It appears that at least some genotypes of groundnut, pigeonpea and rice have the ability to access and take up Fe- and/or Al-bound P (Otani and Ae, 1996b; Subbarao et al., 1997a,b). Genotypic differences have been detected between pigeonpea cultivars for P solubilizing activity of root exudates (Ishikawa et al., 2000; Subbarao et al., 1997a). Similarly, substantial genetic differences have been detected for root growth in chickpea (Ali et al., 2000) and pearl millet (Krishna et al., 1985), and for both P-use efficiency (Batiano et al., 2000; Wani et al., 1990) and response to mycorrhizal colonization (Krishna et al., 1985) in pearl millet. In some cases (e.g. Kaeppler et al., 2000; Wissuwa et al., 1998), QTLs associated with enhanced P uptake have already been mapped.

Availability of reliable screening procedures for accurate phenotyping

Reliable screening procedures for phosphorus uptake in controlled conditions are now available (Kaeppler et al., 2000; Otani and Ae, 1996a,b; Subbarao et al., 1997a,b). Similarly, a system permitting rapid assessment of root volume on large numbers of plants under field conditions has recently been described (van Beem et al., 1998) and is now being evaluated for improving drought tolerance in maize (Bänziger et al., 2000; Mugo et al., 1999). Additional procedures that could be used in phenotypic characterization of factors contributing to genetic variation in P-acquisition ability are described by others in this publication. Among this set of procedures, we believe there will undoubtedly be some that can be used to assess genetic variation in P-acquisition ability that is relevant to the specific causes of non-availability of P in a particular target environment, and can be used on large enough numbers of mapping progenies, with high enough heritability, to permit QTL detection in mapping populations that are segregating for the trait and having adequate marker polymorphism, as has already been done for rice (Wissuwa et al., 1998) and maize (Kaeppler et al., 2000; Reiter et al., 1991).

Availability of molecular biology tools

Extensive research in recent years has demonstrated genomic synteny across the grasses (Devos and Gale, 1997, 2000; Gale and Devos, 1998; Van Deynze et al., 1998). Once suitable mapping populations for P acquisition and related target traits are identified or developed in pearl millet and sorghum, we can begin by attempting marker saturation in genomic regions related to those in which QTLs for P uptake have already been detected in rice (Wissuwa et al., 1998) and maize (Kaeppler et al., 2000). For pearl millet, we already have available a short genetic linkage map based on RFLP and STS markers (Devos et al., 1995; Liu et al., 1994), and this has recently been tied to the linkage maps of foxtail millet and rice (Devos et al., 2000) and hence other well-studied cereals and forage grasses. Levels of marker polymorphism in pearl millet are very high, even between elite inbred parental lines of hybrids adapted to India. Unfortunately, we do not yet have large sets of STMS or other micro-satellite-based markers, but these are under development (Katrien Devos, pers. comm.). In the case of sorghum, several molecular-marker based genetic linkage maps are available (Peng et al., 1999; Subudhi and Nguyen, 2000). Although the level of marker polymorphism in sorghum appears to be much less than that in pearl millet, it should be adequate as more STMS and other micro-satellite-based markers become publicly available over the next couple of years (Bhattramakki et al., 2000; Boivin et al., 1999; Brown et al., 1996; Kong et al., 2000; Tamarino et al., 1997). The required molecular tools for the legumes in ICRISAT's genetic enhancement mandate are currently under development. In case of chickpea, STMS markers that have recently been developed at the University of Frankfurt (Hüttel et al., 1999; Winter et al., 1999) are now permitting detection of polymorphism where little could be detected with previously available molecular marker systems (Jagdish Kumar, pers. comm.). Similar PCR-compatible co-dominant marker systems based on simple sequence repeats are currently under development for pigeonpea in a DFID-supported collaboration between ICRISAT and the University of Birmingham (Paula Bramel, pers. comm.), but still need to be developed for groundnut (SL Dwivedi, pers. comm.).

Current availability of mapping populations

Pearl millet

Several pearl millet mapping populations of moderate size (120–275 progenies) have been developed as sets of F_4 progeny bulks and their F_3 testcrosses derived from individual skeleton-mapped F_2 plants (Hash and Bramel-Cox, 2000; Hash and Witcombe, 1994). These now involve some 10 pairs of genetically diverse inbred lines, of Asian, African and American origin, selected for QTL mapping of disease resistances (Jones et al., 1995), abiotic stress tolerances (Howarth et al., 1997; Yadav et al., 1999, 2000), grain and stover yield and quality components (Hash and Bramel-Cox, 2000; Hash et al., 2001) and morphological markers. Several of these populations have parents of contrasting Indian and West African origin (e.g. PT 732B × P 1449-2; H 77/833-2 × PRLT 2/89-33; 841B × 863B; and W 504 × P 310-17) that could be expected to differ in their ability to be colonized by VAM (Graham and Eissenstat, 1994; Krishna et al., 1985). All ICRISAT pearl millet mapping population parental lines have recently been testcrossed to a set of four genetically and phenotypically diverse elite male-sterile lines to allow ready assessment of differences for such additional secondary target traits as P-acquisition ability in both inbred and hybrid form. Originally conceived as a quick way to assess opportunities for using existing skeleton-mapped populations for mapping components of stover yield and quality (Hash et al., 2002), this line × tester set should also permit rapid evaluation of the potential for using these populations to map QTLs for many other traits including components of improved P acquisition.

Sorghum
Several skeleton-mapped sorghum random inbred line (RIL) sets are also available, with ICRISAT and several other agencies globally. However, the array of tropically adapted genotypes in these is not so broad as those used in developing the pearl millet mapping populations, and the degree of marker polymorphism in these is markedly less than that commonly detected in pearl millet. Thus, there is a need to increase the diversity of tropically-adapted sorghums to be included in future mapping populations (see sorghum recommendations in Ribaut and Poland, 2000).

Groundnut
Adequate mapping populations derived from crosses of cultivars are not yet available due to the low levels of marker polymorphism in cultivated groundnut; however, the cross of a cultivated tetraploid with an artificial allotetraploid (developed from an interspecific cross of two wild diploids) which shows good levels of RFLP marker polymorphism and linkage map development is underway at the University of Georgia (A. Paterson, pers. comm.). There are also sets of progenies derived from backcross introgression of segments from wild relatives into cultivated groundnut (Garcia et al., 1995; Nalini Mallikarjuna, pers. comm.) that could be screened for both phenotypic and marker polymorphism (Tanksley and McCouch, 1997).

Chickpea
In chickpea, several mapping populations have been developed, including derivatives of both cultivated × wild and cultivated × cultivated crosses (Jagdish Kumar, pers. comm.). These are available in the form of RIL sets. RAPD and STMS markers are being used to develop a genetic linkage map that can readily be used for skeleton map transfer to populations based on other crosses. New mapping populations developed with drought tolerance as a primary target trait might also reveal useful differences in P-acquisition ability as one of the drought-tolerance mechanisms being pursued is increased root length and root volume. It is also likely that differences exist for production of acidic root exudates (e.g. Ali et al., 2000; Rengel, 2000a) and for ability to establish successful associations with VAM (Graham and Eissenstat, 1994; Rengel, 2000b), although screening of mapping populations and their parents for these traits in P-limited growing conditions has not yet been taken up in this species.

Pigeonpea
Mapping populations are not yet available for this crop, but are under development (Paula Bramel, pers. comm.).

An aside: Negative consequences of improving phosphorus uptake ability?

Some will express concern that crop genotypes that are more efficient at acquiring P might be less responsive to applied P and, therefore, at least initially, contribute to abuse of the soil ecosystem by farmers using this as an excuse to mine soil P rather than replenish that which is removed with each harvest. However, Buah et al. (1998) have observed that specific selection for high N-use efficiency in grain sorghums did not diminish responsiveness to applied N, and Akintoye et al. (1999) report similar results from maize. In case of improved P-use efficiency proposed here (to be achieved via improved P-acquisition ability), mechanisms based on altered root architecture (rooting depth, degree of branching, intensity of root hair development, ...) and root exudates appear unlikely to be associated with reductions in responsiveness to applied P. However, for mycorrhizae-based mechanisms of improved P acquisition, it is likely that at least in some crop × VAM combinations these might not be expressed when higher levels of applied P were available (Kaeppler et al., 2000; Koide and Schreiner, 1992), and/or in years of limited soil moisture availability (Hafner et al., 1993). Some host plant genotypes having superior ability to establish associations with mycorrhizae under low-P conditions might indeed be less responsive to applied P than those lacking this mechanism of tolerance to low P-availability. This potential lack of responsiveness is likely to be less problematic under conditions, such as those in the Sahel, where subsistence food crops like pearl millet will for the forseeable future be receiving at most a modest application of sparingly soluble rock phosphate, or more often, being used to harvest residual P fertility following fertilization of a prior cash crop such as cotton (*Gossypium* sp.) or groundnut.

A more troubling concern is the observation by Buerkert et al. (1998) that increased P uptake (as a result of application of P or crop residues) had negative effects on the nutritional quality of pearl millet grain produced in the Sahel. The first observation that increased grain yield as a result of increased P uptake was accompanied by a reduction in grain protein content, is not unexpected. As levels of the limiting nutrient for crop growth and grain yield (in this case P) are raised, other factors (nutrients, water, light, . . .) are expected to become limiting. This could be overcome to some degree by accompanying the P application with a modest dose of N (in organic or inorganic form). In any case, any increase in grain availability as a result of increased P uptake will directly contribute to an increase in the availability of carbohydrates (and thereby energy) for consumption by people and their livestock. Human diets that are more protein-deficient are unlikely to result from such changes provided that total per capita caloric intake increases, although the possibility of micronutrient deficiencies can not be ruled out (Welch and Graham, 1999, 2000).

The observation that increased P uptake leads to an accumulation of the anti-nutritional factor phytate (Buerkert et al., 1998), is also less alarming than it might initially appear. This is so despite the fact that in most crops, the vast majority of P present in grain is stored in the form of phytate. Phytate levels are strongly dependent on P fertilization and the total P concentration and it is therefore unlikely to be possible to increase P uptake without some increase in grain phytate concentration (except under carefully controlled laboratory conditions where water, light and nutrient availability might be carefully balanced). Appropriate non-lethal low phytate mutants, such as that recently reported in rice (Larson et al., 2000), could provide a genetic solution, but are not yet available in the orphan crops in ICRISAT's genetic improvement mandate. However, the primary reason that increased phytate accumulation should not be considered alarming is that milling, fermentation, and other traditional steps in preparation of food from grains high in phytate can all greatly reduce the levels of this anti-nutritional factor (Abdalla et al., 1998a,b; Kheterpaul and Chauhan, 1991; Simwemba et al., 1984).

Conclusions

From the above review of the literature, it appears that for at least some of the academically and economically orphaned tropical food and feed grain crops in the crop improvement mandate of ICRISAT, the genetic variation and molecular tools already exist that can permit plant breeders to enhance the P-acquisition component of efficiency in low-nutrient environments, in a manner similar to that demonstrated for drought tolerance in common bean by Schneider et al. (1997). Where not yet already in place, these genetic tools can be expected to become available in the very near future. Therefore, it is our belief that with appropriate, targeted research these tools can permit empirical exploration of the potential for marker-facilitated mapping and manipulation of major genes that can contribute to enhanced P-acquisition ability of these crops. With these tools, delivery of new versions of currently popular high-yielding, high quality, disease resistant crop cultivars, having genetically improved ability to acquire P currently in soils but unavailable for crop growth, could take as little as 5 – 7 years. Of course, sustainable use of such improved cultivars would require that they be used as components of integrated soil fertility management systems.

Acknowledgements

This review was supported, in part, by a research grant to ICRISAT from the Plant Sciences Research Programme of the UK's Department for International Development (DFID-PSP). The opinions presented are those of the authors and are not necessarily those of DFID, Embrapa, ICARDA or ICRISAT.

References

Abdalla A A, El Tinay A H, Mohamed B E and Abdalla A H 1998a Effect of traditional processes on phytate and mineral content of pearl millet. Food Chem. 63, 79–84.

Abdalla A A, El Tinay A H, Mohamed B E and Abdalla A H 1998b Proximate composition, starch, phytate and mineral contents of 10 pearl millet genotypes. Food Chem. 63, 243–246.

Ae N, Arihara J, Okada K, Yoshihara T and Johansen C 1990 Phosphorus uptake by pigeonpea and its role in cropping systems of the Indian subcontinent. Science 248, 477–480.

Akintoye H A, Kling J G and Lucas E O 1999 N-use efficiency of single, double, and synthetic maize lines grown at four N levels in three ecological zones of West Africa. Field Crops Res. 60, 189–199.

Ali M Y, Krishnamuthy L, Saxena N P, Rupela O P, Jagdish Kumar and Johansen C 2002 Scope for genetic manipulation of mineral nutrition in chickpea. *In* Food Security in Nutrient-Stressed Environments: Exploiting Plants' Genetic Capabilities, Proceedings of an International Workshop 27–30 September 1999, Patancheru, India. Developments in Plant and Soil Sciences volume 95. Ed. Adu-Gyamfi J J. Kluwer Academic Publishers, Dordrecht, The Netherlands (in press)

Bänziger M, Mugo S and Edmeades G O 2000 Breeding for drought tolerance in tropical maize – conventional approaches and challenges to molecular approaches. *In* Molecular Approaches for the Genetic Improvement of Cereals for Stable Production in Water-limited Environments. A strategic planning workshop held at CIMMYT, El Batan, Mexico, 21–25 June 1999. Eds. Ribaut J-M and Poland D. pp. 69–72. International Maize and Wheat Improvement Center (CIMMYT). Mexico, DF, Mexico.

Bationo A, Henao J and Anand Kumar K 2002 Phosphorus use efficiency as related to sources of P fertilizers, rainfall, soil, crop management and genotypes in the West African semi-arid tropics. *In* Food Security in Nutrient-Stressed Environments: Exploiting Plants' Genetic Capabilities, Proceedings of an International Workshop 27–30 September 1999, Patancheru, India. Developments in Plant and Soil Sciences volume 95. Ed. Adu-Gyamfi J J. Kluwer Academic Publishers, Dordrecht, The Netherlands (in press)

Bhattramakki D, Dong J, Chhabra A K and Hart G E 2000 An integrated SSR and RFLP linkage map of *Sorghum bicolor* (L.) Moench. Genome 43, 988–1002.

Boivin K, Deu M, Rami J-F, Trouche G and Hamon P 1999 Towards a saturated sorghum map using RFLP and AFLP markers. Theor. Appl. Genet. 98, 320–328.

Bolan N S, Naidu R, Mahimairaja and Baskaran S 1994 Influence of low-molecular-weight organic acids on the solubilization of phosphates. Biol. Fert. Soils 18, 311–319.

Bramel-Cox, P J, Barker T, Zavala-Garcia F and Eastin J D 1991 Selection and testing environments for improved performance under reduced-input conditions. *In* Plant Breeding and Sustainable Agriculture: Considerations for Objectives and Methods. Eds. Sleper D A, Barker T C and Bramel-Cox P J. pp. 39–56. CSSA Special Pub. N.18. Crop Sci. Soc. Am. Madison, Wisconsin, USA.

Breeze V G, Wild A, Hopper M J and Jones L H P 1984 The uptake of phosphate by plants from flowing nutrient solution. II. Growth of *Lolium perenne* L. at constant phosphate concentrations. J. Exp. Bot. 35, 1210–1221.

Brown S M, Hopkins M S, Mitchell S E, Senior M L, Wang T Y, Duncan R R, Gonzalez-Candelas F and Kresovich S 1996 Multiple methods for the identification of polymorphic simple sequence repeats (SSRs) in sorghum [*Sorghum bicolor* (L.) Moench]. Theor. Appl. Genet. 93, 190–198.

Brück H, Payne W A and Sattelmacher B 2000 Effects of phosphorus and water supply on yield, transpirational water-use efficiency, and carbon isotope discrimination in pearl millet. Crop Sci. 40, 120–125.

Buah S S J, Maranville J W, Traore A and Bramel-Cox P J 1998 Response of nitrogen use efficient sorghums to nitrogen fertilizer. J. Plant Nut. 21, 2303–2318.

Buerkert A, Haake C, Ruckwied M and Marschner H 1998 Phosphorus application affects the nutritional quality of millet grain in the Sahel. Field Crops Res. 57, 223–235.

Devos K M and Gale M D 1997 Comparative genetics in the grasses. Plant Mol. Biol. 35, 3–15.

Devos K M and Gale M D 2000 Genomic relationships: the grass model in current research. Plant Cell 12, 637–646.

Devos K M, Pittaway T S, Busso C S, Gale M D, Witcombe J R and Hash C T 1995 Molecular tools for the pearl millet nuclear genome. Int. Sorghum and Millets Newsl. 36, 64–66.

Devos K M, Pittaway T S, Reynolds A and Gale M D 2000 Comparative mapping reveals a complex relationship between the pearl millet genome and those of foxtail millet and rice. Theor. Appl. Genet. 100, 190–198.

Doerge R W, Zeng Z-B and Weir B S 1997 Statistical issues in the search for genes affecting quantitative traits in experimental populations. Statistical Sci. 12, 195–219.

Edwards M 1992 Use of molecular markers in the evaluation and introgression of genetic diversity for quantitative traits. Field Crops Res. 29, 241–260.

Föhse D, Claassen N and Jungk A 1988 Phosphorus efficiency of plants. I. External and internal P requirement and P uptake efficiency of different plant species. Plant Soil 110, 101–109.

Frisch M, Bohn M and Melchinger A E 1999a Minimum sample size and optimal positioning of flanking markers in marker-assisted backcrossing for transfer of a target gene. Crop Sci. 39, 967–975.

Frisch M, Bohn M and Melchinger A E 1999b Comparison of selection strategies for marker-assisted backcrossing of a gene. Crop Sci. 39, 1295–1301.

Gale M D and Devos K M 1998 Plant comparative genetics after 10 years. Science 282, 656–659.

Garcia G M, Stalker H T and Kochert G 1995 Introgression analysis of an interspecific hybrid population in peanuts (*Arachis hypogaea* L.) using RFLP and RAPD markers. Genome 38, 166–176.

Gavito M E and Miller M H 1998 Early phosphorus nutrition, mycorrhizae development, dry matter partitioning and yield of maize. Plant Soil 1999, 177–186.

Graham J H and Eissenstat D M 1994 Host genotype and the formation and function of VA mycorrhizae. Plant Soil 159, 179–185.

Hafner H, George E, Batiano A and Marschner H 1993 Effect of crop residues on root growth and phosphorus acquisition of pearl millet in an acid sandy soil in Niger. Plant Soil 150, 117–127.

Hash C T 2000 Concepts for marker technique applications in Africa. *In* Training Manual for a Seminar held at IITA, Ibadan, Nigeria, 16–17 August 1999. Eds. Haussmann B I G, Geiger H H, Hess D E, Hash C T and Bramel-Cox P. pp. 155–168. International Crops Research Institute for the Semi-Arid Tropics (ICRISAT). Patancheru 502 324, AP, India. http://www.icrisat.org/text/research/grep/homepage/mol/S3_1Hash.pdf

Hash C T and Bramel-Cox P J 2000 Marker applications in pearl millet. *In* Training Manual for a Seminar held at IITA, Ibadan, Nigeria, 16-17 August 1999. Eds. Haussmann B I G, Geiger H H, Hess D E, Hash C T and Bramel-Cox P. pp. 112–127. International Crops Research Institute for the Semi-Arid Tropics (ICRISAT). Patancheru 502 324, AP, India. http://www.icrisat.org/text/research/grep/homepage/mol/S2_5Hash.pdf

Hash C T and Witcombe J R 1994 Pearl millet mapping populations at ICRISAT. *In* Use of Molecular Markers in Sorghum and Pearl Millet Breeding for Developing Countries. Eds. Witcombe J R and Duncan R R. pp. 69–75. Overseas Development Administration. London, UK.

Hash CT, Abdu Rahman M D, Bhasker Raj A G and Zerbini E 2001 Molecular markers for improving nutritional quality of crop residues for ruminants. *In* Molecular Breeding of Forage Crops, Proceedings of the 2nd International Symposium, Molecular Breeding of Forage Crops, Lorne and Hamilton, Victoria, Australia, November 19–24, 2000. Developments in Plant Breeding 10. Ed. Spangenberg G. pp. 203–217. Kluwer Academic Publishers, Dordrecht, The Netherlands.

Hash C T, Yadav R S, Cavan G P, Howarth C J, Liu H, Qi X, Sharma A, Kolesnikova-Allen M A, Bidinger F R and Witcombe J R 2000 Marker-assisted backcrossing to improve terminal drought tolerance in pearl millet. *In* Molecular Approaches for the Genetic Improvement of Cereals for Stable Production in Water-limited Environments. A strategic planning workshop held at CIMMYT, El Batan, Mexico, 21–25 June 1999. Eds. Ribaut J-M and Poland D. pp. 114–119. International Maize and Wheat Improvement Center (CIMMYT). Mexico, DF, Mexico.

Hinsinger P 1998 How do plant roots acquire mineral nutrients? Chemical processes involved in the rhizosphere. Adv. Agron. 64, 225–265.

Howarth C J, Weltzien Rattunde E, Bidinger F R and Harris D 1997 Seedling survival of abiotic stress: Sorghum and pearl millet. *In* Proceedings of the International Conference on Genetic Enhancement of Sorghum and Pearl Millet, Sep 23–27, 1996, Lubbock, Texas. pp. 379–399. INTSORMIL. Lincoln, Nebraska, USA.

Hüttel B, Winter P, Weising K, Choumane W, Weigand F and Kahl G 1999 Sequence-tagged microsatellite site markers for chickpea (*Cicer arietinum* L.). Genome 42, 210–217.

Inthapanya P, Sipseuth, Sihavong P, Sihathep V, Chanphengsay M, Fukai S and Basnayake J 2000 Genotype differences in nutrient uptake and utilization for grain yield production of rainfed lowland rice under fertilised and non-fertilised conditions. Field Crops Res. 65, 57–68.

Ipke F N, Powell J M, Isiramal N O, Wahua T A T and Ngodigha E M 1999 Effects of primary tillage and soil amendment practices on pearl millet yield and nutrient uptake in the Sahel of West Africa. Expl. Agric. 35, 437–448.

Ishikawa S, Adu-Gyamfi J J, Nakamura T, Yoshihara T and Wagatsuma T 2002 Genotypic variability in phosphorus solubilizing activity of root exudates by crops grown in low-nutrient environments. *In* Food Security in Nutrient-Stressed Environments: Exploiting Plants' Genetic Capabilities, Proceedings of an International Workshop 27–30 September 1999, Patancheru, India. Developments in Plant and Soil Sciences volume 95. Ed. Adu-Gyamfi J J. Kluwer Academic Publishers, Dordrecht, The Netherlands (in press)

Jones D L 1998 Organic acids in the rhizosphere – a critical review. Plant Soil 205, 25–44.

Jones E S, Liu C J, Gale M D, Hash C T and Witcombe J R 1995 Mapping quantitative trait loci for downy mildew resistance in pearl millet. Theor. Appl. Genet. 91, 448–456.

Jones N, Ougham H and Thomas H 1997 Markers and mapping: we are all geneticists now. New Phytol. 137, 165–177.

Kaeppler S M, Parke J L, Mueller S M, Senior L, Stuber C and Tracy W F 2000 Variation among maize inbred lines and detection of quantitative trait loci for growth at low phosphorus and responsiveness to arbuscular mycorrhizal fungi. Crop Sci. 40, 358–364.

Kheterpaul N and Chauhan B M 1991 Effect of natural fermentation on phytate and polyphenolic content and in-vitro digestibility of starch and protein of pearl millet (*P. typhoide*um). J. Sci. Food Agric. 55, 189–195.

Kochert G No date. Introduction to RFLP mapping and plant breeding applications. The Rockefeller Foundation International Program on Rice Biotechnology, New York, USA.

Koide R T and Schreiner R P 1992 Regulation of the vesicular-arbuscular mycorrhizal symbiosis. Annu. Rev. Plant Physiol. Plant Mol. Biol. 43, 557–581.

Kong L, Dong J and Hart G E 2000 Characteristics, linkage map positions and allelic differentiation of *Sorghum bicolor* (L.) Moench DNA simple-sequence repeats (SSRs). Theor. Appl. Genet. 101, 438–448.

Krishna K R, Shetty K G, Dart P J and Andrews D J 1985 Genotype dependent variation in mycorrhizal colonization and response to inoculation of pearl millet. Plant Soil 86, 113–125.

Larson S R, Rutger J N, Young K A and Raboy V 2000 Isolation and genetic mapping of a non-lethal rice (*Oryza sativa* L.) *low phytic acid 1* mutation. Crop Sci. 40, 1397–1405.

Li X-L, George E and Marschner H 1991 Phosphorus depletion and pH increase at the root-soil and hyphae-soil interfaces of VA mycorrhizal white clover fertilized with ammonium. New Phytol. 119, 397–404.

Liu C J, Witcombe J R, Pittaway T S, Nash M, Hash C T, Busso C S and Gale M D 1994 An RFLP-based genetic map of pearl millet (*Pennisetum glaucum*). Theor. Appl. Genet. 89, 481–487.

Lynch J P 1995 Root architecture and plant productivity. Plant Physiol. 109, 7–13.

Marschner H and Dell B 1994 Nutrient uptake in mycorrhizal symbiosis. Plant Soil 159, 89–102.

Mohan M, Nair S, Bhagwat A, Krishna T G and Yano M 1997 Genome mapping, molecular markers and marker-assisted selection in crop plants. Mol. Breed. 3, 87–103.

Moreau L, Charcosset A, Hospital F and Gallais A 1998 Marker-assisted selection efficiency in populations of finite size. Genetics 148, 1353–1365.

Moreau L, Monod H, Charcosset A and Gallais A 1999 Marker-assisted selection with spatial analysis of unreplicated field trials. Theor. Appl. Genet. 98, 234–242.

Morris R A, Sattell R R and Christensen N W 1992 Phosphorus sorption and uptake from Sri Lankan Alfisols. Soil Sci. Soc. Am. J. 56, 1516–1520.

Muchhal U S, Pardo J M and Raghothama K G 1996 Phosphate transporters from the higher plant *Arabidopsis thaliana*. Proc. Nat. Acad. Sci. USA 93, 10519–10523.

Mugo S N, Banziger M and Edmeades G O 1999 The effects of divergent selection for root capacitance in maize. Agronomy Abstracts 67.

Otani T and Ae N 1996a Sensitivity of phosphorus uptake to changes in root length and soil volume. Agron. J. 88, 371–375.

Otani T and Ae N 1996b Phosphorus (P) uptake mechanisms of crops grown in soils with low P status. I. Screening of crops for efficient P uptake. Soil. Sci. Plant Nutr. 42, 155–163.

Paterson A H, Tanksley S D and Sorrells M E 1991 DNA markers in plant improvement. Adv. Agron. 46, 39–90.

Payne W A, Hossner L R, Onken A B and Wendt C W 1995 Nitrogen and phosphorus uptake in pearl millet and its relation to nutrient and transpiration efficiency. Agron. J. 87, 425–431.

Pellet D M, Grunes D L and Kochian L V 1995 Organic acid exudation as an aluminum tolerance mechanism in maize (*Zea mays* L.). Planta 196, 788–795.

Pellet D M, Papernik L A and Kochian L V 1996 Multiple aluminum resistance mechanisms in wheat: The roles of root apical phosphate and malate exudation. Plant Physiol. 112, 591.

Peng Y, Schertz K F, Cartinhour S and Hart G E 1999 Comparative genome mapping of *Sorghum bicolor* (L.) Moench using an RFLP map constructed in a population of recombinant inbred lines. Plant Breeding 118, 225–235.

Prioul J-L, Quarrie S, Causse M and de Vienne D 1997 Dissecting complex physiological functions through the use of molecular quantitative genetics J. Exp. Bot. 48, 1151–1163.

Raghothama K G 1999 Phosphate acquisition. Ann. Rev. Plant Physiol. Plant Mol. Biol. 50, 665–693.

Reiter R S, Coors J G, Sussman M R and Gabelman W H 1991 Genetic analysis of tolerance to low-phosphorus stress in maize using restriction fragment length polymorphisms. Theor. Appl. Genet. 82, 561–568.

Rengel Z 2002a Genetic control and manipulation of root exudates. *In* Food Security in Nutrient-Stressed Environments: Exploiting Plants' Genetic Capabilities, Proceedings of an International Workshop 27–30 September 1999, Patancheru, India. Developments in Plant and Soil Sciences volume 95. Ed. Adu-Gyamfi J J. Kluwer Academic Publishers, Dordrecht, The Netherlands (in press)

Rengel Z 2002b Breeding for better symbioses. *In* Food Security in Nutrient-Stressed Environments: Exploiting Plants' Genetic Capabilities, Proceedings of an International Workshop 27–30 September 1999, Patancheru, India. Developments in Plant and Soil Sciences volume 95. Ed. Adu-Gyamfi J J. Kluwer Academic Publishers, Dordrecht, The Netherlands (in press)

Ribaut J-M and Poland D, Eds 2000 Molecular Approaches for the Genetic Improvement of Cereals for Stable Production in Water-limited Environments. A strategic planning workshop held at CIMMYT, El Batan, Mexico, 21–25 June 1999. International Maize and Wheat Improvement Center (CIMMYT). Mexico, DF, Mexico.

Ribaut J M, Hu X, Hoisington D and Gonzalez de Leon D 1997 Use of STSs and SSRs as rapid and reliable preselection tools in a marker-assisted selection-backcross scheme. Plant Mol. Biol. Rep. 15, 154–162.

Ryan P R, Delhaize E and Randall R J 1995 Malate efflux from root apices: Evidence for a general mechanism of Al-tolerance in wheat. Aust. J. Plant Physiol. 22, 531–536.

Schaffert R E, Alves V M C, Parentoni S N and Raghothama K G 2000 Genetic control of phosphorus uptake and utilization efficiency in maize and sorghum under marginal soil conditions. *In* Molecular Approaches for the Genetic Improvement of Cereals for Stable Production in Water-limited Environments. A strategic planning workshop held at CIMMYT, El Batan, Mexico, 21–25 June 1999. Eds. Ribaut J-M and Poland D. pp. 79–85. International Maize and Wheat Improvement Center (CIMMYT). Mexico, DF, Mexico.

Schneider K A, Brothers M E and Kelly J D 1997 Marker-assisted selection to improve drought resistance in common bean. Crop Sci. 37, 51–60.

Sedcole J R 1977 Number of plants necessary to recover a trait. Crop Sci. 17, 667–668.

Simwemba C G, Hoseney R C, Varriano-Marston E and Zeleznak K 1984 Certain vitamin B and phytic acid contents of pearl millet (*P. americanum* (L.) Leeke). J. Agric. Food Chem. 32, 31–32.

Sinclair T R and Vadez V 2002 Physiological traits for crop yield improvement in low N and P environments. *In* Food Security in Nutrient-Stressed Environments: Exploiting Plants' Genetic Capabilities, Proceedings of an International Workshop 27–30 September 1999, Patancheru, India. Developments in Plant and Soil Sciences volume 95. Ed. Adu-Gyamfi J J. Kluwer Academic Publishers, Dordrecht, The Netherlands (in press)

Smith S E, Robson A D and Abbott L K 1992 The involvement of mycorrhizas in assessment of genetically dependent efficiency of nutrient uptake and use. Plant Soil 146, 169–179.

Stam P and Zeven A C 1981 The theoretical proportion of the donor genome in near-isogenic lines of self-fertilizers bred by backcrossing. Euphytica 30, 227–238.

Subbarao G V, Ae N and Otani T 1997a Genotypic variation in iron-, and aluminum-phosphate solubilizing activity of the pigeonpea root exudates under P deficient conditions. Soil Sci. Plant Nutr. 43, 295–305.

Subbarao G V, Ae N and Otani T 1997b Genetic variation in acquisition, and utilization of phosphorus from iron-bound phosphorus in pigeonpea. Soil Sci. Plant Nutr. 43, 511–519.

Subbarao G V, Renard C, Payne W A and Batiano A 2000 Long-term effects of tillage, phosphorus fertilization and crop rotation on pearl millet–cowpea productivity in the West-African Sahel. Expl. Agric. 36, 243–264.

Subudhi P K and Nguyen H T 2000 Linkage group alignment of sorghum RFLP maps using a RIL mapping population. Genome 43, 240–249.

Tamarino G, Tarchini R, Ferrario S, Lee M and Pe' M E 1997 Characterization and mapping of simple sequence repeats (SSRs) in *Sorghum bicolor*. Theor. Appl. Genet. 95, 66–72.

Tanksley S D and McCouch S R 1997 Seed banks and molecular maps: unlocking genetic potential from the wild. Science 277, 1063–1066.

Tiessen H, Abekoe M K, Salcedo I H and Owusu-Bennoah E 1993 Reversibility of phosphorus sorption by ferruginous nodules. Plant Soil 153, 113–124.

Tinker P B 1984 The role of micro-organisms in mediating and facilitating the uptake of plant nutrients from soil. Plant Soil 76, 77–91.

van Beem J, Smith M E and Zobel R W 1998 Estimating root mass in maize using a portable capacitance meter. Agron. J. 90, 566–570.

Van Deynze A E, Sorrells M E, Park W D, Ayres N M, Fu H, Cartinhour S W, Paul E and McCouch S R 1998 Anchor probes for comparative mapping of grass genera. Theor. Appl. Genet. 97, 356–369.

Wani S P, Zambre M A and Lee K K 1990 Genotypic diversity in pearl millet (*Pennisetum glaucum*) for nitrogen, phosphorus and potassium use efficiencies. *In* Plant Nutrition – Physiology and Applications. Ed. van Beusichem M L. pp. 595–601. Kluwer Academic Publishers, Dordrecht, The Netherlands.

Welch R M and Graham R D 1999 A new paradigm for world agriculture: meeting human needs. Productive, sustainable, nutritious. Field Crops Res. 60, 1–10.

Welch R M and Graham R D 2002 Breeding crops for enhanced micronutrient content. *In* Food Security in Nutrient-Stressed Environments: Exploiting Plants' Genetic Capabilities, Proceedings of an International Workshop 27–30 September 1999, Patancheru, India. Developments in Plant and Soil Sciences volume 95. Ed. Adu-Gyamfi J J. Kluwer Academic Publishers, Dordrecht, The Netherlands (in press)

Weir C C 1972 Phosphate studies on the Jamaican bauxite soils. Trop. Agric. (Trinidad) 49, 89–96.

Weir C C 1977 Phosphate fixation in Jamaican latosolic soils. Soil Sci. Soc. Am. Proc. 30, 653–655.

Winter P, Pfaff T, Udupa S M, Hüttel B, Sharma P C, Sahi S, Arreguin-Espinoza R, Weigand F, Muehlbauer F J and Kahl G 1999 Characterization and mapping of sequence tagged microsatellite sites in the chickpea (*Cicer arietinum* L.) genome. Mol. Gen. Genet. 262, 90–101.

Wissuwa M, Yano M and Ae N 1998 Mapping of QTLs for phosphorus-deficiency tolerance in rice (*Oryza sativa* L.). Theor. Appl. Genet. 97, 777–783.

Yadav R S, Hash C T, Bidinger F R, Dhanoa M S and Howarth C J 2000 Identification and utilisation of quantitative trait loci to improve terminal drought tolerance in pearl millet (*Pennisetum glaucum* (L.) R. Br.). *In* Molecular Approaches for the Genetic Improvement of Cereals for Stable Production in Water-limited Environments. A strategic planning workshop held at CIMMYT, El Batan, Mexico, 21–25 June 1999. Eds. Ribaut J-M and Poland

D. pp. 108-113. International Maize and Wheat Improvement Center (CIMMYT). Mexico, DF, Mexico.

Yadav R S, Hash C T, Bidinger F R and Howarth C J 1999 QTL analysis and marker-assisted breeding for traits associated with drought tolerance in pearl millet. *In* Genetic Improvement of Rice for Water-Limited Environments. Eds. Ito O, O'Toole J and Hardy B. pp. 211–223. International Rice Research Institute (IRRI). Los Baños, Philippines.

Zaongo C G L, Hossner L R and Wendt C W 1994 Root distribution, water use and nutrient uptake of millet and grain sorghum on West African soils. Soil Sci. 157, 379–388.

J.J. Adu-Gyamfi (Ed.), Food security in nutrient-stressed environments: exploiting plants' genetic capabilities: 37–42, 2002.

Ex-ante impact assessment and economic analysis of breeding for nutrient efficiency and alternate strategies, a conceptual model and research issues: A socio-economist's view

B. I. Shapiro[1], D. Osborne[1] & M. Ahmed[2]
[1]*Natural Resource Management Program, International Crops Research Institute for the Semi-Arid Tropics (ICRISAT), B P 320 Bamako, Mali*
[2]*International Livestock Research Institute (ILRI), P O Box 5689, Addis Ababa, Ethopia*

Key words: economic analysis, ex-ante impact assessment, low-input systems

Abstract

The stated impact goal of the technology strategies under consideration during this workshop is food security in nutrient-stressed environments. To do a meaningful ex-ante impact assessment of any proposed technology requires first placing that technology in the context of the broader issues of what is needed in developmental terms to achieve such a goal, and the relevant development possibilities of the target population(s). What then are the development possibilities for the arid and semi-arid tropics (ASAT)? Substantial empirical evidence shows that basically two main development pathways are relevant – an extensive path that involves increasing cultivated area or improving extensive livestock production, and an intensive path that results in increasing yields through more intensive use of land and more cash inputs. Extensive, low-input systems will generally work where rainfall is below 500 mm, population pressure is low, and market access is remote. Here attempts at crop intensification are not appropriate since the use of cash inputs is not profitable. The key development challenge in livestock production in the dryer parts of the ASAT is the shift to improved forages. Technological change through introduction of forage production and feeding cereals to livestock is eventually expected to take place throughout the African ASAT as it is already doing in southern Africa and in parts of Asia. Where this extensification process is not yet occurring in dryer areas, land degradation and desertification are taking place. Intensive systems require external inputs and will work where rainfall is high enough (above 500 mm) to make inorganic fertilizer use profitable. This profitability is supported by high population pressure and good access to markets. Are more nutrient-efficient crops relevant where crop intensification is possible? Low cash-input systems that include legumes are possible as a short-term solution, but they limit, but not entirely eliminate the need for purchased inputs. Breeding for nutrient efficiency is one of a set of low-input strategies that include improved use of manure, crop residues, composting, biological nitrogen fixation or the application of rock phosphate. At the present time, the available empirical evidence shows that imported inorganic fertilizers are the only technically efficient and economically profitable way to overcome the soil fertility constraints faced by most farmers in ASAT regions where rainfall is high enough for intensification to take place. The main constraint to yield increases in such areas is not germplasm, but policies that affect the availability of external cash inputs, especially the essential inorganic fertilizers. Policies are needed to encourage inorganic fertilizer imports and sales in the private market, along with public investments in infrastructure that would lead to better-functioning markets. Dissemination of technical information is also required to make farmers more aware of potential benefits and how best to use fertilizers and their complements.

Introduction

The intertwined nature of the natural resource base – particularly soil nutrients and water management techniques, and selection of crop characteristics – needs no argument. Indeed, it is underscored in the ASAT by the efforts that plant breeders have devoted to developing varieties that can succeed on nutrient-poor soils. This focus comes in part from recognition that, for various reasons, ASAT farmers practice low-input agriculture on poor soils, tending not to use many inorganic fertilizer inputs. Assessment of the potential impact of breeding for nutrient efficiency needs to be considered in the context of the broader issues of the development and evolution of ASAT agriculture. These issues in turn become the basis for ex-ante economic analysis of the various parts of the breeding programme and its justification and presentation to donors.

To do a meaningful ex-ante impact assessment of any proposed technology requires first placing that technology in the context of the broader issues of what is needed in developmental terms to achieve such a goal, and the relevant development possibilities of the target population(s). Considering the development needs and possibilities establishes the relevance of the new technologies for making the required impact. Then, the agricultural development possibilities in the target region(s) or the likely development pathways need to be identified, since this determines appropriate technology strategies. These needs and possibilities form the framework for judging the ex-ante economic potential of the various breeding and management strategies to be considered during this workshop. This paper considers the perspective of an economist in assessing the potential impact of breeding for nutrient efficiency and other related low-input strategies in ASAT regions, where such strategies are most relevant. It presents a conceptual model of development pathways and uses this model to raise questions relevant to the potential impact of the technologies under consideration.

Setting the context for ex-ante impact assessment

The stated impact of the technology strategies under consideration is food security in nutrient-stressed environments. In the ASAT, more so than in other agroecological regions of the developing world, this impact goal is relevant. Since it is well established that food security is not obtainable without cash income generation, some degree of commercialization of agricultural production is required in addition to raising production. As the goal is measuring potential impact, and farmers are most concerned with achieving their individual objectives, protection of the natural resource base is not considered, but will be discussed when alternative development strategies that could possible help achieve food security are considered. The example of sub-Saharan Africa (SSA) is used to illustrate how the context of the relevant development needs and possibilities influence ex-ante impact assessment and the questions that need to be asked when evaluating the potential of alternative technology strategies. Although specific conditions in SSA are different from those in Asia, the principles are the same and the implications are also relevant.

Table 1. Population projections for Sub-Saharan Africa (SSA)[a]

Population	1990	2025	Increase year^{-1} (%)
Total (millions)	498	1294	2.8
Urban (%)	29	54	6.9

[a]Source: Adapted from World Bank (1992).

Development needs in the SSA and the ASAT

Even though income levels are not increasing in SSA, population growth and shifts in population to urban areas are significantly increasing the demand for food. This increasing demand is increasing the pressure to intensify and commercialize SSA agriculture. As shown in Table 1, the population of SSA is expected to increase from about 500 million in 1990 to 1.3 billion in 2025. Of all the developing regions, SSA is growing the most rapidly, with an annual population growth rate in the 1990s projected to be 3.1%, while population growth in the developing world as a whole is expected to be 1.9% (World Bank, 1992). Between 1990 and 2025, the population growth rate in SSA will decline slightly to become 2.8% on average, but an average population growth rate of 6.9% in urban areas will result in 54% of the total population being urban by 2025 (from 29% in 1990). This will add significantly to growth in demand for food, as similar trends have already shown in most of Asia and Latin America. To produce a surplus that would enable some commercialization, an annual increase in per capita food production of around 1% on average would have to be achieved in SSA. This would require an increase in output of about 4.0% per year given the current population growth rate of 2.9–3.1%. Even if one argues that population is growing at a lower rate in semi-arid areas, and that such a rate of increase in production not possible there, the same principle of the necessity of increasing production to enable commercialization to achieve greater food security will still apply.

Agricultural development pathways in arid and semi-arid SSA

Experience has shown that there are essentially two agricultural development pathways possible to increase production and cash incomes; an extensive path of increasing cultivated area, and an intensive path of increasing yields. Due to rapid population growth, it is likely that most of the potentially arable land in SSA is already being cultivated, at least in crop/fallow systems. There is thus very little room for expansion of crop land, so the needed output increase will have to come from agricultural intensification on existing land. In Asia, the land constraint is even more pronounced, especially on land that can be irrigated, so there is a similar need to intensify rainfed agriculture.

Relevance to arid and semi-arid SSA

The relevant question now is whether the low-input strategies to be discussed in this workshop, or high-input strategies, or a combination of the two are most relevant to the low and variable rainfed regions of semi-arid SSA. As already stated, the rapidly growing population is also causing disequilibrium in farming systems in arid areas. In semi-arid areas agriculture is carried out mainly on smallholder mixed crop/livestock farms. The region is also experiencing rapid urbanization leading to expanding urban markets for farm output. Growing population pressure, markets, and food demand are causing the value of land, crops, and livestock to increase. As these mixed systems evolve in response, crops and livestock can play important roles in increasing farmers' incomes and in making the systems viable in the long run. The question is, how this can be done? To identify appropriate technology strategies, it is necessary to find out what determines the likely development pathways and how this affects technology research efforts.

Evolution of agricultural systems in semi-arid areas

Based on field research in semi-arid SSA and an extensive review of the literature and development success stories, two main development pathways can be identified. Which pathway ultimately occurs will depend upon the profitability of crop intensification as determined by rainfall amount and distribution, and on the forces of population pressure and market access. The specific focus here is the pathways subsequent to the predominant low-input mixed systems in the semi-arid region. This is also critical for development pathways in arid areas, and is the point at which much of semi-arid SSA finds itself. Therefore, analyzing the next step in the evolution of the systems should be very useful in evaluating the potential for breeding for nutrient efficiency and other low-input strategies. Alternative low-input and intensification strategies are also considered because of their implications for technology research relevant to those being considered.

The rapid population growth in many semi-arid areas is because human and animal disease pressure there is lower than in the humid and sub-humid zones. The first response to increasing population pressure is usually an expansion of cultivated area to maintain per capita crop output (Boserup, 1965, 1981; Ruthenberg, 1983). Fallow periods for maintaining soil fertility on cropland are declining or disappearing in many semi-arid areas, and pasture land is coming under crop cultivation (McIntire et al., 1992; Peyre de Fabreques, 1986). Continuing population growth eventually reduces farm sizes as land becomes more constrained. Land becomes limited in quality and quantity as the fallow system breaks down and cropping is extended onto the less fertile lands previously used for grazing. As yields decline and the land supply becomes even more limited, the potential returns to more intensive production practices increase. Research in semi-arid SSA has shown that the two primary agricultural constraints are agronomic, i.e. water availability and soil fertility, and not socioeconomic, i.e. labour or capital (Sanders et al., 1996). These agronomic constraints need to be addressed simultaneously for successful technology introduction that will lead to sustainable production systems in the long run. In the sub-humid region of West Africa, the use of inorganic fertilizer is already increasing rapidly where rainfall is sufficient for the use of fertilizer on high-value crops to be profitable. Here the land constraint, both quality and quantity, and relative profitability of intensified crop production cause fertilizer to replace or be combined with manure in farmer strategies to maintain soil fertility after the discontinuance of fallows (Sanders et al., 1996). This process is also taking place in the more southerly, relatively higher rainfall areas of the semi-arid region of West Africa, i.e. southern Mali and southwestern Burkina Faso, where production of such nitrogen-fixing legumes as pigeonpea are relevant. When water-retention technologies are practised, use of fertilizers can often be pushed further north into the drier regions (Sanders et al., 1996).

Are low-input strategies appropriate where crop intensification using cash inputs is profitable? Where population pressure is high and the conditions appropriate for intensification, system dynamics are leading to land degradation. Ultimately, there will be a return to extensive livestock production if intensification does not take place. With increasing population pressure, inadequate water availability, and/or low use of inputs due to a poor economic environment (low output and high input prices) make it unprofitable to use fertilizers. So yields and incomes decline. In the short term, possibilities for improved, but low cash-input livestock activities exist here together with improved crop rotations or intercropping with forages and legumes. Such systems result in better feeding of animals, and improved soil fertility management. Low cash-input systems that include legumes are also possible. They would however, limit, but not entirely eliminate the need for purchased inputs, and increasing the availability of one nutrient increases the demand for others. Moreover, the income gains from such low-input systems would be small because of the small farm size and thus would not be sustainable in the long term. Extensive, low-input systems work where both rainfall and population pressure are low. Here crop intensification is not appropriate and little intensification of crops or livestock through use of cash inputs is likely to occur. Improved livestock activities that can substantially increase farmer incomes would ultimately require replacement of crops with livestock activities and eventual consolidation of landholdings to enable improved, but extensive animal production. Where this is not yet occurring, land degradation and desertification is taking place.

Over time, human population will decline. Seasonal migration of most men is already occurring from such areas in West Africa (Painter, 1986; Sutter, 1984). Ultimately as the opportunity costs of farmers' time increases outside of agriculture, they will leave the farms and consolidation of the area into extensive, improved livestock production can take place. Possibilities for improved, but low cash-input livestock activities include improved pastures or intercropping cereals with forages, that results in better feeding of animals, and improved soil fertility management. Such innovations could raise livestock income at drier, semi-arid sites since more animals could be fed and sent to market, thus increasing offtake rates. So, the critical fork in the agricultural development path is where the use of fertilizer on crops becomes profitable. In such areas, fertilizer use needs to be combined with techniques that increase water availability. Without this increase in land productivity, the system declines as soil fertility is exhausted. With fertilizer inputs and evolving markets for livestock products the intensive livestock systems become attractive to farmers.

Alternative strategies to overcome the soil-fertility constraint

Numerous alternatives to imported inorganic fertilizer have been proposed over many years for the ASAT. Most purport to substitute locally produced materials – animal or crop by-products, or such sophisticated production techniques, as *Rhizobium* inoculation – for imported inorganic fertilizer. Many are claimed to be low-cost or low-risk. Inorganic fertilizer can be risky because cash outlays can jeopardize the ability of farm households to make cereal purchases to feed family members when the rains are not adequate and crops fail. Also, input distribution systems often do not function efficiently rendering inorganic fertilizer unavailable at the right time, thus lessening its effectiveness. Sources of organic fertilizer, such as manure, however, are severely limited in most of semi-arid West Africa. For most farmers, sufficient manure is only available for small areas surrounding family compounds (Sanders et al., 1996; Williams et al., 1993). At the present time, most of the crop residues are removed for uses with higher economic value, such as feed, fuel, and building materials. With the disappearance of the fallow system and encroachment onto grazing lands, the large quantities of organic fertilizer required to attain even moderate levels of the basic soil nutrients can not be achieved. At present, manure often replaces wood as a cooking fuel in many regions. The ability to increase the efficiency of manure use, is also limited in most areas of semi-arid SSA. While manure can be managed more effectively by improved corralling and its efficiency increased through composting, the overall increase in terms of essential soil nutrients (nitrogen, phosphorus and potassium) is small (Williams et al., 1993). A recent increase in the relative price of inorganic to organic fertilizer due both to the elimination of input subsidies and currency devaluation has resulted in the expansion of manure use and the development of techniques for improving manure quality in the Sudanian zone of Burkina Faso.

Manure use has thus somewhat replaced inorganic fertilizer there. However, the limited ability to expand the supply of organic fertilizer means the potential for this substitution strategy is very limited. Manure is thus expected to be complementary to inorganic fertilizer rather than a substitute in the long term and will remain so, even under intensification (Sanders et al.,1996; McIntire et al., 1992). Problems associated with inorganic fertilizer use in sandy dune soils can include soil acidification and aluminium toxicity due to low soil organic matter. Nevertheless, ICRISAT results consistently show that combining moderate levels of fertilizer (N 45, P 20, and K 25 kg ha^{-1}, of pure nutrients) leads to higher yields on these sandy dune soils. With higher production, more crop residues can be left on the soil, raising its organic matter content. The potential problems of acidification and aluminium toxicity associated with inorganic fertilizers when there is insufficient organic matter can thus be avoided. On the sandy clay loam soils, meanwhile, it has been shown that inorganic fertilizer use on sorghum has led to yield declines after 12 years. Inorganic fertilizers need to be supplemented with organic fertilizers. An Institut des recherches d'agriculture tropicale/Institut national d'investigation agricole (IRAT/INIA) experiment assumed that despite sorghum yield increases from inorganic fertilizer, farmers would not change their cultural practices and begin to incorporate some of the increased crop residues. Farmers presently use almost all of these residues and often burn what is left. With increased yields, farmers could use part of the residues and leave the rest to decompose for the next year's crop. It would be necessary either to keep animals from grazing these residues, or to find a method of incorporating them into the soil with animal traction.

With foreign exchange constraints discouraging the use of imported fertilizer, some Sahelian governments have been developing their own sources of natural rock phosphate. Recent research in Niger, meanwhile, indicates that high-quality rock phosphate from Tahoua, even when partially acidulated, would only be competitive with imported single superphosphate (SSP) if supplied to farmers at approximately 25% of the price of imported SSP. This implies that at present efficiency levels, rock phosphate is unlikely to become competitive with the imported inorganic sources, even through the CFA was devalued by 50% in 1994 (Mokwunye and Hammond, 1992). Furthermore, adequate costing of potential alternatives needs to include the additional labour, management, and extension costs of labour or skill-intensive activities. Environmentalists have also aroused concern among donors and development agencies, about natural resource degradation in developing countries citing use of inorganic fertilizer as a contributing factor. The environmental impacts of fertilizer are assumed to be analogous to those in the developed countries where high levels of fertilizer use can pose a threat to the environment. However, in SSA only 5–10 kg of fertilizer are used ha^{-1} cropland; compared to 60 kg ha^{-1} in India. These very low levels of fertilizer use contribute to soil degradation by requiring the extension of cropping onto more marginal lands to maintain output once the fallow system breaks down. The traditional extensive cropping systems are not sustainable under increasing population pressure. Per capita cereal production declined in SSA at an average annual rate of over 1% from 1961 to 1991 (Sanders et al., 1996).

At the present time, imported inorganic fertilizers are the only technically efficient and economically profitable way to overcome the soil fertility constraints in the Sudanian and Sahelo-Sudanian zones for most regions and farmers. Some scientists have perpetuated this belief in alternatives to inorganic fertilizers. However, until substantial on-station and on-farm experimentation with alternatives is more successful, other measures of improving soil fertility must be considered as complements to rather than as substitutes for imported inorganic fertilizers.

Potential of low-input solutions to increase food

Where agroecological and market conditions are conducive to intensification, a critical issue is whether further improvements in low-input solutions are viable alternatives to more intensive crop activities that depend on purchased inputs.

Implications for technology development strategies

Breeding for nutrient efficiency is but one of a set of alternative and sometimes complementary strategies to raise production of smallholder agriculture on the impoverished soils of the ASAT. Others include:

- Management of biomass and post-harvest residues
- Fertilizer use (organic and inorganic)
- N-fixing intercrops, woody plants, and enhanced fallows

- Mastery of rainwater runoff to decrease soil loss, increase soil moisture, and even harvest and conserve water
- Techniques of pest and disease management
- Conditions for marketing (prices, transport)

There are no simple solutions to the complex problem of food security in nutrient-stressed environments. Where intensification of crop production can not take place, there will be potential for specialized, commercial livestock activities (cow–calf operations and small ruminants). The key development challenge in livestock production in the semi-arid region is the shift to improved forages, but so far this is only occurring in the intensive, specialized finishing operations for small ruminants and peri-urban dairy operations in West Africa. As the opportunity cost of labor falls sufficiently relative to land, and the price of meat rises relative to cereals, technological change through intensification of forage production and feeding cereals to livestock is expected to take place as it is already doing in southern Africa. Fodder banks could also play an increasing role on-farm over time in these semi-arid zones. Higher levels of inputs for improved forage and hay production can then also be expected for fattening livestock on-farm, and for sale to the specialized finishing and peri-urban dairy operations. It is false to assume that increased use of inorganic fertilizer would lead to degradation of soils in the long term, where rainfall is sufficient for intensification to be profitable. Rapid increases of inorganic fertilizer use will produce improved productivity and result is improved soil quality, again a vital long-term matter. Moreover, the seed varieties now being used for most crops are capable of supporting about a 50% increase in average yield. Many authorities believe that a 100% increase over present low levels is attainable with modest increases in fertilizer use and improved soil fertility management. The main constraint to yield is thus not germplasm, but policies and inputs, such as producer returns, fertilizers, irrigation, and transportation. Some marginal improvements with drought-resistant cultivars and some improved animal health control measures are also feasible but will have small effects on income. Research will need to resolve the technical constraints to these developments taking place. Public investments in infrastructure (roads and services) will also be required to lead to better-functioning markets. Furthermore, African governments will need to further improve the already more conducive policy environment that has come about in recent years. The realization of the potential benefits from crop and livestock development will depend mainly on support by African governments, not only in the form of public policies and investments, but also as support for private business. To maintain even the current level of support from donor agencies, these agencies will need to see this type of commitment to be able to overcome the fatigue of their constituencies for development. In addition, inter-government policy decisions needed to ensure more rational use of land resources, and promoting regional integration and trade will have to be made a reality.

References

Boserup E 1965 The Conditions of Agricultural Growth: The Economics of Agrarian Change under Pressure. Aldine, New York, USA.

Boserup E 1981 Population and Technological Change: A Study of Long-Term Trends. University of Chicago Press, Chicago, IL, USA.

McIntire J, Bourzat D and Pingali P 1992 Crop-Livestock Interaction in Sub-Saharan Africa. World Bank Washington, DC, USA.

Mokwunye A U and Hammond L L 1992 Myths and science of fertilizer use in the Tropics. *In* Myths and Science of Soils in the Tropics. Eds. R Lal and P A Sanchez. Soil Sci. Am. and Am. Soc. of Agron SSA Special Publication No. 29. Madison, WI, USA.

Peyre de Fabreques B 1986 The rangeland-livestock sector of Niger. Dept. Agric. Econ., Purdue Univ., West Lafayette, IN, USA (mimeo).

Ruthenberg H 1983 Farming Systems in the Tropics. Oxford University Press Oxford, UK.

Sanders J H, Shapiro B I and Ramaswamy S 1996 The Economics of Agricultural Technology in Semiarid Sub-Saharan Africa. Johns Hopkins University Press, Baltimore, MA, USA.

Sutter J W 1984 Peasants merchant capital and rural differentiation: A Nigerian Hausa case study. Ph.D. dissertation. Cornell University Ithaca, NY, USA (unpublished).

Williams T O, Powell J M and Fernandez-Rivera S 1993 Manure utilization, drought cycles and herd dynamics in the Sahel: Implications for cropland productivity. *In* Proceedings of the Nutrient Cycling Workshop, ILCA, 26–30 November 1993. pp 89–113. International Livestock Centre for Africa, Addis Ababa, Ethiopia.

World Bank 1992 World Development Report 1992: Development and the Environment. Oxford University Press, for the World Bank, Washington, DC, USA. 308 pp.

J.J. Adu-Gyamfi (Ed.), *Food security in nutrient-stressed environments: exploiting plants' genetic capabilities*: 43–46, 2002.

Breeding for low-nutrient environments: Is it sustainable?

J. M. Peacock
International Center for Agricultural Research in the Dry Areas (ICARDA), Arabian Peninsula Regional Program (APRP), PO Box 13979, Dubai, United Arab Emirates

Introduction

This conference set out to address the important issue of how physiological mechanisms of nutrient uptake can best be combined with genetic options to improve the adaptation of crops to low-nutrient availability, thereby enhancing productivity of nutrient-poor soils in the semi-arid tropics. The first session, attempted to address the question from the viewpoint of three interrelated disciplines; physiology, breeding, and socio-economics.

The crop physiologists' view The physiologist's task was to give some perspectives about opportunities for physiological improvement that might result in increased yields in low-nutrient environments. Interestingly, physiologists went about their task by posing six related and important questions.

1. What are the essential physiological roles of nitrogen (N) and phosphorus (P) in crop growth?
2. Can the need for N and P in plant growth be decreased?
3. Can the allocation of N and P to grain be improved to achieve increased grain yield?
4. What is the quantitative yield limitation based on N and P accumulation?
5. Can N and P uptake from the soil be increased in low-nutrient environments?
6. What is the opportunity for legumes in low-nutrient environments?

The authors did an admirable job in addressing these and came to some pertinent conclusions (Sinclair and Vadez, 2000).

In summary, after listing all the important physiological roles of N and P they conclude that both N and P are essential and the need for both elements can not be decreased. Low levels of either invariably mean lower yields. It seems clear that little opportunity exists to increase N recovery in low-nutrient environments, apart from adding more N, and we learn later that this is sometimes neither economical nor environmentally acceptable. Alternatively, having longer-duration varieties and improving the plant architecture could provide a larger leaf area and duration and thus more storage (Sinclair and Sheehy, 1999). But, low-nutrient conditions are often associated with low water and high temperature conditions and as such, long-duration varieties and larger leaf areas are often not well adapted.

One approach suggested is to increase N uptake by the use of more legumes in the farming systems. However, Wood et al. (1997) showed that growing chickpea and lentil in a wheat–legume rotation did not provide a net input of N. These data also indicated that the inclusion of these grain legumes in rotation with wheat may reduce the rate of depletion of soil N compared with a continuous wheat crop, but it is unlikely to yield a positive N balance. It must be concluded therefore that these particular grain–legume rotation systems do not offer the immediate prospect of a sustainable cropping system. The case is likely to be the same for many other cereal–legume rotations and may account for the reason why in the Mediterranean region, grain legumes are grown on only 3.5% of the crop land (Buddenhagen, 1990).

On the other hand, several short-term options can be investigated for increasing P accumulation and the authors look closely at roots; long root hairs are highlighted and although selection for these could be time-consuming, this seems a possible route to take. The interaction of root hairs with mycorrhiza or 'managing mycorrhiza' must be regarded as an essential area for further research.

In summary, the physiologists point to concentrating on trying to increase the P uptake by plants with the help of the plant breeders and molecular biologists.

It is therefore an appropriate point in the summary to turn to the views of applied geneticists, but before doing so I would like to quote from a section taken from Ghazanfer and Fisher (1998). "Endomycorrhiza (in desert plants) increase absorption of nutrients, particularly phosphorous and bacterial nitrogen fixation may be an important nutrient aid in the vascular plants of dune areas. Bowers (1982) lists several genera, including *Aristada* and *Artemesia* that have associated nitrogen-fixing bacteria. There is growing evidence that the rhizosheaths, characteristic of the root systems of desert perennial grasses, such as *Stipagrostis plumosa, Cenchrus ciliaris*, and *Coelachyrum piercie* may be important aids both to nitrogen nutrition and water absorption. Nitrogen-fixing bacteria have been associated with the cylindrical root sheaths formed of matted root hairs and sand grains held by the secreted mucilage (Danin, 1991; 1996a, b)".

Under P scarcity in soil, P influx into roots is immensely affected by root hairs; nearly 90% of root P uptake could be accounted for by root hairs (Föhse et al., 1988). Prolific root hair formation can increase P exploitation efficiency of tomato plants. 'Cottony root' (crt) hair trait that increases P absorption is such an example (Hochmuth et al., 1985). It is well known that plant tolerance to low-P in soil can be significantly influenced by the mycorrhizal component in the roots because vascular–arbuscular mycorrhiza (VAM) can extract P more efficiently at lower concentrations through better soil exploitation and absorption (Krishna, 1997).

Earlier at ICRISAT, this group (Ae et al., 1990) demonstrated the importance of secretions (piscidic acid) from pigeonpea in acidifying the soil to increase P uptake. Indeed, many acids have been identified viz. acetic, glycolic, malonic, oxalic, and formic in the root exudates of a number of plants (Dakora and Philips, 2000; Fox and Comerford, 1990; Vancura and Hovadik, 1965).

These root exudates, as shown by many (Dakora and Phillips, 2000) play a crucial role in enabling desert grasses to survive under very low-nutrient conditions, usually with little water. Now the ICRISAT breeders and crop improvement scientists have a problem, or better stated a challenge; how to make their mandate crops more productive under low-nutrient conditions. If one considers the case of pearl millet (*Pennisetum glaucum*) and the farmers in the Sahel and northern India. *Pennisetum glaucum* is well-adapted in Rajasthan, as is the related forage grass, *Cenchrus ciliaris*; the two species are growing alongside each other. Krishna and his colleagues again at ICRISAT (1985) showed that pearl millet in the presence of VAM was able to take up more P. Genes that code for transporter proteins for all inorganic nutrients and some of the minor nutrients have been cloned from plants (Chrispeels et al., 1999). The question is, is the desert grass system more effective in taking up P and if so, can we identify the genes responsible and get them into *Pennisetum glaucum*?

The plant breeders' view

In recognizing that the nitrogen route may be a dead end, the breeders quite rightly focus their attention on how to genetically enhance the crops' ability to access the small amounts of available P and/or increase the amounts of available P in the soil solution. Genotypic differences in P use efficiency (PUE) due to mycorrhiza occur in many crops (Smith et al., 1992) but the genetics of this phenomenon are still not clearly understood. Many reports support the possibility of increasing the VAM component in roots, and its effect on P efficiency of the host genotype via selection and breeding (Manske, 1990; Smith et al., 1992).

The breeders go straight to the core of the problem, and address the issue of how the genetic enhancement of root traits (Lynch, 1995), including root-induced changes in the chemistry of the rhizosphere, can enhance nutrient availability (Hinsinger, 1998). They argue, like others (Dakora and Philips, 2000; Dunlop and Phung, 2000) that this will be critical to improving uptake of soil nutrients, particularly P. They also argue that molecular biology, combined with Mendelian and quantitative genetics in quantitative trait locus (QTL) mapping and marker-assisted selection (MAS), provide powerful new tools to facilitate efficient genetic manipulation by plant breeders of complex traits. They examine current opportunities for genetically manipulating the ability of crop plants to more efficiently take up essential soil nutrients, using as examples P and ICRISAT's mandate crops–chickpea, groundnut, pearl millet, pigeonpea, and sorghum. They conclude as do others (Dunlop and Phung, 2000) that for at least some of these important food crops, the genetic variation and molecular tools already exist, or can be expected to become available in the very near future. With appropriate, targeted research, these tools can permit us to empirically explore the potential for marker-facilitated mapping and manipulation of major genes that can contribute to the enhanced P uptake ability of these crops.

The authors show that there is overwhelming evidence that associations of plant roots with VAM and the exudation of organic acids and enzymes from plant roots can contribute to significant increases in the supply of P to plant roots. They therefore have two sets of target traits for assessment and possible utilization by plant breeders and physiologists.

If we turn to the conclusions drawn by the physiologists on identifying genes associated with roots exudates, then this is very encouraging, and is clearly a route to be followed. However, each set of traits is complex, and difficult to measure, but despite this the authors argue that cultivars with a genetically improved ability to take up P could take as little as 5–7 years to develop. This time frame will also be dependent to some extent on the availability of reliable screening procedures for roots, but experience shows that this has never been easy or really mastered.

It is over 15 years since some of these ideas on root exudates were being discussed and worked out (Krishna et al., 1985). With so many new breeding and molecular tools now available the way forward is clearer, and it is a matter of the physiologists, breeders, agronomists, molecular biologists, and soil scientists finishing the job. The subsequent papers in this workshop set out to show us the way.

The socio-economists' view

The socio-economist posed three important research and development questions pertinent to the problem of low-nutrient soils.

1. What are the development needs in the arid and semi-arid tropics (ASAT)?
2. How are agricultural systems likely to evolve and what are the implications for technology research strategies?
3. What are the critical research questions related to alternative technology strategies?

In order to address these important questions an example from Sub-Saharan Africa (SSA) was used.

It is clear that agricultural systems in SSA are in crisis, in response to increasing population and growing markets, and that with the current population increases (approx. 3% per year) the region will require a per capita food production increase of about 4.0% per year. It is also recognized that the two major constraints to primary production in SSA are water availability and soil fertility, and that these should be addressed together and not in isolation from each other.

A simple model is presented that indicates which pathway could be followed as a function of population pressure and access to markets. The result of the model is also strongly affected by rainfall. The model allows crop improvement scientists to focus on the priority issues as they relate to improving crop production in low-nutrient soils.

It can be seen that in the driest areas, low-input strategies with traditional crops are not viable because of population pressure and that therefore, extensive livestock activities may have to replace existing systems. Here, the focus would be on improving forages and pastures rather than food or feed crops. In contrast, in the wettest areas the best scenarios are to use inorganic fertilizers, that are clearly profitable. It is the transition zone that requires the most research input and is probably the climatic zone that this workshop must address.

Good local cultivars such as Sadoré Local already exist in the Sahelian zone of SSA. What is now required is to find ways to breed similar varieties that can better utilize and obtain P, and at the same time do not enhance land degradation through the mining of soil nutrients and the total depletion of soil organic matter. Land degradation must be avoided at all costs.

In conclusion, economists admit that there are no silver bullets, and that a combination of technologies will be required. Breeding solutions will need to be combined with crop and water management strategies. Higher productivity, more sustainable systems in the ASAT will require location-specific strategies, that will be dependent on population, markets, and rainfall. Two of the listed technologies, the use of more legumes in the ecosystems, and the use of physiological and molecular biological tools to improve P uptake through plant breeding are addressed.

References

Ae N, Arihara J, Okada K, Yoshihara T and Johansen C 1990 Phosphorus uptake by pigeonpea and its role in cropping systems of the Indian subcontinent. Science 248, 477–480.

Bowers J E 1982 The plant ecology of inland sand dunes in western North America. J. Arid Environ. 5, 199–220.

Buddenhagen I W 1990 Legumes in farming systems in Mediterranean climates. *In* The Role of Legumes in Farming Systems of the Mediterranean Areas. Eds. A E Osman, M H Ibrahim and M A Jones. pp 39. Kluwer Academic Publishers, Dordrecht, the Netherlands.

Chrispeels M J, Crawford N M and Schroeder J I 1999 Proteins for transport of water and mineral nutrients across the membranes of plant cells. Plant Cell 11, 61–675.

Dakora F D and Phillips D A 2000 Root exudates as mediators of mineral acquisition in low-nutrient environments. Plant Soil 245, 35–47

Danin A 1991 Plant adaptation in Desert Dunes. J. Arid Environ. 21, 193–212.

Danin A 1996a Adaptations of *Stipagrostis* spp. to Desert Dunes. J. Arid Environ. 34, 279–311.

Danin A 1996b Plants of Desert Dunes. Springer, Berlin, Germany.

Dunlop J and Phung T 2000 Transporter genes to enhance nutrient uptake: Opportunities and challenges. Plant Soil 245, 115–122.

Föhse D, Claasen N and Jungk A 1988 Phosphorus efficiency in plants 1. External and internal P requirements and P uptake efficiency of different plant species. Plant Soil 110, 101–109.

Fox T R and Comerford N B 1990 Low-molecular weight organic acids in selected forest soils of the southeastern USA. Soil Sci. Soc. Am. J. 54, 1139–1144.

Ghazanfar S A and Fisher M 1998 Vegetation of the Arabian Peninsula. Kluwer Academic Publishers, Dordrecht, The Netherlands. 352 pp.

Hinsinger P 1998 How do plant roots acquire mineral nutrients? Chemical processes involved in the rhizosphere. Adv. Agron. 64, 225–265.

Hochmuth G J, Gabelman W H and Gerloff G C 1985. A gene affecting tomato root morphology. Hort. Sci. 20, 1099–1101.

Krishna K R 1997 Phosphorus efficiency of tropical and semi-dryland crops. *In* Accomplishments and Future Challenges in Dryland Soil Fertility Research in the Mediterranean Area. Ed. J Ryan. pp 343–363. International Center for Research in the Dry Areas (ICARDA), Aleppo, Syria.

Krishna K R, Sheety K G, Dart P J and Andrews D J 1985 Genotype dependent variation in mycorrhyzal colonization and response to inoculation of pearl millet. Plant Soil 86, 113–125.

Lynch J P 1995 Root architecture and plant productivity. Plant Physiol.109, 7–13.

Manske G G B 1990 Genetical analysis of the efficiency of VA mycorrhiza with spring wheat. 1. Genotype differences and a reciprocal cross between efficient and non-efficient variety. *In* Genetic Aspects of Plant Mineral Nutrition. Eds. N El-Bassam, M Dambroth and B C Loughman. pp 397–405. Kluwer Academic Publishers, Dordrecht, The Netherlands.

Shapiro B I, Osbourne D and Ahmed M 2000 Ex-ante impact assessment and economic analysis of breeding for nutrient efficiency: The socio-economists' view. *In* Food Security in Nutrient-Stressed Environments: Exploiting Plants' Genetic Capabilities, Proceedings of an International Workshop 27–30 September 1999, Patancheru, India. Developments in Plant and Soil Sciences volume 95. Ed. J J Adu-Gyamfi. pp 37–42. Kluwer Academic Publishers, Dordrecht, The Netherlands.

Sinclair T R and Sheehy J E 1999 Erect leaves and photosynthesis in rice. Science 283, 1456–1457.

Sinclair T R and Vadez V 2000 Physiological traits for crop yield improvement in low N and P environments. Plant Soil 245, 1–15.

Smith S E, Robson A D and Abbot L K 1992 The involvement of mycorrhizas in assessment of genetically dependent efficiency of nutrient uptake and use. Plant Soil 146, 169–179.

Vancura V and Hovadik A 1965 Root exudates of plants. II. Composition of root exudates of some vegetables. Plant Soil 22, 21–32.

Wood M, Philbeam C, McNeil A and Harris H 1997 Nitrogen cycling in a dryland cereal-legume rotation system. *In* Accomplishments and Future Challenges in Dryland soil Fertility Research in the Mediterranean Area. Ed. J Ryan. pp 71–78. International Center for Research in the Dry Areas (ICARDA), Aleppo, Syria.

J.J. Adu-Gyamfi (Ed.), Food security in nutrient-stressed environments: exploiting plants' genetic capabilities: 47–57, 2002.

Breeding for low-nutrient environments

Y.-G. Zhu[1,2,3], C. Huang[4], N. K. Howes[5] & S. E. Smith[1,2,3]
[1]*Research Centre for Eco-environmental Sciences, Chinese Academy of Sciences, Beijing, China*
[2]*Department of Soil and Water, Waite Campus, The University of Adelaide, Glen Osmond, SA 5064, Australia*
[3]*Department of Plant Science, Waite Campus, The University of Adelaide, Glen Osmond, SA 5064, Australia*
[4]*Field Crops Pathology Unit, South Australia Research and Development Institute (SARDI), Glen Osmond, SA 5064, Australia*
[5]*Corresponding author**

Key words: plant breeding, marker-assisted selection, molecular approach, nutrient efficiency, soil

Abstract

To feed the continuously growing world population requires sustainable food production in both nutrient-stressed environments (low-fertility soils) and fertile soils. In many low-fertility soils, the limiting factor for plant growth is often the bioavailability of the nutrient of concern rather than its total amount in the soil. The traditional approach to correcting low-fertility soils has been fertilizer application. This is now being challenged by economic viability and the negative environmental impacts of nutrient loss from soil to ground water. Modifying the plant to suit the soil is emerging as an useful alternative. Differences in plant nutrient acquisition/efficiency exist among different species and/or genotypes. This genetic variation allows possible improvement of crop cultivars for nutrient-stressed environments by manipulating related plant traits through either conventional or molecular plant breeding (including marker-assisted selection and genetic engineering). This paper examines strategies employed by various plant species/cultivars to acquire nutrients with low bioavailability and the genetic control of these strategies. Current approaches to breeding for low-nutrient environments are also discussed in order to identify new research directions and initiatives for future development of genetic resources and crop cultivar improvement.

Introduction

With the mounting pressure from the ever-increasing population throughout the world, sustainable production of adequate and high-quality food has now been prioritised by many international organizations and governmental agencies. Management of soil fertility is a key component of any food production system. Low-fertility soils are distributed widely in many parts of the world, and their utilization/management will play an important role in feeding the population. Faced with nutrient-deficient soils, the traditional approach is to change the soil to fit the plant, i.e. to apply fertilizers and other soil amendments to increase soil fertility and/or to make soil nutrients available for plant uptake. Application of chemical fertilizers however, means extra costs for farmers and could also lead to greater risks of off-site water pollution by nitrogen (N) and phosphorus (P) from agricultural land. Furthermore, soil deficiency of some micronutrients such as iron (Fe) and manganese (Mn) can not be easily alleviated by fertilizer application. Even where soil nutrient mining could be a serious problem, and application of fertilizer is essential for sustainable crop growth, breeding nutrient-efficient cultivars means better utilization of applied fertilizers, and thus, higher economic returns. Therefore, a plant breeding approach should still be considered as an important component of a management system that includes fertilizer application.

On the other hand it is worth noting that, in many cases a lack of nutrients is due largely to their low bioavailability rather than to their total contents in soils. Taking P as an example, the total soil-P is often 100-fold more than the fraction of soil-P readily

* FAX No: +86-10-62923563.
E-mail: ygzhu@mail.rcees.ac.cn

available to plant uptake (Al-Abbas and Barber, 1964). This discrepancy is due to the differential availability of the different pools of P in soil, and their rate of exchange with P in the soil solution. In general, the soil chemistry of P controls its availability, but some activities of plant roots and soil micro-organisms do help to maintain the P concentration in soil solution by inducing P release from solid phase (Comerford, 1999; Schachtman et al., 1998). The same situation may well be true for such other nutrients as Fe, zinc (Zn), Mn, and copper (Cu) etc. It has therefore been postulated that the large discrepancy between total and bioavailable nutrient could be viewed as the result of a 'genetic deficiency' in plants (Graham, 1984), and that there is considerable scope for breeding cultivars with increased nutrient efficiency.

The exploitation of plant genetic variation in nutrient efficiency has been increasingly emerging as a viable alternative approach to crop production in low-nutrient environments in terms of better resource utilization and environmental quality. Furthermore, varietal differences in plant tolerance to low-nutrient soils have been observed for many elements and on many crop species. For example, since rye is exceptionally high in Zn efficiency, genes controlling Zn efficiency could be transferred to wheat through a breeding program to develop Zn-efficient wheat cultivars (Cakmak et al., 1997). Therefore, breeding nutrient-efficient cultivars is necessary and feasible. In this paper, an overview of recent achievements in this field, is provided in order to identify new research directions and initiatives for the development of genetic resources.

Table 1. Plant traits related to acquisition of nutrients with low availability in the soil

Category	Plant traits
Increasing absorptive surface	Formation of proteoid root
	Root hair development
	Shoot/root ratio
	Change of root architecture
	Formation of mycorrhiza
Uptake kinetics/molecular structure	Kinetic parameters: K_m, V_{max}, and C_{min}
	Ion transporters/channels
Increasing bioavailability in the rhizosphere	Acidification due to net extrusion of H^+
	Change of redox potential
	Excretion of organic chelates
	Production of reductase

Definitions of nutrient efficiency

Nutrient efficiency is considered broadly as the ability of a genotype to grow in a soil deficient in that nutrient (very low bioavailability) and produce more and/or better quality biomass (e.g. micronutrient densities), especially of the harvestable component than other genotypes (Graham, 1988). From an agronomic aspect, the definition of nutrient efficiency is likely to be specific to particular agricultural systems, soils, and plant products (e.g. grain or total above-ground biomass). Plant nutrient efficiency contains two physiological components, i.e. uptake efficiency and utilization efficiency (El Bassam, 1998).

The term uptake efficiency is used to describe the ability of a plant to absorb a particular nutrient per unit of absorbing tissue (root length, area, weight, etc.) usually by exploiting more soil volume or different sources of nutrients within the same volume. Uptake efficiency is directly related to plant root physiology, morphology, and architecture and to symbioses with mycorrhizal fungi or other soil microorganisms.

Utilization efficiency is a term used to describe how efficiently the plant produces dry matter per unit of nutrient taken up, a process mainly regulated internally. Utilization efficiency is influenced by the amount of nutrient of concern required for enzyme synthesis and activation, and also by the mobility of the nutrient within plant, e.g. the movement of nutrients from senescent leaves to newly formed tissues. However, the quest for utilization efficiency may be problematic for some elements (e.g. P and micronutrients), because high utilization efficiency effectively means low tissue concentration, and hence affects the nutritional quality of foodstuffs produced (Graham et al., 1999) and maybe the quality of seed germination.

Definitions of nutrient efficiency vary widely, and in some cases may mislead in the quest for increased productivity with efficient nutrient utilization. For example, a cultivar may have higher P utilization efficiency (biomass per unit P taken up) but this does not mean that it can acquire insoluble P more efficiently. Several authors have reported that different measures (definitions) of P efficiency can result in plant species or genotypes having different ranking for P efficiency

(Föhse et al., 1988; Gourley et al., 1994; Kemp and Blair, 1991). From the agronomic point of view, a combination of total P uptake and plant dry matter accumulation (or harvestable dry matter yield) under low external P supply would be appropriate. Such high dry matter accumulation and/or total P removal could be due to more efficient uptake or more efficient utilization. Nevertheless, an ad hoc approach to identification of efficient genotypes may not reveal the mechanisms of efficiency and in consequence the predictive value of the information with respect to conditions – (soil type etc.) under which the genotype will perform well – will be limited and the search for genes relevant to plant improvement programmes would be difficult.

Plant adaptation to low-nutrient environments

Wild plants have evolved various strategies to ensure their survival in response to low soil fertility, including either slow (Type I) and ruderal species (Type II) (Marschner, 1995). Slow-growing plants often have high root/shoot ratios, and low maximum potential growth rate, even when the nutrient supply is high (Hommels et al., 1989). Plants of this type can store nutrients (i.e. luxury consumption and higher internal concentration) for later utilization when supply is limited. Another characteristic of some Type I plants is their large seed reserves of nutrients used in seedling establishment (Stock et al., 1990). Ruderal species are more immediately responsive to external nutrient supply and can easily display visual deficiency symptoms. This type of plant has relatively low nutrient storage capacity.

Apart from these differing growth strategies, plant species have developed various specific mechanisms to acquire nutrients from different forms in the soil. These mechanisms can be both morphological and physiological, and are important for crop production systems under nutrient-stressed environments. For example, in adaptation to N-limited environments the non-mycorrhizal sedge, *Eriophorum vaginatum*, has evolved specific systems to take up organic N; at least 60% of the N demand of this species can be ascribed to amino acid uptake (Chapin et al., 1993). Such root-induced changes in the rhizosphere as chelation, acidification, and redox potential, etc. are major physiological responses of plants to nutrient deficiency (Marschner, 1995). While the formation of symbioses between plant roots and soil microorganisms can either add nutrients to the soil (N_2-fixation) or enhance plant uptake (mycorrhizae). Morphological and physiological traits related to plant adaptation to low-nutrient environments are summarized in Table 1, detailed discussion follows.

Plant characteristics related to nutrient efficiency and their genetic control

For the purpose of breeding programs, identification of plant traits important for nutrient uptake and utilization efficiency is essential, as these will allow the development of marker-assisted selection for particular traits, whether they are quantitatively or qualitatively inherited. Identification of such traits is also useful in understanding the mechanisms of nutrient efficiency. This understanding will lead to the identification of genes controlling nutrient efficiency, and to the possible manipulation of these genes to produce nutrient-efficient crop cultivars.

Root system: size and architecture

The acquisition of soil-based resources is the major function of the root systems of terrestrial plants. Such root morphological characteristic, as architecture and branching, size, density, and length of hairs will profoundly affect plant acquisition of nutrients from soil. The size and/or rate of growth of root systems into unexploited soil is particularly important for the aquisition those nutrients with low mobility, such as P. The importance of root size in nutrient uptake is well represented in mechanistic models for nutrient uptake (Barber and Silberbush, 1984). Experimental results have also shown that efficient P-accumulation by some crop cultivars is related to large root systems. The spatial arrangement of the root system (root architecture) is another important factor in determining P uptake. For example, P-efficient genotypes of common bean often have a vigorous and highly branched root systems with large numbers of apices (Lynch and van Beem, 1993).

The presence of root hairs can substantially increase the root surface area available for ion uptake and therefore increase the nutrient efficiency of plants (Gahoonia and Nielsen, 1998). Inclusion of root hairs in mechanistic models greatly improved the prediction of P influx/uptake of seven plant species (e.g. Föhse et al., 1991). Stimulation of root hair growth in *Arabidopsis thaliana* by low-P supply was observed

Table 2. Simulation of root hair growth of *Arabidopsis thaliana* by P availability[a]

Transfer treatment	Root hair length (mm)		
	Before	After	*P*-value
High P→low P	0.470±0.162	0.913±0.275	<0.1
Low P→high P	0.995±0.228	0.521±0.141	<0.01

[a] Bates and Lynch, 1996.

in a Phytagel-based experiment (Table 2) (Bates and Lynch, 1996). An increase in the abundance of root hairs in response to Fe deficiency has also been observed (Romheld and Kramer, 1983).

Some plants, such as white lupin, brassicas, and members of the Proteacea form proteoid roots under P-deficient conditions, this enhances their P uptake capabilities (Keerthisnghe et al., 1998). Under constant P-supply, Keerthisnghe et al. (1998) demonstrated that with external P in the range of 1–10 mmol m^{-3}, proteoid roots accounted for over 50% of the total roots biomass, but that the formation of proteoid root was greatly suppressed when external P supply was over 25 mmol m^{-3}. Proteoid roots can increase P-uptake both directly via the enhanced absorptive area, and indirectly through the excretion of citric acid which acts as a chelating agent (see above).

Genetic control of plant root characteristics has been widely discussed (Caradus and Woodfield, 1998), and it is generally accepted that this variation is largely multi- or polygenic. Caradus (1995) has given a detailed review on the heritability of such root traits, as size, length, weight, volume, surface area, root number, etc. Since plant root systems are complex, with a range of functional characteristics, the importance of specific root trait(s) in terms of nutrient uptake is likely to vary widely among plant species. In some cases, root hair development is the dominant factor, but in others it may be root architecture. Parameters reflecting root trait(s) crucial for P efficiency, e.g. taproot diameter, rate of root growth, root plasticity (i.e. root/short ratio), density of root hairs, and proportion of fibrous roots to total root dry weight, are needed as solution criteria. Further information is therefore required to quantify the relationship between parameters relating to root system functions and plant nutrient efficiency, so that individual or combinations of root trait(s) can be identified for the purpose of identifying the genes responsible for nutrient efficiency.

Rhizosphere modification

Root-induced rhizosphere modification is a key factor affecting plant uptake of less-mobile nutrient ions such as P and many micronutrients. Johansen et al. (1995) suggested that genetic manipulation of the rhizosphere modification processes (largely root exudation), and of root-associated microorganisms are key targets in the search for more-efficient use of P by plants; and that this may also be true for micronutrients.

Exudation of organic acids from plant roots has two possible effects: (1) reduction in rhizosphere pH that will increase the bioavailability of P and micronutrients in soils with high pH, such as calcareous soils; (2) chelation of metal ions by organic anions thereby releasing fixed nutrients from solid phase into the soil solution. A wide range of organic anions exuded from nutrient-stressed plant roots have been identified. Generally speaking, root exudation of organic acids is not specific to a particular nutrient. For example, exudation of phytosiderophores has been identified as a key process for Fe acquisition by gramineous species, but deficiencies of Zn (Hopkins et al., 1998) or Cu (Gries et al., 1998) could also induce the release of phytosiderophores from root cells. On the other hand, rhizosphere changes induced by the deficiency of one nutrient may increase the uptake of other nutrients. For example, mechanisms leading to plant Fe-efficiency may also result in increased uptake of other micronutrients in addition to Fe (e.g. Mench and Fargues, 1994). Welch (1995) therefore pointed out that root exudation is possibly an integral part of a more basic and complex array of stress-response mechanisms operating in nutrient acquisition by roots. As in many cases deficiencies of micronutrients and P may coexist in the same soil, it could therefore be useful to link root exudation of organic acids with plant uptake efficiency for multiple elements. This would allow the development of crop cultivars with broad adaptation to low-fertility environments.

Root activities, such as absorption of cations/anions, net excretion of organic acids (as stated above) and N_2-fixation, etc., can induce changes of pH in the growth medium (Marschner, 1995). Rhizosphere acidification is in some cases a response to P stress by plant roots. For example, net proton excretion under P-deficient conditions has been observed in tomato (Heuwinkel et al., 1992), white lupin (Dinkelaker et al., 1989), wheat (Fan and Liu, 1991), and rice (Fan and Liu, 1991; Kirk and Saleque, 1995). The enhanced H^+ excretion under P stress is beneficial for

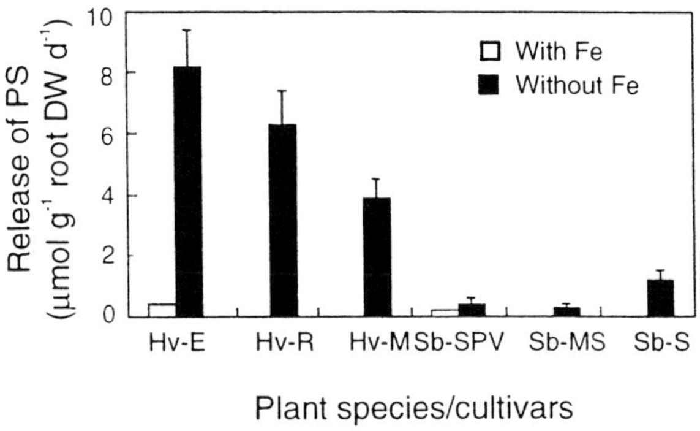

Figure 1. Genetic variation in release of phytosiderophores (PS) in response to Fe-deficiency. Plant age 19 days. Hv, *Hordeum vulgare*; Sb, *Sorghum bicolor*. Other letters represent following cultivars: E, Europa; R, Rikuzen-Mugi; M, Minori-Mugi; SPV, SPV-393; MS, MS 35-1; S, Sudangrass (Romheld and Marschner, 1990).

the plant to mobilize insoluble P in soils of high pH, while this mechanism (reduction of pH in the rhizosphere) may not be important in acidic soils where the dominant P forms are Fe- and Al-P. Kirk and Saleque (1995) reported that solubilization of fixed P by reduced pH accounted for over 80% of the P absorbed by rice plants. Reduction in pH in the rice rhizosphere was mainly due to H^+ directly released from the roots or produced in Fe^{2+} oxidation by root-released O_2. The mechanism of enhanced H^+ excretion in response to P stress is not fully understood, but proton/P co-transporters may contribute to the pH change of the growth medium in response to P deprivation or deficiency (Dunlop and Gardiner, 1993).

Genetic variation in root exudation of organic acids in response to nutrient stress has been widely recognized. For example, different cultivars of barley and sorghum differ substantially in their ability to release phytosiderophores in response to Fe deficiency (Figure 1). In a hydroponic experiment, the Al- and Fe-P solubilizing activity of root exudates from P-deficient pigeonpea was also shown to vary genotypically (Subbarao et al., 1997), whereas root exudates from P-sufficient plants had only negative effects on the solubilization of Fe- or Al-P. The solubilizing activity was found to be negatively correlated with tissue-P concentrations, suggesting that this ability was possibly regulated by the internal P status, which is in accordance with the general concept that P uptake is regulated by internal P status (Schachtman et al., 1998). The exudation of organic acids may also depend on the plant growth stage (Otani and Ae, 1997), and should be taken into account when a bioassay is to be designed to screen for this plant trait. The ability of plant roots to acidify the rhizosphere in relation to P efficiency has also been found to differ genotypically in some wheat cultivars (Zhu et al., unpublished data).

Vesicular arbuscular mycorrhizae (VAM)

The role of VAM in plant acquisition of nutrients, such as P and Zn has been well established (Smith and Read, 1997). It is generally recognized that the action of arbuscular mycorrhizal associations to improve uptake of immobile nutrients is not by dissolving relatively unavailable forms of soil nutrients, but by allowing more efficient uptake of nutrient ions from soil solutions and more effective exploitation of the soil volume by the external mycelium (Li X et al., 1991).

Mycorrhizal responsiveness has been shown to vary markedly with host plant species and genotypes (e.g. Bryla and Koide, 1990; Hetrick et al., 1993, 1995; Khalil et al., 1994; Koide et al., 1988). It is generally recognized that modern soil management pratices and plant breeding under fertile soil conditions may inadvertently limit or reduce the effect of this symbiotic contribution to the nutrition of crop plants (Hetrick et al., 1993, 1995; Smith et al., 1993), because high levels of soil-P are often found to reduce the colonization of VAM (e.g. Bruce et al., 1994) or minimize their beneficial effects. Landraces of spring wheat were observed to be more responsive to mycorrhiza than high-yielding modern varieties (Manske, 1989). However, according to their results on oats and tomato, Koide and his co-workers (Bryla and Koide 1990; Koide and Li, 1991; Koide et al., 1988) tended to support the hypothesis that in infertile soils improved cultivars are more responsive to VAM than wild types. They argued that wild types have greater inherent tolerance to soil infertility, and therefore the beneficial effect of mycorrhizal infection is less important than in cultivated forms. On the other hand, Khalil et al. (1994) reported that considerable variability existed in the mycorrhizal dependency of maize and soybean cultivars when grown in P-deficient soils, and that this variability extended across both improved and unimproved cultivars. In general, mycorrhizal responsiveness is related to the inherent rate of growth, the plasticity of root/shoot ratios, root architecture, and root hair development. For example, plants with rapid growth rate and high tissue P concentration are probably more dependent on mycorrhizae (Koide and Li, 1991). In barley plants, Baon et al. (1993) found that the responsiveness of different cultivars to mycorrhizal infection was positively correlated with their response to P application, indicating the importance of VAM in plant P efficiency.

Absorption mechanisms at plasma membrane/molecular levels

Transport of ions across the plasma membrane is a key step in plant acquisition of nutrients from external environments. Absorption mechanisms mainly refer to the kinetic characteristics of uptake processes and the types of plasma membrane proteins involved. Parameters such as K_m (affinity), V_{max} (maximum uptake rate), and C_{min} (minimal ion concentration in the solution below which the plant can not take up) have been frequently measured and used to characterize the mechanisms of plant uptake (e.g. high- and low-affinity transporters) of a particular nutrient ion and nutrient efficiency (Smith et al., 1998). However, it seems that the relationship between these parameters and plant nutrient efficiency has not as yet been well established since in most cases, the soil-based processes have a much more important potential role than the membrane-associated processes in plants. Bloom (1994) argued that N uptake may depend more on root architecture and distribution than on kinetic parameters, and that the physiology of N uptake/assimilation may be similar in wild and cultivated taxa of barley and in older and newer cultivars (Jackson and Koch, 1997). In the case of P, there is also evidence to show that P-absorption potential does not differ consistently among wild and cultivated *Hordeum* species (Chapin et al., 1989). While most species/cultivars remain unexamined in terms of absorption kinetics, further efforts to link nutrient uptake characteristics to plant nutrient efficiency should be encouraged.

Research on the molecular basis of plant nutrient uptake has lead to the identification of a number of ion transporters, such as high-affinity K uptake transporter and Zn transporters (see Fox and Guerinot, 1998 for general review), and P transporters (e.g. Liu et al., 1998; Smith et al., 1997). Molecular studies of such transporters are mainly to elucidate the mechanisms of nutrient uptake at the plasma membrane level, and are often carried out in hydroponic systems at extreme concentrations (i.e. either at starvation or at mm range) of the nutrients of concern. This is not an appropriate reflection of field conditions. It is therefore necessary to take different approaches if the physiological significance of individual transport genes is to be resolved (Smith et al., 1998). Understanding the structures of genes encoding for these transporters and their functions is still at an early stage, and to date no solid relationship between the types/expression of ion transporters and plant nutrient efficiency at whole-plant level has been revealed. Further discussion in this topic is covered by another presentation in this publication (Dunlop and Phung, 2000).

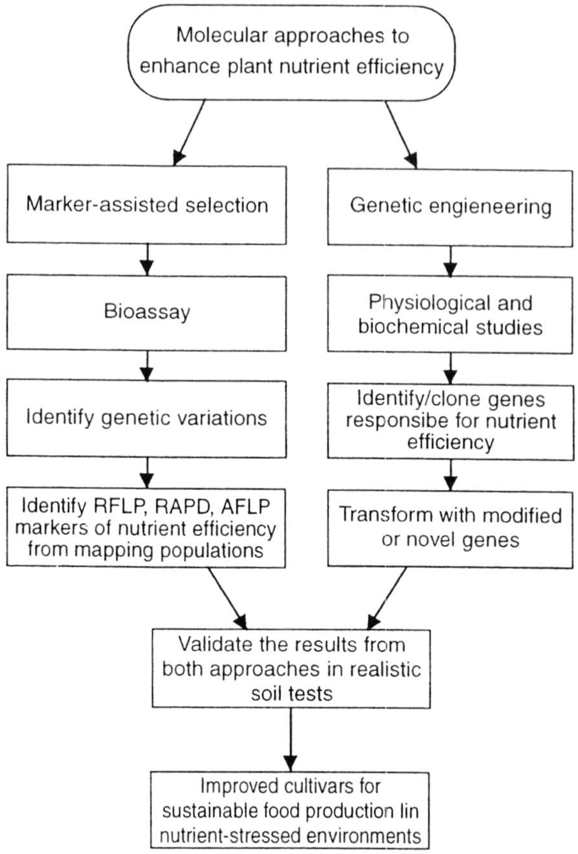

Figure 2. Molecular approaches to improve plant nutrient efficiency in low-fertility soils (modified from Schachtman and Barker, 1999).

Strategies for breeding nutrient-efficient cultivars

As indicated in the Introduction, improvement of crop production by altering such soil conditions as low fertility has limitations in terms of environmental impacts and economical and practical viability. However, the existence of plant traits responsible for nutrient efficiency in different species/ecotypes raises the possibility of genetic improvement of crop species in terms of mineral nutrition. It has now been well documented that plant traits related to nutrient efficiency are generally under genetic control. For example, micronutrient

Table 3. Zinc (Zn) efficiency (based on yield) rankings of some wheat cultivars (out of 30 tested) grown on Zn-deficient soil in two seasons at two sites in South Australia (based onn yields with no Zn application)[a]

Cultivars	Lameroo		Yeelanna	
	1988	1989	1988	1989
Excalibur	1	5	2	2
RAC629	3	2	13	3
Aroona	2	20	1	19
Schomburgk	12	13	11	8
Warigal 5RL	14	12	6	20
Warigal	22	16	14	25
Takari	5	25	23	15
Gatcher	23	27	27	27
Durati	29	29	25	29

[a]Data from Graham et al., 1992).

efficiency is controlled by a major and several minor genes (see Graham et al., 1999, for general review), it is therefore possible to incorporate these genes into a locally adapted cultivar, or to genetically modify the local cultivars. Molecular approaches to breeding crop cultivars with high nutrient efficiency are illustrated in Figure 2.

Bioassay system

One of the key issues in breeding programs for nutrient-efficient crop cultivars is how to screen P efficiency realistically and effectively in a way that will be relevant to field conditions. As discussed earlier, the ranking of genotypes may vary with the definition of efficiency, the harvestable plant parts, and soil type. Experimental designs should be based on both the plant and soil of concern, so that results can reveal both utilization and uptake efficiencies. However, in most current screening systems there are still limitations in the definition of mechanisms of nutrient efficiency. For example, in a study to conduct linkage analysis and quantitative trait loci (QTL) mapping of tolerance to low-P stress in maize, Reiter et al. (1991) used a bioassay system with sand-alumina mixture as growth medium that simulated diffusion-limited P supply from soils where P was primarily fixed as Al-P. It is also possible to design systems where Fe/calcium (Ca)-P or organic P are the dominant forms, thus targeting particular soil conditions where different mechanisms (such as chelation of organic anions, pH change, exudation of phosphatases, or mycorrhizal colonization) are likely to be important. In such systems identified traits were related largely to P-uptake efficiency. In contrast, in a study on rice, Ni et al. (1998) employed a hydroponic system, in which the mechanism of internal P efficiency was detected, but not the uptake efficiency, because P supplied in solution (whether sufficient or insufficient) is freely available, unlike the situation in the soil. In the case of P, it is suggested that a bioassay with dual treatments using defined sources of P that represent diffusion-limited and freely available forms of P in the soil could reveal factors relating both to plant uptake and internal P (utilization) efficiencies. This strategy may well be applicable to other nutrients. Other limitations in designing effective bioassays, such as environment×genotype interactions, and interactions with other nutrients are frequently encountered by both plant breeders and plant physiologists (RD Graham, personal communication). Table 3 shows that the ranking of Zn efficiency in wheat cultivars could vary substantially with experimental sites and year. In consequence, any results from pot assays must be validated in field situations if they are to be used effectively in breeding programs.

Parameters to be evaluated in screening systems mainly include biomass production, internal concentration/total uptake of the nutrient of concern, tillering, root length/size, and root exudation of H^+ and organic acids. Tillering and dry weight have been used effectively in identifying molecular markers for tolerance to P-deficiency in rice (Ni et al., 1998) and in maize (Reiter et al., 1991). On the other hand, measurement of phytosiderophore production has been suggested as useful in selection for Fe-efficiency in wheat cultivars (Hansen et al., 1996). Suitable parameters used to select cultivars with enhanced nutrient efficiency would also be very useful in unravelling candidate mechanisms underlying the efficiency of a specific nutrient and to identify gene(s) encoding the efficiency.

Genetic engineering

Soil–root interactions could broadly involve a wide range of factors, i.e., root architecture, root exudation, rhizosphere acidification, and plant–soil microbe interactions – particularly VAM (Smith et al., 1999), ion transporters/channels, and factors controlling their expression and activity. Genetic engineering of plant characters related to nutrient efficiency has emerged as a widely recognized alternative in recent years. For example, considerable effort has been directed to work on molecular aspects of ion transporters (e.g. Fox and

Guerinot, 1998). For the purpose of understanding the mechanisms, it is important to elucidate the molecular basis of plant acquisition of nutrient ions. However, expectations of the potential for manipulation of ion transporters/channels to enhance nutrient efficiency should not be too high. The process of nutrient uptake is often limited by the mobility of ions in the soil and their diffusion to the root surface, rather than by the kinetics of uptake across the plasma membrane. With many cloned genes already to hand, the next challenge is to understand their expression pattern and regulation, and to establish the relationship between the expression of these genes and plant nutrient efficiency and rates of uptake in realistic soil tests.

For root exudation of organic acids, processes involved in both the synthesis and the efflux of organic acids in response to a particular nutrient stress must be considered. Several genes, that are probably involved in the biosynthesis of phytosiderophores and response to Fe-deficiency have been cloned (Mori, 1997), but little is known about the regulation of gene expression. Cloning and characterization of the genes involved in the biosynthesis pathways, and the transporter genes of the Fe^{3+}-phytosiderophore complex, together with the regulation of these genes under Fe-deficiency are required for improvement of Fe-efficiency in grass species. This strategy is probably also applicable to other nutrients. Exudation of organic anions may not be specific to a single nutrient, indicating that enhancement of this trait would improve the plant efficiencies of multiple elements. Studies on root exudation of organic anions and/or H^+ therefore deserve considerable efforts in the future.

Genetic manipulation of mycorrhizal symbiosis is an another option for improvement of plant nutrient efficiency, particularly P and Zn. Smith et al. (1993) suggested that the interactions between mycorrhizal infection and P efficiency should be considered at two levels: (a) mycorrhizae may act as general modifiers of efficiency regardless of the extent to which the plants are colonized and in some mycorrhizae-dependent plants infection may change the ranking of genotypes and (b) the extent of infection is also under genetic control (e.g. Hetrick et al., 1995), and shows considerable variability between genotypes in some species. Lackie et al. (1988) and Mercy et al. (1990) demonstrated that selection for high incidence of mycorrhizal infection is realistic and could be achievable. However, mycorrhizal responsiveness and/or mycorrhiza-assisted P-uptake by plants may not be closely correlated with colonization rate. It is therefore a pre-requisite to analyse the contribution of VAM to plant P efficiency under a range of soil-P concentrations. Inclusion of a wide range of germplasm would be helpful to the identification of mycorrhizal responsive gene(s) of various crop species, and to the development of molecular markers for this trait. Studies into genetic control of mycorrhizal colonization, e.g., mutants with highly reduced mycorrhizal colonization, will greatly advance the molecular basis of the control of VAM symbiosis (e.g. Barker et al., 1998).

Marker-assisted selection (MAS)

The advent of molecular-marker technology has led to the development of genetic maps for a number of crops to identify QTLs related to nutrient efficiency. Measuring the efficiency of nutrient uptake is difficult, due to its variation under different environmental conditions. Molecular markers based on plant genotype may overcome some of these difficulties (Schachtman and Barker, 1999). Marker assisted selection (MAS) does not require detailed knowledge of the plant physiological and molecular basis of desired traits, and it will reduce the time and cost of selecting nutrient-efficient progeny, especially in a backcrossing program. To date only limited MAS application in breeding programmes aimed at increasing the nutrient-efficiency of crop cultivars has been reported. For example, molecular markers have been identified for P efficiency in maize based on sand culture (Reiter et al., 1991) and for internal P efficiency in rice using a hydroponic system (Ni et al., 1998; Wissuwa et al., 1998). Markers for P, Zn, and Mn efficiency in wheat and barley are being developed at the Cooperative Research Centre for Molecular Plant Breeding, Australia. It is expected that this approach has great potential, particularly in breeding for nutrient efficiency, that is almost certainly controlled by multiple genes.

The use of doubled haploid (DH) populations provides an option for improved selection if molecular marker technology is not available, because all progeny of DH lines are genetically identical, thus allowing easy replication of nutrient-uptake experiments. However, care must be taken to reduce variations to the extent possible as in ripening time, seed size in such parameters or nutrient status, with the availability of molecular markers, DH populations provide further advantages in characterizing segregating populations, especially for characteristics (nutrient efficiency in this case) controlled by multiple genes.

Conclusion

The existence of genetic variation in plant nutrient acquisition and adaptation to low-nutrient environments provides the rationale for breeding crop cultivars for such situations, and to secure food supply in a sustainable fashion. Screening strategies should be devised on the basis of ecophysiological understanding of plant nutrient efficiency. QTLs will be established by MAS using reliable bioassay and appropriate parameters, that may overcome some of the current limitations in breeding for low-nutrient environments. Recent developments in molecular biology and gene technologies to genetically modify plants provide further opportunities for enhancement of nutrient efficiency, but this requires more information before the approach can be justified.

References

Al-Abbas A H and Barber S A 1964 A soil test for P based upon fractionation of soil P: I. Correlation of soil P fractions with plant-available P. Soil Sci. Soc. Am. Proc. 28, 218–221.

Baon J, Smith S and Alston A 1993 Mycorrhizal responses of barley cultivars differing in P efficiency. Plant Soil 157, 97–105.

Barber S A and Silberbush M 1984 Plant root morphology and nutrient uptake. *In* Root, Nutrient and Water Influx, and Plant Growth. Eds. S A Barber and D R Bouldin. pp 65–87. Am. Soc. Agron. Madison, WI, USA.

Barker S, Stummer B, Gao L, Dispain I, O'Connor P and Smith S E 1998 A mutant in *Lycopersicon esculentum* Mill. with highly reduced VA mycorrhizal colonisation-isolation and preliminary characterisation. Plant J. 15, 791–797.

Bates T and Lynch J 1996 Stimulation of root hair elongation in *Arabidopsis thaliana* by low P availability. Plant Cell Environ. 19, 529–538.

Bloom A 1994 Crop acquisition of ammonium and nitrate. *In* Physiology and Determination of Crop Yield. Eds. K J Boote, T R Sinclair and G M Paulsen. pp 303–309. ASA, CSA, SSSA, Madison, WI, USA.

Bruce A, Smith S E and Tester M 1994 The development of mycorrhizal infection in cucumber: effects of P supply on root growth, formation of entry points and growth of infection units. New Phytol. 127, 507–514.

Bryla D and Koide R 1990 Role of mycorrhizal infection in the growth and reproduction of wild vs. cultivated plants. II. Eight wild accessions and two cultivars of *Lycopersicon esculentum* Mill. Oecologia 84, 82–92.

Cakmak I, Derici R, Torun B,Tolay I, Braun H J and Schlegel R 1997 Role of rye chromosomes in improvement of zinc efficiency in wheat and triticale. *In* Plant Nutrition for Sustainable Food Production and Environment. Eds. T Ando, K Fujita, T Mae, H Matsumoto, S Mori, J Sekiya. pp 237–241. Proceedings of the XIII International Plant Nutrition Colloquium, 13–19 September 1997. Tokyo, Japan. Developments in Plant and Soil Sciences, Volume 78. Kluwer Academic Publishers, Dordrecht, The Netherlands.

Caradus J R 1995 Genetic control of P uptake and P status in plant. *In* Genetic Manipulation of Crop Plants to Enhance Integrated Nutrient Management in Cropping Systems. 1. Phosphorus. Proceedings of an FAO/ICRISAT Expert Consultancy Workshop 15–18 March 1994, ICRISAT Asia Center, Patancheru, India. Eds. C Johansen, K K Lee, K K Sharma, G V Subbarao and E A Kueneman. pp 55–74. International Crops Research Institute for the Semi-Arid Tropics, Patancheru, Andhra Pradesh, India.

Caradus J and Woodfield D 1998 Genetic control of adaptive root characteristics in white clover. Plant Soil 200, 63–69.

Chapin F I III, Groves R and Evans L 1989 Physiological determinants of growth rate in response to P supply in wild and cultivated *Hordeum* species. Oecologia 79, 96–105.

Chapin F I III, Moilanen L and Kielland K 1993 Preferential use of organic nitrogen for growth by non-mycorrhizal arctic sedge. Nature 361, 150–153.

Comerford N B 1999 Soil bioavailability. *In* Phosphorous in Plant Biology: Regulatory Roles in Molecular, Cellular, Organismic, and Ecosystem Processes. Proceedings 12th Annual Penn State Symposium in Plant Physiology, 28–30 May 1998. Current Topics in Plant Physiology, Volume 19. Eds. J P Lynch and J Deikman. pp 136–147. Am. Soc. Plant Physiol. Rockville, MD, USA.

Dinkelaker B, Romheld V and Marschner H 1989 Citric acid excretion and precipitation of calcium citrate in the rhizopshere of while lupin (*Lupinus albus* L.). Plant Cell Environ. 12, 285–292.

Dunlop J and Gardiner S 1993 Phosphorus uptake, proton extrution and membrane electropotentials of P-deficienct *Trifolium repens* L. J. Exp. Bot. 44, 1801–1808.

El Bassam N 1998 A concenpt of selection for 'low input' wheat varieties. Euphytica 100, 95–100.

Fan X and Liu Z 1991 Phosphorus supply and excretion of H^+ by rice and wheat roots. Acta Phytophysiol. Sinica 17, 125–132.

Föhse D, Claassen N and Jungk A 1988 P efficiency of plants: I. External and internal P requirement and P uptake efficiency of different plant species. Plant Soil 110, 101–109.

Föhse D, Claassen N and Jungk A 1991 P efficiency of plants: II. Significance of root radius, root hairs and cation-anion balance for P influx in seven plant species. Plant Soil 132, 261–272.

Fox T C and Guerinot M L 1998 Molecular biology of cation transport in plants. Ann. Rev. Plant. Physiol. Mol. Biol. 49, 669–696.

Gahoonia T and Nielsen N 1998 Direct evidence on participation of root hairs in P (^{32}P) uptake from soil. Plant Soil 198, 147–152.

Gourley C, Allan D and Russelle M 1994 Plant nutrition efficiency: A comparison of definitions and suggested improvement. Plant Soil 158, 29–37.

Graham R 1984 Breeding for nutritional characteristics in cereals. Adv. Plant Nutr. 1, 57–102.

Graham R D 1988 Development of wheats with enhanced nutrient efficiency: Progress and potential. *In* Wheat Production Constraints in Tropical Environmenta. Ed. A R Klatt. pp 305–320. Centro Internacional de Mejoramiento de Maïz y Trigo, El Batan, Mexico.

Graham R, Senadhira D, Beebe S, Iglesias C and Monasterio I 1999 Breeding for micronutrient density in edible portions of staple food crops: conventional approaches. Field Crops Res. 60, 57–80.

Graham R D, Ascher J S and Hynes S C 1992 Selecting zinc-efficient cereal genotypes for soils of low zinc status. Plant Soil 146, 241–250.

Gries D, Klatt S and Runge M 1998 Copper-deficiency-induced phytosiderophore release in the calcicole grass *Hordelymus europaeus*. New Phytol. 140, 95–101.

Hansen N C, Jolley V D, Berg W A, Hodges M E and Krenzer E G 1996 Phytosiderophore release related to susceptibility of wheat to iron deficiency. Crop Sci. 36, 1473–1476.

Hommels C H, Kuiper P J C and Tanczos O G 1989 Luxury consumption and specific utilization rates of three macro-elements in two *Taraxacum* microspecies of contrasting mineral ecology. Physiol. Plant. 77, 569–578.

Hetrick B, Wilson G and Cox T 1993 Mycorrhizal dependence of modern wheat cultivars and ancestors: a synthesis. Can. J. Bot. 71, 512–518.

Hetrick B, Wilson G, Gill B and Cox T 1995 Chromosome location of mycorrhizal responsive genes in wheat. Can J. Bot. 73, 891–897.

Heuwinkel H, Kirkby E A, Le Bot J and Marschner H 1992 Phosphorus deficiency enhances molybdenum uptake by tomato plants. J. Plant Nutr. 15, 549–568.

Hopkins B G, Whitney D A, Lamond R E and Jolley V D 1998 Phytosiderophore release by sorghum, wheat and corn under zinc deficiency. J. Plant. Nutr. 21, 2623–2637.

Jackson L and Koch R 1997 The ecophysiology of crops and their wild relatives. *In* Ecology in Agriculture. Ed. L Jackson. pp 3–37. Academic Press, San Diego, CA, USA.

Johansen C, Subbarao G V, Lee K K and Sharma K K 1995 Genetic manipulation of crop plants to enhance integrated nutrient management in cropping systems – the case of phosphorus. *In* Genetic Manipulation of Crop Plants to Enhance Integrated Nutrient Management in Cropping Systems – 1. Phosphorus: Proceedings of an FAO/ICRISAT Expert Consultancy Workshop, 15–18 March 1994, ICRISAT Asia Center, Patancheru, India. Eds. C Johansen, K K Lee, K K Sharma, G V Subbarao and E A Kueneman. pp 9–30. International Crops Research Institute for the Semi-Arid Tropics, Patancheru, Andhra Pradesh, India.

Keerthisinghe G H P, Ryan P R and Delhaize E 1998 Effect of P supply on the formation and function of proteoid roots of white lupin (*Lupinus albus* L.). Plant Cell Environ. 21, 467–478.

Kemp P and Blair G 1991 P efficiency in pasture species, VI: A comparison of Italian ryegrass, phalaris, red clover and white clover over time. Austr. J. Agric. Res. 42, 541–558.

Khalil S, Loynachan T E and Tabatabai M A 1994 Mycorrhizal dependency and nutrient uptake by improved and unimproved corn and soybean cultivars. Agron. J. 86, 949–958.

Kirk G J D and Saleque M 1995 Solubilisation of P by rice plants growing in reduced soil: prediction of the amount solubilized and the resultant increase in uptake. Eur. J. Soil Sci. 46, 247–255.

Koide R T and Li M 1991 Mycorrhizal fungi and the nutrient ecology of three old field annual plant species. Oecologia 85, 403–412.

Koide R T, Li M, Lewis J and Irby C 1988 Role of mycorrhizal infection in the growth of and reproduction of wild vs. cultivated plants. I. Wild vs. cultivated oats. Oecologia 77, 537–543.

Lackie S M, Bowley S R and Peterson R L 1988 Comparison of colonisation among half-sib families of *Medicago sativa* L. by *Glomus veriforme* (Daniels and Trappe) Berch. New Phytol. 108, 477–482.

Li X, George E and Marschner H 1991 P depletion and pH decrease at the root–soil and hyphae–soil interfaces of VA mycorrhizal white clover fertilised with ammonium. New Phytol. 119, 397–404.

Liu H, Trieu A, Blaylock L and Harrison M 1998 Cloning and characterisation of two P transporters from *Medicago truncatula* roots: regulation in response to P and to colonisation by arbuscular mycorrhizal (AM) fungi. Mol. Plant-Microbe Interact. 11, 14–22.

Lynch J P and van Beem J 1993 Growth and architecture of seedling roots of common bean genotypes. Crop Sci. 33, 1253–1257.

Manske G 1989 Genetical analysis of the efficiency of VA mycorrhiza with spring wheat. Agric. Ecosyst. Environ. 29, 273–280.

Marschner H 1995 Mineral Nutrition of Higher Plants. 2nd edition. Academic Press, London. 889 pp.

Mench M and Fargues S 1994 Metal uptake by iron-efficient and inefficient oats. Plant Soil 165, 227–233.

Mercy M A, Shivashankar G and Bagyaraj D J 1990 Mycorrhizal colonisation in cowpea is host dependent and heritable. Plant Soil. 121, 292–294.

Mori S 1997 Reevaluation of the genes induced by iron deficiency in barley roots. Soil Sci. Plant Nutr. 43, 975–980.

Ni J J, Wu P, Senadhira D and Huang N 1998 Mapping QTLs for phosphorus deficiency tolerance in rice (*Oryza sativa* L.). Theor. Appl. Genet. 97, 1361–1369.

Otani T and Ae N 1997 The exudation of organic acids by pigeonpea roots for solubilizing iron- and aluminum-bound P. *In* Plant Nutrition for Sustainable Food Production and Environment. Proceedings of the XIII International Plant Nutrition Colloquium, 13–19 September 1997. Tokyo, Japan. Developments in Plant and Soil Sciences, Volume 78. Eds. T Ando, K Fujita, H Matsuomoto, S Mori and J Sekiya. pp 325–326. Kluwer Academic Publishers, Dordrecht, The Netherlands.

Reiter R, Coor J, Sussman M and Gabelman W 1991 Genetic analysis of tolerance to low-P stress in maize using restriction fragment length polymorphisms. Theor. Appl. Gen. 82, 561–568.

Römheld V and Kramer D 1983 Relationship between proton efflux and rhizodermal transfer cells induced by iron deficiency. Z. Pflanzenphysiol. 113, 73–83.

Römheld V and Marschner H 1990 Genotypical differences among graminaceous species in release of phytosiderophores and uptake of iron phytosiderophores. Plant Soil 123, 147–153.

Schachtman D and Barker S 1999 Molecular approaches for increasing the micronutrient density in edible portions of food crops. Field Crops Res. 60, 81–92.

Schachtman D P, Reid R and Ayling S 1998 P uptake by plants: From soil to cell. Plant Physiol. 116, 447–453.

Smith F A, Smith S E and Reid R 1998 Membrane and nutrition: opportunities for integration and progress. *In* 6th International Symposium on Genetics and Molecular Biology of Plant Nutrition. Risø National Laboratory, Elsinore, Denmark.

Smith F W, Ealing P, Dong B and Delhaize E 1997 The cloning of two *Arabidopsis* genes belonging to a phosphate transporter family. Plant J. 11, 83–92.

Smith S, Robson A and Abott L K 1993 The involvement of mycorrhizas in assessment of genetically dependent efficiency of nutrient uptake and use. Plant Soil 146, 169–179.

Smith S E and Read D 1997 Mycorrhizal Symbiosis, 2nd edition. Academic Press, London, UK. 605 pp.

Smith S E, Rosewarne G M, Ayling S M, Dickson S, Schachtman D P, Barker S J, Reid R J and Smith F A 1999 Phosphate transfer between vesicular-arbuscular mycorrhizal symbionts: Insights from confocal microscopy, microphysiology and molecular studies. *In* Phosphorus in Plant Biology: Regulatory Roles in Molecular, Cellular, Organismic and Ecosystem Processes. Proceedings 12th Annual Penn State Symposium on Plant Physiology, 28–30 May 1999. Current Topics in Plant Physiology, Volume 19. Eds. J P Lynch and J Deikman. pp 111–123. Am. Soc. Plant Physiol. Rockville, MD, USA.

Stock W D, Pate J S and Delfs J 1990 Influence of seed size and quality on seedling development under low nutrient conditions in five Australian and South African members of *Proteaceae*. J. Ecol. 78, 1005–1020.

Subbarao G, Ae N and Otani T 1997 Genetic variation in acquisition, and utilization of phosphorus from iron-bound P in pigeonpea. Soil Sci. Plant Nutr. 43, 511–519.

Welch R W 1995 Root exudate modification of rhizosphere nutrient availability and nutrient uptake by roots. *In* Genetic Manipulation of Crop Plants to Enhance Integrated Nutrient Management in Cropping Systems: 1. Phosphorus. Proceedings of an FAO/ICRISAT Expert Consultancy Workshop, 15–18 March 1994, ICRISAT Asia Center, Patancheru, India. Eds. C Johansen, K K Lee, K K Sharma, G V Subbarao and E A Kueneman. pp 97–106. International Crops Research Institute for the Semi-Arid Tropics, Patancheru, Andhra Pradesh, India.

Wissuwa M, Yano M and Ae N 1998 Mapping of QTLs for phosphorus-deficiency tolerance in rice (*Oryza sativa* L.). Theoret. Appl. Genet. 97, 777–783.

J.J. Adu-Gyamfi (Ed.), *Food security in nutrient-stressed environments: exploiting plants' genetic capabilities*: 59–66, 2002.

Genetic adaptation of crop plants to low-nutrient environments: morphological and ecophysiological characteristics of adaptation

H. Nakano[1,4], J. J. Adu-Gyamfi[2] & T. Nakamura[3]
[1]*Hokkaido National Agricultural Experimental Station (HNAES), Shinsei, Memuro, Kasaigun, Hokkaido 052-0071, Japan.* [2]*International Crops Research Institute for the Semi-Arid Tropics (ICRISAT), Sabo Bakin Zuwo Road, PMB 3491, Kano, Nigeria.* [3]*Japan International Research Center for Agricultural Sciences (JIRCAS), 1-2 Ohwashi, Tsukuba, Ibaraki 305-8686, Japan.* [4]*Corresponding author**

Key words: Alfisol, biomass, leaf, mineral stress, N utilization efficiency, Vertisol

Abstract

One hundred genotypes each of sorghum and pigeonpea were screened for their adaptation to soils low in nitrogen (N) and phosphorus (P) on an Alfisol and a Vertisol. The tested genotypes were classified by growth duration, growing habit, plant shape, germplasm collection site, and end use, and results were analyzed by classified groups. A wide range of genetic variation in grain and dry matter yields was observed. Dry matter yield depended on growth duration. For sorghum to attain maximum dry matter yield, growth duration should exceed 80 days, and to attain the maximum growth rate, a main stem longer than 250 cm, large and thin leaves (mean leaf area over 320 cm^2, specific leaf area (SLA) larger than 250 cm^2 g), and a leaf/stem weight ratio below 0.6 are required. Total accumulated plant N was positively correlated to dry matter yield but not to N-use efficiency of accumulated N. Results suggest that the ability of roots to absorb soil-N was comparatively more important for biomass productivity under low-N conditions than N-use efficiency. Biomass production in the growth stage and at harvest were not positively correlated, because some hybrids and forage varieties grew vigorously in the late growth stage. This suggests that improvement of growth activities in both early- and late-maturing types should be integrated when breeding for adaptation to low-N soils. Such morphological characteristics as large thin leaves, long main stems, and low leaf/stem weight are required for future ideotypes of sorghum and pigeonpea adapted to low-N and low-P conditions.

Introduction

Improving the adaptability of crop varieties to low soil fertility is an effective way to increase the crop productivity of low-input agriculture in the semi-arid tropics. Many studies have been made to screen genotypes adapted to low soil-nitrogen (N) and phosphorus (P) and to investigate their mechanisms of adaptation, to low-N conditions in wheat (Spanakakis and Viedt, 1990), maize (Lafitte and Edmeades, 1994), and sorghum (Franca, 1981), and to low-P in rice (Koyama et al., 1973) and bean (Gabelman and Gerloff, 1983). To improve the selection efficiency of breeding programmes, it is necesary to identify secondary traits associated with grain yield and biomass productivity in low-nutrient environments that can be measured easily in field-based evaluation. The objectives of this study were to: (1) estimate genetic variation in sorghum and pigeonpea in biomass and grain productivity in soils low in N and P, and (2) to identify plant morphological and physiological characteristics that correlated with biomass productivity and grain productivity under low-nutrient conditions.

Materials and methods

Screening tests were done on germplasm provided by the Genetic Resources and Enhancement Program of the International Crops Research Institute for the Semi-Arid Tropics (ICRISAT). Genotypic differences in grain yield and biomass productivity were investigated in sorghum grown without application of N or P fertilizer in Alfisol fields and in pigeonpea grown without application of P fertilizer in Alfisol and Vertisol fields at ICRISAT, Patancheru, India. In each field test 100 sorghum or pigeonpea genotypes classified into groups by maturity type, growing habit, plant shape, germplasm collection site, and end use were tested. Responses to low soil-N or -P and to application of N and P fertilizer were analyzed by genotypic differences and by differences among the groups. In 1996, sorghum and pigeonpea were sown in a low-N Alfisol field (RCE21). Pigeonpea was sown in a low-P Vertisol field (BR1E). The N treatments were; no N application (N0) or N applied at 100 kg N ha^{-1} in two splits (N 100) for sorghum. Similarly, two P treatments were; no applied P (P0) and 20 kg P ha^{-1} (P20). All plots received a basal application of 30 kg K ha^{-1}. In 1997, the same genotypes of sorghum were sown in a low-P Alfisol field (RM19). Treatments were 0 or 30 kg P ha^{-1}. Similar genotypes of pigeonpea used in 1996 were tested again with the same treatments in a low-P Vertisol field (BR3B). All screening tests were done by split plot design with 3 replications.

The test fields were chosen through a survey of soil analysis of their N and P status. For soil diagnosis, 9 soil cores to 30-cm depth were collected in each field in the dry season (January). They were mixed, dried, and sieved. The electrical conductivity (EC), mineral N, Total N, Olsen-P, calcium (Ca)-bound P, aluminium (Al)-bound P and iron (Fe)-bound P of the samples were determined. For comparison with soils at ICRISAT, soil from farmers' fields was collected from upland crop fields on the Deccan Plateau in the dry season (April). Soils were sampled and chemically analysed by the same procedure as those from ICRISAT.

Results

N and P status, and responses of sorghum and pigeonpea to N and P fertilizer Soil P and N status of screened ICRISAT and farmers' fields are shown in Table 1. Soil from farmers' fields showed low values in total N, mineral N, electrical conductivity (EC) and Olsen P. The EC ranged from 0.07 to 0.30 m mho^{-1} cm; total N ranged from 0.017 to 0.059%; mineral N from 3.9 to 11.8 mg kg^{-1}; and Olsen-P from 0.8 to 6.1 mg kg^{-1}. Almost all the values of the test fields at ICRISAT were distributed in the range of the farmers' field soils. Only the Olsen-P (6.7 mg kg^{-1}) value of the RCE21 Alfisol field was higher than those of the farmers' field soils. Sorghum and pigeonpea genotypes grew well without P fertilizer application and did not respond to P fertilizer application in RCE21. This result must be due to the high P fertility of the REC21 field.

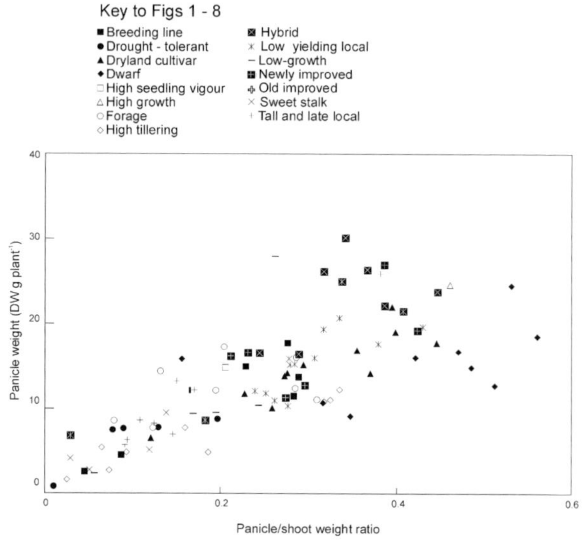

Figure 1. Relationship between panicle weight and panicle/shoot weight ratio at harvest in sorghum genotypes grown without N or P fertilizer in a low-N Alfisol. Figures 1 to 8 based on results under the same conditions in the same field, symbols indicate the same genotype groups.

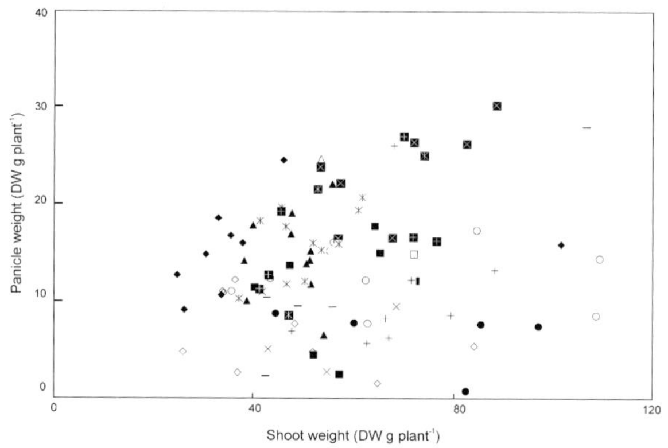

Figure 2. Relationship between shoot weight and panicle weight at harvest in sorghum genotypes.

Adaptability of sorghum genotypes to low N availability

A wide range of genetic variation in grain yield was observed in the responses of sorghum genotypes to the low soil N Alfisol. Harvest Index (HI; panicle weight /shoot weight ratio) also varied widely (Figure 1). Dwarf varieties, dryland cultivars, hybrids, and low-yielding local varieties had comparatively higher HIs.

Table 1. Nutritional status of soil from farmers' upland crop fields in Mahabubnagar district, Andhra Pradesh, India and test fields at the International Crops Research Institute for the Semi-Arid Tropics, Patancheru, 1995–97

Cropping[a]		EC (m mho cm^{-1})	Mineral N (mg kg^{-1})	Total N (mg kg^{-1})	Olsen P (mg kg^{-1})	Ca-P (mg kg^{-1})	Al-P (mg kg^{-1})	Fe-P (mg kg^{-1})
Mahabubnagar								
/ /S+C/–/	0.21		8.00	534	5.40			
/ /C+Pi/–/C		0.08	9.30	468	0.80			
S/–/C/–/S+Pi		0.07	5.90	172	1.90			
/ /C /–/S+Pi		0.07	11.80	590	2.80			
S/–/C/–/C		0.10	7.90	433	1.30			
/G/G/G/		0.08	3.90	307	6.00			
ICRISAT (field number)								
Alfisol	RCE21	0.04	2.90	310	6.70	250	–	130
	RM19	0.05	4.90	520	1.50	120	–	50
Vertisol	BR1E	0.15	3.80	430	2.90	47	–	56
	BR3B	0.17	5.80	540	3.80	110	–	85

[a]Cropping; 95 rainy season / 95 dry season / 96 rainy season / 96 dry season / 97 rainy season.
//: Unknown or before cropping. /–/ = no crops C=castor, S=sorghum, Pi=pigeonpea, G=groundnut, S+Pi=intercropped sorghum and pigeonpea, -: trace.

But, drought-tolerant varieties, sweet stalked varieties, and tall late local varieties had lower HIs. All the genotypes were distributed on a regression line between panicle weight and HI in Figure 1. However, owing to low biomass productivity, the panicle weight of dwarf varieties was low despite their high HIs. Some hybrids and improved varieties had high panicle weights in addition to high HIs because of their high biomass productivity. No positive relationships were observed between shoot and panicle morphology of genotypes included in the screening test (Figure 2). However positive trends between shoot and panicle weights were observed within each genotypic group except for dwarf varieties and hybrids, e.g., low-yielding local varieties, tall and late local varieties, newly improved varieties, and old improved varieties. This result suggested that improvement in biomass productivity was also effective in increasing grain yield in many sorghum genotypes. High dry matter yields were observed in genotypes of forage varieties, hybrids, drought-tolerant varieties, and tall, late local varieties (Figure 2).

Total dry matter yield might be influenced by growth duration (Figure 3). However, some forage varieties, hybrids and, drought-tolerant varieties had higher shoot weights than others of similar growth durations. Growth rate, expressed as shoot dry matter (DW) accumulated per day, also varied widely among the tested genotypes (Figure 4). The growth rate until harvest ranged from 0.3 to 1.4 g DW plant^{-1}

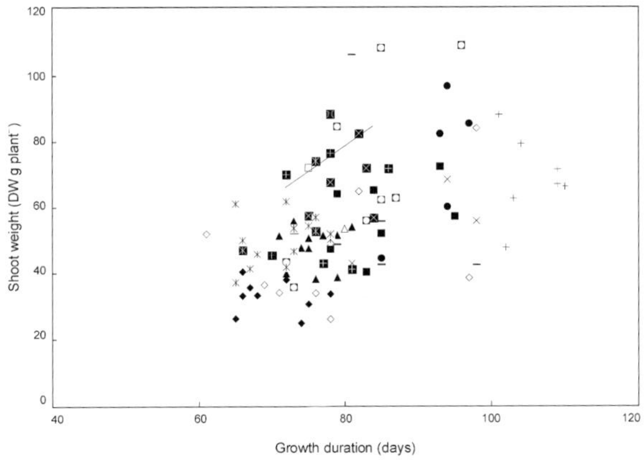

Figure 3. Relationship of growth duration and shoot weight at harvest in sorghum genotypes.

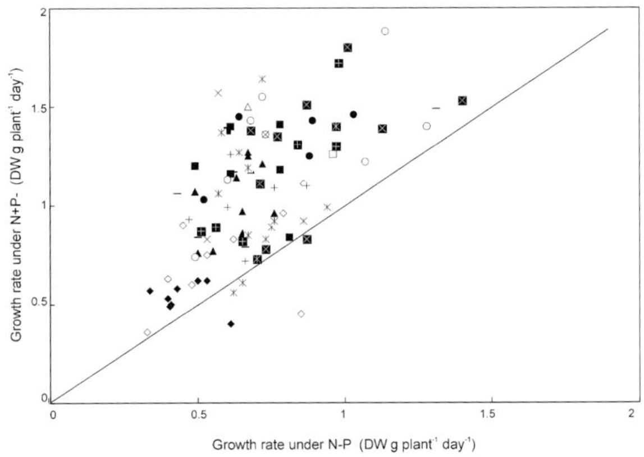

Figure 4. Effect of N fertilizer application on growth rate of shoot at harvest in sorghum genotypes.

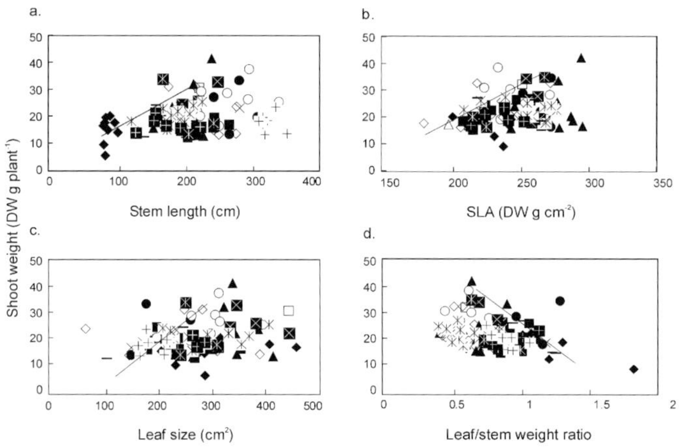

Figure 5. Relationships of main stem length, specific leaf area (SLA), leaf size, and leaf/stem weight ratio shoot biomass measured in late August in sorghum genotypes.

day^{-1}without application of N-fertilizer. Dwarf varieties and almost all the high-tillering varieties had extremely low values. Some genotypes of hybrids, forage varieties, and drought-tolerant varieties showed high values but tall, late local varieties had low or medium values. This result indicated that the high dry matter yields of the hybrids, forage varieties and drought-tolerant varieties were due not only to their long growth durations, but also to their high growth rates. The high dry matter yields of the tall, late local varieties at harvest was solely due to their long growth duration. In terms of relationships between biomass productivity and ecological or morphological characteristics in the tested genotypes, growth duration was positively related to shoot weight (biomass productivity). In order to attain the maximum level of dry matter yield the growth duration of sorghum genotypes should be longer than 80 days (Figure 3).

Plant height, leaf size, and specific leaf area [SLA; leaf area cm^2/unit leaf dry weight (DWg)] were examined in terms of their relationship with biomass productivity. Under low-N conditions the shoot weight of most of the 100 genotypes at the booting stage ranged from 5 to 40 DW g plant^{-1}. At harvest some genotypes of forage varieties, hybrids, drought-tolerant varieties, and dryland cultivars had high shoot weights. Plant height, leaf size, and SLA did not show conspicuous positive correlations with shoot weight at this stage in the tested genotypes. However, in order to attain a maximum shoot weight, the stem length should be more than 250 cm (Figure 5a), SLA should be larger than 250 cm^2 g^{-1} (Figure 5b), and leaf area should be more than 320 cm^2 (Figure 5c). The values of these

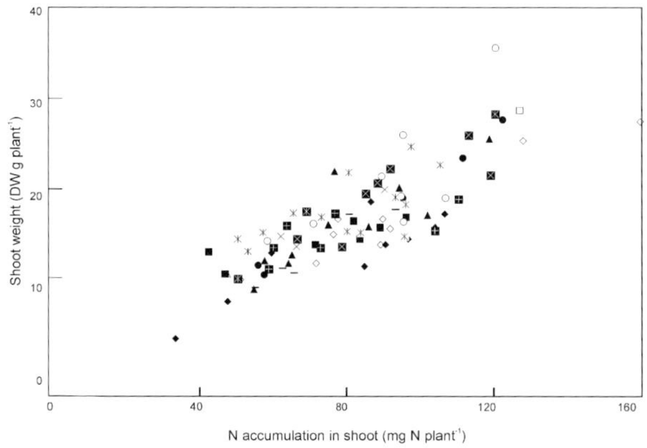

Figure 6. Relationship between N accumulation in shoot and shoot weight in late August in sorghum genotypes.

traits were increased by application of N fertilizer. On the other hand, the leaf/stem ratio tended to be negatively correlated to shoot weight (Figure 5d) and the values were decreased by N fertilizer application. The highest shoot weight at the booting stage was observed in a genotype collected from a dryland area whose leaf/stem weight ratio was 0.6.

In addition to morphological characteristics, other factors control the biomass productivity of sorghum genotypes. Nitrogen uptake was closely correlated to shoot weight at the booting stage (Figure 6) and N accumulation was high in some genotypes of hybrids, dryland cultivars, drought-tolerant varieties, and forage varieties. Responses to N fertilizer application in N uptake also differed among tested genotypes but the ratio of N uptake under N+ to that under N– did not depend on N uptake ability under N– conditions in the same way as crop growth rate did (Figure 4).

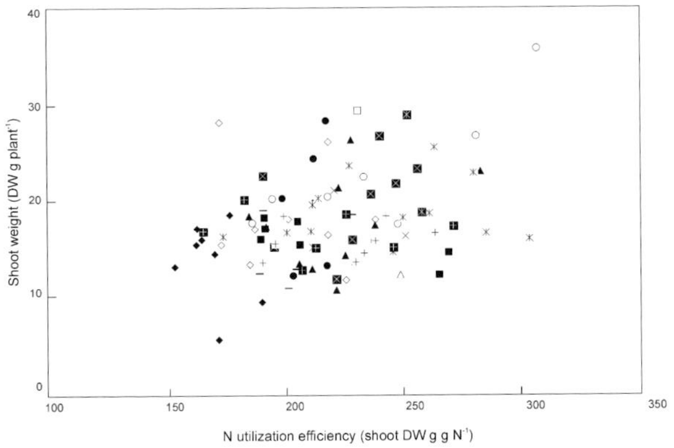

Figure 7. Relationship between use efficiency of N accumulated in shoot, and shoot weight in late August in sorghum genotypes.

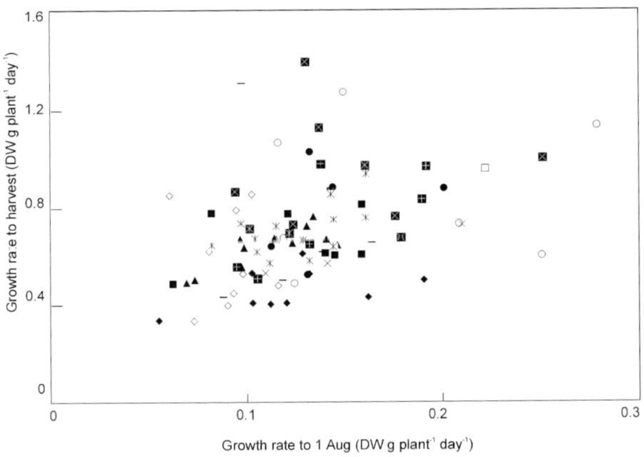

Figure 8. Relationship between growth rate of shoot in early growth stage and that at harvest in sorghum genotypes.

Although the tested genotypes differed widely in N-use efficiency (shoot weight DWg/ accumulated N g in shoot), a positive correlation between N-use efficiency and shoot weight was not observed (Figure 7). The biomass productivity of sorghum genotypes in low-N soil might mainly depend on their N-uptake ability rather than the utilization efficiency of accumulated N in a plant.

The relationship between shoot growth rate in early growth stages and that at harvest was not clear, because some hybrids and forage varieties grew vigorously in the late growth period (Figure 8). This discordance between values of the growth parameter measured at two different stages suggested that growth during each period was decided by independent mechanisms.

Adaptability of sorghum genotypes to low P availability

Analysis for low-P adaptability was done by a similar approach to that used for low-N adaptability. The adaptability of the same 100 genotypes to low-P conditions also varied widely. However, there were no genotype groups in which all members grew well without P application in the low-P soil (RM19 Alfisol), although dwarf and high-tillering varieties grew poorly. In terms of relations between plant morphology and biomass productivity in low-P treatments; large thin leaves and low leaf/stem weight ratio, similar to those in low-N conditions were required to attain high shoot weights. Some genotypes with large leaves were observed in drought-tolerant varieties, hybrids, and sweet-stalked varieties. However a positive relationship between shoot and panicle weights was not identified, even in analysis within each genotype group under low-P conditions. Further, seedling vigor evaluated 15 days after sowing, and shoot weight at the grain-filling stage did not show a clear positive relationship.

Adaptability of pigeonpea genotypes to low-P availability

Dry matter yield of pigeonpea at harvest ranged from 5 to 74 g DW plant^{-1} in 100 genotypes cultivated without P-fertilizer in a low-P Vertisol (BR1E). Responses of the genotypes to P-fertilizer application in terms of biomass production varied widely. Increases in dry matter yield following P application in the tested genotypes did not depend on their performance in biomass production under low-P conditions as it did in the screening tests for low-N and -P adaptability of sorghum genotypes.

Pod yield was closely related to dry matter yield in the tested pigeonpea genotypes (Figure 9), HI (pod weight: shoot weight ratio) was positively related to pod yield, although the correlation was influenced by short-duration determinate (DT) varieties, short-duration (NDT) varieties, and varieties from arid areas (Figure 10). Pod yield was low due to low shoot weights in spite of the high pod/shoot ratios in these varieties, because their growth duration was comparatively short (only 180 days). However, in the other genotypes (except for three groups), dry matter yield at harvest did not depend on growth duration that ranged from 180 days to 320 days. Further, no positive relationships were observed between shoot weight at flowering stage and that at harvest.

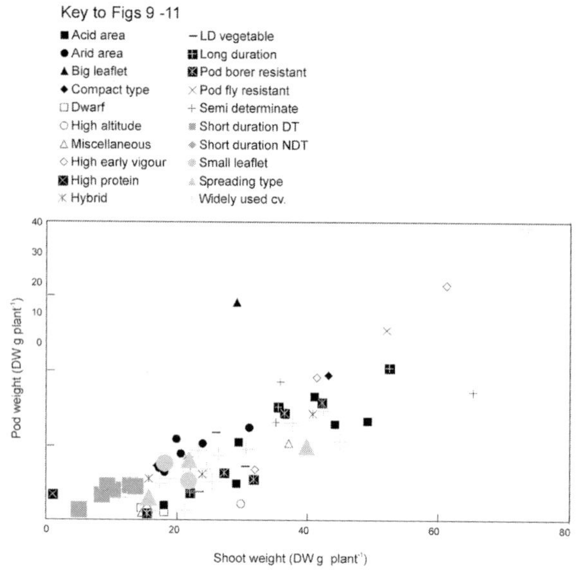

Figure 9. Relationship between shoot weight and pod weight at harvest in pigeonpea genotypes. Figures 9 to 11 based on results under the same conditions in the same field, symbols indicate the same genotype groups.

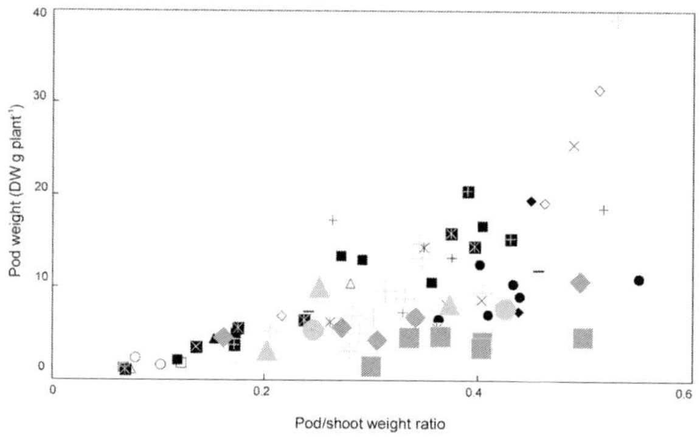

Figure 10. Relationship between pod weight and pod/shoot weight ratio at harvest in pigeonpea genotypes.

Relationships between pigeonpea biomass productivity and plant morphology were examined. In the tested genotypes at the pod-setting stage, main stem length ranged from 54 to 128 cm, branch numbers from 0.3 to 9.5 plant^{-1}and total node numbers from 18 to 39 plant^{-1}. They were positively correlated to pod weight under low-P condition (Figures 11a). SLA and leaf size showed trends of positive correlation with shoot weight under low-P (Figures 11b and 11c). Shoot weight was increased through the increase of leaf size and SLA by P-fertilizer application. Semi-determinate varieties and long-duration vegetable varieties had large leaves. Leaf/stem weight ratio correlated negatively to pod weight at this stage (Figure 11d). By the application of P fertilizer, although total leaf weight per plant increased, leaf/stem ratio decreased. In order to attain the maximum level of dry matter production, SLA should be more than 240 cm^2 g^{-1}, single trifoliate leaf size should be more than 80 cm^2, and leaf/stem ratio should be lower than 0.6.

Discussion

In the SAT, Alfisols and Vertisol are the major soil types. Screening for the adaptability of sorghum and pigeonpea genotypes to low -N and P fertility in both soils were conducted on fields that had not received N or P fertilizer for several years. In this study in the RCE21 Alfisol field, pigeonpea and sorghum grew well without P application and P responses were not observed in either crop. This result might be due to the high –P soil content of the field, Ca-P (250 mg kg^{-1}) for sorghum and Fe-P (130 mg kg^{-1}) for pigeonpea (Ae et al., 1991) Response to P-fertilizer was conspicuous in sorghum genotypes in the RM19 Alfisol field of low Ca-P (120 mg kg^{-1}) and in pigeonpea genotypes in Vertisol fields of low Fe-P (47 mg kg^{-1} in BR1E and, 110 mg kg^{-1}, in BR3B). Response to N-fertilizer was pronounced in sorghum genotypes in the RCE21 Alfisol field. The poor N status of the field was indicated with low mineral N: 2.9 mg kg^{-1}, low total N: 0.031%, and a low EC value: 0.04 m mho cm^{-1}. These were the same levels as those of farmers' fields that had not received chemical fertilizers for several years.

Biomass productivity under low fertility conditions was investigated. The contributions of HI and biomass productivity to the crop yield of genotypes depends on crop species and soil fertility (Kawano, 1986). Under low-fertility conditions, biomass productivity mainly determined grain yield in rice (Kawano et al., 1974), maize (Yamaguchi, 1974), and soybean (Buzzell and Buttery, 1977). In sorghum genotypes grown without N fertilizer in a low-N Alfisol, HI affected grain yield. However, the comparative importance of HI to grain yield potential was determined in positively diverse genotypes. Because genotypes were extremely different in plant shape and panicle morphology, e.g., dwarf varieties and varieties with panicles of broom-corn type were included in the test. In separate analysis of each genotype group, biomass productivity tended to be positively correlated to grain yield in sorghum. On the other hand, Sheldrake and Narayana (1979) pointed out that the relatively low yield of pigeonpea was due to a restricted partitioning of dry matter into pods

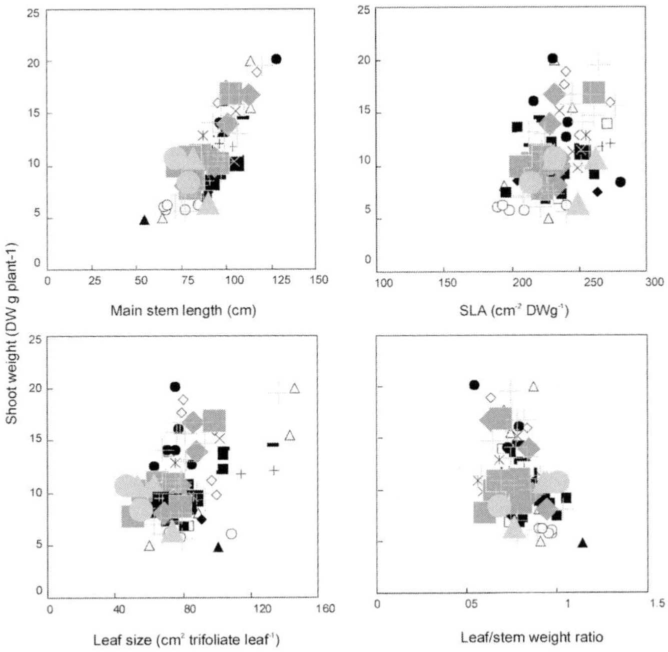

Figure 11. Relationships between main stem length, SLA, leaf size, and leaf/stem weight ratio, and shoot biomass at pod-setting stage in late September in pigeonpea genotypes.

that may be related to their perennial nature. However, both biomass productivity and HI affected pod yield in many pigeonpea varieties in this study. Further improvement in biomass productivity seemed more important than that of HI to increase grain yield potential in short-duration pigeonpea varieties. In both of sorghum and pigeonpea, it seemed that improvement in biomass productivity under low-input conditions was required for increase in grain yield potential.

Generally, the dry matter yield of genotypes is decided by their growth duration in many crops. However, in this study, except for short-duration pigeonpea, whose growth duration was less than 180 days, growth duration was not correlated to dry matter yield. This might be due to the very dry conditions after maturity in short-duration pigeonpeas. On the other hand, the dry matter yield of sorghum genotypes under low-N conditions was positively correlated with their growth duration (from 60 to 118 days). In sorghum genotypic differences were also observed in crop growth rate (dry matter production day^{-1} plant^{-1}).

Thorough comparative studies of local and improved varieties on genotypic adaptation to soil fertility, crop species were classified into genotypes adapted to low-fertility conditions and those adapted to high fertility. Generally, local varieties belonged to the former and improved varieties belonged to the latter. It was theoretically explained that their adaptation mechanism was based on their differing morphological characteristics (Tsunoda, 1959, 1962). Local varieties adapted to low fertility have large, thin leaves (Tsunoda, 1959), they have long stems, but their leaf/stem ratio is not very low (Tsunoda, 1962). These characteristics are adaptations to increase the photosynthetic rate per unit leaf weight. The plant height of N-stressed plants was shown to be a good indicator of biomass productivity (Lafitte and Edmeades, 1988; Tsunoda, 1962). In this study, sorghum genotypes of high biomass productivity under low-N conditions had leaves of larger area, and higher SLA, longer stems and lower leaf/stem ratios than a critical value in each trait. Similar characteristics were required for sorghum genotypes and pigeonpea genotypes adapted to low-P soils. This result corresponded to those on morphological characteristic adaptation to low fertility reported in other crops.

Based on relationships between biomass productivity and plant morphological characteristics, crop varieties have been classified into local varieties adapted to low fertility, and improved modern varieties with high response to fertilizer application (Tsunoda 1959, 1962). However, sometimes, modern varieties prove to be superior even under low-input conditions (Schinkel and Mechelke, 1990). In this study, some sorghum hybrids and improved varieties recorded high dry matter yields in low-N and low-P soil, and some forage and drought-tolerant varieties also had high biomass productivity in low-N (Figure 4). Some local sorghum varieties (low-yielding local forage, and sweet stalked variety) had high biomass productivity in low-P soil. Breeding material for low-P adaptation was available in local pigeonpea germplasm (Figure 9).

N uptake ability and N-use efficiency of accumulated plant-N differ among crop varieties (Franca, 1981; Maraville et al., 1980). In this study, N content and N-use efficiency were examined in sorghum genotypes grown in low-N soil. N uptake and N-use efficiency ranged widely in the tested genotypes. However, although N content was positively correlated to dry matter yield, N-use efficiency was not positively related to dry matter yield. This result suggested that the improvement of traits related to efficient N uptake would be more effective in increasing biomass productivity in low-N soils than N-use efficiency in sorghum.

Sorghum genotypes showing high N uptake rates as young plants usually had higher uptake rates as older plants (Franca, 1981). Total plant biomass and N content at anthesis are also positively associated with grain yield under conditions of N deficiency in maize and sorghum (Muchow, 1988; Muchow and Davis, 1988). These results suggested that biomass productivity during the whole growing period could be predicted by growth activities at an early stage. However, the early growth rates of sorghum genotypes under low-N conditions were not correlated to these at harvest, because some hybrids and improved varieties grew actively late in the growing period. Similar discrepanices between early growth and that in late growing stages was observed in sorghum and pigeonpea genotypes grown in low-P soils. Various traits associated with N or P uptake ability and biomass production at each growth stage should be evaluated, and should be integrated into breeding program to improve low-N and -P adaptability in these crops.

Acknowledgment

We thank ICRISAT Genetic Resources and Enhancement Program for providing sorghum and pigeonpea seeds.

References

Ae N, Arihara J and Okada K 1991 Phosphorus uptake mechanisms of pigeonpea grown in Alfisols and Vetisols. *In* Phosphorus nutrition of Grain Legumes in the Semi-Arid Tropics. Eds. C. Johansen, K K Lee and K L Sarawat. pp 91–104, International Crops Research Institute for the Semi-Arid Tropics (ICRISAT), Patancheru, Andhra Pradesh, India.

Buzzell R I and Buttery B R 1977 Soybean harvest index in hillplots. Crop Sci. 17, 968–970.

Franca G E 1981 Differences in dry-matter yield and the uptake, distribution, and use of nitrogen by sorghum genotypes. Ph.D. Thesis, Univ. of Nebraska, Lincoln, Nebraska, U S A. Diss. Abstr. 41,4018B.

Gabelman W H and Gerloff G C 1983 The search and interpretation of genetic controls that enhance plant growth under deficiency levels of a macronutrient. Plant Soil 72, 335–350.

Kawano K 1986 Yield breeding under limited- input condition. *In* Proceeding of the 28th symposium of Jap. Breeding Sci. Soc. pp 4–10. Tokyo, Japan.

Kawano K, Gonzales H and Lucena M 1974 Intraspecific competition, competition with weeds, and spacing response in rice. Crop Sci. 14, 841–845.

Koyama T, Chammek C and Snitwongse P 1973 Varietal differences of Thai rice in the irresistance to phosphorus deficiency. Tech. Bull. TARC No. 4, 1–32.

Lafitte H R and Edmeades G O 1994 Improvement for tolerance to low soil nitrogen in tropical maize I. Selection criteria. Field Crop Res. 39, 1–14.

Maraville J W, Clark R and Ross W M 1980 Nitrogen efficiency in grain sorghum. J. Plant Nutr. 2, 577–589.

Muchow R C 1988 Effect of nitrogen supply on the comparative productivity of maize and sorghum in a semi-arid tropical environment. III. Grain yield and nitrogen accumulation. Field Crops Res. 18, 31–34.

Muchow R C and Davis 1988 Effect of nitrogen supply on the comparative productivity of maize and sorghum in a semi-arid tropical environment. II . Radiation interception and biomass accumulation. Field Crops Res. 18, 17–30.

Schinkel B and Mechelke W 1990 A method to estimate the prospect of specific breeding for nutrient efficiency. *In* Genetic aspects of plant mineral nutrition. Eds. N El Bassam, M Dambroth and B C Loughman. pp 449–456, Kruwer Academic Publishers, Dordrecht, The Netherlands.

Sheldrake A R and Narayana A 1979 Growth, development and nutrient uptake in pigeonpea (*Cajanus cajan*). J. Agric. Sci. Camb 92, 513–526.

Spanakakis A and Viedt A 1990 Performance of winter wheat cultivars under reduced nitrogen conditions. *In* Genetic Aspects of Plant Mineral Nutrition. Eds. N El Bassam, M Dambroth and B C Loughman. pp 465–473, Kluwer Academic Publishers, Dordrecht, the Netherlands.

Tsunoda S 1959 A developmental analysis of yielding ability in varieties of field crops. II. The assimilation system of plants as affected by the form, direction and arrangement of single leaves. Jap. J. Breeding 9, 237–244.

Tsunoda S 1962 A developmental analysis of yielding ability in varieties of field crops. IV. Quantitative and spatial development of the stem-system. Jap. J. Breeding 12, 49–56.

Yamaguchi J 1974 Varietal traits limiting the grain yield of tropical maize. IV. Plant traits and productivity of tropical varieties. Soil Sci. Plant Nutr. 20, 287–304.

J.J. Adu-Gyamfi (Ed.), Food security in nutrient-stressed environments: exploiting plants' genetic capabilities: 67–79, 2002.

Genotypic variability and physiological characteristics of crop plants adapted to low-nutrient environments

J. J. Adu-Gyamfi[1,4], S. Ishikawa[1], T. Nakamura[2] & H. Nakano[3]
[1]*International Crops Research Institute for the Semi-Arid Tropics (ICRISAT), Sabo Bakin Zuwo Road, PMB 3491, Kano, Nigeria.* [2]*Japan International Research Center for Agricultural Sciences (JIRCAS), 1-2 Ohwashi, Tsukuba, Ibaraki 305-8686, Japan.* [3]*Hokkaido National Agricultural Experimental Station (HNAES), Shinsei, Memuro, Kasaigun, Hokkaido 082-0071, Japan.* [4]*Corresponding author**

Key words: low-P, pigeonpea , P-sufficient, P efficiency, sorghum, translocation

Abstract

The efficiency of crop improvement for nutrient-stressed environments can be enhanced if adaptive plant traits (morphological, physiological, and developmental) can be identified. A series of field and greenhouse (solution culture) experiments were conducted to screen large collections (98 lines each) of sorghum and pigeonpea accessions from different agro-ecological systems for their ability to produce high yields in both nitrogen (N) and phosphorus (P)-sufficient and-deficient conditions, to assess the extent of genotypic variation among cultivars, identify physiological and agronomic traits, and investigate and prioritize important mechanisms responsible for these adaptation of the crops to low-N and low-P environments. Fifteen genotypes each of sorghum and pigeonpea (hybrids and improved, local, drought-tolerant, and breeding lines selected from an initial screening of 98 genotypes) were evaluated for their growth characteristics, and N and P concentrations at the early vegetative stage, booting, and final harvest in low-N or low-P Alfisols. To compare the uptake, translocation, and distribution of P, a ^{32}P isotope was used. For sorghum; the hybrids, improved varieties, and forage lines produced higher biomass and grain yields than the farmers' varieties and other lines used. Hybrid pigeonpea and germplasm accessions from arid and high-altitude areas were the most adapted to both low-P and sufficient-P soil conditions Further selection of a sub-sample of 15 genotypes each and their evaluation under both field and solution-culture conditions clearly showed that hybrids and newly improved cultivars produced the highest biomass and grain yield in both low-P and sufficient-P conditions.

An important trait of sorghum observed in this study is the rapid and efficient translocation and remobilization of P from leaves and stem reserves to support growth and grain filling in P-stressed conditions. A ^{32}P study showed that in low-P conditions, P-efficient genotypes translocated more P from the roots to flag leaves as indicated by a high specific radioactivity. When P was withdrawn from the nutrient solution after booting, the contributions from leaf- and stem-P offered a powerful resource for grain filling for sorghum, indicating that stem-P reserves could therefore play an important role in grain-P and total grain mass under P-deficient conditions. For pigeonpea, the ability to re-translocate P from old leaves to the growing parts before senescence could be considered an important trait when breeding for P-efficient genotypes. Future research directions are suggested.

Introduction

The sustained productivity of agricultural ecosystems of the semi-arid tropics (SAT) where some 850 million people (one sixth of humanity) reside is constrained by low soil availability of nutrients, especially phosphorus (P) and nitrogen (N), the limited availability of inorganic fertilizers, and the poor crop response to P inputs on many soils. The situation is further exacerbated by low and unreliable rainfall, and limited groundwater resources. Although crop improvement in these regions has resulted in the release of hybrids and improved new cultivars, yields of traditional coarse grains (millet and sorghum) and legumes

* E-mail: j.j.adugyamfi@cgiar.org

(groundnuts, pigeonpea, cowpea, and chickpea) are still very low.

Low P bioavailability is the major cause of the declining crop productivity in the SAT countries of West Africa (van Kuelen and Bremen, 1990). Total P in most soils of the West African SAT ranges from 25 to 350 mg kg^{-1}, and available P (1 to 30 mg kg^{-1}) is seldom adequate for optimal plant growth (Bationo et al., 1989, 2000). In addition, the concentration of inorganic P (Pi) in soil solution, the form in which is most readily accessed by plants rarely exceeds 10 μM (Bieleski, 1973). Plants must therefore have specialized transporters at the root/soil interface to extract Pi from solutions of micromolar concentrations in addition to other mechanisms for transporting Pi across the membrane between the intercellular compartments, where the concentration of Pi may be 100-fold higher than in the external solution (Schachtman et al., 1998).

In the SAT regions where total P is low (below 50 mg kg^{-1} P) and the available P levels are far below the critical levels needed for most crops, replenishment of P in the soil is necessary. This entails the addition of P through; (1) the application of inorganic-P fertilizers, (2) the use of crops that can effectively utilize locally applied rock phosphate, and (3) the biological cycling of P through legume intensification. In regions where total P is high, but is poorly acquired by crops, physiological modifications of crop varieties that would increase P acquisition is a viable alternative to increasing available P levels. The use of crop plants that can exploit P by solubilizing calcium (Ca), aluminum (Al) or iron (Fe)-bound P, or modifying its root morphology to increase its absorbing surface area, is desirable (Ae et al., 1990; Dakora and Philips, 2002; Hoffland, 1992; Ishikawa et al., 2002; Rengel, 1999, 2002). Thus, optimizing P nutrition of crops involves combining better soil and crop management practices to enhance soil fertility restoration with genetic strategies to improve plants' ability to tolerate to low-P soils.

There is an obvious need for increased emphasis on improving the adaptation of crop plants to soil conditions for low-P available soils by; selecting and breeding crops that require only moderate external inputs to express their maximum genetic potential, improving the nutrient acquisition abilities of existing crops, and/or optimizing nutrient cycling via the use of locally available plant nutrient sources like rock phosphate. Available literature suggests that producing crops adapted to low-P soils is a major challenge that will require a long-term research horizon and substantial research investments. However, there have been successes in screening crop genotypes for general vigour and tolerance to low-P soils. Promising information is available on possible low-P tolerance mechanisms in pigeonpea (Adu-Gyamfi et al., 1989, 1991; Nakano et al., 2002; Otani et al., 1996; Subbarao et al., 1997), common bean (Lynch, 1999; Lynch and Beebe, 1995; Schettini et al., 1987; Snapp and Lynch, 1996; Vadez et al., 1999), sorghum (Maranville et al., 1980), pearl millet (Hafner et al., 1993; Payne et al., 1995), lotus (Hart and Jessop, 1983), rape (Hoffland, 1992), white clover (Caradus et al., 1993; Dunlop et al., 1990; Mackay et al., 1990), maize (da Silva et al., 1993), and rice (Ray et al., 1996; Yano and Sasaki, 1997; Wissawa et al., 1998).

Reliable screening methods to identify agronomical useful traits, and an understanding of how they operate at the physiological, biological, and genetic levels should be very valuable for initiating an efficient plant breeding programme to optimize P acquisition by crops in low-P environments. Current research is unraveling mechanisms of genetic control of these traits, and techniques are being developed to transfer relevant genes across genotypes and species to enhance P-deficiency tolerance by rice (Yano and Sasaki, 1997; Wissawa et al., 1998) and to develop detailed molecular linkage maps for rice. The detection and mapping of putative quantitative trait loci (QTLs) for P-deficiency tolerance in rice represents a crucial step that could eventually lead to the identification of genes controlling P uptake and use efficiency or for the identification of linked markers to be used in marker-assisted selection.

The objective of the study was to initially screen large collections (98 lines each) of sorghum and pigeonpea accessions from different agro-ecological systems for their ability to produce high yields in both N-and P-sufficient and-deficient conditions, assess the extent of genotypic variation among cultivars, identify physiological and agronomic traits, and investigate candidate mechanisms of P-deficiency tolerance.

Materials and methods

A series of field and greenhouse (solution culture) experiments were conducted on sorghum and pigeonpea lines from 1996 to 1998.

Field experiments: initial screening for adaptation to low-P

The study was conducted during the rainy and postrainy seasons of 1996 and 1997 at ICRISAT, Patancheru (17° 38′N and 78° 21′E), India. Ninety-eight genotypes each of sorghum and pigeonpea accessions from the genebank at ICRISAT were grown on a medium-deep Alfisol (Ferric Luvisols: Udic Rhodustalf) and Vertisol (Typic Pellustert). The general physico-chemical soil characteristics have been described by (El-Swaify and Caldwell, 1991). The Alfisol used had an average available moisture storage capacity of 80 mm, pH (H_2O) of 6.2. The nutrient concentrations (mg kg^{-1}) in the 0–30 cm soil depth were; mineral N 2.9, total N 310, total P 208, available P (Olsen) 6.7. The pH (H_2O) of the Vertisol was 8.0 and the nutrient concentrations in the 0–30 cm soil depth (in mg kg^{-1}) were: mineral N 5.8, total N 540, total P 280, and available P (Olsen) 3.8. The sorghum was grown on the Alfisol and the pigeonpea on the Vertisol. The experiments were split-plot design. The N treatments on the Alfisol were: no N application (N0) or N applied at 100 kg N ha^{-1} (N100) in 2 splits to the sorghum. On the Vertisol, there were two P treatments: no P application (P0), or P applied at 40 kg P ha^{-1} (P40) as superphosphate banded at sowing. At final harvest, ranging from 90 to 150 days after sowing (DAS), the aboveground materials were sampled, separated into leaves, stem, and head (pods for pigeonpea), dried, and weighed. The head/pods were further separated into seeds and rachis/husk. Shoot weight, grain weight, and harvest index were estimated.

Evaluation of genotypic differences in response to P application

Fifteen genotypes each of sorghum and pigeonpea classified as hybrids, improved varieties, farmers' varieties, drought-tolerant lines, and breeding lines selected from the initial screening of 98 genotypes were evaluated for their growth characteristics and response to P applications on an Alfisol at ICRISAT Patancheru during the rainy seasons of 1997 and 1998. Physical and chemical characteristics of the top 30 cm of the soil used were: EC 0.06 mS cm^{-2}, pH (H_2O) 6.9, bulk density 1.6 g cm^{-2}, and organic carbon (C) 3.1 g kg^{-1}. The nutrient concentrations (in mg kg^{-1}) were total N 560, total P 200, Al-P 6.9, Ca-P 23.0, Fe-P 52, available P (Olsen) 6.5, 12.7 (Troug P) and 4.7 (Bray II). The same field was used for both sorghum and pigeonpea experiments.

The experiments were factorial set out in a split-plot design with 3 replications. For the sorghum trials, the main treatments were CSH 5, CSH 9, CSH 11 (hybrids), ICSV 112, ICSV 745, SPV351 (improved varieties), FSRP, Aima, Khundijowar (farmers' varieties) Nyanjang, Kesarijola, Sanchisan (drought-tolerant lines), and ICSV 93053, 296 B, ICSV 89057 (breeding lines). Similarly, the 15 genotypes for the main plots in the pigeonpea trial were ICPH 8, ICPH 73 , ICPH 526 (hybrids), ICPL 87, ICPL 85010, ICPL 88032 (improved varieties), UPAS 120, Manak, Field Collection (farmers' varieties) ICPL 88039, ICPL 84023, ICPL 89021(drought-tolerant lines) and ICPL 161, ICPL 7120, C11 (breeding lines). The sub-plot treatments were; no P application (P0), or P applied at 40 kg P ha^{-1} (P40). All the P fertilizer was applied as superphosphate by opening the furrows and banding at 5–8 cm soil depth at sowing. For the sorghum, N was applied at 50 kg N ha^{-1} in 2 splits to all plots. No N was applied to the pigeonpea plots.

Four seeds of pigeonpea and sorghum were sown at a spacing of 60 × 10 cm and seedlings thinned to one at 8 DAS. Five samplings were done; 3 at the vegetative stage, 1 at flowering, and 1 at final harvest. The final harvest of mature sorghum ranged from 90 to 130 days except for Kesarijola that matured at 160 days. Similar samplings were done for pigeonpea. With the exception of ICPL 161 and ICPL 7120, and C11 that matured at 200 DAS, all the other genotypes matured within 120–160 DAS. The aboveground parts were cut at harvest, and separated into leaves, stem, leaves, pods/head. After drying to constant weight these were weighed, dried samples were ground to pass through a 1-mm sieve, thoroughly mixed, and a portion of the ground samples used for chemical analysis.

Greenhouse experiments

P uptake and translocation by sorghum in solution culture

Seeds of the 15 genotypes of sorghum (most of them used for the field experiment) were pre-germinated in paper pots filled with Alfisol. The 15 genotypes included 3 hybrids, 3 improved varieties, 3 farmers' varieties, 2 drought-tolerant lines, 2 Chinese lines, and 2 breeding lines. The seedlings were transplanted at 8 DAS into six 600-L tubs containing nutrient solution. The composition of the nutrient solution was (μM): N 1000 ($NaNO_3$), K 250 (K_2SO_4:KCl [1:1]), Ca 370

($CaCl_2$. H_2O), Mg 280 ($MgSO_4$), Fe 4 (EDTA), Mn 0.9 ($MnCl_2$. $4H_2O$), Zn 0.15 ($ZnSO_4$. $7H_2O$), B 4 (H_3BO_3), Cu 0.16 ($CuSO_4$. $5H_2O$), Mo 0.1 (MoO_4. $2H_2O$). Three P treatments; 4, 20, and 100 (μM P) were initiated 3 days after transplanting. The nutrient solutions were well aerated, the pH of the solution was adjusted to 5.5 daily, and the solution renewed every 5–6 days.

The tubs were covered with an aluminum sheet with 100 holes (10 rows and 10 columns). The holes were spaced at 20 × 10 cm allowing for 10 seedlings in 10 rows. Two tubs were used for each P treatment and the 15 genotypes were randomized and transplanted in single rows for each genotype. Sorghum hybrid CSH 9 was used for the 2 border rows in all the treatments. The plants were grown in the 600-L nutrient solution for 20 days. Seven uniform plants from each of the 15 genotypes and the 3 treatments were carefully transplanted into 5-L plastic buckets. There were 315 plastic buckets (7 plants, 15 genotypes, and 3 P treatments). The nutrient solutions in the buckets were well aerated, pH adjusted to 5.5 daily, and renewed every 3 days. All the plants were allowed to grow to the booting stage. At booting (50–70 DAS) 3 plants were harvested, and separated into leaves, stem, and roots. Leaf area was measured using a leaf area meter (Hayashi-Denko Co. Ltd.). The total root length was measured with a COMAIR root length scanner (line intersection method; Newman, 1966). The separated plant parts were dried at 70 °C for 48 h and then ground. The rest of the plants were allowed to grow in the same nutrient solution but without P (-P) till maturity. The treatment was to ensure that plants utilize and remobilize the P accumulated during booting for grain production. At maturity, 3 plants were harvested and similar measurements to those at the booting stage were taken. The contributions of P from the different plant parts to grain were estimated as the difference in amounts of P in plant parts at harvest and at booting.

P translocation using ^{32}P

Three genotypes of sorghum: CSH 9 (hybrid), 296 B (breeding line), and FSRP (farmers' variety) were used to study the uptake kinetics and translocation efficiency. Seedlings were grown in the 600-L tub containing nutrient solution at 3 P treatments (1, 10 and 100 μM P). To estimate the rate of translocation, 40 day-old plants from the 600-L tub were transferred to 20-L pots containing nutrient solution treated with two rates of P, 10 and 100 μM P, and were grown till flowering. To estimate the short-term translocation of absorbed P from roots to the other plant parts, 6 plants from each genotype were used. The solution in the pots was labeled with ^{32}P and the plants were allowed to absorb it for 2 h. After 2 h, the plants were thoroughly washed and transferred to 2 non-labeled pots containing 10 and 100 μM P, and the distribution of the absorbed P to different plant parts were traced by sampling the plants at 1, 3, and 6 h after transferring plants to ^{32}P-free solution. The amount of ^{32}P labeled-P in plant parts were estimated using a liquid scintillation counter (Beckman LS 6500) after acid digestion (Adu-Gyamfi et al., 1991).

Analytical methods and data analysis

Nitrogen concentration in plant parts was estimated by the indophenol color formation (Chaykin, 1969) after micro-Kjeldahl digestion and P estimation by the modified molybdate acid method (Murphy and Riley, 1962). All data were analyzed using Genstat 5 Release 4.1 (Copyright 1998, Lawes Agricultural Trust, Rothamstead Experimental Station) following standard analysis of variance procedures (Gomez and Gomez, 1984).

Estimation of P efficiencies

P efficiencies were estimated as:

(a) uptake efficiency [mg/unit root dry mass or mg /unit root length] (Barrow, 1975).
(b) incorporation efficiency [shoot DM/unit nutrient in shoot] (Blair and Cordero, 1978)
(c) utilization efficiency [shoot DM/unit nutrient uptake] (Blair and Cordero, 1978)
(d) P efficiency ratios [kg grain/ kgP in shoot; kg grain/ kgP grain] (Batten, 1993)

Results

Evaluation of genotypic differences in growth and grain yield

Grain weight, shoot weight, and harvest index of the 98 genotypes of sorghum and pigeonpea are shown in Tables 1 and 2. The hybrids, the tall and late, and recently improved varieties of sorghum produced more grain and higher shoot weights than the other varieties in N-sufficient and low-N conditions. Harvest index was highest in the tall and late varieties, the dwarf exotic lines from USA, and the hybrids, irrespective

Table 1. Genotype comparisons of grain weight, shoot weight, and harvest index (HI) of 98 genotypes of sorghum grown in the field without N application (N0) and at 100 kg N ha^{-1}(N100). All the accessions were selected from the genebank at ICRISAT Patancheru, Andhra Pradesh India, rainy seasons 1996 and 1997

Classification	Entries	Grain wt (g^{-1} plant)		Shoot wt (g^{-1} plant)		HI	
		N0	N100	N0	N100	N0	N100
Breeding line	6	4.8	16.1	29.6	69.1	0.18	0.24
Dwarf exotic line from USA	9	4.5	8.4	15.9	29.3	0.29	0.29
Variety collected in dryland	11	4.1	14.2	22.3	54.0	0.19	0.27
Drought-tolerant	5	4.7	14.4	24.9	55.2	0.19	0.25
Quick early growth	1	4.0	12.8	33.1	65.4	0.12	0.20
Forage line	8	3.3	10.8	24.4	56.2	0.15	0.21
High growth index from breeding	1	5.6	19.5	26.4	70.3	0.21	0.28
High-tillering line	9	3.9	11.4	19.3	42.9	0.20	0.27
Hybrid	8	7.2	20.9	30.0	65.2	0.24	0.32
Low growth index from breeding	5	3.8	18.6	25.4	75.7	0.15	0.25
Local	2	4.8	13.1	29.5	56.1	0.17	0.23
Low-yielding cultivar	11	4.2	12.0	19.9	42.6	0.21	0.28
Recently improved variety	7	5.6	14.5	24.8	53.3	0.22	0.27
Old improved variety	4	3.4	12.6	18.4	47.2	0.19	0.28
Sweet-stalk line	6	5.5	16.0	22.4	53.8	0.26	0.30
Tall and late	5	7.2	19.4	22.1	51.8	0.33	0.38
Means		4.5	14.7	23.8	55.5	0.21	0.27

Factors	Grain wt		Shoot wt		HI	
	P values	LSD@0.01	*P* values	LSD@0.01	*P* values	LSD@0.01
Cultivars (CV)	<0.001	0.63	<0.001	1.9	<0.001	0.01
N levels (P)	<0.001	4.32	<0.001	10.9	<0.001	0.08
CV × N	<0.01	6.11	<0.01	14.9	0.497	NS

of N treatment (Table 1). Pigeonpea accessions collected from arid, high-altitude areas, high-protein lines, and the hybrids produced higher grain weights and higher harvest indices than the other lines at low-P and P-sufficient conditions. Generally, most varieties performed well in both low-P and P-sufficient conditions.

Genotypic differences in response to N and P

Biomass, grain yield, harvest index, and P accumulation by sorghum were higher in the hybrids and the improved cultivars than in the farmer and drought-tolerant lines (Table 3). A similar trend was observed for pigeonpea except that the medium-duration breeding lines produced the highest biomass and grain yield in low-P and P-sufficient conditions (Table 4). Phosphorus incorporation (shoot DM per kg P in shoot) and utilization (kg shoot DM per kg P applied) efficiencies were highest in pigeonpea hybrids and breeding lines and decreased with P application (Figure 1). The P efficiency ratios (kg grain per grain P in shoot or kg grain per kg P in grain) decreased with P application, but there were no significant differences among the genotypes (Figure 1).

Greenhouse experiments

P uptake and translocation

Shoot and grain weights increased with P supply in solution (Table 5). Shoot DMs for the hybrids and improved varieties were higher than for the other genotypes. The high shoot DM in the drought-tolerant lines was due to a contribution from a late-maturing line (Kesarijola) that produced a high stem weight in both low-P and P-sufficient conditions. Grain weight was slightly higher in the hybrids than the other cultivar groups at low-P, but the difference was more

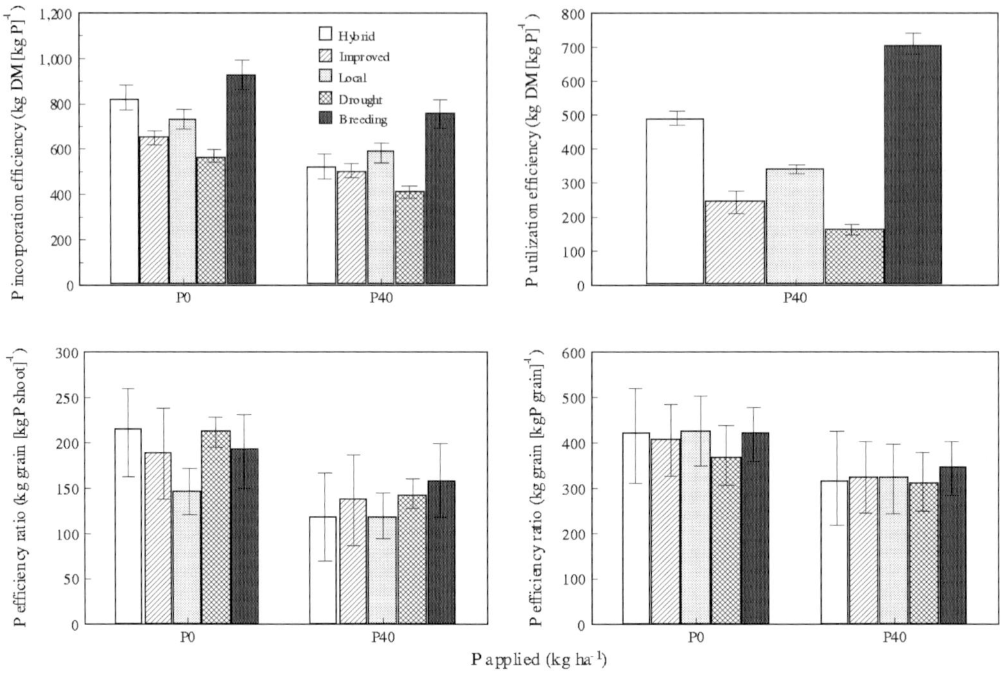

Figure 1. Genotypic differences in phosphorus efficiencies of field-grown pigeonpea at final harvest, ICRISAT Patancheru, Andhra Pradesh, India 1998 rainy season. Data represent average values for 3 genotypes each of hybrid, improved, local, drought-tolerant and breeding lines. Error bars represent ± standard error of mean (n = 3).

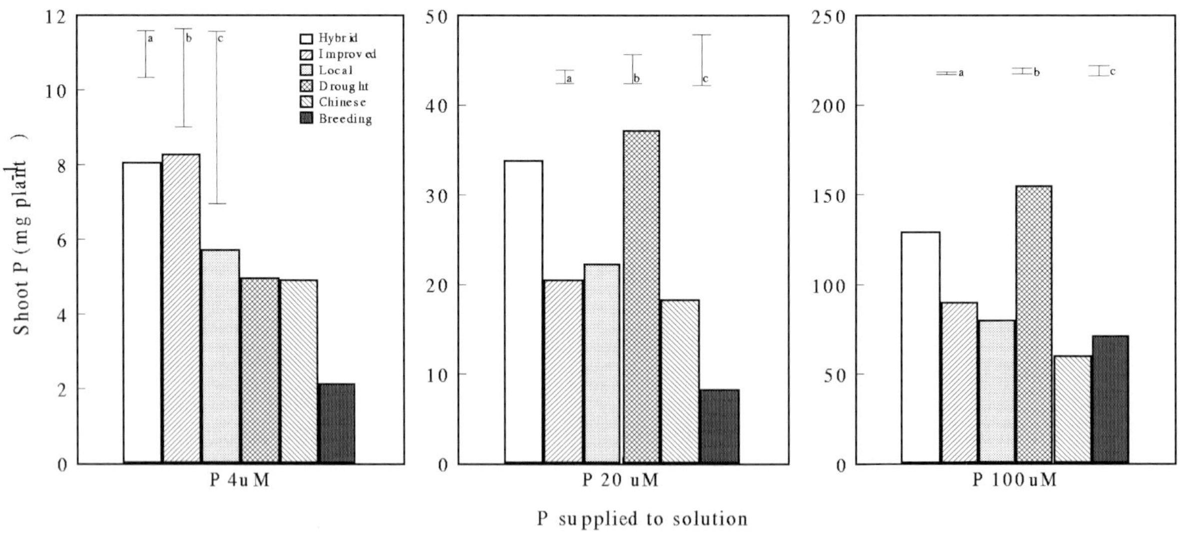

Figure 2. Genotypic differences in phosphorus uptake efficiency by sorghum grown in solution culture at P concentrations of 4, 20 and 100 μM. Data represent average values for 3 genotypes each of hybrid, improved, local, and 2 genotypes each of drought tolerant, Chinese, and breeding lines greenhouse experiment, ICRISAT Patancheru, Andhra Pradesh, India 199/98. Bars indicate LSD (< 0.05) values when comparing (a) cultivars, (b) phosphorus levels and (c) cultivar × phosphorus interaction.

Table 2. Genotype comparisons of grain weight, shoot weight, and harvest index (HI) of 98 genotypes of pigeonpea grown in the field without P application (P0) and at 40 kg P ha^{-1} (P40). All the accessions were selected from the genebank at ICRISAT Patancheru, Andhra Pradesh India, rainy seasons 1996 and 1997

Classification	Entries	Grain wt (g^{-1} plant)		Shoot wt (g^{-1} plant)		HI	
		P0	P40	P0	P40	P0	P40
Acid areas	8	2.7	3.3	26.5	52.1	0.11	0.07
Arid areas	8	7.2	12.7	33.2	57.9	0.21	0.21
Big leaflet	1	–	–	28.0	28.0	–	–
Compact type	3	2.6	5.1	35.1	70.7	0.08	0.09
Dwarf	2	4.1	6.9	20.0	39.2	0.19	0.17
High altitude	5	–	–	24.3	41.9	–	–
High diversity	4	0.4	1.1	30.1	56.8	0.01	0.01
High early vigour	6	5.4	8.3	39.1	66.1	0.13	0.12
High protein	3	5.5	7.7	28.3	50.1	0.17	0.13
Hybrid	2	5.7	11.2	23.2	37.0	0.25	0.30
LD vegetable	2	4.2	1.4	32.7	64.7	0.15	0.02
Long duration	3	0.3	0.2	30.8	51.8	0.01	0.00
Pod borer tolerant	1	3.2	10.2	30.1	74.6	0.11	0.14
Pod fly tolerant	3	2.5	1.0	42.1	71.3	0.08	0.02
Semi-determinate	9	5.9	10.8	22.2	33.5	0.27	0.33
Short-duration (determinate)	2	4.3	8.5	32.3	54.0	0.14	0.17
Short-duration (non-determinate)	2	4.3	2.9	8.1	40.7	0.35	0.16
Small leaflet	6	5.9	10.3	28.6	54.8	0.22	0.18
Spreading type	3	3.9	8.4	25.4	42.0	0.13	0.20
Widely used cultivar	25	4.5	6.3	35.7	59.8	0.13	0.11
Means	–	4.0	6.4	28.8	52.3	0.15	0.14

Factors	Grain wt		Shoot wt		HI	
	P values	LSD@0.01	*P* values	LSD@0.01	*P* values	LSD@0.01
Cultivars (CV)	<0.001	0.84	<0.001	3.5	<0.001	0.01
P levels (P)	<0.001	3.28	<0.001	18.3	<0.001	0.04
CV × P	<0.001	4.70	0.937	NS	0.002	0.06

evident when plants were supplied with 20 μM P (Table 5).

Shoot P accumulation followed a similar trend to that of the dry weight (Figure 2). Surprisingly, the Chinese lines that have been reported to have high N-use efficiency (Maranville and Madhavan, 2000) accumulated very little P in the shoot. With P supply, P uptake efficiency increased, but P incorporation efficiency decreased. No significant cultivar group differences were observed in P incorporation efficiency, except for the drought-tolerant lines that showed higher values than the other genotypes in low-P conditions (Figure 3). Similar to the observations from the field, there were no significant differences in P efficiency ratio between genotypes (data not shown).

The translocation and redistribution of P from plant parts of sorghum to grain after booting is shown in Table 6. When grown in nutrient solution without P after booting, most of the P in grain was translocated from leaves and stem. When supplied with low-P (20 μM P), more P was translocated from the leaves and stem of the hybrid, drought-tolerant, and Chinese lines to the reproductive parts, than in the farmer's variety and breeding lines. Thus, reserves in stems during pre-anthesis plant assimilation was an important source of P for grain filling.

P translocation using ^{32}P

The specific and total radioactivities in plant parts of sorghum grown in 10 μm P solution is shown are

Table 3. Grain weight and harvest index of 15 entries selected from 98 sorghum genotypes at ICRISAT Patancheru, Andhra Pradesh India, rainy seasons 1996 and 1997

Classification	Entries	Grain wt (g^{-1} plant)		HI		Grain wt (g^{-1} plant)		HI	
		N0	N100	N0	N100	P0	P40	P0	P40
Hybrid	3	7.6	21.2	0.31	0.33	14.7	20.1	0.34	0.40
Improved	3	10.3	25.0	0.28	0.34	11.1	14.6	0.24	0.29
Local	4	4.3	11.7	0.19	0.27	8.9	10.3	0.30	0.31
Drought	3	5.9	20.5	0.25	0.31	6.6	8.0	0.15	0.14
Breeding lines	2	5.7	22.0	0.20	0.31	10.2	7.1	0.24	0.21
Means	–	6.8	20.1	0.25	0.31	10.3	12.0	0.25	0.27
Factors		Grain wt		HI		Grain wt		HI	
		P values	LSD@0.01	*P* values	LSD@0.01	*P* values	LSD@0.01	*P* values	LSD@0.01
Cultivars (CV)		<0.001	3.82	<0.001	0.026	<0.001	3.89	<0.001	0.019
N/P levels (P)		<0.001	6.11	<0.001	0.047	<0.001	6.72	0.001	0.046
CV × N/P		<0.001	8.22	0.010	0.083	0.040	10.01	0.100	NS

Table 4. Shoot, grain dry weights, and grain P content in different cultivars of pigeonpea at different levels of P nutrition at ICRISAT Patancheru, Andhra Pradesh India, rainy seasons 1996 and 1997

Classification	Entries	Shoot wt (Mg ha^{-1})		Grain wt (Mg ha^{-1})		Grain P (kg P ha^{-1})	
		P0	P40	P0	P40	P0	P40
Hybrid	3	13.4	19.1	3.5	4.4	8.3	14.2
Improved varieties	3	7.9	9.8	2.3	2.6	5.8	8.1
Local varieties	3	9.6	13.6	1.9	2.7	4.4	8.5
Drought-tolerant varieties	3	5.0	6.6	1.8	2.3	5.0	7.3
Breeding line	3	17.4	28.3	3.6	6.0	8.6	17.1
Means	–	10.6	15.5	2.6	3.6	6.4	11.0
Factors		Shoot wt		Grain wt		Grain P content	
		P values	LSD@0.01	*P* values	LSD@0.01	*P* values	LSD@0.01
Cultivars (CV)		<0.001	1.49	<0.001	0.45	<0.001	1.26
P levels (P)		<0.001	4.56	<0.001	1.10	<0.001	3.12
CV × P		0.100	6.40	0.104	NS	0.060	4.44

Table 7. The specific activity of ^{32}P was highest in flag leaves followed by leaves and stem. The hybrid, CSH 9 translocated more P from roots to flag leaves than the non-hybrids. Similarly, more P accumulated in leaves, stem, and head in the hybrid, than in the non-hybrids. In low-P conditions, a higher proportion of the P was distributed to leaves and stem in the hybrids, than in the non-hybrids. The sequential translocation of P from roots to leaves (particularly flag leaves) by the hybrid compared to the non-hybrids accounted for its high biomass and high P content.

Discussion

Screening for genotypic differences by crops in response to low-N and-P conditions

The traditional coarse cereals (pearl millet and sorghum) and legumes (groundnuts, pigeonpea, and cowpea) are generally grown in areas where stresses of various types persist. The most common stress that affects grain and forage production is nutrient stress, particularly low-P. Genotype screening and selection

Table 5. Shoot and grain weights (g plant^{-1}) at final harvest in different cultivars of sorghum green house experiment at ICRISAT Patancheru, Andhra Pradesh, India 1998

Classification	Entries	Shoot wt (g^{-1} plant) P levels (μM)			Grain wt (g^{-1} plant) P levels (μM)		
		4	20	100	4	20	100
Hybrid	3	15.3	52.5	95.1	1.2	9.5	31.8
Improved varieties	3	13.0	33.9	62.7	1.0	4.8	19.6
Local varieties	3	9.7	30.9	45.0	1.0	4.5	13.8
Drought-tolerant varieties	2	12.8	64.1	113.5	0.6	3.1	11.7
Chinese lines	2	8.6	21.7	33.8	0.7	4.6	11.2
Breeding lines	2	3.1	11.6	46.0	0.0	1.7	14.9
Means	–	10.4	35.8	66.0	0.7	4.7	17.2

Factors	Shoot wt		Factors	Grain wt	
	P values	LSD@0.01		*P* values	LSD@0.01
Cultivars (CV)	<0.001	0.67	Cultivars (CV)	<0.001	0.11
P levels (P)	<0.001	1.39	P levels (P)	<0.001	0.47
CV × P	<0.001	2.42	CV × P	<0.001	0.80

Table 6. Contributions of translocated P from leaves, stem, and roots to grain after P supply was halted at booting stage in sorghum greenhouse experiment, ICRISAT Patancheru, Andhra Pradesh, India 1998

P levels	Classification	Leaf	Stem (mg P plant)$^{-1}$	Root	Total Change	Total Grain
20 μM	Hybrid	–8.1	–6.5	–0.8	–15.4	16.7
	Improved	–4.5	–4.2	–0.8	–9.6	9.9
	Local	–5.7	–1.8	–1.0	–8.5	11.1
	Drought-tolerant line	–6.7	–1.4	–0.6	–8.7	8.6
	Chinese cultivar	–5.7	–5.5	0.5	–10.8	11.7
	Breeding line	–0.9	–2.0	0.7	–2.1	3.6
100 μM	Hybrid	–33.7	–37.0	–2.7	–73.3	70.8
	Improved	–12.4	–37.6	–1.8	–51.8	51.3
	Local	–11.6	–29.6	–1.9	–43.1	42.5
	Drought-tolerant line	–19.9	–40.8	19.7	–41.0	40.3
	Chinese cultivar	–17.8	–15.3	0.3	–32.8	34.8
	Breeding line	–10.5	–24.1	–4.2	–38.8	37.9

for tolerance to low-N and -P soil conditions is emerging as an important strategy to increase productivity in nutrient-stressed environments.

There is a lot of controversy about the performance of landraces and farmer-selected varieties compared to hybrids and newly improved varieties in low-nutrient conditions. Various studies on wild, landraces, old, intermediate, hybrids, and improved genotypes of cereals showed that the new cultivars showed higher yield potential than the old cultivars or landraces under low levels of mineral inputs, and were also efficient responders to N and P fertilization. In wheat (El Bassam, 1998), reported that the present-day cultivars have higher yield potential than the local cultivars under

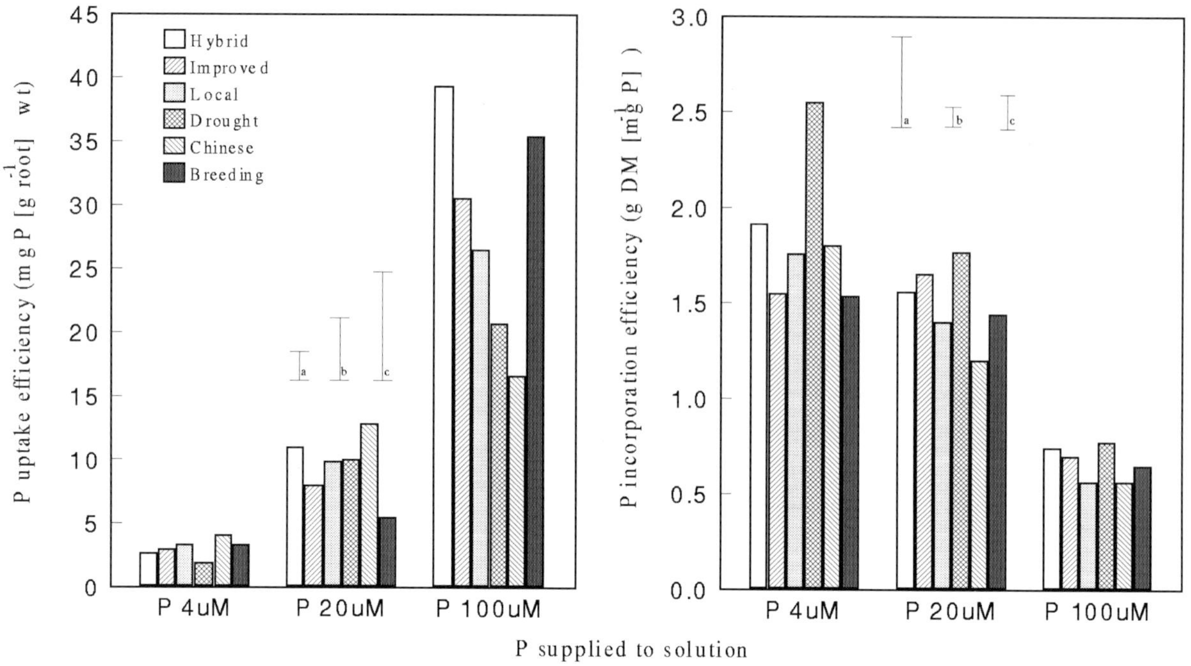

Figure 3. Shoot P accumulation by sorghum grown in solution culture at P concentrations of 4, 20 and 100 μM. greenhouse experiment, ICRISAT Patancheru, Andhra Pradesh, India 1997/98. Data represent average values for 3 genotypes each of hybrid, improved, local, and 2 genotypes each of drought-tolerant, Chinese and breeding lines. Bars indicate LSD (< 0.05) values when comparing (a) cultivars, (b) phosphorus levels and (c) cultivar × phosphorus interaction.

low-P and sufficient-P conditions. On the contrary (Bationo et al., 1989; Payne et al., 1995), reported that local varieties, landraces, and farmer-selected local lines of sorghum and pearl millet are better adapted to low-N and low-P soils than improved varieties and hybrids.

There are many biotic and abiotic factors that influence the adaptation of crop plants to low-nutrient environments. Investigations into low-P input selection based on single treatments at a single location (in this study) may not be adequate. Screening should take into consideration the interactions of nutrients, water, the chemical and physical characteristics of the soil type, temperature and photoperiod, and the crop production system. Wild varieties and landraces of sorghum and pearl millet that are adapted to the harsh Sahelian regions of high temperature, low available nutrients (N, P, K, Zn, and other micronutrients) coupled with shallow soils, low rainfall (< 400 mm a year), and pest-prone environments have acquired specific traits that would not be expressed in other agro-environmental conditions. But in practice, it is difficult to include all these treatments and their interactions in a single screening experiment. The topic of environmental characterization has been extensively discussed (Cooper and Hammer, 1996).

Soil physical properties and screening crops using an artificial medium are criteria for consideration in screening for adaptation to low-nutrient environments. For example in the Sahel region the soils (sandy silicaeous isohyperthemic Psammentic Paleustalf) used for pearl millet and sorghum production are 91% sand, 5.3% clay, and 0.17% organic carbon. Sand-alumina culture systems have been used extensively to evaluate genotypic variability for P acquisition in tomato (Coltman et al., 1985), lettuce (Buso and Bliss, 1988), dry beans (Pierera and Bliss, 1987), maize (da Silva et al., 1993), and pearl millet, (Payne et al., 1995), but may not be representative of such field conditions.

Reliable screening for crop cultivars using an appropriate bioassay system that reflects the type of soil in which crops will be grown, and the criteria for evaluation is a prerequisite to breeding for cultivars with enhanced tolerance to low-nutrient environments. Identifying the appropriate selection environment for

Table 7. Specific and total radio activity at flowering in plant parts of sorghum grown in solution culture, greenhouse experiment, ICRISAT Patancheru, Andhra Pradesh, India 1998

Plant part	Genotype	Specific Radioactivity × 10^3 dpm (g sample)$^{-1}$	Total radioactivity × 10^3 dpm
Leaf	CSH 9	349.0	4673.6
	296 B	25.2	285.1
	FSRP	293.1	2063.1
Flag leaf	CSH 9	858.7	721.9
	296 B	247.8	96.0
	FSRP	673.9	714.4
Stem	CSH 9	218.9	8859.6
	296 B	82.0	1811.0
	FSRP	360.8	8979.5
Head	CSH 9	196.4	2097.8
	296 B	7.9	42.8
	FSRP	90.1	1386.5

both evaluation and genetic resources and to select within breeding population is highly essential.

Physiological and agronomic characteristics of crops adapted to low-nutrient environments

Above-ground traits and nutrient use efficiency

Breeders often have little information on the physiological basis of adaptations of crop plants to low-nutrient environments, thus leaving them without clear ideas of how to further exploit genetic material. The efficiency of crop improvement for nutrient-stressed environments can be enhanced if adaptive plant traits (morphological, physiological, and developmental) can be identified. Agronomic traits (grain dry weight, total plant weight, grain N, grain P) and physiological traits (PEPcase, RuBisCo, NADP – malic enzymes, and carbon isotope ratio of leaves) are known to influence N use efficiencies by sorghum (Maranville and Madhavan, 2000; Youngquist et al., 1992).

Although there were small differences in P uptake efficiency among genotypes, large and significant differences in incorporation and utilization efficiencies were observed among the genotypes and crop species (Figures 1 and 3). The difficulties in devising specific selection criteria for P efficiency in breeding programmes lies in the definitions of P use efficiency. Blair (1993) listed 18 operational definitions of P efficiency of both uptake and utilization. Selection and genetic manipulation of crops based on P use efficiency need to be made in consideration of the end product, be it for forage or for grain production.

Remobilization of pre-anthesis dry mater for grain yield

An important trait observed in this study is the rapid and efficient translocation and remobilization of P from leaves and stem reserves to support growth and grain filling in P-stressed conditions (Table 6). Results from a ^{32}P study showed that in low-P conditions, P-efficient genotypes translocated more P from the roots to flag leaves than P-inefficient genotypes as indicated by a high specific radioactivity (Table 7) than in P-efficient genotypes. Stem-P reserves could therefore play an important role in grain-P and total grain mass under P-deficient conditions. Table 6 shows that when the P supply was withdrawn from the nutrient solution after booting, the contributions from leaf- and stem-P offered powerful reserves for grain filling in sorghum. Improving grain- filling capacity from the stem reserves in wheat has been reported as an important breeding target (Blum, 1998). Blum et al. (1994) reported that the capacity for maintaining large reserves in stems appears to be a genetically controlled constitutive trait that may be linked to assimilate partitioning during stem elongation, and to the developmental characteristics of the stem. Winter wheat cultivars with fairly high yield potential also had good grain filling from stem reserves (Hossain et al., 1990).

For pigeonpea, pre-anthesis P mobilization from stem to grain is not efficient due to the woody nature of the stem (Adu-Gyamfi et al., 1991). Phosphorus stored in pigeonpea stems is hardly translocated to grain. The ability to re-translocate P from old leaves to the growing parts before senescence could be considered an important trait in breeding for P-efficient pigeonpea genotypes (Adu-Gyamfi et al., 1991).

Below-ground traits

Root morphological traits and modifying the rhizosphere through the release of carboxylate anions have been extensively discussed in these proceedings and reviewed elsewhere (Hinsinger, 1998; Jones, 1998; Schachtman et al., 1998). Much has been written about root exudates and heritability of such root traits as size, length, weight, volume, surface area, number, density of root hairs, diameter, growth rate, and energy cost for respiration, but in contrast too little field analysis has been done.

Future research directions

Integrated approach: Plant breeders / Crop physiologists / Statisticians / Crop modelers

The potential contribution of physiological research to genetic improvement of crop plants has long been expounded, but the traditional separation of physiological and breeding research tends to persist. Ideally, the development of a new selection procedure (the adoption of a new screening method, the survey of genotypic variation for a putative trait, the demonstration of its worth in terms of effect on low-nutrient availability, and genetic analysis to establish its heritability) should be achieved through collaborative physiological and breeding research (Cooper and Hammer, 1996). A common behaviour model for physiologists has been to advocate to breeders the selection of particular traits on the basis of studies with a limited range of germplasm, often conducted in isolation from the key objectives of the breeding programme and with limited number or no information on heritability and / or efficiency of the putative trait. Recently, an increasing number of crop improvement programmes have successfully integrated breeding and physiological research and involved statisticians, crop modelers, and socio-economists at the initial and planning stages of the research. This approach is commended.

References

Adu-Gyamfi J J , Fujita K and Ogata S 1989 Phosphorus absorption and utilization efficiency of pigeonpea (*Cajanus cajan* (L) Millsp.) in relation to dry matter production and dinitrogen fixation. Plant Soil 119, 315–324.

Adu-Gyamfi J J, Fujita K and Ogata S 1991 Competition for P among plant parts in the early and medium duration cultivars of pigeonpea (*Cajanus cajan* (L) Millsp.). Plant Soil. 136, 163–169.

Ae N, Arihara J, Okada K, Yoshihara T and Johansen C 1990 Phosphorus uptake by pigeonpea and its role in cropping systems of the Indian subcontinent. Science 248, 477–480.

Barrow N J 1975 The response of phosphate of two annual pasture species. II. The specific rate of uptake of phosphate, its distribution and use for growth. Aust. J. Agric. Res. 26, 145–156.

Bationo A, Henao J and Anand Kumar K 2002 Phosphorus use efficiency as related to sources of P fertilizers, rainfall, soil, crop management, and genotypes in the West African semi-arid tropics. *In* Food Security in Nutrient-Stressed Environments: Exploiting Plants' Genetic Capabilities, Proceedings of an International Workshop 27–30 September 1999, Patancheru, India. Developments in Plant and Soil Sciences volume 95. Ed. J J Adu-Gyamfi. pp 145–154. Kluwer Academic Publishers, Dordrecht, The Netherlands.

Bationo A, Christianson C B and Baethgen W E 1989.Plant density and nitrogen fertilizer effects on pearl millet production in Niger. Agron. J. 82, 290–295.

Batten G D 1993 A review of phosphorus efficiency in wheat. *In* Genetic Aspects of Plant Mineral Nutrition. Eds. P J Randall, E Delhaise, R A Richards and R Munns. pp 215–220. Kluwer Academic Publishers, Dordrecht The Netherlands.

Bieleski R L 1973 Phosphate pools, phosphate transport and phosphate availability. Ann. Rev. Plant Physiol. 24, 225–252.

Blair G 1993 Nutrient efficiency – what do we really mean? *In* Genetic Aspects of Plant Mineral Nutrition. Eds. P J Randall, E Delhaise, R A Richards and R Munns. pp 205–213. Kluwer Academic Publishers Dordrecht, The Netherlands.

Blair G J and Cordero S 1978 The phosphorus efficiency of three annual legumes. Plant Soil 50, 387–398.

Blum A 1998 Improving wheat grain filling under stress by stem reserve mobilization. Euphytica 100, 77–83.

Blum A, Simmena B, Mayer J, Golan G and Shipiler L 1994 Stem reserve mobilization supports wheat grain filling under heat stress. Aust. J. Pl. Physiol. 21, 771–781.

Buso G S C and Bliss F A 1988 Variability among lettuce cultivars grown at two levels of available phosphorus. Plant Soil 11, 67–74.

Caradus J R, Mackay A D, Wewala G S, Dunlop J and Hart A L 1993 Inheritance of phosphorus response in white clover (*Trifolium repens* L.). *In* Genetic Aspects of Plant Mineral Nutrition. Eds. P J Randall, E Delhaise, R A Richards and R Munns. pp 251–260. Kluwer Academic Publishers, Dordrecht, The Netherlands.

Chaykin S 1969 Assay of nicotinamide deaminase. Determination of ammonium by the indophenol reaction. Anal. Biochem. 32, 357–382

Coltman R R, Gerloff G C and Gabelman W H 1985 Differential tolerance of tomatoes strains to maintain and deficient levels of phosphorus. J Am. Soc. Hort. Sci. 110, 140–144.

Cooper M and Hammer G L 1996 Synthesis of strategies for crop improvement. *In* Plant Adaptation and Crop Improvement. Eds. M Cooper and G L Hammer. pp 591–623. CAB International, ICRISAT, IRRI, Wallingford, Oxon, UK.

da Silva A E, Gabelman W H and Coors J G 1993 Screening maize inbred lines for tolerance to low-P stress. *In* Genetic Aspects of Plant Mineral Nutrition. Eds. P J Randall, E Delhaise, R A Richards and R Munns. pp 233–239. Kluwer Academic Publishers, Dordrecht, The Netherlands.

Dakora F D and Philips D A 2002 Root exudates as mediators of mineral acquisition in low-nutrient environments. Plant Soil 245, 35–47.

Dunlop J, Lambert M G, van den Bosch J, Caradus R, Hart A L, Wewale G S, Mackay A D and Hay M J M 1990 A programme to breed a cultivar of *Trifolium repens* L for more efficient phosphate. *In* Genetic Aspects of Plant Mineral Nutrition. Eds. N El Bassam, M Dambroth and B C Loughman. pp 547–552. Kluwer Academic Publishers, Dordrecht, The Netherlands.

El Bassam N 1998 A concept of selection for 'low input' wheat varieties. Euphytica 100, 95–100.

El-Swaify S A and Caldwell R C 1991 Potential crop productivity of Alfisols and Vertisols in the semi-arid tropics. *In* Phosphorus Nutrition of Grain Legumes in the Semi-Arid Tropics. Eds. C Johansen, K K Lee and K L Sahrawat. pp. 3–16. International Crops Research Institute for the Semi-Arid Tropics, Patancheru, Andhra Pradesh, India.

Gomez K A and Gomez A A 1984 Statistical Procedures for Agricultural Research. 2nd edn. John Wiley and Sons, New York, USA. 680 pp.

Hafner H, George E, Bationo A and Marschner H 1993 Effect of crop residues on root growth and phosphorus acquisition of pearl millet in an acid sandy soil in Niger. Plant Soil 150, 117–127.

Hart A L and Jessop D 1983 Phosphorus fractions in trifoliate leaves of white clover and lotus at various levels of P supply. N. Z. J. Agric. Res. 26, 357–361.

Hinsinger P 1998 How do plant roots acquire mineral nutrients? Chemical processes involved in the rhizosphere. Adv. Agron. 64, 225–265.

Hoffland E 1992 Quantitative evaluation of the role of organic acid exudation in the mobilization of rock phosphate by rape. Plant Soil 140, 279–289.

Hossain A B S, Sears R G, Cox T S and Paulsen G M 1990 Dessication tolerance and its relationship to assimilate partitioning in winter wheat. Crop Sci. 30, 622–627.

Ishikawa S, Adu-Gyamfi J J, Nakamura T, Yoshihara T and Wagatsuma T 2002 Genotypic variability in phosphorus solubilizing activity of root exudates by pigeonpea grown in low-nutrient environments. Plant Soil 245, 71–81.

Jones D L 1998 Organic acids in the rhizosphere – a critical review. Plant Soil 205, 25–44.

Lynch J 1999 The role of nutrient-efficient crops in modern agriculture. *In* Nutrient Use in Crop Production. Ed Z Rengel. pp 241–264. The Haworth Press, New York, USA.

Lynch J P and Beebe S E 1995 Adaptation of beans (*Phaseolus vulgaris* L.) to low phosphorus availability. HortSci. 30, 1165–1171.

Mackay A D, Caradus J R, Dunlop J, Wewale G S, Mouat M C H, Lambert M G, Hart A L and van den Bosch J 1990. Response to phosphorus of a world collection of white clover cultivars. *In* Genetic Aspects of Plant Mineral Nutrition. Eds. N El Bassam, M Dambroth and B C Loughman. pp 553–558. Kluwer Academic Publishers, Dordrecht, The Netherlands.

Maranville J W and Madhavan S 2002 Physiological adaptations of nitrogen use efficiency in sorghum. Physiological adaptations for nitrogen use efficiency in sorghum. Plant Soil 245, 25–34.

Maranville J W, Clark R B and Ross W M 1980 Nitrogen efficiency in grain sorghum. J. Plant Nutr. 2, 577–589.

Murphy J and Riley J P 1962 A modified single solution method for the determination of phosphorus in natural waters. Anal. Chim. Acta 27, 31–36.

Nakano H, Adu-Gyamfi J J and Nakamura T 2002 Genetic adaptation of crop plants to low-nutrient environments: Morphological and ecophysiological characteristics of adaptation. *In* Food Security in Nutrient-Stressed Environments: Exploiting Plants' Genetic Capabilities, Proceedings of an International Workshop 27–30 September 1999, Patancheru, India. Developments in Plant and Soil Sciences vol 95. Ed. J J Adu-Gyamfi. pp 59–66. Kluwer Academic Publishers, Dordrecht, The Netherlands.

Newman E J 1966 A new method of estimating the total root length of a root in a sample. J. Appl. Ecol. 3, 139–145.

Otani T, Ae N and Tanaka H 1996 Phosphorus (P) uptake mechanisms of crops grown in soils with low P status. I. Screening of crops for efficient P uptake. Soil Sci. Plant Nutr. 42, 155–163.

Payne W A, Hossner L R, Onken A B and Wendt C W 1995 Nitrogen and phosphorus uptake in pearl millet and its relation to nutrient and transpiration efficiency. Agron. J. 87, 425–431.

Pereira P A A and Bliss F A 1987 Nitrogen fixation and plant growth of common bean (*Phaseolus vulgaris L.*). at different levels of phosphorus availability. Plant Soil 104, 79–84.

Ray J D, Yu L, McCough S R and Chamoux M C 1996 Mapping quantitative trait loci associated with root penetration ability in rice (*Oryza sativa* L.). Theor. Appl. Genet. 92, 627–636.

Rengel Z 1999 Physiological mechanisms underlying differential nutrient efficiency of crop genotypes. *In* Mineral Nutrition of Crops: Mechanisms and Implications. Ed. Z Rengel. pp 227–265. The Haworth Press, New York, USA.

Rengel Z 2002 Genetic control of root exudates. Plant Soil 245, 59–70.

Schachtman D P, Reid R J and Ayling S M 1998 Phosphorus uptake by plants: From soil to cell. Plant Physiol. 116, 447–453.

Schettini T, Gabelman W H and Gerloff G C 1987 Incorporation of phosphorus efficiency from exotic germplasm into agriculturally adapted germplasm of common bean *(Phaseolus vulgaris* L.) *In* Genetic Aspects of Plant Mineral Nutrition. Eds. W H Gabelman and B C Loughman. pp 559–568. Martinus Nijhoff Publishers, Dordrecht, The Netherlands.

Snapp S S and Lynch J P 1996 Phosphorus distribution and remobilization in bean plants as influenced by phosphorus nutrition. Crop Sci. 36, 929–935.

Subbarao G V, Ae N and Otani T 1997 Genotypic variation in iron- and aluminum-phosphate solubilizing activity of pigeonpea root exudates under P deficient conditions. Soil Sci. Plant Nutr. 43, 295–305.

Vadez V, Lasso J H, Beck D P and Drevon J J 1999 Variability of N_2-fixation in common bean (*Phaseolus vulgaris* L.) under P deficiency is related to P use efficiency - N_2-fixation tolerance to P deficiency. Euphytica 106, 231–242.

van Keulen H and Breman H 1990 Agricultural development in the West African Sahel region: a cure against land hunger? Agric. Ecosyst. Environ. 32: 177–197.

Wissuwa M, Yano M and Ae N 1998 Mapping of QTLs for phosphorus-deficiency tolerance in rice (*Oryza sativa* L.). Theor. Appl. Genet. 97, 777–783.

Yano M and Sasaki T 1997 Genetic and molecular dissection of quantitative traits in rice. Plant Mol. Biol. 35, 145–153.

Youngquist J B, Bramel-Cox P and Maranville J W 1992 Evaluation of alternative sreening criteria for selecting nitrogen-use efficient genotypes in sorghum. Crop Sci. 32, 1310–1313.

J.J. Adu-Gyamfi (Ed.), Food security in nutrient-stressed environments: exploiting plants' genetic capabilities: 81–90, 2002.

Physiological adaptations for nitrogen use efficiency in sorghum†

Jerry W. Maranville[1,3] & S. Madhavan[2]
[1]*Department of Agronomy and Horticulture, University of Nebraska, Lincoln, NE 6853-0817, U.S.A.*
[2]*Department of Biochemistry, University of Nebraska, Lincoln, NE 68583-0817, U.S.A.* [3]*Corresponding author**

Received 10 May 2000. Accepted in revised form 2 June 2001

Key words: assimilation efficiency indices, carbon isotope discrimination, carboxylating enzymes, immunoblots, nitrogen stress, nitrogen use efficiency

Abstract

Known high nitrogen utilization efficiency (NUE_I, biomass per unit plant N) China lines of sorghum, China 17 and San Chi San, were compared with relatively low NUE_I U.S. lines, CK60 and Tx623, for both their physiological and biochemical adaptations to tolerate an imposed N stress in the greenhouse. Assimilation efficiency indices (ACi) were significantly greater for the China lines than the U.S. lines at both low and high soil nitrogen levels by about two-fold. Chlorophyll levels in leaves of high NUE_I lines were lower at both soil N treatments. Immunoblots of leaf extracts of sorghum subjected to N stress indicated reduced levels of both phosphoenolpyruvate carboxylase (PEPcase) and ribulose 1,5-bisphosphate carboxylase (Rubisco) while NADP-malic enzyme levels, in general, appear not to be affected. However, NUE_I China line, China 17, retained a significantly greater PEPcase activity than the less-NUE_I U.S. lines, and also the NUE_I China line San Chi San, when grown under N stress conditions. This suggests that PEPcase and enzymes associated with phosphoenolpyruvate synthesis, perhaps, are significant factors in maintaining relatively high photosynthesis under N stress. Carbon isotope ratios of leaves from sorghum genotypes, as indicated by $\delta^{13}C$ values, became less negative when sorghum plants were grown under N stress, but a genotypic variation either at a low or high N was not observed.

Abbreviations: ACi – assimilation efficiency index; CER – carbon exchange rate; N – nitrogen; NADP-ME – NADP-malic enzyme; NUE_I – nitrogen use efficiency (biomass per unit plant nitrogen); PEPcase – phosphoenolpyruvate carboxylase; Rubisco – ribulose 1,5-bisphosphate carboxylase

Introduction

Effective improvement in obtaining more yield from any increment of applied nitrogen (N), or in sustaining some productivity at moderate to extreme soil N stresses, can occur through breeding most rapidly if metabolic/physiological processes of N use efficiency or N stress tolerance are understood. Definitions of N use efficiency vary. Moll et al. (1982) and Onken et al. (1985) recognized that two major components, uptake efficiency and utilization efficiency, contribute to variation in N use efficiency. Uptake efficiency has been defined as the amount of N taken up from that available in the soil which is largely dependent on the root system and its ability to extract N (Eghball and Maranville, 1993).

The subject of N use efficiency quantification has been evaluated by Cassman et al. (1998) in their research with rice (*Oryza sativa* L.) at the International Rice Research Institute (IRRI). They argue that incremental efficiency, or change in yield for each change in applied N rate from that of the control, is a more valid approach than the simpler definitions of yield per unit N applied or per unit plant N. Physiological efficiency then was defined as $_{\Delta}Y/_{\Delta}N_p$ or change in yield for each change in plant N resulting from the applied inputs. Moll et al. (1982) referred to 'utilization efficiency' as grain produced per unit of plant

* FAX No: +1-402-472-3654; E-mail: jmaranville1@unl.edu
†Contribution of the Nebraska Exp. Stn., Lincoln, NE as paper no. 12810.

This paper previously appeared in *Plant and Soil* **245**, 2002, pp. 25–34.

N. However, Maranville et al. (1980) argued that both forage and grain are important economic components and defined NUE for each separately. Their NUE_1 was total biomass per unit plant N (inverse of total plant N concentration), and NUE_2 was grain yield per unit plant N. Siddiqi and Glass (1981) felt these 'quotients' were oversimplifications and did not place enough emphasis on yield. They proposed that N use efficiency should be defined as the ratio of biomass to tissue concentration rather than content. Field (1983) termed the measure of carbon exchange rate (photosynthesis) per unit of leaf N as 'instantaneous' N use efficiency. However, Alagarswamy et al. (1988) found that his concept failed to explain overall NUE differences (Maranville's NUE_1), but that NUE_1 was more related to partitioning of N into additional leaf area.

Physiological N use efficiency can be influenced by morphological and anatomical features (Longstreth and Nobel, 1980). Youngquist et al. (1992) attempted to establish alternative screening criteria for NUE_1 or NUE_2 from morphological traits that would require no or only limited N analysis and save time. They found that the best predictor of NUE_1 was total dry weight/total grain N, and that harvest index was a good predictor of NUE_2 in sorghum.

These definitions assume that improved N use efficiency will occur by (1) maintaining constant grain or biomass yield while reducing plant or leaf N, (2) increasing grain or biomass yield at a constant level of plant or leaf N, or (3) reducing levels of plant or leaf N per unit increase in grain or biomass yield. Regardless of the method employed, it is implied that photosynthesis must be maintained or increased while N supply is reduced. The mechanism likely involves entities critical to both N and carbon metabolism. Sage et al. (1987) showed that total leaf N was correlated with both CO_2 assimilation and the activity of the carboxylating enzymes ribulose-1,5-bisphosphate carboxylase/oxygenase (Rubisco) and phosphoenolpyruvate carboxylase (PEP case). Since 40–50% of the soluble N in leaves of C_4 plants can be contained in these two enzymes plus pyruvate orthophosphate dikinase (Lawlor et al., 1987; Sugiyama et al., 1984), the correlation is not surprising. Under suboptimal N levels, Rubisco can be preferentially synthesized over other enzymes, and can contain up to 50% of the total soluble N (Sugiyama et al., 1984). Maranville and Settimi (1989) proposed that observed differences of C_4 crops for NUE_1 at low leaf N levels may lie in the comparative ability to mobilize N from carboxylating enzymes with minimal loss of catalytic function, thus maintaining photosynthesis at levels which sustain growth as N supply is reduced.

Carbon isotope ratios, $\delta^{13}C$ values, reflect the integrated balance of stomatal diffusion and carboxylation during photosynthesis. In C_4 species, carbon isotope discrimination reflects the initial carboxylation reaction of PEPcase fixing CO_2 into C_4 acids and the refixation by Rubisco of CO_2 released into the bundle sheath cells through decarboxylation of those C_4 acids. Thus, $\delta^{13}C$ values or Δ ($\delta^{13}C$ source–$\delta^{13}C$ product) when calculated, are integral measurements of various photosynthetic properties of a leaf until its harvest. It is known from several studies that a variety of physical and physiological factors influence carbon isotope ratios of C_3 plants (Madhavan et al., 1991), and carbon isotope ratios in these plants are negatively correlated with water use efficiency. Very few studies have been made thus far to correlate $\delta^{13}C$ values of C_4 plants with either N use efficiency or water use efficiency. This is due to the fact that in C_4 plants there is a disadvantage of a much reduced signal as the slope of the relationship between Δ and Pi/Pa (ratio of intercellular to ambient CO_2 partial pressure) is much less, unlike in C_3 plants (Henderson et al., 1998). Nitrogen availability and/or utilization of available N efficiently under conditions of stress will significantly affect the functions of both PEP case and Rubisco, or their activity ratio. By determining the Δ of the leaf dry matter, Ranjith et al. (1995) have suggested that greater relative investment of N in Rubisco activity resulted in a higher N use efficiency in N stressed sugarcane.

This study was conducted to further investigate the role that specific enzymes and other physiological traits including carbon isotope discrimination may play in plants having high N use efficiency defined as biomass per unit plant N (NUE_1).

Materials and methods

Several greenhouse experiments were conducted on sorghum lines differing in NUE (g biomass g^{-1} N) at the University of Nebraska-Lincoln, using the same protocol for each repeated experiment. Two of these lines were originated in China (China 17 and San Chi San), and identified as day length insensitive with marked N stress tolerance (produced satisfactory grain and biomass yield under severe soil N limitation) by Dr Paula Bramel-Cox, ICRISAT, (personal communication). These were compared to two U.S. derived sorghum lines, CK60 and Tx623, known to have markedly less NUE_1 and NUE_2 than the China lines from previous tests on the field. Plants were grown in 9 L pots containing a mixture of 80% sand and 20% soil to insure low N in the medium (1 mg kg^{-1}). Treatments consisted of low and high N for the four genotypes in eight replications by providing low N pots with full nutrient except N, and high N pots with full nutrient (Hoagland and Arnon, 1950) at each watering. When low N plants were at the 7–9 leaf growth stage, 0.7 m*M* of N was added to insure plants would continue growing through seed fill when experiments were terminated.

Chlorophyll concentration of leaves was monitored using a Minolta SPAD meter which was standardized to convert the readings to concentrations using the method of Markwell et al. (1995). Physiological measurements were made at anthesis after low N plants had developed distinguishable N deficiency symptoms compared to high N plants. Carbon exchange rates (CER) were determined on the 3rd or 4th leaf from the top (including flag leaf) using a LI-6200 portable photosynthesis system (LI-COR, Inc., Lincoln, NE) equipped with a 1.0 L leaf chamber. This portable photosynthesis system is calibrated to calculate CO_2 assimilation rates, stomatal resistance to water vapor and internal CO_2 concentration using equations derived from von Caemmerer and Farquhar (1981). Assimilation efficiency indices (ACi) were also determined using the same equipment, and the method described by Settimi and Maranville (1998) was used to deduce carboxylation activity. The chamber containing the specified leaf was placed beneath a light source providing a constant 1800 μmol photons m^{-2} s^{-1} (400–700 nm) and excluded light in the infrared region to minimize heat buildup. Measurements were made between 1000 and 1400 h. Each leaf required about 20 min for an ACi measurement. The initial slope value of CO_2 assimilation (A) plotted against internal leaf CO_2 concentration (Ci) gives an estimate of activity. Assimilation values less than 50% of the maximum and their Ci values were used in the regression analysis. The regression analysis is the ACi as μmol CO_2 m^{-2} s^{-1} versus μmol CO_2 mol^{-1}. The ACis were then compared using a Student Newman-Kuels test (Zar, 1984).

Leaves that were non-destructively used for the measurements of CER and ACi were also sampled for the determination of amount and activities of different enzymes and carbon isotope ratio measurements. Several leaf punches (0.75 cm^2) from these leaves were obtained, fixed instantaneously in liquid N and stored at −62 °C until the performance of various analyses.

Enzyme assays

Crude extracts were prepared by grinding 4–5 leaf punches in ice-cold 100 m*M* Tris-HC1 buffer, pH 7.5 containing 10 m*M* $MgCl_2$, 14 m*M* 2-mercaptoethanol, 1 m*M* EDTA and 5% glycerol (v/v). The homogenate was filtered through four layers of cheesecloth and the filtrate was centrifuged at 27 000 *g* for 15 min. The supernatant fraction was used immediately for both PEPcase and NADP-malic enzyme activity determinations.

PEPcase activity was measured according to the procedures described by Iglesias and Andreo (1989). The standard assay medium in a total volume of 1 mL consisted of 50 m*M* Tris-HC1, pH 8.0, 5 m*M* $MgCl_2$, 0.15 m*M* NADH, 10 m*M* sodium bicarbonate ($NaHCO_3$), 2 IU malic dehydrogenase, 4 m*M* phosphoenolpyruvate and an aliquot of the leaf extract containing ~20 μg of total soluble protein. The enzyme activity was followed by monitoring the NADH oxidation at 340 nm in a CARY 2200 spectrophotometer at 30 °C.

NADP-malic enzyme activity was also measured as per the procedures of Iglesias and Andreo (1989). The standard assay medium contained in 1 mL, 50 m*M* Tris-HC1, pH 8.0, 10 m*M* $MgCl_2$, 0.5 m*M* $NADP^+$, an aliquot of the leaf extract containing ~20 μg of total soluble protein and 4 m*M* L-malate with which the reaction was started. The enzyme activity was followed by NADPH production at 340 nm in a CARY 2200 spectrophotometer at 30 °C.

For the determination of Rubisco activity, procedures of Ranjith et al. (1995) were followed with slight modifications. Briefly, 2–3 leaf punches were ground in a minimal volume of ice-cold extraction buffer containing 100 m*M* bicine, pH 8.1, 7.5 m*M* dithiothreitol (DTT), 1 m*M* disodium ethylenediamine tetracetic acid (Na_2EDTA) and 10 g L^{-1} polyethylene glycol (PEG, 3600) (w/v), in a mortar with pestle. The homogenate was centrifuged at 27 000 *g* for 15 min and the supernatant was used as the enzyme source after determining the total soluble protein content by the method of Bradford et al. (1976). Prior to the start of the Rubisco assay, aliquots of various sample extracts were preincubated (1:1) with an activation medium containing 100 m*M* bicine buffer, pH 8.1, 20 m*M* $MgCl_2$ and 20 m*M* ($NaHCO_3$) at 25 °C for 30 min. The reaction was initiated by the addition of

200 μL of activated enzyme extract to 800 μL of an assay mixture containing 100 mM bicine buffer, pH 8.1, 5 mM DTT, 0.4 mM ribulose bisphosphate, 0.1 mM Na_2EDTA, 20 mM $MgCl_2$, 10 mM $NaHCO_3$ and 111kBq $NaH^{14}CO_3$, and thoroughly mixed. The reaction was carried out at 25 °C for 2 min and terminated by the addition of 0.4 mL of 3M formic acid in methanol. The reaction vials were kept in an oven at 50 °C overnight to remove the remaining free CO_2 and dry the contents. Radioactivity was measured by scintillation counting in a Packard Tri-Carb 1600 TR Liquid Scintillation Analyzer, after adding 5 mL of scintillation fluid. Controls for each set of assays were made by the addition of inactivated enzyme extracts (0.4 mL of 3M formic acid in methanol was added to 200 μL of activated enzyme extracts) to the assay mixtures.

Protein determination

Leaf soluble protein concentrations were determined spectrophotometrically at 595 nm by the method of Bradford (1976) using bovine serum albumin (BSA) as the standard.

Immunoblotting procedures

SDS-PAGE in 7.5% gels (Ready-Gels, Bio-Rad) were performed according to the method of Laemmli (1970). Proteins were transferred electrophoretically to 0.45 μm nitrocellulose membranes (Schleicher & Schuell, Inc., Keene, NH) using a trans-blot apparatus (Mini Protean II, Bio-Rad) (Towbin et al., 1979). Membranes were blocked overnight at 4 °C, by incubating in phosphate buffered saline (PBS), (10 mM NaH_2PO_4, pH 7.5 and 150 mM NaC1) containing 0.5 mL L^{-1} Tween 20 and 50 g L^{-1} non-fat milk powder. Prior to incubating with respective antibodies for Rubisco, PEPcase and NADP-ME the blots were washed twice for 5 min in PBS containing Tween 20. Blots were incubated with respective primary antibodies at 1:5000 for 1 h and then washed three times with PBS containing 0.5 mL L^{-1} Tween 20. All blots were then incubated with a secondary antibody (1:25 000) conjugated with Horse radish peroxidase. Detection of second antibody utilized the Super Signal chemiluminescent system (Pierce Chemical Co., Rockford, IL).

Carbon isotope ratio measurements

Carbon isotope ratio measurements were made following published procedures (Madhavan et al., 1991). Briefly, leaf punches that were dried in an oven for 48 h at 60 °C were ground to a fine powder. A subsample of this leaf powder (2–3 mg) was analyzed for carbon isotope ratio using an elemental analyzer (Heraeus, CHN-O Rapid) interfaced with an automated trapping box system and a Finnigan Delta-S isotope ratio mass spectrometer. The isotope ratio of each sample was determined by comparison with a previously calibrated working standard with a known δ^{13}C value (precision ± 0.2‰) relative to PDB (Pee Dee Belemnite international standard). The ^{13}C content of CO_2 is given as an isotope ratio, R,

$$R = [^{13}CO_2]/[^{12}CO_2]$$

and δ^{13}C is given by,

$$\delta^{13}C = 1000[(R_{\text{sample}}/R_{\text{standard}}) - 1]$$

Results

A comparative field test to confirm the NUE_1 of lines was made in 1996, and showed that the China lines had higher values than the U.S. lines. The China lines averaged approximately 94 g biomass g^{-1} N while the U.S. lines averaged 85 g g^{-1} when grown with N stress. This compared to 77 g g^{-1} and 76 g g^{-1}, respectively, for these contrasting groups at normal soil N.

Chlorophyll concentration of the measured leaves indicated that N stress was evident for all lines when grown at the low N treatment in the greenhouse (Table 1). The U.S. lines had greener leaves at both high and low N treatments compared to the China lines. Line CK60 lost only 32% of its leaf chlorophyll compared to 45–50% for the other lines. Leaf chlorophyll as measured by the SPAD meter was highly correlated to leaf N with correlation coefficients ranging from 0.89 to 0.98 for the four lines; however, the slope of the relationship was steeper for CK60 than the other lines (data not shown). Thus, the relative measure of leaf chlorophyll was a direct indicator of leaf N levels for these lines. Even though chlorophyll and thus leaf N levels were higher for the U.S. lines, this did not translate into higher CER values (photosynthesis) at either N level (Table 1). San Chi San had a higher CER than the other lines at the low N treatment, and there was no difference among lines at the high N level.

Assimilation efficiency indicies (A/Ci slopes) were significantly higher for the China lines than the U.S.

Table 1. Leaf chlorophyll concentration, carbon exchange rate (CER), and assimilation efficiency indicies (A/Ci) at anthesis for China and U.S. sorghum lines grown in the greenhouse at high and low N supply measured on the third leaf from the top

Line	High N			Low N		
	[Chl] μmol m^{-2}	CER μmol CO_2 m^{-2} s^{-1}	A/Ci slope	Chl μmol m^{-2}	CER μmol CO_2 m^{-2} s^{-1}	A/Ci slope
China 17	523 b[a]	24.1 a	0.833 a	283 c	13.4 b	0.187 a
San Chi San	494 b	23.4 a	0.572 b	279 c	16.6 a	0.160 a
CK60	620 a	22.2 a	0.389 c	419 a	11.2 b	0.070 b
Tx623	723 a	22.6 a	0.329 c	347 b	13.4 b	0.089 b
Mean	590[b]	23.1[b]	0.531[b]	332	13.7	0.127

[a]Values within a column followed by the same letter are not significantly different ($P \leq 0.05$).
[b]Nitrogen treatment is significantly different from the low N treatment ($P \leq 0.01$).

Table 2. NADP-Malic enzyme activities ($\bar{x}$±S.E.) in leaves of two China and two U.S. derived sorghum lines grown in the greenhouse at two N levels. Values are calculated on either a protein or leaf area basis

Line	High N		Low N	
	Protein nmol min^{-1} mg^{-1}	Area nmol min^{-1} cm^{-2}	Protein nmol min^{-1} mg^{-1}	Area nmol min^{-1} cm^{-2}
China 17	2662±331	16.9±0.05	338±20	0.50±0.03
San Chi San	1979±161	17.9±0.64	2468±256	3.90±0.14
CK60	927±126	24.8±4.25	1936±146	6.27±0.47
Tx623	1477±89	25.1±1.37	2547±350	4.81±1.37
Mean	1761±177	21.2±1.58	1822±193	3.87±0.50

Table 3. PEPcase enzyme activities ($\bar{x}$±S.E.) in leaves of two China and two U.S. derived sorghum lines grown in the greenhouse at two N levels. Values are calculated on either a protein or leaf area basis

Line	High N		Low N	
	Protein nmol min^{-1} mg^{-1}	Area nmol min^{-1} cm^{-2}	Protein nmol min^{-1} mg^{-1}	Area nmol min^{-1} cm^{-2}
China 17	29.1±8.3	3.65±1.57	41.1±5.6	0.59±0.05
San Chi San	17.1±1.6	1.50±0.15	8.1±0.9	0.14±0.02
CK60	44.4±6.0	11.58±1.53	10.8±2.4	0.35±0.07
Tx623	59.4±2.8	10.24±0.33	7.5±1.2	0.15±0.02
Mean	37.5±4.7	6.74±0.90	16.9±2.5	0.31±0.04

Table 4. Rubisco enzyme activities ($\bar{x}$±S.E.) in leaves of two China and two U.S. derived sorghum lines grown in the greenhouse at two N levels. Values are calculated on either a protein or leaf area basis

Line	High N		Low N	
	Protein nmol min^{-1} mg^{-1}	Area nmol min^{-1} cm^{-2}	Protein nmol min^{-1} mg^{-1}	Area nmol min^{-1} cm^{-2}
China 17	13.1±0.4	1.52±0.07	1.5±0.6	0.06±0.01
San Chi San	16.4±2.9	2.20±0.09	2.0±0.2	0.10±0.02
CK60	19.8±1.1	5.03±0.31	3.2±0.3	0.16±0.02
Tx623	23.0±1.2	4.48±0.08	2.4±0.1	0.11±0.01
Mean	18.1±2.1	3.3±1.7	2.3±0.36	0.11±0.02

lines at both N levels (Table 1) indicating a stronger carboxylation activity. China 17 was particularly notable in that it had the highest ACi slope at either N level. These high assimilation efficiency indices for China 17 were attained at comparatively low leaf N levels as indicated by the low chlorophyll concentrations with low soil N supply.

Tables 2, 3 and 4 show the activities of NADP-Malic enzyme, PEPcase and Rubisco, respectively. NADP-ME activity was higher in China lines than in the U.S. lines when calculated on a protein basis at high N treatment, whereas at low N, China 17 had the lowest activity. However, when compared on an area basis, NADP-ME activity in the U.S. lines was much higher than China lines under both N treatments (Table 2). There was no difference in activity between N levels if based on protein, but when based on area, there was an 82% reduction from the high to the low N treatment. The China line, China 17, appeared to retain greater PEPcase activity than San Chi San and the U.S. lines when N stress was imposed (Table 3). China 17 had activity that actually was increased with low N while San Chi San lost 53% activity, CK60 lost 76% and Tx623 lost 87% activity when based on protein. When based on area, all lines had reduced activity at the low N treatment. The average loss of activity on an area basis for the average of the China lines was 87% compared to a loss of 98% for the average of the U.S. lines.

Rubisco activity (Table 4) was greater in the U.S. lines than the China lines at both N levels. On a protein or area basis, the China lines lost a similar percentage of activity (95%) as the U.S. lines when grown with N stress, which was different than PEPcase in China 17 where it retained a greater percentage of activity.

There were no differences in the levels of NADP-Malic enzyme between genotypes or between N levels (Figure 1B). There were differences, however, among lines for the carboxylation enzymes, PEPcase (Figure 1A) and Rubisco (Figure 1C), and there was less of each of these two enzymes at the low N level. The China 17 line appeared to retain more PEP case protein at the low N treatment than either San Chi San or the U.S. lines perhaps indicating a preferential allocation of N in PEP case under N stress.

Leaf carbon isotope ratios which are dependent largely upon carbon isotope discriminations (δ^{13}C, ‰ values), occurring during photosynthesis, were not significantly different among the four genotypes at any growth stage up to flowering within high or low N treatments (Table 5). However, leaf δ^{13}C values became consistently less negative (0.4‰) under N stress. The effects of N stress on foliar δ^{13}C values became more apparent during the boot and anthesis stages of plant development than in their juvenile stages of growth. Leaf δ^{13}C values were consistently less negative (0.8‰) during later stages of plant development (Table 6), but there was still no difference among the lines.

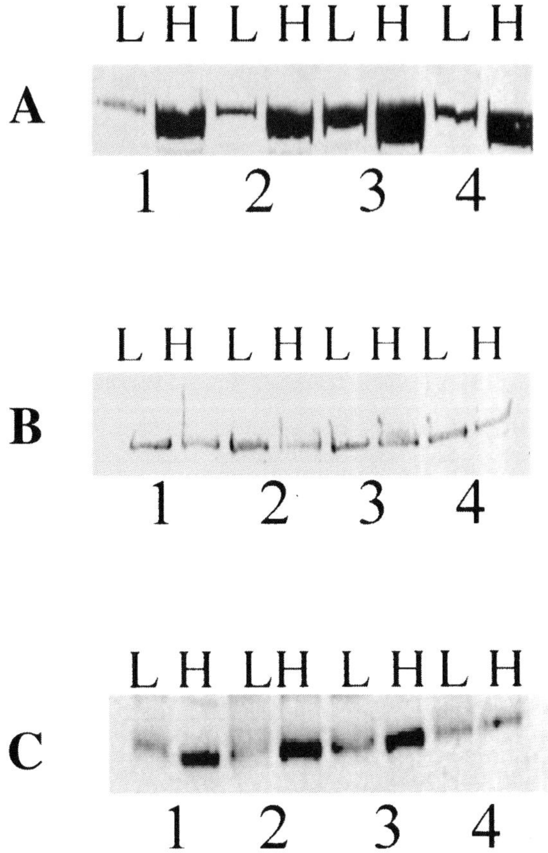

Figure 1. Immunodetection of phosphoenolpyruvate carboxylase (A), NADP-malic enzyme (B) and Rubisco (C) in greenhouse grown sorghum lines subjected to low and high N treatments. For panels A and B, 1, 2, 3 and 4 represent sorghum genotypes, San Chi San, CK60, China 17 and Tx623 respectively. For Panel C, 1, 2, 3 and 4 represent sorghum genotypes, Tx623, China 17, CK60 and San Chi San, respectively. Letters L and H in all panels indicate low or high N treatments. All lanes in the gel were loaded with identical amounts (10 μg) of total extractable soluble protein from crude leaf extracts. Note that the amounts of both carboxylating enzymes, PEPcase and Rubisco are lower in low N treated plant extracts, whereas no appreciable variation can be observed for NADP-malic enzyme levels between treatments in all genotypes.

Table 5. Leaf carbon isotope ratios ($\bar{x}$±S.E.) at different stages of plant development in four sorghum genotypes grown under low and high nitrogen treatment

Genotype	N Treatment	$\delta^{13}C$, (‰) at stage of development		
		Vegetative	Boot	Anthesis
China 17	Low	−13.8±0.3	−12.6±0.1	−12.7±0.1
	High	−13.8±0.2	−12.9±0.1	−13.1±0.1
San Chi San	Low	−13.7±0.2	−12.2±0.2	−12.4±0.1
	High	−13.7±0.4	−12.6±0.1	−12.8±0.1
TX 623	Low	−13.6±0.2	−12.2±0.1	−12.3±0.1
	High	−13.9±0.2	−12.8±0.1	−12.8±0.1
CK 60	Low	−13.4±0.1	−12.5±0.1	−12.4±0.1
	High	−13.6±0.3	−12.8±0.1	−12.8±0.2
Mean	Low	−13.6±0.17	−12.38±0.21	−12.45±0.17
	High	−13.8±0.13	−12.78±0.13	−12.88±0.15

Table 6. Carbon isotope fractionation ratios for leaves of plants after anthesis from China and U.S. lines grown in the greenhouse at two N levels. Each mean is from six measurements

Line	High N $\delta^{13}C$ ‰	Low N $\delta^{13}C$ ‰	Mean $\delta^{13}C$ ‰
China 17	−14.3 a[a]	−13.5 a	−13.9
San Chi San	−14.1 a	−13.2 a	−13.6
CK60	−14.2 a	−13.4 a	−13.8
Tx623	−14.4 a	−13.6 a	−14.0
Mean	−14.3[b]	−13.4	

[a] Values within a column followed by the same letter are not significantly different ($P \leq 0.05$).
[b] Nitrogen treatment is significantly different from the low N treatment ($P \leq 0.05$).

Discussion

We conducted a series of greenhouse experiments comparing both high NUE and low NUE sorghums to determine biochemical/physiological attributes associated with the trait. By determining leaf N and chlorophyll contents from plants grown at two different soil N levels (adequate and stress), we showed that the Minolta SPAD chlorophyll meter values when converted to chlorophyll concentration were highly correlated with leaf N (r^2=0.92–0.98). This provided the opportunity to determine the leaf N of plants in this experiment nondestructively and with a high degree of accuracy.

There is evidence that carboxylating enzymes are linked closely to biomass production and that preferential allocation of N to PEPcase from Rubisco occurs especially under high N conditions (Sugiyama et al., 1984). Since Rubisco activity is correlated with the initial slope of the CO_2 response of photosynthesis in C_3 plants (van Caemmerer and Farquhar, 1981), there has been speculation that the total carboxylation enzyme capacity may also be estimated in this manner (Settimi and Maranville, 1998). Our results showed that the slope values for CO_2 assimilation versus internal leaf [CO_2] were markedly higher for the high NUE China lines than the less N efficient U.S. lines under N stress indicating that carboxylation ability was indeed related to greater biomass production at reduced leaf [N]. There was no difference in CER on a leaf area basis among the genotypes when grown at high N supply, but the China line San Chi San had significantly greater CER at low N supply. The other China line (China 17) was not higher than U.S. lines for CER, but when comparing leaf chlorophyll concentrations among lines, both China lines were on average 26.6% less than the U.S. lines. Given the high correlation between chlorophyll and leaf N of these genotypes, it can be assumed the leaf N of China lines was much lower than U.S. lines. If a comparison of CER values were made on a leaf N basis rather than an area basis, the China lines would have a marked advantage in photosynthesis at low N supply compared to the U.S. lines.

Major carbon assimilating enzymes in C_4 species include PEPcase, pyruvate orthophosphate dikinase (PPDK) and Rubisco (Sugiharto et al., 1990), while NADP-malic enzyme is the major decarboxylating enzyme in NADP-ME C_4 subtypes. In this study, we have attempted to determine both the levels and activities of these enzymes, except PPDK, in plants that were subjected to low and high N treatments. As demonstrated in earlier studies (Ranjith et al., 1995; Sugiharto et al., 1990; Sugiyama et al., 1984), PEP-case levels and activity were considerably reduced in sorghum genotypes under N stress. However, PEP case activity was higher particularly in China 17, when grown with N stress. Additionally, immunoblots of leaf extracts from China 17 genotype showed greater amounts of this protein than in U.S. lines where plants were N stressed. This may further suggest that there may be some preferred allocation of leaf N to PEPcase under N stress to maintain this enzyme activity.

Levels and activity of the enzyme Rubisco also declined when sorghum plants were grown under N stress, as has been found in other C_4 species (Ranjith et al., 1995; Sugiharto et al., 1990; Sugiyama et al., 1984). However, Rubisco activity was found to be much higher in U.S. lines than in China lines. This does not coincide with the observed ACi values, which has been suggested to be an estimator of Rubisco activity in C_3 crops. Rubisco enzyme activity was much higher in U.S. lines which had lower ACi values.

Although there was significant reduction in NADP-malic enzyme activity on a leaf area basis when plants were grown under N stress, there was no consistent patterns of activity among the China or U.S. lines to indicate involvement of this enzyme in observed NUE_I pattern.

Carbon isotope ratios

Carbon isotope ratios, $\delta^{13}C$ values, reflect the integrated balance of stomatal diffusion and carboxylation during photosynthesis. Carbon isotope discrimination, particularly in C_4 species, reflects the initial carboxylation reaction of PEPcase fixing CO_2 into C_4 acids and the refixation by Rubisco of CO_2 released into the bundle sheath cells through decarboxylation of those C_4 acids. In addition, it is also dependent on the ratio of intercellular to ambient CO_2 partial pressure, P_i/P_a, and leakiness, Φ (the measure of CO_2 not fixed by Rubisco but leaks back into mesophyll) (Farquhar, 1983). Carbon isotope discrimination in C_4 plants is thus calculated by using the equation proposed by Farquhar (1983) and Henderson et al. (1992):

$$\Delta = a + b_4 + \Phi(b_3 - s) - a)P_i/P_a,$$

where a is the fractionation during diffusion of CO_2 in air (4.4‰), b_4 is the combined fractionation during dissolution of CO_2 to bicarbonate and subsequent fixation by PEPcase (−5.2‰), b_3 is the fractionation associated with Rubisco carboxylation (29‰) and s is the fractionation during leakage of CO_2 from the bundle sheath to the mesophyll (∼1.8‰) (Ranjith et al., 1995).

In this study, no significant differences in foliar $\delta^{13}C$ values among genotypes were imminent at all stages. However, within genotypes there were consistent, significant difference in leaf $\delta^{13}C$ values between high and low N treatments. $\delta^{13}C$ became consistently less negative within plants that were N stressed than those which were not. The trend towards a less negative foliar $\delta^{13}C$ value within genotypes appeared to be progressive with the developmental stage of the plant. This trend towards a less negative leaf $\delta^{13}C$ value is also consistent from our earlier studies on maize (Madhavan et al., 1996). However, in contrast to our results, Ranjith et al. (1995) have reported an increased Δ (more negative $\delta^{13}C$) with N stress in sugarcane leaves. They have suggested that this may be due to a decreased partitioning of N to Rubisco under N stress resulting in an increased Φ and reduced photosynthetic efficiency.

Further, Ranjith et al. (1995) have suggested that this increase in Δ (0.5–0.7 between N stress and non-stressed plants) in sugarcane leaves may be due to a reduced Rubisco:PEPcase activity ratio, leading to an increased Φ in these plants. Results from our study with sorghum (another NADP-ME C_4 sub-type as sugarcane), also showed a reduced Rubisco:PEPcase activity ratio (0.1–0.8 between N stressed and non-stressed plants) with N stress but with a less negative $\delta^{13}C$ value (a difference of ∼0.8‰ between N stress and non-stressed) for these plants. In order for the $\delta^{13}C$ to become more negative, there should be a greater extent of CO_2 leakage from bundle sheath cells. This is because under steady-state photosynthesis, bundle sheath CO_2 will become enriched in ^{13}C due to the discrimination by Rubisco against ^{13}C. The leakage of this ^{13}C enriched CO_2 will result in more negative $\delta^{13}C$ values (Sasakawa et al., 1989). Sasakawa et al. (1989) also suggested that variations exist in the level of Rubisco, and thus its activity need not necessarily change the variation in foliar $\delta^{13}C$ values in C_4 species.

However, $\delta^{13}C$ values of kernels became more negative (∼0.4‰) with N stress, in all four sorghum genotypes selected for this study (data not presented). This observation is consistent with the observations of Wong and Osmond (1991) who reported that N stress resulted in a more negative $\delta^{13}C$ value for root dry matter in *Echinochloa frumentacea*, another C_4

plant. Our results suggest that there may be fractionations associated with carbon partitioning during mobilization.

There was a consistent increase in $\delta^{13}C$ as plants were subjected to N stress. This result is interesting in that the response to N stress is opposite to that of plants to water or salinity stress. Bowman et al. (1989) found that $\delta^{13}C$ values were decreased in *Zea mays* (L.) and *Andropogon glomeratus* when water or salinity stress was imposed. This was also observed in other C_4 species (Meinzer et al., 1994). It is generally interpreted that changes in bundle sheath leakage account for differences in $\delta^{13}C$ values as plants undergo drought or salt stress and the stomates close. In N stress when water is non-limiting, the stomates remain open and no or little restriction to atmospheric CO_2 occurs. The increase in $\delta^{13}C$ with N stress then must reside in the differential fixation of CO_2 by PEPcase and Rubisco which reside in different physical proximities within the bundle sheath/mesophyll area.

Although carbon isotope discrimination does not appear to account for the genetic differences observed between the two groups of sorghums for NUE1, it may provide a means to distinguish characteristics of response to N stress which appear opposite of the response to water or salinity.

Conclusions

Greenhouse studies of sorghum genotypes varying in NUE_I indicated a strong correlation between N stress and both physiological and biochemical determinants of photosynthetic efficiency. The observation that the N use efficient China line, China 17, retained significantly more active PEPcase than other sorghum lines, when grown under N stress, suggests that retention of an active PEP case and enzymes associated with phosphoenolpyruvate synthesis may play a key role in maintaining their photosynthetic efficiency. However, a re-examination of this suggestion is necessary because San Chi San had a moderately high CER and ACi at low N supply but not differ from U.S. lines in its PEPcase activity. Carbon isotope ratio measurements of sorghum genotypes grown under N stress indicated that their foliar $\delta^{13}C$ values became less negative when compared to plants that were not N stressed. This observation is significant because foliar $\delta^{13}C$ values of C_4 plants tend to become more negative when grown under water stress, suggesting that this methodology may become a tool of choice to differentiate N stress from water stress in sorghum genotypes.

Acknowledgements

The authors are grateful to Dr R. J. Spreitzer, Department of Biochemistry, University of Nebraska, for assistance in Rubisco assays and to Patricia Echtenkamp for the enzyme and immunoblot assays. Research was funded in part by USAID Grant No. LAG-G-00-96-90009-00 to the International Sorghum and Millet Cooperative Research Support Program (INTSORMIL).

References

Alagarswamy G, Gardner J C, Maranville J W and Clark R B 1988 Measurement of instantaneous nitrogen use efficiency among pearl millet genotypes. Crop Sci. 28, 681–685.

Bowman W D, Hubick K T, von Caemmarer S and Farquhar G D 1989 Short-term changes in leaf carbon isotope discrimination in salt and water stressed C_4 grasses. Plant Physiol. 90, 162–166.

Bradford M M 1976 A rapid and sensitive method for the quantification of microgram quantities of protein utilizing the principle of protein-dye binding. Anal. Biochem. 72, 248–254.

Cassman K G, Peng S, Olk D C, Ladha J K, Reichardt W, Doberman A and Singh U 1998 Opportunities for increasing nitrogen-use efficiency from improved resource management in irrigated rice systems. Field Crops Res. 56, 7–39.

Eghball B and Maranville J W 1993 Root development and nitrogen influx of corn genotypes grown under combined water and nitrogen stresses. Agron. J. 85, 147–152.

Farquhar G D 1983 On the nature of carbon isotope discrimination in C_4 species. Aust. J. Plant Physiol. 10, 205–226.

Field C 1983 Allocating leaf nitrogen for the maximization of carbon grain: Leaf age as a control on the allocation program. Oecologia 45, 341–347.

Henderson S A, von Caemmerer S and Farquhar G D 1992 Short-term measurements of carbon isotope discrimination in several C_4 species. Aust. J. Plant Physiol. 15, 263–285.

Henderson, S A, von Caemmerer S, Farquhar G D, Wade L and Hammer G 1998 Correlation between carbon isotope discrimination and transpiration efficiency in lines of the C_4 species *Sorghum bicolor* in the greenhouse and the field. Aust. J. Plant Physiol. 25, 111–123.

Hoagland D R and Arnon D I 1950 The water-culture methods used in the study of plant nutrition. Tech. Comm. No. 22 (2nd ed). Commonwealth Agricultural Bureaux, Farnham Royal. Bucks, England.

Iglesias A A and Andreo C S 1989 Purification of NADP-malic enzyme and phospho*enol*pyruvate carboxylase from sugar cane leaves. Plant Cell Physiol. 30, 399–405.

Laemmli U 1970 Cleavage of structural proteins during the assembly of the head of bacteriophage T4. Nature 227, 680–685.

Lawlor D W, Boyle F A, Kendall A C and Keys A J 1987 Nitrate nutrition and temperature effects on wheat; enzyme composition, nitrate and total amino acid content of leaves. J. Exp. Bot. 35, 378–392.

Longstreth D J and Nobel P S 1980 Nutrient influences on leaf photosynthesis. Plant Physiology 65, 541–543.

Madhavan S, Treichel I and O'Leary M H 1991 Effects of relative humidity on carbon isotope fractionation in plants. Botanica Acta 104, 292–304.

Madhavan S, Swanson S T, Blackmer T M, Pinkerton T S and O'Leary M H 1996 Relationship between nitrogen nutrition and photosynthetic carbon isotope fractionation in corn. Plant Physiol. 11(s), 366.

Maranville J W, Clark R B and Ross W W 1980 Nitrogen efficiency in grain sorghum. J. Plant Nutr. 2, 577–589.

Maranville J W and Settimi J R 1989 Remobilization of nitrogen from carboxylating enzymes and its implication for N-use efficiency in maize. Agronomy Abstracts, p. 117. American Society of Agronomy, Madison, WI.

Markwell J, Osterman J C and Mitchell J L 1995 Calibration of the Minolta SPAD-502 leaf chlorophyll meter. Photosynth. Res. 46, 467–472.

Meinzer F C, Zvi P and Saliendia N Z 1994 Carbon isotope discrimination, gas exchange, and growth of sugar cane cultivars under salinity. Plant Physiol. 104, 521–526.

Moll R H, Kampreth E J and Jackson W A 1982 Analysis and interpretation of factors which contribute to efficiency of nitrogen utilization. Agron. J. 74, 562–564.

Onken A B, Lavelle M J and Peterson G C 1985 Improving nutrient use efficiency in sorghums. *In* Proceedings of the 40th Annual Corn and Sorghum Research Conference. Ed. D Wilkinson. American Trade Association. Washington, D.C.

Ranjith S A, Meinzer F C, Perry M H and Thom M 1995 Partitioning of carboxylase activity in nitrogen stressed sugarcane and its relationship to bundle sheath leakiness to CO_2 photosynthesis and carbon isotope discrimination. Aust. J. Plant Physiol. 22, 903–911.

Sage R F, Pearcy R W and Seeman J R 1987 The nitrogen use efficiency of C_3 and C_4 plants. Plant Physiol. 85, 355–359.

Sasakawa H, Sugiharto B, O'Leary M H and Sugiyama T 1989 $\delta^{13}C$ values in maize leaf correlate with phosphoenolpyruvate carboxylase levels. Plant Physiol. 90, 582–585.

Settimi J R and Maranville J W 1998 Carbon dioxide assimilation efficiency of maize leaves under nitrogen stress at different stages of plant development. Commun. Soil Sci. Plant Anal. 29, 777–792.

Siddiqi M Y and Glass A D M 1981 Utilization index: a modified approach to the estimation and comparison of nutrient utilization efficiency in plants. J. Plant Nutr. 4, 289–302.

Stiborova M 1988 Phospho*enol*pyruvate carboxylase: the key enzyme of C_4-photosynthesis. Photosynthetica 22, 240–263.

Sugiharto B, Miyata K, Nakamoto H, Sasakawa H and Sugiyama T 1990 Regulation of expression of carbon-assimilating enzymes by nitrogen in maize leaf. Plant Physiol. 92, 963–969.

Sugiyama T, Mizuno M and Hayashi M 1984 Partitioning of nitrogen among ribulose-1,5-bisphosphate carboxylase/oxygenase, phospho*enol*pyruvate carboxylase, and pyruvate orthophosphate dikinase as related to biomass productivity in maize seedlings. Plant Physiol. 75, 665–669.

Towbin H, Staehelin T and Gordon J 1979 Electrophoretic transfer of proteins from polyacrylamide gels to nitrocellulose sheets: procedures and some applications. Proc. Natl. Acad. Sci. USA 76, 4350–4354.

van Caemmerer S and Farquhar G D 1981 Some relationships between the biochemistry of photosynthesis and the gas exchange of leaves. Planta 153, 376–387.

Wong, S C and Osmond C B 1991 Elevated atmospheric partial pressure of CO_2 and plant growth. III. Interaction between *Triticum aestivum* (C_3) and *Echinochloa frumentacea* (C_4) during growth in a mixed culture under different CO_2, N nutrition and irradiance treatments with emphasis on below ground responses estimated using the $\delta^{13}C$ value of root biomass. Aust. J. Plant Physiol. 18, 137–152.

Youngquist J B, Bramel-Cox P and Maranville J W 1992 Evaluation of alternative screening criteria for selecting nitrogen-use efficient genotypes in sorghum. Crop Sci. 32, 1310–1313.

Zar J H 1984 Biostatistical Analysis. Prentice Hall, Englewood Cliffs, NJ.

J.J. Adu-Gyamfi (Ed.), *Food security in nutrient-stressed environments: exploiting plants' genetic capabilities*: 91–101, 2002.

Photosynthesized carbon translocation and distribution of crops adapted to low-nutrient environments

K. Fujita[1], M. I. Chaudhary[1], A. Fujita[1], Y. Kai[1], M. Takayanagi[1], H. El Shemy[1] & J. J. Adu-Gyamfi[2]
[1]*Faculty of Applied Biological Science, Hiroshima University, 1-4-4 Kagamiyama, Higashi-Hiroshima 739-8528, Japan*
[2]*International Crops Research Institute for the Semi-Arid Tropics (ICRISAT), PMB 3491, Kano, Nigeria*

Key words: grain legume, leaf expansion, phosphorus, ^{13}C tracer

Abstract

Reduction in plant leaf area and growth as a consequence of phosphorus (P) limitation have been attributed to the direct effects of lack of P on the expansion of individual leaves, and to reduced production of the assimilates required for growth. A series of field and pot experiments was conducted to assess the effect of low-P supply on photosynthesis, photoassimilate partitioning, leaf area expansion, and on the ethanol-soluble (cytosol) and cell wall fractions (pectin hemicellulose and cellulose). The crops used were pigeonpea (*Cajanus cajan*), and mashbean (*Vigna aconitifolia*). Phosphorus levels were 5 and 200 kg P ha^{-1} for the field experiments and 5 and 250 kg P ha^{-1} for the pot experiments. Leaves were exposed to $^{13}CO_2/^{12}CO_2$ (99.5 at.% ^{13}C) for 2 h in natural light conditions using a steady state $^{13}CO_2/^{12}CO_2$ feeding system. The carbon (C) content and the ^{13}C abundance in ethanol and water-soluble cellulose and hemicellulose fractions were determined using a mass spectrometer (MAT 252 Finnigan, San Jose, California, USA). Biomass and leaf area expansion were severely decreased by low P, however, there was genotypic variation among the pigeonpea cultivars used. Low-P treatment resulted in a decrease in P concentration in whole leaves and cell walls but an increase in the cytosol fraction. P-limitation resulted in the partitioning of more ^{13}C assimilates to roots and less to leaves and stems. For pigeonpea, a hybrid (ICPH 8) that is more adapted to low-P conditions than the non-hybrids (ICPL 87 and UPAS 120) recorded a relatively higher amount of ^{13}C in leaves and stems than the non-hybrids. These results suggest that at low-P availability there is increased partitioning of photoassimilates to roots and decreased partitioning to growing leaves. The synthesis into hemicellulose and cellulose in leaves is suppressed, causing an inhibition of leaf area expansion. Leaf area expansion is an important trait affecting crop productivity. Variability among crop plants in the translocation of photoassimilates to leaves and the other growing parts at low-P availability provides scope for genetic manipulation of this trait to improve the adaptation of crops to P-stressed environments.

Introduction

Leaf area development is an important factor determining crop productivity. It is largely inhibited under P-deficiency conditions resulting in substantial yield losses. All the authors' previous experiments showed that low-P application resulted in the inhibition of leaf area development. The negative effect of sub-optimal P nutrition on leaf development becomes more severe in leguminous plants in which nodules also need sufficient P for nitrogen (N) fixation. Radin and Eidenbock, (1984) reported that P deficiency inhibited leaf expansion by decreasing the hydraulic conductance of the root system in young cotton (*Gossypium hirsutum* L.) plants. Israel and Rufty, (1988) showed that leaf development was inhibited due to reduced allocation of plant biomass to leaves and an increased allocation of biomass to roots in P-deficient plants. In addition to the reduced photosynthetic surface area, carbohydrate utilization efficiency is decreased in P-deficient plants.

Starch concentration is increased in P-stressed soybean (Israel and Rufty, 1988), clover, and lotus plants (Hart and Jessop, 1984).

The increase in starch concentration in low-P tissue is reflected in the limitation of such sink activities as leaf expansion in shoots (Radin and Eidenbock, 1984). Moreover, higher triose-P/inorganic P ratios in chloroplasts favour starch synthesis at the expense of sucrose synthesis due to the activation of ADP-glucose pyrophosphorylase (Stitt and Quick, 1989). Therefore, under P stress plant undergo a number of physiological and biochemical changes that are ultimately reflected in growth inhibition.

The direct effect of P nutrition on either cell wall extensibility or the threshold turgor for growth could also account for decreased leaf expansion (Radin and Eidenbock, 1984). Cytokinin contents are decreased in low-P plants (Horgan and Wareing, 1980) and the implication of cytokinin in the control of cell wall extensibility has been reported by Thomas et al. (1981). Puliga et al. (1996) suggested that soil-drying treatment caused production of ABA in plants and there was a distinct relationship between abscisic acid (ABA) accumulation and decrease in leaf growth. Since leaf development includes both the production and expansion of leaf cells and these processes need synthesis and expansion of cell walls.

Previous studies showed cell wall dry-matter weight was decreased in field-grown mashbean plants that received low applications of P (Chaudhary and Fujita, 1998a). Similarly cell wall P and N contents were lower in low-P than control plants, indicating that synthesis of cytoplasmic and cell wall components might be negatively affected.

Leaf growth involves the production and expansion of leaf cells, with the latter suggested to be of overriding importance (Ferris and Taylor, 1994; Gardner et al., 1995). Expansive growth is co-regulated by the rate at which cell walls loosen and extend, and the rate at which water and solutes are accumulated by the growing cell (Lockhart, 1965). The cell wall consists of two continuous interpenetrating systems, one of which is the cellulose fibrils and the other the continuous system of microcapillary spaces. These spaces may be filled with hemicellulose and other organic substances (Bailey, 1957). The properties of the cell wall are determined by those of cellulose and other substances which may add some characteristics or slightly alter those of the cellulose. Indeed, a putative wall-loosening enzyme, xyloglucan endotransglycosylase (XET) (Fry et al., 1992) has been shown to be correlated with increased growth in a number of plant cells and organs (Passioura and Fry, 1992) and may contribute to the cell-wall loosening that occurrs in rapidly expanding leaves. The increase in leaf growth was linked to an increase in the cell-wall loosening enzyme, XET (Ranasinghe and Taylor, 1996). This enzyme is thought to cut and rejoin xyloglucan molecules spanning adjacent microfibrils, some of which are load-bearing and form 'tethers' (Smith and Fry, 1991).

Rapid growth in primary leaves of *Phaseolus vulgaris* exposed to elevated CO_2 occurred as a result of enhanced cell expansion as cell-wall loosening and extensibility were promoted (Ranasinghe and Taylor, 1996). The effect on cell walls has been reported to be an indirect one, following enhanced photosynthesis and an increase in the concentration of soluble carbohydrates available for both metabolic reactions and to provide the substrates for further plant growth.

It is known that cell wall relaxation and turgor pressure are prerequisites for cell elongation (Cosgrove, 1987). For instance, when the amount of hemicellulose in *Graminiae* is reduced by degradation with 1,3:1,4-β-glucanase, the cells elongate by turgor pressure (Sakurai, 1991). This suggests that cell wall degradation plays a vital role in cell elongation. The objectives were to assess the effect of low-P supply on photosynthesis, photoassimilate partitioning, leaf area expansion, and on the ethanol (EtOH)-soluble (cytosol), and cell wall fractions (pectin, hemicellulose, and cellulose) of pigeonpea and mashbean.

Materials and methods

Experiment 1

Mashbean (*Vigna aconitifolia*) was grown in P-deficient (Granite Regosol; see Ogata et al., 1988 for some of the chemical characteristics of the soil) soil at P application levels of 5 and 200 kg P ha^{-1} in the form of superphosphate. A basal dose of 14 kg ha^{-1} N as $(NH_4)_2SO_4$ and 150 kg ha^{-1} K as K_2SO_4 was applied. Soil pH of the experimental field was adjusted to approximately 6.0 by the application of dolomitic lime at the rate of 1600 kg ha^{-1}.

Plant growth and P content

Plant shoots were cut at ground level and nodulated roots were dug out carefully. Data on leaf number,

Table 1. Effect of low phosphorus (P) on leaf area and biomass production of three pigeonpea genotypes at 78 DAS

Plant part	ICPH 8		ICPL 87		UPAS 120		LSD
	Control	Low P	Control	Low P	Control	Low P	(0.05)
Leaf area	861	46.3	710	203	669	156	134
(cm^2 $plant^{-1}$)	(100)	(54)	(100)	(29)	(100)	(23)	
Leaf number	17	12	15	6	11	9	1.6
(number $plant^{-1}$)							
Average leaf area	48.7	22.1	28.7	12.7	26.5	16.2	10.79
(cm^2 $leaf^{-1}$)							
Leaf thickness	2.56	3.13	3.29	4.23	3.67	3.91	0.96
(mg dry weight cm^{-2})							
Dry weight (g $plant^{-1}$)							
Shoots	4.52	2.50	4.26	1.44	4.62	0.99	2.20
Roots	0.98	1.04	1.40	0.74	1.38	0.51	ns[a]
Whole plant	5.55	3.54	5.66	2.18	6.00	1.50	2.89
P amount	23.99	6.84	17.26	2.75	13.67	2.22	2.57
(mg $plant^{-1}$)	(100)	(29)	(100)	(16)	(100)	(16)	
P uptake activity	24.3	6.6	12.3	3.7	9.9	4.4	3.46
(mg root DW^{-1})	(100)	(27)	(100)	(30)	(100)	(44)	

[a] ns=non significant.

total leaf area, shoot, and roots+nodule dry weight, and N and P contents of various plant parts were collected.

Leaf area expansion and P content

Leaves immediately before initiation of laminae unfolding were removed and fixed by dipping into liquid N. The leaf area was measured by an auto leaf area meter (Model AAM-5, Hayashi Denco Co. Ltd., Osaka, Japan) and the leaves oven-dried at 65 °C for more than 3 days. The dried leaves were ground and digested in a mixture of concentrated H_2SO_4 and nitric acid and their P content was determined by the molybdenum blue method as described by Ogata et al. (1988).

Photosynthetic rate

The rate of photosynthesis (Po) of intact plants was determined with a portable Infrared Gas Analyzer (Model LI-6200, LI-COR, Co. Ltd., Lincoln, NE). The photosynthetic active radiation (PAR) was more than 1700 μmol m^{-2} s^{-1}. Photosynthetic rates of terminal leaflets of second and third trifoliate leaves from the uppermost positions in all the cultivars were determined between 10.00 and 12.00 h.

Experiment 2

Seeds of three genotypes of pigeonpea, ICPH 8, ICPL 87, and UPAS 120 were grown in pots of soil consisting mainly of Granite Regosols, that received basal dressings of 150 kg N ha^{-1} and 139 kg K ha^{-1}. Phosphorus was applied at 250 kg P ha^{-1} (control P) and at 5 kg P ha^{-1} (low-P). The soil pH was adjusted to about 6.5 with dolomitic lime prior to fertilizer application.

Three plants from each replication were harvested at 30 days after sowing (DAS) during the vegetative growth stage. The harvested plants were separated into leaves and stem at upper (about five leaves from the uppermost leaf) and lower parts, and roots. They were dried at 70 °C in a forced-air oven for 3 days, and their dry weights taken.

Fractionation and measurement of sugars in each fraction

Cell walls prepared from the upper leaves were fractionated according to the method described by Sakurai

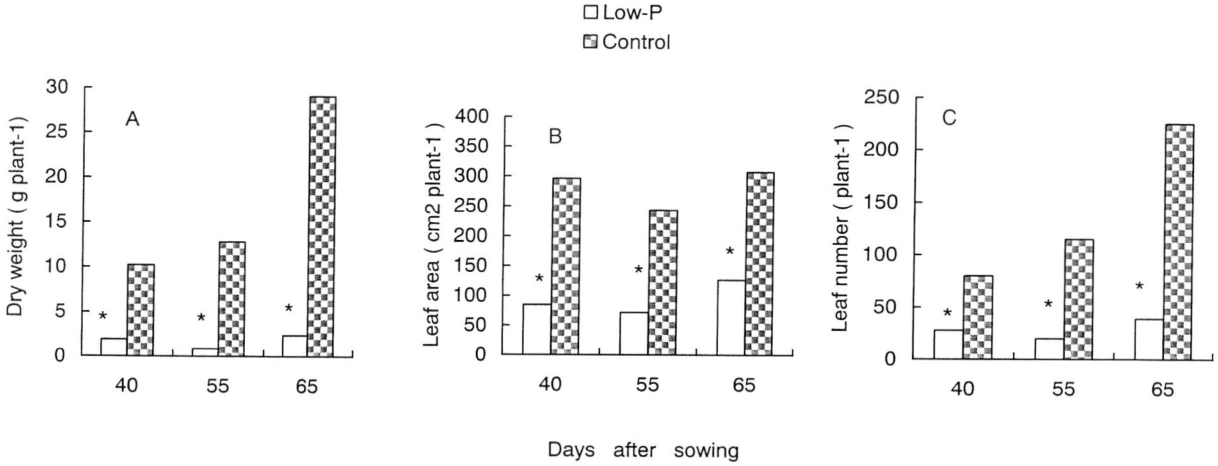

Figure 1. Effect of phosphorus (P) application levels on: (a) dry-matter weight, (b) leaf area and (c) leaf number in field-grown mashbean plants at 40, 55 and 65 days after sowing (DAS). *significant at P=0.05

et al. (1987). Analyses of cell wall components were conducted. Fresh leaves were fixed in liquid N and preserved at −20 °C until analysis. The EtOH extract was designated the EtOH-soluble fraction and consisted of neutral sugars including glucose, fructose, and sucrose in the cytosol. The leaves were rehydrated and homogenized in liquid N. The homogenate was centrifuged at 1000×*g* for 10 min. The resultant supernatant was designated the water-soluble (WS) fraction. This fraction also consisted of neutral sugars in the cytosol, but was collected after denaturation of the proteins with trichloro acetic acid (TCA) (5%, w/v). The wall residue was washed with deionized water, acetone, and a methanol:chloroform mixture (1:1, v/v) and air dried. The dried cell wall material was treated with 2 units ml^{-1} porcine pancreatic α-amylase (type 1-A, Sigma) in a Na-acetate buffer (pH 6.5, 50 mM) for 2 h to remove starch. Proteins were removed by treating the residue with 200 mg ml^{-1} pronase (Actinasee Kaken Kagaku Co., Tokyo, Japan) in Na-phosphate buffer (pH 6.5, 50 mM) containing 5% ELOH for 18 h. The pectic fraction was extracted from the wall residue (cell wall), by treating with 50 mM EDTA at 95 °C for 15 min. The hemicellulose fraction was extracted with a solution of 17.5% NaOH containing 0.02% $NaBH_4$ for 18 h. The alkaline-insoluble fraction was designated cellulose. The hemicellulose fraction was neutralized with glacial acetic acid.

$^{13}CO_2$ feeding

A $^{13}CO_2$ feeding experiment was conducted to determine changes in the labeled assimilates in the various plant parts and in the cell wall composition in the upper developing leaves for a short period (72 h after the introduction of $^{13}CO_2$). The experiment was carried out as described by Sawada et al. (1995). A leaf was exposed to $^{13}CO_2$ feeding 78 DAS. The leaf was enclosed in a transparent plastic bag to which a gas with air balance containing 36 Pa $^{13}CO_2$ (99.5 at.% ^{13}C) was flowed. The leaf was then allowed to assimilate the $^{13}CO_2$ under natural light conditions for 2 h. During $^{13}CO_2$ assimilation, the PAR was above 1700 μmol m^{-2} s^{-1}. Five plants were harvested at 72 h after the end of $^{13}CO_2$ assimilation. Harvested plants were immediately separated into the $^{13}CO_2$-fed leaf, leaves and stem above or below the fed leaf, leaves and stem below the fed leaf, and roots. Upper leaves were dipped into liquid N and preserved at −20 °C until analysis. Other plant parts were separately dried at 70 °C in a forced-air oven for 3 days. The plant parts were weighed, and ground to fine powder with a vibrating sample mill (Model TI-100 Heiko Co. Ltd., Iwaki, Fukushima, Japan).

^{13}C analysis

The ^{13}C abundance in the ground plant materials, EtOH, and water-soluble fraction, cellulose and hemicellulose fractions was determined with a mass spectrometer (Delta Plus, Finnigan, San Jose, California, USA). The ^{13}C atom (%) excess in the plant sample was calculated as the difference in ^{13}C atom (%) between the sample and standard pure chemical, glycine. The amount of labeled C in the plant sample was calculated as described by Sawada et al. (1995).

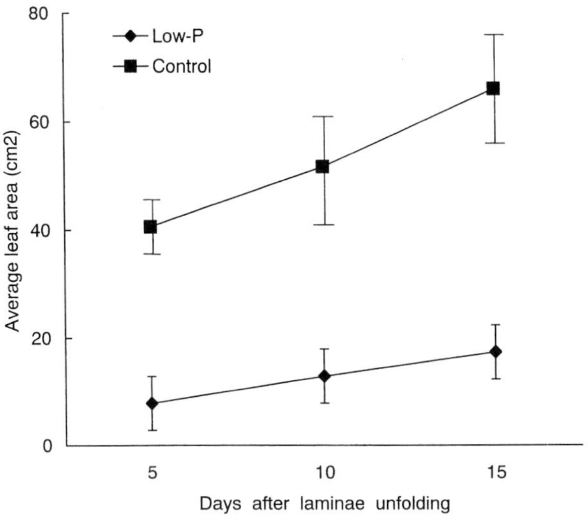

Figure 2. Effect of phosphorus (P) application on average leaf area of mashbean plants 5–15 days after laminae unfolding.

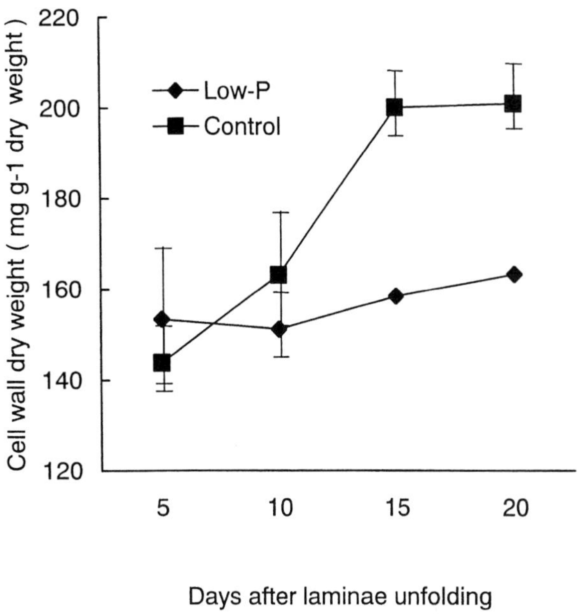

Figure 3. Effect of phosphorus (P) application level on cell walls dry weight in leaves of field-grown mashbean plants. Vertical bars indicate standard deviation (not visible when smaller than symbol).

Statistical analysis
All the experiments were conducted with three to four replications. The effect of treatments was assessed according to the expected mean squares given by McIntosh (1983). For mean separation, the treatment sum of squares was partitioned by the method of orthogonal contrasts. The coefficients of variation for all response variables were below 10%.

Results

Experiment 1

Biomass production and leaf area
Throughout the experimental period, whole plant dry weight was extremely reduced in low-P mashbean plants compared to the control plants and at the final sampling it was only 8% of the control (Figure 1 and Table 1). Total leaf area was adversely affected in low-P plants as found by their 26% reduction at 40 DAS, 30% at 55 DAS, and 41% at 65 DAS (Figure 1).

The negative impact of P deficiency was reflected in a smaller average leaf size in low-P leaves than those in control plants as found in unfolding laminae (Figure 2). Cell wall content was also less in P-deficient plants (Figure 3).

Phosphorus content in various fraction in leaves
The P content of the cell walls in developing leaves of low-P plants was lower than in control plants (Figure 4). The P content in the methanol-soluble extract of low-P developing leaves was lower than those of control plants indicating more input of P into organic compounds in control plants (Figure 5). In parallel to a decreased P content in the developing low-P leaves, their cell wall P content decreased (data not shown).

Experiment 2

The photosynthetic rate (Po) in all cultivars decreased at low P, mainly due to stomatal conductance and leaf internal CO_2 concentration (Table 2).

Sugar concentration in the upper leaves increased at low-P in all the cultivars.

^{13}C partitioning among various plant parts
At low-P, ^{13}C atom (%) excess in roots increased, whereas that in the upper leaves and stem decreased. This tendency was more pronounced in ICPL 87 and UPAS 120 than in ICPH 8 (Table 3).

No significant difference in ^{13}C export rate was observed in spite of P level and cultivars (Figure 6). The ^{13}C partitioning to developing upper leaves was higher than that to roots in control plants, while at low-P, in the upper leaves it tended to decrease, but increased in the roots (Figure 7). The degree of changes induced by the P-deficiency treatment was less in ICPH 8, while it was considerable in ICPL 87 and UPAS 120, particu-

Table 2. Effect of low phosphorus (P) on photosynthesis, stomatal conductance, leaf internal CO_2 concentration, and transpiration of pigeonpea genotypes

	ICPH 8		ICPL 87		UPAS 120	
	Control	Low P	Control	Low P	Control	Low P
Photosynthesis rate (μ mole CO_2 m^{-2} s^{-1})	13.90	8.84[a]	11.69	6.57 ns[b]	10.87	6.07[a]
Stomatal conductance (μ mole CO_2 m^{-2} s^{-1})	0.155	0.135 ns	0.214	0.088 ns	0.212	0.057[a]
Leaf internal CO_2 conc. (μ IL^{-2})	183	221 ns	247	190 ns	254	164[a]
Transpiration (μ mol)	2.51	2.35 ns	3.44	1.73	3.72	1.17[a]

[a] Significant at P=0.05.
[b] ns=not significant.

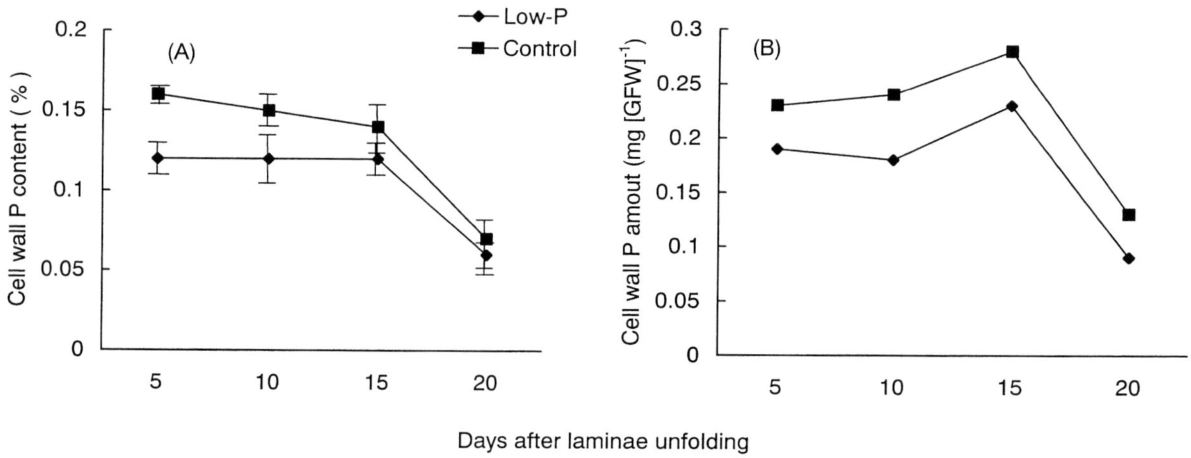

Figure 4. Effect of phosphorus (P) application level on (A) cell wall P content, and (B) cell wall P amount in leaves of field-grown mashbean plants. Vertical bars indicate standard deviation (not visible when smaller than symbol).

larly the changes in ^{13}C partitioning to roots was most markedly obvious.

^{13}C partitioning into various fractions in the developing upper leaves

Relative ^{13}C atom (%) excess tended to be higher in the cytosol and such cell wall fractions as hemicellulose and cellulose (Table 4). These changes decreased in ICPH 8, however the reverse was true for other cultivers.

^{13}C partitioned more to the cell wall components than to cytosol in spite of P nutrient status and cultivars in developing upper leaves (data not shown).

Discussion

Effect of low P on biomass production and leaf area expansion

Throughout the experimental period, whole plant dry weight was extremely reduced in low-P plants compared to the control. Total leaf area was adversely affected in low-P plants (Table 1). Leaf numbers in low-P plants were 35% of the control at 40 DAS and 17% at 55 and 65 DAS suggesting that leaf initiation is sensitive to low-P conditions (Lynch et al., 1991). The negative impact of P deficiency was reflected in a smaller average leaf size in low-P leaves than in the control (Table 1). Furthermore, cell wall P concentration was maintained stable at low-P, while it increased with time in the control plants (Figure 4). These results indicate that both leaf initiation and leaf expansion are

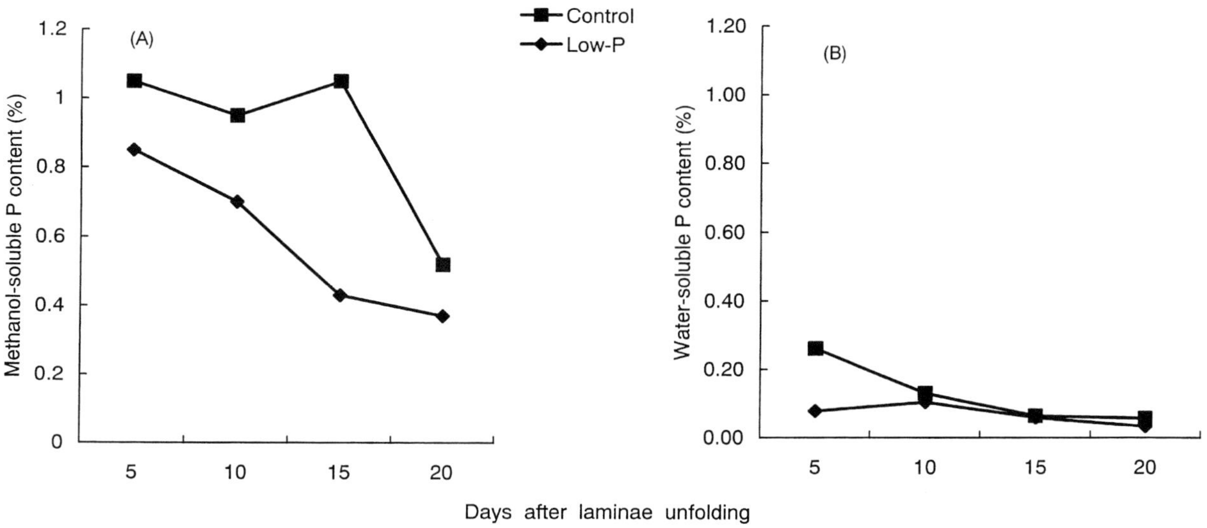

Figure 5. Effect of phosphorus (P) application levels on (A) methanol-soluble P, and (B) water-soluble P contents of leaves of field-grown mashbean plants.

adversely affected by P-deficiency, particularly due to inhibition of cell wall synthesis leading to a reduction in whole plant leaf area. Therefore, severe reduction in the growth of pigeonpea genotypes under low-P application suggests that leaf area is the single most important factor determining dry-matter production.

A proportional decrease in average leaf area and P content was found in the developing leaves under low-P conditions, suggesting that P is involved in the expansion of leaves. The P concentration in the cell wall was decreased in leaves of P-deficient plants. These data suggest that reduced P transport from lower matured leaves to upper developing leaves in P-deficient plants might have played an inhibiting role in the initiation and expansion of new leaves, resulting in fewer leaf numbers and smaller leaf size.

The authors attempted to identify the factors controlling leaf expansion in relation to synthesis and loosening of cell walls, and P nutrition in plant parts. A lower dry weight of the cell walls in the developing leaves in low-P plants than in control plants suggests that suppression of synthesis of cell wall material is responsible for impairment of leaf expansion in the low-P plants (Figure 3). Sustained cell enlargement in plants requires irreversible expansion of the cell wall, uptake of water, and the synthesis of new wall components. Changes in the structural properties of the cell wall, resulting in the modification of wall extensibility underlie growth responses. Leaf growth involves both the production and expansion of leaf cells, and in leaves of dicotyledons cell division and expansion normally occur simultaneously during much of leaf expansion. Leaf cell expansion may be studied by analyzing the biophysical components of growth in a model proposed by Lockhart (1965) for the elongation of single cells. The model shows that expansive growth is co-regulated by the rate at which cell walls loosen and extend, and by the rate at which water and solutes are accumulated by the growing cell. Taylor and Ranasinghe (1996) reported that leaf growth in primary leaves of beans (*Phaseolus vulgaris*) exposed to elevated CO_2 occurs as a result of increased cell expansion as cell wall loosening and extensibility are promoted. The results of the present study indicate that the implication of cell walls in leaf growth is an indirect one and input of substrates is a prerequisite for leaf expansion to take place.

It has been reported that cell wall synthesis is necessary in addition to changes in the mechanical properties of the cell wall for long-term sustained growth (Brummer and Hall, 1987). On the basis of their experimental results Hoson and Masuda (1991) reported that cell wall synthesis is essential for the elongation of intact rice coleoptiles. Galactose inhibits the synthesis of the cell wall polysaccharides by suppressing the formation of UDP-glucose, a key precursor for the synthesis of polysaccharides in oat coleoptiles (Inouhe et al., 1987). A lower P content in the cell walls in the developing leaves of low-P plants than in control plants was indicated. The P contents in the methanol-soluble extract of low-P developing leaves was lower than that of control plants indicating more input of

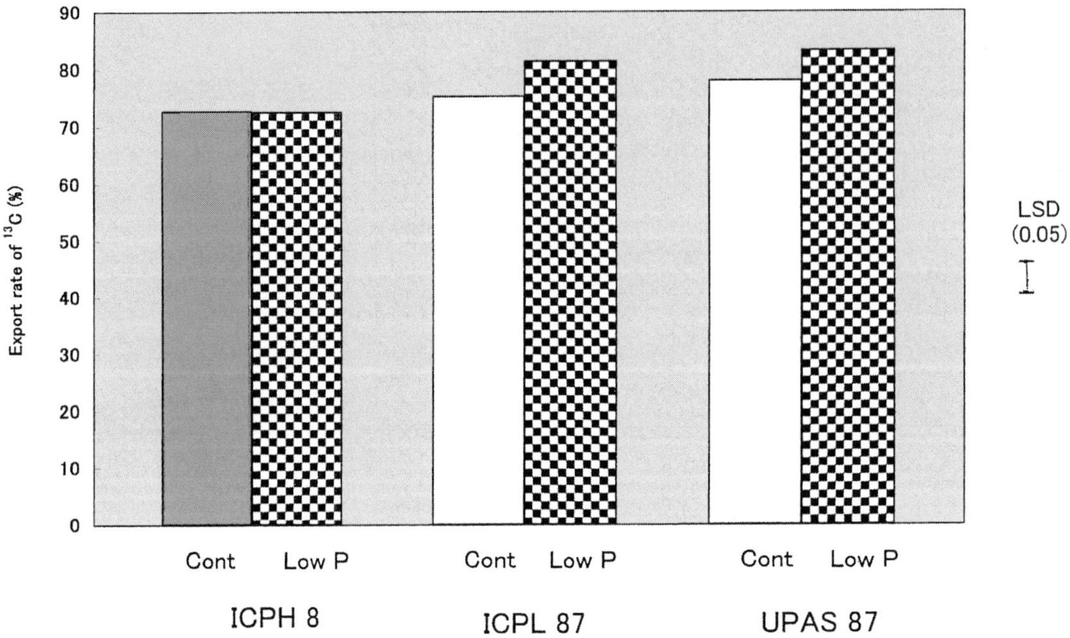

Figure 6. Export rate of ^{13}C in three pigeonpea cultivars grown at low phosphorus (P) concentration.

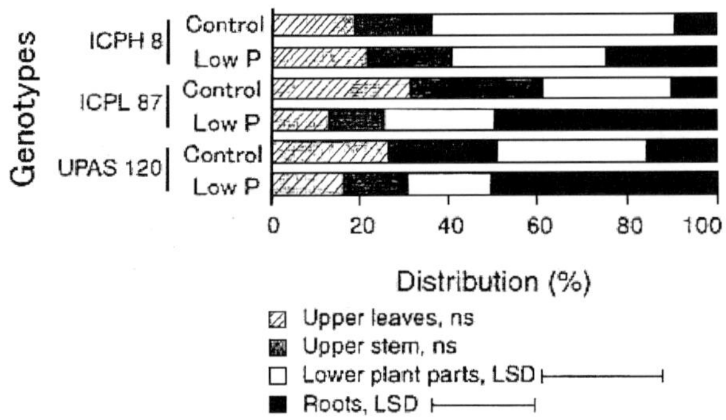

Figure 7. Distribution of ^{13}C in various plant parts of three pigeonpea cultivars grown at low phosphorus (P) concentration.

P into organic compounds in control plants (Figure 5). However, in matured leaves a sharp reduction in the P content was indicated both in low-P and control plants. A leaf during its ontogeny functions first as a metabolic sink and later, before it has expanded to its final size, as a metabolic source. Therefore, a similar phenomenon could be responsible for transport of P from a matured leaf to newly developing leaves. As the leaf developed, a declining trend in P content was found. Though P content was higher in developing leaves of control plants than those in low-P plants, it sharply decreased in both when they attained maturity. These results indicate that having reached maturity, leaves release most of their P content for the newly developing tissues. Bieleski (1968a) showed that even under P stress some amount of Pi is required for continued metabolic activity of the cell. In another experiment (1968b) he demonstrated that P-deficient *Spirodela* plants lost their vacuolar Pi pool but retained their metabolic Pi.

A higher water-soluble content of P was found in the developing leaves of low-P than control plants showing the incorporation of less P into the organic fraction of low-P leaves (Figure 5). These results suggest that low-P application inhibited leaf expansion through reduction in the synthesis of cell wall components and, perhaps lesser incorporation of P into organic compounds. In parallel to a decreased-P content in the developing low-P leaves, their cell wall P content decreased (Figure 4). Under P deficiency, there can be a considerable increase in phosphatase activity in the apoplast (Bieleski and Johnson, 1972) and it results in the release of Pi, making it available for growth (Knyple, 1978).

The results suggest under suboptimal P supply conditions, leaf expansion is inhibited by reduction in the synthesis of cell walls. The reduced P supply from relatively matured tissues to newly initiating leaves might have been responsible for more growth reduction in P-deficient than in control plants.

Table 3. Effect of low phosphorus (P) on ^{13}C atom (%) excess[a] in various plant parts of three pigeonpea genotypes 3 days after feeding with $^{13}CO_2$

Plant part	ICPH 8		ICPL 87		UPAS 120		LSD
	Control	Low P	Control	Low P	Control	Low P	(0.05)
^{13}C fed leaf	0.252	0.591	0.591	0.361	0.351	0.288	ns[a]
Upper leaves	0.023	0.079	0.100	0.049	0.082	0.040	0.042
Upper stem	0.055	0.177	0.219	0.144	0.177	0.106	ns
Lower leaves and stem	0.032	0.067	0.065	0.074	0.023	0.042	ns
Root	0.033	0.083	0.057	0.120	0.037	0.091	ns

[a]ns=not significant.

Table 4. Effect of low P on relative ^{13}C atom (%) excess[a] in various fractions of upper leaves of that pigeonpea genotypes 3 days after introduction of $^{13}CO_2$

Leaf fraction	ICPH 8		ICPL 87		UPAS 120	
	Control	Low P	Control	Low P	Control	Low P
EtOH+water-soluble	1.61	0.89	0.99	1.39	0.91	1.38
Pectin	0.13	0.03	0.06	0.06	0.05	0.05
Hemicellulose	1.30	0.66	0.68	0.86	0.62	1.05
Cellulose	1.57	0.87	0.97	1.16	0.88	1.45
LSD (0.05)	0.374		0.56		0.681	

a. Relative ^{13}C atom % excess= $\frac{^{13}C \text{ atom (\%) excess in each fraction}}{^{13}C \text{ atom (\%) excess in leaves above } ^{13}CO_2\text{-fed leaf}}$

Effect of low P on partitioning of ^{13}C

The leaf area and biomass production decreased in the low-P treatment, and the reduction was less in ICPH 8 than in other genotypes, indicating that tolerance to low-P conditions differs among pigeonpea genotypes and is higher in ICPH 8 than in others in this experiment.

This genotypic difference to P deficiency may be related to tolerance of both source activity consisting of leaf area and Po and sink activity to low-P. It has been reported that under suboptimal P supply conditions, leaf area decreases due both to decrease in leaf number and leaf size in mashbean (Chaudhary and Fujita, 1998a,b), and in common bean (Lynch et al., 1991). On the other hand, there were slight but significant differences in response of Po to P deficiency among pigeonpea genotypes (Table 2) and mashbean (Chaudary and Fujita, 1998a,b). Po decreased to about 75% of that of the control plant at low-P in pigeonpea, which is almost the same figure as reported by Rao and Terry (1995) in sweet potato. Low-P treatment restricts the growth of the shoot, especially the area expansion of leaves, but has only minor effects on photosynthate production per unit area (Rao et al., 1990).

Palmer et al. (1996) reported that low N reduced leaf area expansion not through a reduced turgor pressure but probably through changes in cell wall properties. Radin and Boyer (1982) have found that N has an effect similar to that of P on roots. According to these results, it is unlikely that in low-P plants changes in turgor would be the sole factor inhibiting leaf expansion at low-P. On the other hand, cell enlargement involves substantial synthesis of new cellulose materials (Rodriguez et al., 1998). For example, a rapid expansion of husk leaf area was accompanied by synthesis of cell wall materials (Sato et al., 1997).

Plants respond to P stress by altering the partitioning of photosynthate among plant parts and by altering the allocation of carbon to starch, sucrose, and other structural carbohydrates such as cellulose and hemicelluloses (Rao et al., 1990; Rao and Terry 1995).

Our data showed that ^{13}C partitioning to developing leaves in the upper position on the main stem decreased at low-P, whereas it increased to roots, and these changes were less in tolerant pigeonpea cultivars like ICPH 8 (Figure 7). These results suggest that when sink activity in developing leaves for photosynthate is impaired by P deficiency, more photosynthate becomes available to roots.

It is assumed that at low-P, sink activity for photosynthate is more severely impaired than source activity. This is supported by the findings that at low-P leaf sugar concentration increased in all the cultivars. In other words, it infers that sink activity of developing young leaves plays a crucial role on determining the adaptation of crops to low-P. Based on these discussions, ICPH 8, the hybrid, is more tolerant to low-P than the other cultivars tested, mainly due to the tolerance of its developing leaves.

When competition for photosynthate is intense between the growing leaves and stem and roots under high population density and adverse climate conditions, it is possible that, at the expense of photoassimilate supply to the growing shoot parts, roots utilize it in P-deficient conditions.

References

Bailey I W 1957 Aggregation of microfibrils and their orientations in the secondary wall of coniferous tracheids. Am. J. Bot. 44, 415–418.

Bieleski R L 1968a Levels of phosphate esters in *Spirodela*. Plant Physiol. 43, 1297–1308.

Bieleski R L 1968b Effect of phosphorus deficiency on levels of phosphorus compounds in *Spirodela*. Plant Physiol. 43, 1309–1316.

Bieleski R L and Johnson P N 1972 The external location of phosphatase activity in phosphorus-deficient *Spirodela oligorrhiza*. Aust. J. Biol. Sci. 25, 707–720.

Brummer B and Hall J L 1987 Rapid cellular responses to auxin and the regulation of growth. Plant Cell Environ. 10, 523–543.

Chaudhary M I and Fujita K 1998a Comparison of phosphorus deficiency effects on the growth parameters of mashbean, mungbean and soybean. Soil Sci. Plant Nutr. 44, 19–20.

Chaudhary M I and Fujita K 1998b Responses of dinitrogen fixation and photosynthetic activity in mashbean (*Vigna aconitifolia*) cultivars under low P conditions. Soil Sci. Plant Nutr. 44, 31–41.

Cosgrove D J 1987 wall relaxation and the driving forces for call expansive growth. Plant Physiol. 84, 561–564.

Ferris R and Taylor G 1994 Elevated CO_2, water relations and biophysics of leaf extension in four chalk grassland herbs. New Phytol. 127, 297–307.

Fry S C, Smith R C, Renwick K F, Martin D J, Hodge S K and Matthew K J 1992 Xyloglucan endo transglycosylase, a new wall loosening enzyme activity from plants. Biochem. J. 282, 821–828.

Gardner S D L, Bosac C and Taylor G 1995 Leaf growth of hybrid poplar following exposure to elevated CO_2. New Phytol. 131, 81–90.

Hart A L and Jessop D 1984 Diurnal variation in starch and inorganic phosphate concentrations in leaves of white clover and lotus. N. Z. J. Agric. Res. 27, 1–3.

Horgan J N and Wareing P F 1980 Cytokinins and the growth response of seedlings of *Betula pendula* Roth. and *Acer pseudoplatanus* L. to nitrogen and phosphorus deficiency. J. Exp. Bot. 31, 525–532.

Hoson T and Masuda Y 1991 Role of polysaccharides synthesis in elongation growth and cell wall loosening in intact rice coleoptiles. Plant Cell Physiol. 32, 763–769.

Inouhe M, Yamamoto R and Masuda Y 1987 UDP-glucose level as a limiting factor for IAA-induced cell elongation in *Avena* coleoptile segments. Physiol. Plant. 69, 49–54.

Israel D W and Rufty Jr T W 1988 Influence of phosphorus nutrition on phosphorus and nitrogen utilization efficiencies and associated physiological responses in soybean. Crop Sci. 954–960.

Knyple J S 1978 Reversal of the symptoms of phosphate deficiency in *Spirodela* by RNA and adenosine monophosphates. Z. Pflanzenphysiol. 90, 265–277.

Lockhart J A 1965 An analysis of irreversible plant cell elongation. J. Theor. Biol. 8, 264–275.

Lynch J, Andre L and Epstein E 1991 Vegatative growth of common bean in response to phosphorus nutrition. Crop Sci. 31, 380–387.

McIntosh M S 1983 Analysis of combined experiments. Agron. J. 75, 153–155.

Ogata S, Adu-Gyamfi J J and Fujita K 1988 Effect of phosphorus and pH on dry matter production, nitrogen fixation and critical phosphorus concentration in pigeonpea (*Cajanus cajan* (L.) Millsp.) Soil Sci. Plant Nutr. 34, 55–64.

Palmer S J, Berridge D M, McDonald A J S and Davis W J 1996 Control of leaf expansion insunflower (*Helianthus annuus* L.) by nitrogen nutrition. J. Exp. Bot. 47, 259–268.

Passioura J B and Fry S C 1992 Turgor and cell expansion: beyond the Lockhart equation. Aust. J. Plant Physiol. 19, 565–576.

Puliga S, Vazzana C and Davies W J 1996 Control of crops leaf growth by chemical and hydraulic influences. J. Exp. Bot. 47, 529–537.

Radin J W and Eidenbock M P 1984 Hydraulic conductance as a factor limiting leaf expansion of phosphorus-deficient cotton plants. Plant Physiol. 75, 372–377.

Ranasinghe S and Taylor G 1996 Mechanism for increased leaf growth in elevated CO_2. J. Exp. Bot. 47, 349–358.

Rao I M and Terry N 1995 Leaf phosphate status, photosynthesis, carbon partitioning in sugarbeet. IV. Changes with time following increased supply of phosphate to low phosphate plants. Plant Physiol. 107, 1313–1321.

Rao I M, Freeden A L and Terry N 1990 Leaf phosphate status, photosynthesis and carbon partitioning in sugarbeet. III. Diurnal changes in carbon partitioning and carbon export. Plant Physiol. 92, 29–36.

Rodriguez D, Zubillaga M M, Polshuk E L, Keltjens W G, Goudriaan J and Lavado R S 1998 Leaf area expansion and assimilate production in sunflower (*Helianthus annuus* L.) growing under low phosphorus conditions. Plant Soil 202, 133–147.

Sakurai N 1991 Cell wall functions in growth and development – A physical and chemical point of view. Bot. Mag. Tokyo 104, 235–251.

Sakurai N, Tanaka S and Kuraishi S 1987 Changes in wall polysaccharides of squash (*Cucurbita maxima* Duch.) hypocotyls under water stress condition. 1. Wall sugar composition and growth as affected by water stress. Plant Cell Physiol. 28, 1051–1058.

Sato H, Sakurai N, Sendo S, Saneoka H, Nobuyasu H and Fujita K 1997 Factors affecting leaf area development in husk leaf of flint corn. Crop Sci. 37, 1826–1831.

Sawada O, Itoh J and Fujita K 1995. Characteristics of photosynthesis and translocation of ^{13}C-labelled photosynthate in husk leaves of sweet corn. Crop Sci. 35, 480–485.

Smith R C and Fry S C 1991 Endo-transglycosylation of xyloglucans in plant cell suspension cultures. Biochem. J. 279, 529–535.

Stitt M and Quick W P1989 Photosynthetic carbon partitioning: its regulation and possibilities for manipulation. Physiol. Plant. 77, 633–641.

Taylor G and Ranasinghe S 1996 Mechanism for increased leaf growth in elevated CO_2. Exp. Bot. 47, 349–358.

Thomas J, Ross C W, Chastain C J, Koomanoff N, Hendrix J E and van Volkenburgh E 1981 Cytokinin-induced wall extensibility in excised cotyledons of radish and cucumber. Plant Physiol. 85, 331–334.

J.J. Adu-Gyamfi (Ed.), *Food security in nutrient-stressed environments: exploiting plants' genetic capabilities:* 103–110, 2002.

Varietal differences in root growth as related to nitrogen uptake by sorghum plants in low-nitrogen environment

Takuji Nakamura[1,5], Joseph J. Adu-Gyamfi[2], Akiko Yamamoto[3], Satoru Ishikawa[2], Hiroshi Nakano[4] & Osamu Ito[1]

[1]*Japan International Research Center for Agricultural Sciences (JIRCAS), 1-1 Ohwashi, Tsukuba, Ibaraki 305-8686, Japan.* [2]*International Crops Research Institute for the Semi-Arid Tropics (ICRISAT), Sabo Bakin Zuwo Road, PMB 3491, Kano, Nigeria.* [3]*Experimental Farm of Tohoku University, Tohoku University, Naruko, Tamatsukuri-gun, Miyagi 989-6711, Japan.* [4]*Hokkaido National Agricultural Experimental Station (HNAES), Shinsei, Memuro, Kasaigun, Hokkaido 052-0071, Japan.* [5]*Corresponding author**

Key words: hybrid, local, N absorption rate, semi-arid tropics (SAT), sorghum, specific N absorption rate in roots

Abstract

Hybrid (*Sorghum bicolor* (L.) Moench cv CSH 9) and local variety (*S. bicolor* cv. FSRP) of sorghum were grown in an Alfisol field and in pots of Alfisol in semi-arid tropical India, and the productivity of sorghum in low N condition was studied in relation to N absorption by plant. The grain yield of the hybrid (CSH 9) was higher than that of the local variety (FSRP) irrespective of the rate of N application. The reduction in hybrid grain yield was less than that for the local variety when N was not applied. After booting stage hybrid absorbed more N than the local variety. To estimate the role of roots in N absorption, the characteristics of N absorption were parameterized. Relative growth rate was highly correlated with relative N absorption rate in plant. N absorption was regulated by root activity (specific N absorption rate per unit root mass and length). Specific N absorption rate in roots at physiological maturity was higher in the hybrid than in local variety in low N conditions. Therefore, the smaller reduction of hybrid yield in low-N conditions was ascribed to high-N absorption rate in its roots, and the high ability of roots to absorb N after booting stage.

Introduction

Sorghum is widely grown in the semi-arid tropics (SAT), where various types of stress exit. Nutrient – especially nitrogen (N) and phosphorus- and drought stresses are the most important factors affecting productivity (Marschner, 1995; Osaki et al., 1992). Nitrogen is the most limiting nutrient in Alfisol of the SAT because of their low organic matter, high losses of nitrogen through various mechanisms, and immobilization in the soil (Das and Chatterjee, 1982; Khanna et al., 1982; Prasad and Thomas, 1982).

The development and release of hybrids and high-yielding varieties of sorghum have led to a substantial increase in the yield of sorghum under N-limiting condition in SAT areas (Doggett et al., 1970; Martin, 1970; Rachie, 1970; Rao, 1982), suggesting that some hybrids and improved varieties are adapted to limited-N condition.

The ability of crop plants to use the mineralized-N depends on the root system development and the rate at which they absorb N (Lee et al., 1996; Rao et al., 1993). It was reported that, in vegetative stage, the rate of N accumulation by plant was regulated by the N absorption rate in several cereal crops (Osaki et al., 1995; Shinano et al., 1994). However, these results were observed in adequate N level, not low-N level, and the effect of low-N environment on the relationship between N absorption and production and N accumulation in plant was still unclear. Thus, it was assumed that effective N absorption from soils is one of the mechanisms by which sorghum is adapted to low-N conditions and that sorghum hybrids could have the ability to absorb more nutrients from low-nutrient soils in SAT than the local varieties.

* FAX No: +81-298-38-6354. E-mail: takuwan@jircas.affrc.go.jp

This paper previously appeared in *Plant and Soil* **245**, 2002, pp. 17–24.

Table 1. The size and N concentration of the seeds sown in the field and pot experiments, ICRISAT, Patancheru, India, rainy season, 1996 and 1997

	Seed size ($mg\ seed^{-1}$)		N concentration ($g\ N\ kg^{-1}$)	
CSH9	27.6	(0.6)	11.4	(0.6)
FSRP	31.1	(0.3)	15.4	(0.7)

Numbers in parentheses are standard errors

A number of genotypes of sorghum were evaluated in low-N condition to assess the extent of genotypic variation among cultivars and identify physiological traits. Biomass, grain yield and N uptake in plant were highest in the hybrids and improved cultivars compared to local varieties in sorghum (Adu-Gyamfi et al., 2002; Nakano et al., 2002). Based on their results, we selected a hybrid (CSH 9) and local variety (FSRP), which have been commonly cultivated in farmers' field in India. Field and pot experiments were conducted to compare the physiological basis of growth and dry matter production in a hybrid (CSH 9) and a local variety (FSRP) of sorghum in low-N condition by examining the relationship between root growth and N-absorption rate.

Materials and methods

Field experiment

The experiment was conducted during 1996 rainy season on an Alfisol at ICRISAT, Patancheru, India. Hybrid sorghum (*Sorghum bicolor* L. Moench) cv. CSH 9 and local variety FSRP were sown on 25 June 1996. The size and N concentration of the seeds used in field experiment is shown in Table 1. The timings of growth stage (booting and growth period, etc.) in both genotypes were almost same at previous experiments (Adu-Gyamfi et al., 2002; Nakano et al., 2002). Spacing was 60 × 10 cm. Two levels of N fertilizer, 0 [0N] and 100 kg N ha^{-1} [100N] as urea were applied before sowing. All plots received a uniform basal application of 20 kg P ha^{-1} as single superphosphate. The experiment layout was a randomized block design with three replications. The size of each plot was 5 × 6 m. Annual rainfall was 1063 mm, with 1000 mm between June and October. The mean temperature was 26.8 °C between June and October.

Samples (six to 10 plants) were taken at about 21-day intervals after emergence. Samples were separated into leaves, stems, roots, and panicles. Samples were oven-dried at 80 °C for 48 h, ground, dry weights determined, and a portion of the ground material used for N analysis. In addition, samples (80–110 plants) were taken at final harvest and grain yield was determined.

A monolith sampling methods was used to collect roots in the soil to 60 cm depth. Roots were washed in mesh net in the tap water to remove the soils, and the total root lengths were determined with a COMAIR root scanner (Devi et al., 1996).

Pot experiment

A hybrid (CSH 9) and local variety (FSRP) of sorghum were grown in 10-l pots (one plant per pot) contained 10 kg Alfisol soil during 1997 rainy season in the greenhouse at ICRISAT, Patancheru, India. The size and N concentration of the seeds used in pot experiment is shown in Table 1. Nitrogen was supplied by basal application of urea at two levels; 0 g N pot^{-1} (-N) and 2 g N pot^{-1} (+N) (Shinano et al., 1991). Phosphorus was applied at 2 g P_2O_5 pot^{-1} as single superphosphate. Pots were irrigated with water during the experiment to maintain 60% maximum water holding capacity. The mean temperature of greenhouse during experiment was 25 °C. The experiment layout was a randomized block design with three replications.

Plants were sampled at vegetative, booting, physiological maturity, and final harvest stages. Twelve plants were sampled at each sampling time. Samples were separated into leaves, stems, roots, and panicles. The dry weight samples were determined after they had been oven-dried at 80 °C for 48 h. Samples were ground and a portion of the ground material used for N analysis.

Measurement

Nitrate in the soil was determined in dried soil after extraction with KCl (Keeney and Nelson, 1982), total N by Kjeldahl's method. The N-status and pH of soil used in field and pot experiments is shown in Table 2. The N concentration in samples was analyzed by indophenol color formation (Chaykin, 1969) after digestion with a hydrogen peroxide–sulfuric acid mixture (Singh et al., 1984).

N absorption rate in the plant (AR) was calculated from total amount of N absorbed in whole plant using the following equation:

$$AR = (N_1 - N_2)/(t_1 - t_2),$$

Table 2. Nitrate-N concentration and total N in field soil and pot soil before fertilizer application, ICRISAT, Pantancheru, India, rainy seaseon 1996 and 1997, respectively

		Field		Pot	
$pH(H_2O)$		6.26	(0.11)	6.69	(0.02)
NO_3-N	(mg kg^{-1} soil)	2.93	(0.99)	5.83	(0.24)
Total N	(g kg^{-1} soil)	0.60	(0.09)	0.39	(0.01)

Numbers in parentheses are standard errors.

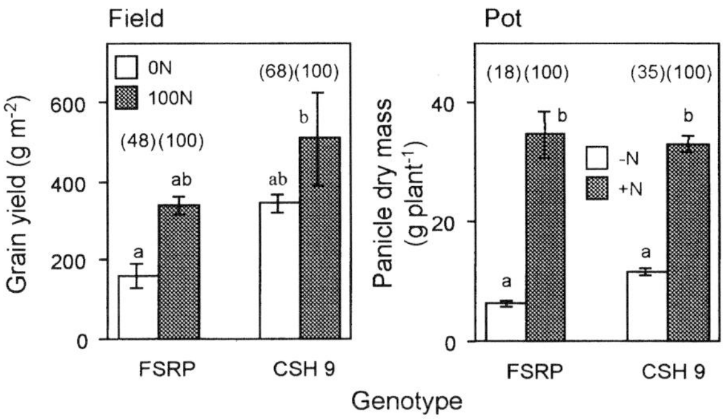

Figure 1. Grain yield in the field experiment, and panicle dry mass in pot experiment of a sorghum hybrid (CSH 9) and a local variety (FSRP), ICRISAT, Patancheru, India, rainy season 1996 and 1997, respectively. Values in parentheses are the relative values compared to the 100N treatment in the field experiment and +N treatment in pot experiment. The vertical bars indicate ±SE. Bars with the same letter indicate no significant difference at p=0.05 level.

where: N_1 is the total N absorbed in the plant at t_1, N_2 the total at t_2; t is the number of days after emergence. The relative N absorption rate in whole plant (RAR) was calculated as AR divided by total dry mass in whole plant.

To estimate the role of root in N absorption, RAR was studied with parameter analysis related to root activity and distribution ratio of dry mass to roots as follows:

$$\mathrm{RAR} = \mathrm{SAR} \times \mathrm{FWR}, \qquad (1)$$

where: SAR is specific absorption rate of N per roots mass, and FWR is the distribution ratio of dry mass to roots (the ratio of root dry mass to whole plant dry mass). In addition, SAR was parameterized into specific absorption rate of N per root length (SARL) and specific root length (SRL) as follows:

$$\mathrm{SAR} = \mathrm{SARL} \times \mathrm{SRL} \qquad (2)$$

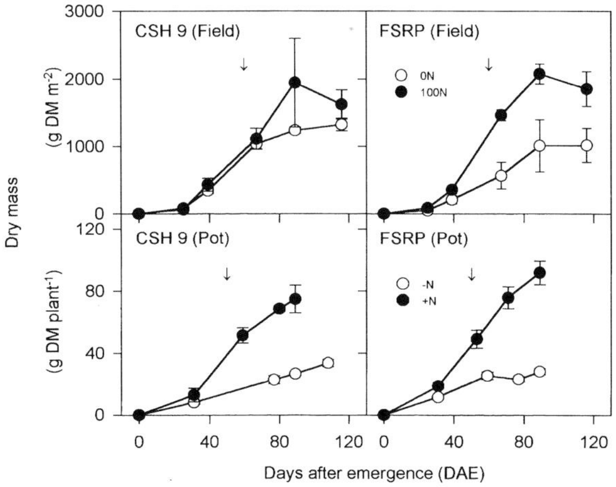

Figure 2. Amount of dry mass accumulated in whole plant at successive growth stages of a sorghum hybrid (CSH 9) and a local variety (FSRP) in field and pot experiments, ICRISAT, Patancheru, India, rainy season 1996 and 1997, respectively. The arrow in the figure indicates the booting time. The vertical bars indicate ±SE.

Results

Yield

In the field experiment, the grain yield of the hybrid CSH 9 was higher than that of local variety FSRP (Figure 1). The grain yield of 0N treatment was 68% of 100N grain yield in the hybrid, while in the local variety, grain yield of 0N treatment was 48% of grain yield of the 100N treatment. In the pot experiment, the dry mass of panicle of −N was 35% of panicle dry mass of +N in hybrid, while dry mass of panicle of −N was 18% of panicle dry mass of +N in local variety. There was the smaller reduction in yield and panicle dry mass in the hybrid than in the local variety when no N was applied.

Accumulation of dry mass and N in plants

In the field experiment, the growth in 100N was larger than that in 0N in hybrid and local variety (Figure 2). The difference in the growth between 100N and 0N treatments in hybrid was smaller than that in local variety. In the pot experiment, the difference in the growth between +N and −N treatments also was smaller in hybrid than in local variety. Dry mass of whole plant in hybrid increased after the booting stage in −N treatment, while dry mass of whole plant in local variety remained constant after the booting stage in −N treatment.

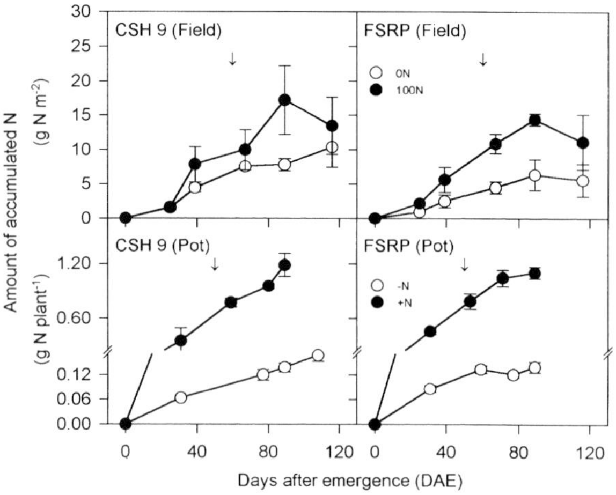

Figure 3. Amount of N accumulated in whole plant of a sorghum hybrid (CSH 9) and a local variety (FSRP) at successive growth stages in field and pot experiments, ICRISAT, Patancheru, India, rainy season 1996 and 1997, respectively. The arrow in the figure indicates the booting time. The vertical bars indicate ±SE.

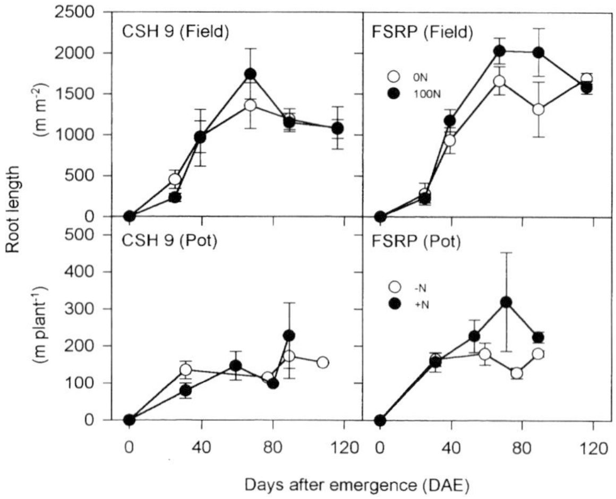

Figure 5. Root length at successive growth stages of a sorghum hybrid (CSH 9) and a local variety (FSRP) in field and pot experiments, ICRISAT, Patancheru, India, rainy season 1996 and 1997, respectively. The vertical bars indicate ±SE.

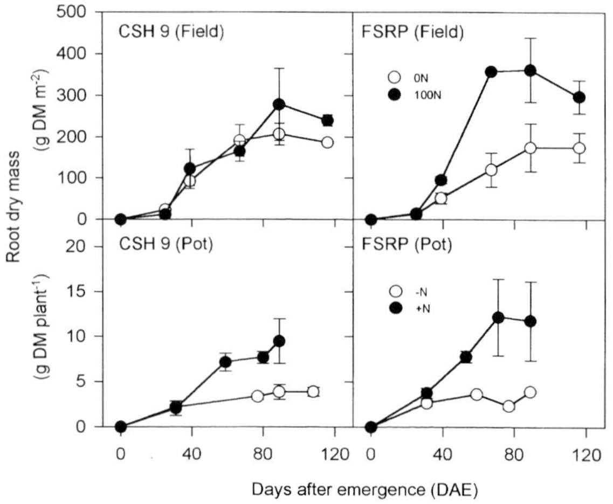

Figure 4. Root dry mass at successive growth stages of a sorghum hybrid (CSH 9) and a local variety (FSRP) in field and pot experiments, ICRISAT, Patancheru, India, rainy season 1996 and 1997, respectively. The vertical bars indicate ±SE.

In the field experiment, the amount of N absorbed by hybrid plants increased after the booting stage in both N treatments, but the amount of N absorbed by plants of local variety in 100N treatment remained constant after booting (Figure 3). The amount of N absorbed in plants of local variety at 0N in the field experiment did not increase after booting stage. In the pot experiment, the amount of N absorbed by hybrid plants increased continuously after booting stage in both N treatments, but the increase of amount of N absorbed in plants of local variety was slowed down after booting stage in +N treatment. The amount of N absorbed in plant of local variety did not increase after booting stage in −N treatment.

Root growth

Root dry mass increased with fertilizer-N, while root length was not much affected by fertilizer-N in hybrid and local variety (Figures 4 and 5). The reduction of dry mass and length of roots were larger in local variety than that in hybrid when no N was applied in the field and pot experiments.

Relative N absorption rate (RAR), specific absorption rate of N per root weight and length (SAR, SARL), distribution ratio of dry mass to root (FWR), specific root length (SRL)

In the field and pot experiments, FWR was not correlated to RAR, but SAR was highly correlated with RAR in both genotypes (Figure 6). In addition, SRL was not closely correlated with SAR, but SARL was highly correlated with SAR (Figure 7). In Table 3, regression between relative growth rate (RGR) and each parameter that depend on N absorption were shown. RGR was correlated with RAR, SAR, FWR and SARL in both experiments.

Table 3. Correlation of relative N absorption rate in whole plant (RAR), specific absorption rate of N per roots mass (SAR), distribution ratio of dry mass to roots (FWR), specific absorption rate of N per root length (SARL), and specific root length (SRL) with relative growth rate in plant

	RAR	SAR	FWR	SARL	SRL
Field experiment	0.94**	0.77**	0.61*	0.81**	0.76**
Pot experiment	0.82**	0.76**	0.65**	0.52*	0.30

*Significant at 5% level; **significant at 1% level.

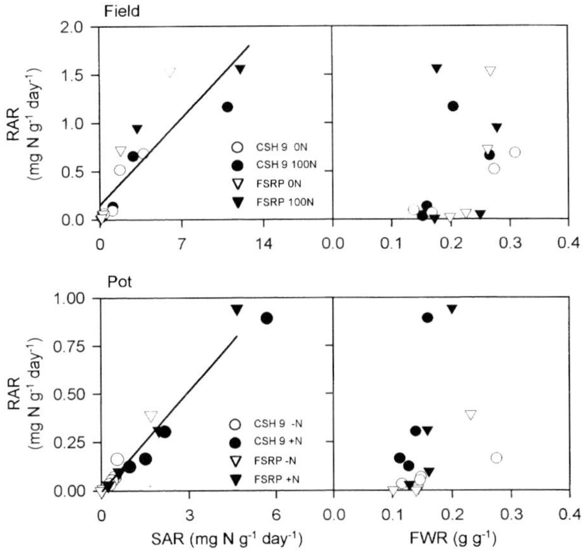

Figure 6. Relationship between relative absorption rate of N (RAR), specific absorption rate of N per root mass (SAR), and the ratio of root dry mass to whole plant dry mass (FWR) of a sorghum hybrid (CSH 9) and a local variety (FSRP) at successive growth stages in field and pot experiments, ICRISAT, Patancheru, India, rainy season 1996 and 1997, respectively. The regression slope is represented by RAR=0.128SAR+0.150 (r^2=0.771, p<0.01) in the field experiment and RAR=0.174SAR-0.006 (r^2=0.955, p<0.01) in the pot experiment.

RAR, SAR, and SARL at the physiological maturation stage

The RAR, SAR, and SARL tended to be higher in the hybrid than local variety irrespective of N level in both field and pot experiments at physiological maturity (Table 4). The ability of roots to absorb N remained higher in the hybrid than local variety during maturation.

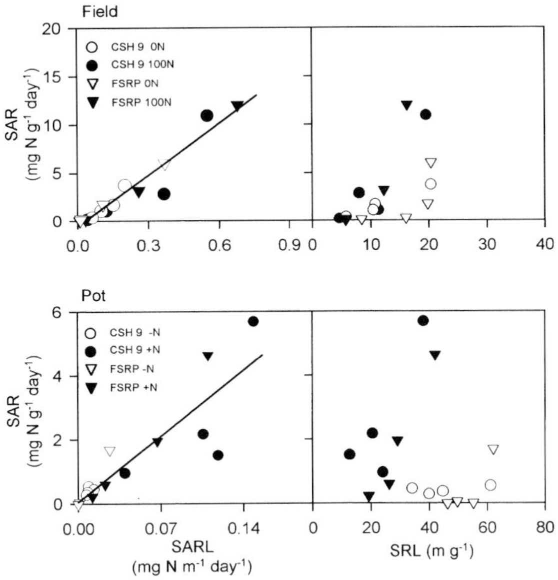

Figure 7. Relationship between specific absorption rate of N per root mass (SAR), specific absorption rate of N per root length (SARL) and specific root length (SRL) of a sorghum hybrid (CSH 9) and a local variety (FSRP) at successive growth stages in field and pot experiments, ICRISAT, Patancheru, India, rainy season 1996 and 1997, respectively. The regression slope is represented by SAR=17.9SARL–0.597 (r^2=0.927, p<0.01) in the field experiment and SAR=29.3SARL+0.058 (r^2=0.774, p<0.01) in the pot experiment.

Discussion

The present study focused on the physiological mechanism of sorghum for adaptation to limited N conditions, for which a little information is available. The yields of hybrid (CSH 9) were higher than that of local variety (FSRP) in field experiment (Figure 1). The reduction in yield of the local variety was larger than that of hybrid when no N was applied in both field and pot experiments. These data suggest that the hybrid was more adapted to limited-N conditions than local variety. It is known that seed vigor and viability are important components influencing seedling establishment, crop growth, and productivity (TeKrony and Egli, 1991; Welch, 1986) and that seeds with low nutrient reserves are disadvantaged unless sown to soils with adequate available nutrients for seedling establishment and growth after germination (Welch, 1986, 1999). The size and N concentration of the seeds sown in our experiments were larger in local variety than in hybrid (Table 1). Thus, the difference in the results between hybrid and local variety would not be due to the size and N concentration of the seeds used in our experiments. The amount of N absorbed by plants

Table 4. Relative absorption rate of N of sorghum in whole plant (RAR), specific root absorption rate of N per root weight (SAR), and specific root absorption rate of N per root length (SARL) at physiological maturity of sorgum hybrid (CSH 9) and local variety (FSRP) in field and pot experiments, ICRISAT, Patancheru, India, rainy season 1996 and 1997, respectively

Genotype	N treatment	RAR (μg N g^{-1} day^{-1})	SAR (μg N g^{-1} day^{-1})	SARL (μg N m^{-1} day^{-1})
Field experiment				
CSH 9	0N	64.7^{b}	377^{b}	62.8^{c}
FSRP	0N	21.6^{a}	133^{a}	12.6^{ab}
CSH 9	100N	30.8^{ab}	202^{ab}	43.2^{bc}
FSRP	100N	9.3^{a}	53.2^{a}	9.5^{a}
Pot experiment				
CSH 9	−N	52.4^{xy}	368^{xy}	8.2^{xy}
FSRP	−N	0.8^{x}	8.3^{x}	0.2^{x}
CSH 9	+N	165^{yz}	1507^{y}	118^{y}
FSRP	+N	98.2^{y}	613^{xy}	23.3^{xy}

Field experiment and pot experiments analyzed separately. Values followed by the same letter in the same column are not significantly different at P=0.05.

increased in hybrids after the booting stage (Figure 3). Furthermore, in low-N condition, the amount of N absorbed in plants increased in hybrids, but did not increase in the local variety after booting stage. The N absorption rate remained high in the hybrid, especially in low-N conditions, and this continuous absorption of N may have contributed to the higher yields of the hybrids (Osaki et al., 1991, 1995). Thus, it is suggested that the ability of roots to absorb N after the booting stage play an important role in controlling grain and dry matter production in sorghum.

Root dry mass was affected by N-treatments but root length was not much (Figures 4 and 5). This observation is consistent with that of Lee et al. (1996), who also reported that 0N application reduced root biomass, not root length. Myers (1980) reported that shoot biomass varied with fertilizer levels, but root biomass did not, suggesting that the ratio of root biomass to whole plant dry mass is affected by fertilizer levels. Brown et al. (1987) observed fertilizer-N had little effect on the ratio of roots to total plant weight of barley, which indicated that root biomass production responded to fertilizer in a similar way to above-ground biomass production. In this study, the similar responsiveness of the root biomass and the above-ground biomass production would be supported by little difference of the range of FWR in fertilizer-N levels (Figure 6).

The plants showed a close relationship between amount of absorbed N in plant and dry mass in plant and plant growth was substantially regulated by the amount of absorbed N in plant (Nakamura et al., 1997; Osaki et al., 1992; Shinano et al., 1994). Thus, N absorption rate may affect the plant growth rate. To evaluate the role of roots in N absorption, each parameter was estimated by based on Eqs. (1) and (2) and correlation between each parameter and relative growth rate (RGR) was analyzed (Table 3). RAR, SAR, FWR and SARL was closely related to RGR in both experiment, especially regression coefficients of RAR-RGR and SAR-RGR were higher than the others, which indicated plant growth rate would be regulated by N absorption rate. Nutrient absorption is controlled by root size in plant (Haynes et al., 1991; Iwama, 1988). On the contrary, it is reported that nutrient absorption by plant is controlled by root activity (e.g., nutrient absorption rate of roots) (Hilbert, 1990; Takahashi et al., 1991). In seven crops, root biomass was not related to nitrogen absorption or dry mass production especially during maturation (Osaki et al., 1995, 1997). SAR was strongly correlated with RAR, and SARL was strongly correlated with SAR (Figures 6 and 7). These results indicated that N absorption rate per root dry mass and length regulated N absorption rate in whole plant.

RAR, SAR, SARL at the physiological maturity were higher in the hybrid than in the local variety irrespective of N treatments (Table 4). In the hybrid, N was absorbed continuously after booting stage (Figure 3). This ability of high N absorption in hybrid was due to its high SAR and SARL (Figures 6 and 7). High-N absorption by roots after the booting stage in sorghum would be an important physiological trait for high productivity in low-N environment.

Acknowledgement

We sincerely thank Drs. K. Okada and T. Kanno, Japan International Research Center for Agricultural Sciences, for critical and invaluable comments on the manuscript.

References

Adu-Gyamfi J J, Ishikawa S, Nakamura T and Nakano H 2002 Genotypic variability and physiological characteristics of crop plants adapted low-nutrient environment. *In* Food Security in Nutrient-Stressed Environments: Exploiting Plants' Genetic Capabilities. Ed. J J Adu-Gyamfi. pp. 42-43. Summary and Recommendations of an International Workshop, September 1999, ICRISAT, Patancheru, India.

Brown S C, Keatinge J D H, Gregory P J and Cooper P J M 1987 Effect of fertilizer, variety and location on barley production under rainfed conditions in Northern Syria. 1. Root and shoot growth. Field Crop Res. 16, 53–66.

Chaykin S 1969 Assay of nicotinamide deaminase. Determination of ammonia by the indophenol reaction. Anal. Biochem. 32, 375–382.

Das S C and Chatterjee R K 1982 Chemical composition of soils of India, *In* Review of Soil Research, Eds. N S Randhawa et al. pp 83-109. Proceeding of 12th International Congress of Soil Science, Indian Society of Soil Science, Division of Soil Science and Agricultural Chemistry, February 1982, New Delhi, India

Devi G, Ito O, Matsunaga R, Tobita S, Rao T P, Lakshmi N V and Lee K K 1996 Simulating root system development of short-duration pigeonpea. Exp. Agric. 32, 1–13.

Doggett H, Curtis D L, Laubscher F X and Webster O J 1970 Sorghum in Africa. *In* Sorghum Production and Utilization. Eds. J S Wall and W M Ross. pp. 288–327. The AVI Publishing Company, Westport, CT, USA.

Haynes B, Koide R T and Elliott G 1991 Phosphorus uptake and utilization in wild and cultivated oats (*Avena* spp.). J. Plant Nutr. 64, 1105–1118.

Hilbert D W 1990 Optimization of plant root:shoot ratios and internal nitrogen concentration. Ann. Bot. 88, 91–99.

Iwama K 1988 Difference in root growth of potato among years and cropping seasons. Jpn J. Crop Sci. 57, 346–354.

Keeney D R and Nelson D W 1982 Nitrogen-inorganic forms. *In* Methods of Soil Analysis, Part 2. Eds. A L Page et al. pp. 643–698. Madison, WI USA.

Khanna S S, Pathak A N and Saxena S N 1982 Phosphorus, *In* Review of Soil Research, Eds. N S Randhawa et al. pp. 322–330. Proceeding of 12th International Congress of Soil Science, Indian Society of Soil Science, Division of Soil Science and Agricultural Chemistry, February 1982, New Delhi, India

Lee K K, Singh P, Wani S P, Rego T J, Trimurtulu N and Monteith J L 1996 Effect of fertilizer nitrogen and irrigation on root growth, and water uptake with special reference to postrainy season sorghum. *In* Dynamics of Roots and Nitrogen in Cropping Systems of Semi-Arid Tropics. Eds. O Ito et al. pp. 262–271. Japan International Research Center for Agricultural Sciences, Tsukuba, Japan.

Marschner H 1995 Mineral Nutrition of Higher Plants, 2nd edition, Academic Press, London. 889 pp.

Martin J H 1970 History and classification of sorghum *Sorghum bicolor* (Linn.) Moench. *In* Sorghum Production and Utilization. Eds. J S Wall and W M Ross. pp. 1–27. The AVI Publishing Company, Westport, CT, USA.

Myers R J K 1980 The root system of a grain sorghum crop. Field Crops Res. 3, 53–64.

Nakamura T, Osaki M, Koike T, Hanba Y, Wada E and Tadano T 1997 Effect of CO_2 enrichment on carbon and nitrogen interaction in wheat and soybean. Soil Sci. Plant Nutr. 43, 789–798.

Nakano H, Adu-Gyamfi J J and Nakamura T 2002 Genotypic adaptation of crop plants to low-nutrient environment: morophological and ecophysiological characteristics of adaptation. *In* Food Security in Nutrient-Stressed Environments: Exploiting Plants' Genetic Capabilities. Ed. J J Adu-Gyamfi pp. 40–41. Summary and Recommendations of an International Workshop, September 1999, ICRISAT, Patancheru, India.

Osaki M, Morikawa K, Shinano T, Urayama M and Tadano T 1991 Productivity of high-yielding crops. II. Comparison of N, P, K, Ca, and Mg accumulation and distribution among high-yielding crops. Soil Sci. Plant Nutr. 37, 445–454.

Osaki M, Shinano T and Tadano T 1992 Carbon-nitrogen interaction in field crop production. Soil Sci. Plant Nutr. 38, 553–564.

Osaki M, Shinano T, Matsumoto M, Ushiki J, Mori M and Tadano T 1995 Productivity of high-yielding crops. V. Root growth and specific absorption rate of nitrogen. Soil Sci. Plant Nutr. 41, 635–647.

Osaki M, Shinano T, Matsumoto M, Zheng T and Tadano T 1997 A root–shoot interaction hypothesis for high productivity of field crops. Soil Sci. Plant Nutr. 43, 1079–1084.

Prasad R and Thomas J 1982 Nitrogen, *In* Review of Soil Research. Eds. N S Randhawa et al. pp. 309-322, Proceeding of 12th International Congress of Soil Science, Indian Society of Soil Science, Division of Soil Science and Agricultural Chemistry, February 1982, New Delhi, India

Rachie K O 1970 Sorghum in Asia. *In* Sorghum Production and Utilization. Eds. J S Wall and W M Ross. pp. 328–381. The AVI Publishing Company, Westport, CT, USA.

Rao N G P 1982 Transforming traditional sorghums in India. *In* Sorghum in the Eighties. Eds. L R House et al. pp. 39–59. Proceedings of the International Symposium on Sorghum, November 1981, ICRISAT, Patancheru, A.P., India.

Rao T P, Ito O and Matsunaga R 1993 Differences in uptake kinetics of ammonium and nitrate in legume and cereals. Plant Soil 154, 67–72.

Shinano T, Osaki M and Tadano T 1991 Comparison of reconstruction of photosynthesized ^{14}C compounds incorporated into shoot between rice and soybean. Soil Sci. Plant Nutr. 37, 419–417.

Shinano T, Osaki M, Yamada S and Tadano T 1994 Comparison of root growth and nitrogen absorbing ability between Gramineae and Legumunosae during the vegetative stage. Soil Sci. Plant Nutr. 40, 485–495.

Singh U, Sahrawat K L, Jambunathan R and Burford J R 1984 The use of hydrogen peroxide for the digestion and determination of total nitrogen in chickpea (*Cicer arientinum* L.) and pigeonpea (*Cajanus cajan* L.). J. Sci. Food Agric. 35, 640–646.

Takahashi Y, Chinushi T, Nakano T, Hagino K and Ohyama T 1991 Effect of placement of coated urea fertilizer on root growth and rubidium uptake activity in soybean plant. Soil Sci. Plant Nutr. 37, 735–739.

TeKrony and Egli 1991 Relationship of seed vigor to crop yield. Crop Sci. 31, 816–822

Welch R M 1986 Effects of nutrient deficiencies on seed production and quality. Adv. Plant Nutri. 2, 205–247.

Welch R M 1999 Importance of seed mineral nutrient reserves in crop growth and development. *In* Mineral Nutrition of Crops. Ed. Z Rengel pp. 205–226. Food Products Press, Binghamton, NY, USA.

J.J. Adu-Gyamfi (Ed.), Food security in nutrient-stressed environments: exploiting plants' genetic capabilities: 111–121, 2002.

Genotypic variability in phosphorus solubilizing activity of root exudates by pigeonpea grown in low-nutrient environments

S. Ishikawa[1,4,5], J. J. Adu-Gyamfi[1], T. Nakamura[2], T. Yoshihara[3], T. Watanabe[3] & T. Wagatsuma[4]
[1]*International Crops Research Institute for the Semi Arid Tropics (ICRISAT), Sabo Bakin Zuwo Road, PMB 3491, Kano, Nigeria.* [2]*Japan International Research Center for Agricultural Sciences (JIRCAS), 1-2 Ohwashi, Tsukuba, Ibaraki, 305-8686, Japan.* [3]*Faculty of Agriculture, Hokkaido University, Sapporo, Hokkaido, 060-8589, Japan.* [4]*Faculty of Agriculture, Yamagata University, Tsuruoka, Yamagata, 997-8555, Japan.* [5]*Corresponding author**

Received 10 May 2000. Accepted in revised form 24 October 2000

Key words: Alfisols, phosphorus deficiency, phosphorus solubilization, pigeonpea genotype, piscidic acid, root exudates

Abstract

A pot experiment confirmed that pigeonpea could efficiently utilize various sources of phosphorus (P) (aluminium phosphate, iron phosphate and apatite), irrespective of genotype. A qualitative assay method for iron (Fe)-P solubilizing activity showed that root exudates collected from P-deficient pigeonpea contained Fe-P solubilizing substances and that they were released mainly from root tips. Citric, malic, malonic, succinic and piscidic acids were identified in root exudates. Citric and piscidic acids release from roots was increased by low-P treatment in all the genotypes tested. The release rates of citric and piscidic acids were affected by the P concentration of shoots rather than that of roots. The pigeonpea roots released approximately 5–100 times more piscidic acid than citric acid depending on P stress status, plant age and genotype. When organic acids were added to Alfisols, citric acid was most capable of mobilizing P from the soil, followed by piscidic acid and malic acid. No correlation was found between genotypic variability in the release rates of citric and piscidic acids from the roots under low-P treatment at hydroponic culture and in the growth and P uptake of plants on Alfisols. Although citric and piscidic acids released from pigeonpea roots may play a partial role in solubilizing unavailable insoluble P in soils, the releases were thought to be an unsatisfactory strategy for explaining genotypic variation in low P availability of pigeonpea.

Introduction

Many tropical and subtropical soils with high levels of iron (Fe) and aluminium (Al) oxides have extremely high capacities to immobilize phosphorus (P) and render it unavailable for uptake by plants. However, it has been found that several plant species are able to directly modify the rhizosphere in order to acquire and utilize unavailable soil-P compounds (Jones, 1998; Rengel, 1999). The release of organic acids from roots that is induced by P deficiency is an effective way for plants to acquire sparingly soluble P from soils. Investigations using white lupin (*Lupinus albus* L.) showed that the amount of citrate excreted under P-deficiency was as much as 12% of the total plant dry weight (Gardner et al., 1983), and that proteoid roots exuded larger amounts of citrate and malate than normal roots (Johnson et al., 1996). Hoffland (1992) reported that P uptake from rock phosphate by rape (*Brassica napus* L.) increased by acidifying part of the rhizosphere by exuding of citric and malic acids. Ohwaki and Hirata (1992) compared the differences in carboxylic acid exudation among six tropical leguminous crops subjected to P starvation, and found that larger amounts of citric and malic acids were exuded

* FAX No: +81-235-28-2812. Tel.: +81-235-28-2917.
E-mail: isatoru@tds1.tr.yamagata-u.ac.jp

This paper previously appeared in *Plant and Soil* **245**, 2002, pp. 71–81.

from the roots of chickpea than those exuded by other leguminous crops. Thus, citric and malic acids were identified as the primary carboxylic acid released from the roots of many plant species under P-starvation.

Pigeonpea [*Cajanus cajan* (L.) Millsp] is an important leguminous crop and a protein source for people living in semi-arid tropics (SAT). Ae et al. (1990) reported that pigeonpea is more able to utilize Fe-P than several cereals and other leguminous crops because it can exude piscidic acid from its roots. Piscidic acid is a new substance reported in root exudates of P-deficient plants, it differs from such carboxylic acids as citric and malic acids because it is a phenolic compounds (Toshima et al., 1999; Yoshihara et al., 1971). The few reports on genotypic variation in P-solubilizing activity of pigeonpea root exudates in low-P environments (Subbarao et al., 1997a, b) did not show a correlation between genotypic differences in the release rates of carboxylic acids (including piscidic) from pigeonpea roots in response to low-P environments, or in the acquisition of P from soils where it was scarce.

The purpose of this study was to analyze the P acquisition mechanisms implicated in organic acids released from the roots of pigeonpea genotypes to provide information useful in breeding crop ideotypes that could utilize unavailable P more efficiently than the present pigeonpea genotypes. The exploitation of plant genetic variation in the acquisition of unavailable insoluble P in soils is emerging as an alternate approach to improve productivity in low-nutrient environments. If successful, it would result in low-input and sustainable agriculture systems in the SAT.

Materials and methods

Comparison of P utilization from different sources of P among pigeonpea genotypes

A pot experiment was carried out at the International Crops Research Institute for the Semi-Arid Tropics (ICRISAT), Patancheru, Andhra Pradesh, India. The experiment was a randomized complete block design with four replicates. Four pigeonpea genotypes consisting of a hybrid (ICPH 8), a improved variety (ICPL 87), a local variety (Manak) and a drought-tolerant line (ICPL 88039) were obtained from an initial screening of 15 genotypes (Adu-Gyamfi et al., 2000). The genotypes used were of extra-short-duration (maturity period < 110 d) or short-duration (maturity period < 120–140 d). Four different forms of fertilizers; aluminium phosphate ($AlPO_4$), iron phosphate ($FePO_4{\cdot}H_2O$), apatite [$Ca_5OH(PO_4)_3$] and single superphosphate (SSP) were applied at sowing; all provided 40 mg P kg^{-1} soil. There was also a control (no P applied; P_0). To remove the water-soluble P fraction in the $AlPO_4$ and $FePO_4{\cdot}H_2O$, they were washed in distilled water several times and dried in an oven at 80 °C before mixing with the soil. Each of the P sources with 50 mg N kg^{-1} soil were mixed with 10 kg soil (Alfisol) and the mixtures used to fill pots. Five seeds of pigeonpea were sown in each pot and adequate deionized water was supplied when required. The seedlings were thinned to 1 per pot after emergence. Plants were harvested at the beginning of the flowering stage (60 days after sowing). Plants were separated into leaves, stems and roots, dried to a constant weight and weighed. The P concentrations in the tissues were determined by a molybdenum-ascorbic acid method (Murphy and Riley, 1962) after digestion with H_2SO_4 and H_2O_2 (2:5 v/v). Available P in the soil was estimated using Troug and Bray II, and inorganic and insoluble P forms (Al-P, Ca-P and Fe-P) were estimated by the method of Sekiya (1983) before sowing and after harvest.

Plant preparation for collection of root exudates

In another experiment, the seeds of the same pigeonpea varieties were sown in sand and grown for 14 days with deionized water only. Seedlings were washed from the sand culture and carefully transferred to two 600-L tubs containing nutrient solution of the following composition (μM): N 1000 (NH_4NO_3), K 500 ($KCl{:}K_2SO_4$ [1:1]), Ca 750 ($CaCl_2{\cdot}H_2O$), Mg 500 ($MgSO_4{\cdot}H_2O$), Fe 18 ($FeSO_4{\cdot}7H_2O$), B 46 (H_3BO_3), Mn 0.9 ($MnCl_2{\cdot}4H_2O$), Cu 0.16 ($CuSO_4{\cdot}5H_2O$), Zn 0.15 ($ZnSO_4{\cdot}7H_2O$), Mo 0.1 ($NaMoO_4{\cdot}2H_2O$) and Co 0.001 ($CoSO_4$). The P concentrations in nutrient solution were 2 or 80 μM P as KH_2PO_4. To simplify the terminology, 2 and 80 μM P were designated as low-P and high-P. The 600-L tubs was covered with an aluminium sheet with 100 holes (10 rows and 10 columns). One tub was used for each P treatment and the four genotypes were randomized and transplanted in two rows per genotype. The solutions were continually aerated, refreshed with new solutions once a week and their pH was adjusted to 5.3 every day. These seedlings were grown in a greenhouse regulated at a constant temperature of 30 °C under natural light conditions. After 15 days, the P concentration in the low-P treatment was increased to 4 μM to prevent the plants from dying of severe P deficiency. At 15 or 30 days after transplanting, four uniform seedlings of each genotype were taken from the tub, their seedlings were washed thoroughly in distilled water and transferred to a 1-L plastic container of 1 mM $CaCl_2$ solution at pH 5.3. Root exudates were collected for 24 h under aeration in a growth chamber at 27 °C, 16-/8-h day/night cycles, 70 W m^{-2} light intensity and 70% relative humidity. The solutions containing root exudates were passed through filter paper (No. 6) and concentrated to approximately 5 mL using a rotary evaporator at 40 °C. Samples were stored at −20 °C until organic acid analysis.

To examine the change in the release rate of organic acid from roots depending on change in internal P concentration, seedlings pre-exposed to low-P solution for 15 days were transferred to low-P or high-P nutrient solutions and incubated for 7 days. The collection of root exudates and determination of the P concentrations in shoots and roots were done as described above.

In order to examine the effect of temperature on the release of organic acid from the roots, four seedlings grown in low-P solution for 15 days were transferred to a 1-L plastic container containing 1 mM $CaCl_2$ solution. The containers were put into two water baths that were continuously maintained at 25 °C or 35 °C and incubated in a growth chamber at 25 °C for 24 h. The solution containing root exudates was concentrated as above and stored until organic acid analysis. All experiments on organic acid analysis were repeated twice.

Assay method for Fe-P solubilizing activity

In order to examine the Fe-P solubilizing activity by root exudates, a rapid and simple assay method was developed by modifying the method of Ma et al. (1997) for Al-chelating assay. Filter paper of adequate size (No.1) was immersed in 5 mM $FeCl_3$ solution for 10 min, then soaked in 1/15 M phosphate buffer (pH 6.8) for 10 min. The paper was washed in distilled water for several seconds and dried at 80 °C. Each of the organic acids (citric, malic, succinic and piscidic acids) were dissolved in 0.5 M acetate buffer (pH 4.5) and 2 μL of each organic acid concentration (0–0.5 M) dropped onto the paper. After 3 min, the paper was washed in distilled water for several seconds and placed in 4% ammonium molybdate solution. After 5 min, the paper was taken from the solution and covered with a new paper, which had been permeated with 1.6% ascorbic acid solution, that turned the lower paper blue. White spots on the blue lower paper were visible if solution samples were able to solubilize Fe-P and phosphates were removed by washing with distilled water from the lower paper. Root exudate samples collected from roots treated with low-P for 30 days were also tested using the same method.

Identification of the release portion of Fe-P solubilizing substances

A nondestructive technique was used to identify the part of the root that could release Fe-P solubilizing substances by modifying the method of Zheng et al. (1998) for Al-chelating substances. The roots of Manak seedling that was exposed to low-P for 30 days in hydroponic culture were put between filter papers permeated with $FePO_4$ solution. Wet sponges containing distilled water were put above and below the filter papers to prevent the roots drying out. The P-stressed seedling was incubated in a growth chamber at 27 °C. After 24 h, the seedling was removed, and the lower paper was washed in distilled water for several seconds. The paper was stained with ammonium molybdate and ascorbic acid solution as described above.

Determination of organic acids

All the root exudate samples collected from pigeonpea genotypes grown in solution culture as described above were passed through a cation exchange resin consisting of Dowex 50WX8-400 (200–400 mesh, H^+ form) and then through an anion exchange resin consisting of Dowex 1X8-400 (200–400 mesh, $HCOO^-$ form). The anionic fraction was eluted with 5 M HCOOH and evaporated to dryness using a rotary evaporator at 40 °C. The residue was dissolved in 1 mL dilute H_2SO_4 and passed through a membrane filter (0.45 μm). The organic acid (citric, malic, malonic, succinic and piscidic) concentrations in root exudates were determined by HPLC equipped with an ion-exclusion column. Piscidic acid [(+)-(2*S*, 3*R*)-] was synthesized from 4-hydroxyphenyl-pyruvic acid (Toshima et al., 1999) and used as a standard.

Extraction efficiency of unavailable insoluble P in Alfisol by organic acids

The amount of phosphate released into solution by each organic acid (citric, malic, succinic and piscidic

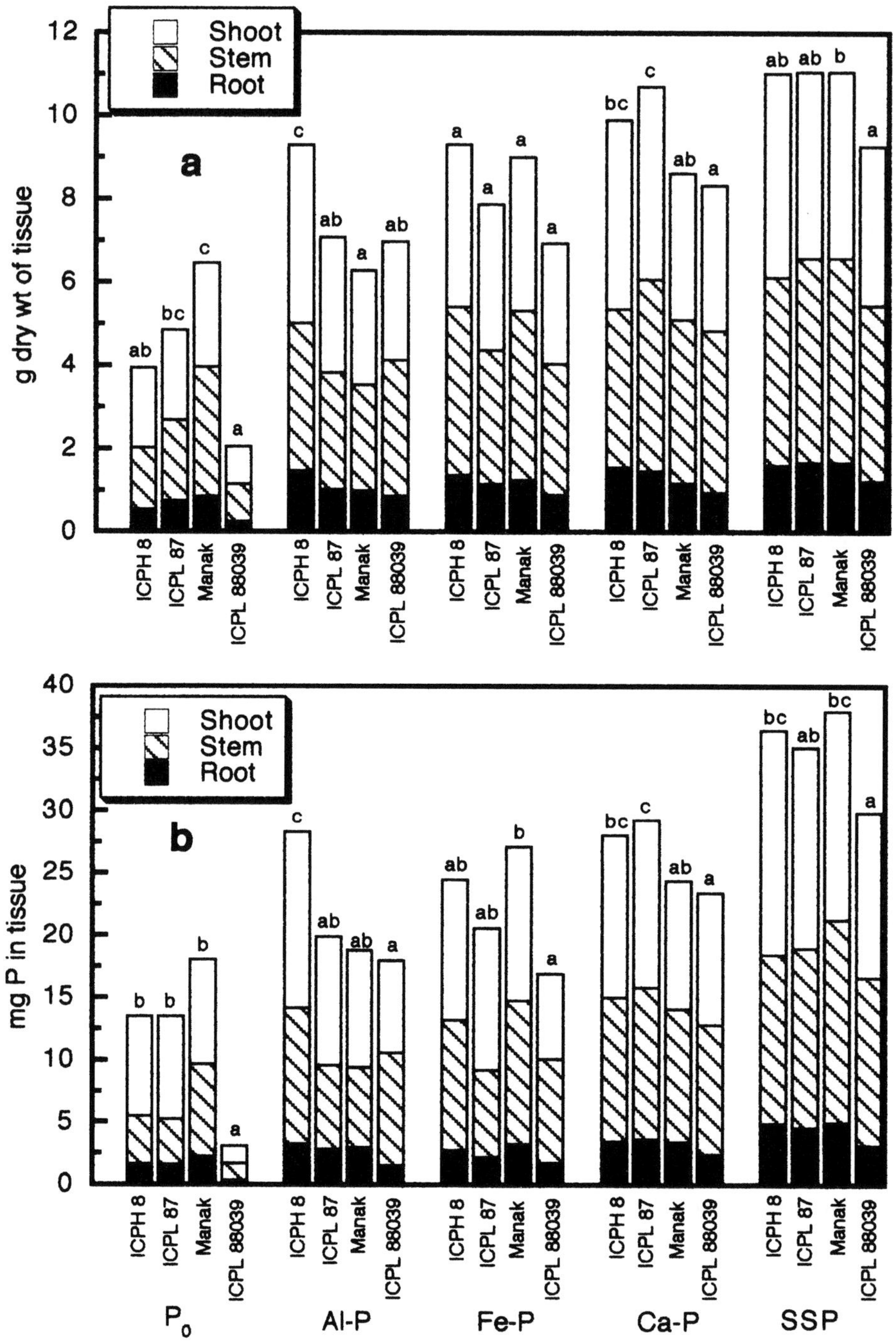

Figure 1. Comparison of dry weight (a) and P content (b) of plant tissues grown on an Alfisol applied with different P sources; $AlPO_4$ (Al-P), $FePO_4 \cdot H_2O$ (Fe-P), apatite (Ca-P), single superphosphate (SSP) and no P added (P_0), among four pigeonpea genotypes. Plants were harvested at the beginning of the flowering stage. Each column represents the mean of four replications. Different letters indicate statistical difference (P<0.05) among genotypes in each treatment. Duncan's Multiple Range Test was performed to separate the means.

acid) was assessed by extracting 1 g of Alfisol with 10 mL of each organic acid solution (pH 4.5) at concentrations ranging between 0 and 5 mM according to Jones and Darrah (1994). After shaking for 60 min, samples were centrifuged at 2000 *g* for 10 min and the solutions were passed through filter paper (No. 6). As organic acids disturb the developing color of P, the solution samples were evaporated to dryness on a hot plate (180 °C), digested by HNO_3 and $HClO_4$ (5:1 v/v), and re-evaporated. The P concentration in the digests were estimated by the molybdenum-ascorbic acid method.

Results

The characteristics of the Alfisol used in this pot experiment before and after application of P from different sources are shown in Table 1. There was significantly more Fe-bound P in Alfisol without P fertilizer application, followed by Ca-P and Al-P. The pH was decreased slightly by applying SSP and was not affected by the other P sources. Approximately 70–80% of $Ca_5OH(PO_4)_3$ added was detected as available P by both methods, whereas approximately 40–45% of $AlPO_4$ and $FePO_4 \cdot H_2O$ added was detected as available P by Bray II method. Most $Ca_5OH(PO_4)_3$ added was detected in Ca-P fraction, but the added $AlPO_4$ was hardly detected in the Al-P fraction and nor $FePO_4 \cdot H_2O$ in the Fe-P fraction.

The total dry weight and total P content of the plant tissues in all the genotypes except for Manak were in the order of $P_0 <$ Al-P, Fe-P $<$ Ca-P $<$ SSP and those of Manak were in the order of P_0, Al-P $<$ Ca-P $<$ Fe-P $<$ SSP (Figure 1a, b). Similar results were reported by Fujita et al. (1995). The dry weight and P contents of each tissue from Manak in the P_0 and Fe-P treatments were higher than the other genotypes. In all treatments, ICPL 88039 had the lowest the dry weight and P content in all its tissues, and this difference was significant under P_0 application. The P available in soils after cultivating pigeonpea genotypes under P_0 application showed an increasing trend compared to the soil before cultivation and the soil without any plants for 60 days but supplied with deionized water (Figure 2).

The Fe-P on filter paper was hardly solubilized by low pH at 4.5 and acetate buffer (Figure 3a). Citric acid had the highest Fe-P solubilizing activity among organic acids tested (Figure 3a–f). Two μL of 5 mM of citric acid, i.e. 10 nmol was the minimum detectable

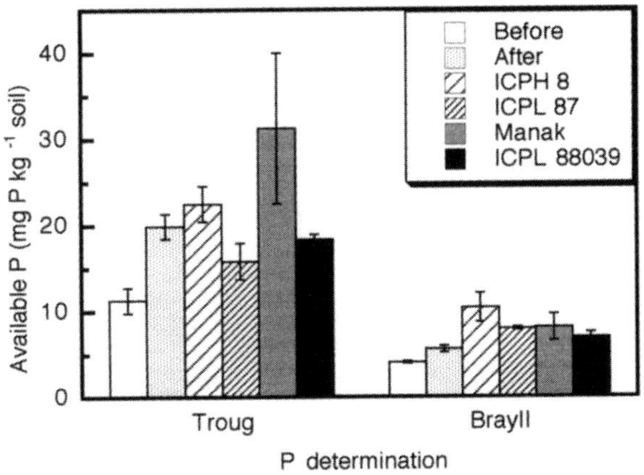

Figure 2. Available P in an Alfisol before and after cultivation. After cultivation of pigeonpea genotypes on an Alfisol without application of P (P_0), available P was determined by both Troug and Bray II methods. The data were compared with P available in the soil before cultivation (Before) and that of a pot of soil retained without any plants for 60 days supplied with water (After). The columns and vertical bars represent the means and the standard errors in four replications.

concentration (Figure 3b). Different detectable limits were observed among organic acids. Twenty nmol of malic acid and piscidic acid were detectable (data not shown), but 1 μmol of succinic acid (Figure 3f) was hardly detectable. A white spot on the filter paper was observed when Manak seedlings were grown in low-P conditions, indicating that root exudates contained Fe-P solubilizing substances (Figure 3g). The solubilizing activity of Fe-P was hardly detected in the other genotypes (data not shown). The extensive white spots were observed at approximately 2-cm from root-tips of Manak, indicating that Fe-P solubilizing substances was released mainly from root-tips (Figure 3h).

The release rates of citric and piscidic acids were increased with low-P treatment for both 15 and 30 days in all the genotypes except for some data of Manak (Table 2). There was approximately 3-fold difference in the citric acid release between the low-P and high-P treatments after 15 days in all genotypes except for ICPL 87. The piscidic acid released by genotypes following low-P treatment for 15 days was slightly higher than that of high-P treatment except for Manak. The organic acid release after 30 days decreased as compared with that after 15 days because of increased external P concentration at low-P treatment. Nevertheless, approximately 2-fold difference in piscidic acid release was observed between treatments at 30 days. There was no significant differences in the amounts of succinic, malic and malonic acids released from

Table 1. Chemical characteristics of Alfisols mixed with various P sources in a pot experiment. Values are means (n=3) with standard error in parentheses

P sources	pH	Available P (mg kg^{-1} soil)		Inorganic P (mg kg^{-1} soil)		
	(H_2O)	Troug	Bray II	Al-P	Ca-P	Fe-P
P_0	6.2	11.2 (1.47)	4.1 (0.13)	2.1 (0.41)	13.5 (1.07)	34.7 (0.36)
SSP	6.0	18.7 (0.42)	13.4 (1.55)	1.2 (0.95)	13.9 (2.91)	41.5 (2.79)
$AlPO_4$	6.2	11.9 (1.98)	22.0 (1.62)	2.1 (1.26)	10.2 (0.81)	52.6 (4.81)
$Ca_5OH(PO_4)_3$	6.2	42.4 (0.48)	31.0 (1.85)	0.8 (0.55)	41.7 (2.91)	46.7 (3.12)
$FePO_4 \cdot H_2O$	6.2	13.4 (1.05)	20.3 (0.49)	3.1 (1.24)	17.9 (1.85)	50.0 (4.19)

Table 2. Effect of external P concentration and the treatment period on the release rates of citric and piscidic acids from the roots of pigeonpea genotypes

Genotype	P treatment (μM P)	Period (days)	Citric acid	Piscidic acid
			(μmol g^{-1} root dry wt 24 h^{-1})	
ICPH 8	2	15	0.182 (0.01)	0.942 (0.01)
	80		0.067 (0.05)	0.739 (0.13)
	low P/high P		2.72	1.27
	4	30	0.091 (0.01)	0.425 (0.06)
	80		ND[a]	0.183 (0.03)
	low P/high P			2.32
ICPL 87	2	15	0.097 (0.05)	0.875 (0.34)
	80		0.071 (0.04)	0.674 (0.11)
	low P/high P		1.37	1.30
	4	30	0.114 (0.02)	0.364 (0.04)
	80		0.040 (0.04)	0.190 (0.02)
	low P/high P		2.85	1.92
Manak	2	15	0.178 (0.03)	0.801 (0.06)
	80		0.071 (0.07)	1.01 (0.07)
	low P/high P		2.50	0.79
	4	30	0.102 (0.01)	0.403 (0.03)
	80		0.127 (0.07)	0.220 (0.06)
	low P/high P		0.80	1.83
ICPL 88039	2	15	0.190 (0.01)	1.05 (0.09)
	80		0.063 (0.06)	1.00 (0.37)
	low P/high P		3.01	1.05
	4	30	0.108 (0.01)	0.404 (0.01)
	80		0.097 (0.04)	0.230 (0.00)
	low P/high P		1.11	1.76

[a]ND = not detectable.

The release rates of citric and piscidic acids were expressed as the amount (μmol) of the release to total root dry weight (g). The external P concentration in the low-P treatment was increased to 4 μM to prevent the plants from dying of severe P deficiency after 15 days. Values are means (n=2) with standard error in parentheses. The values of low P/ high P express the ratio of the release rates in low P treatment (2 or 4 μM P) to that in high P (80 μM P) treatment.

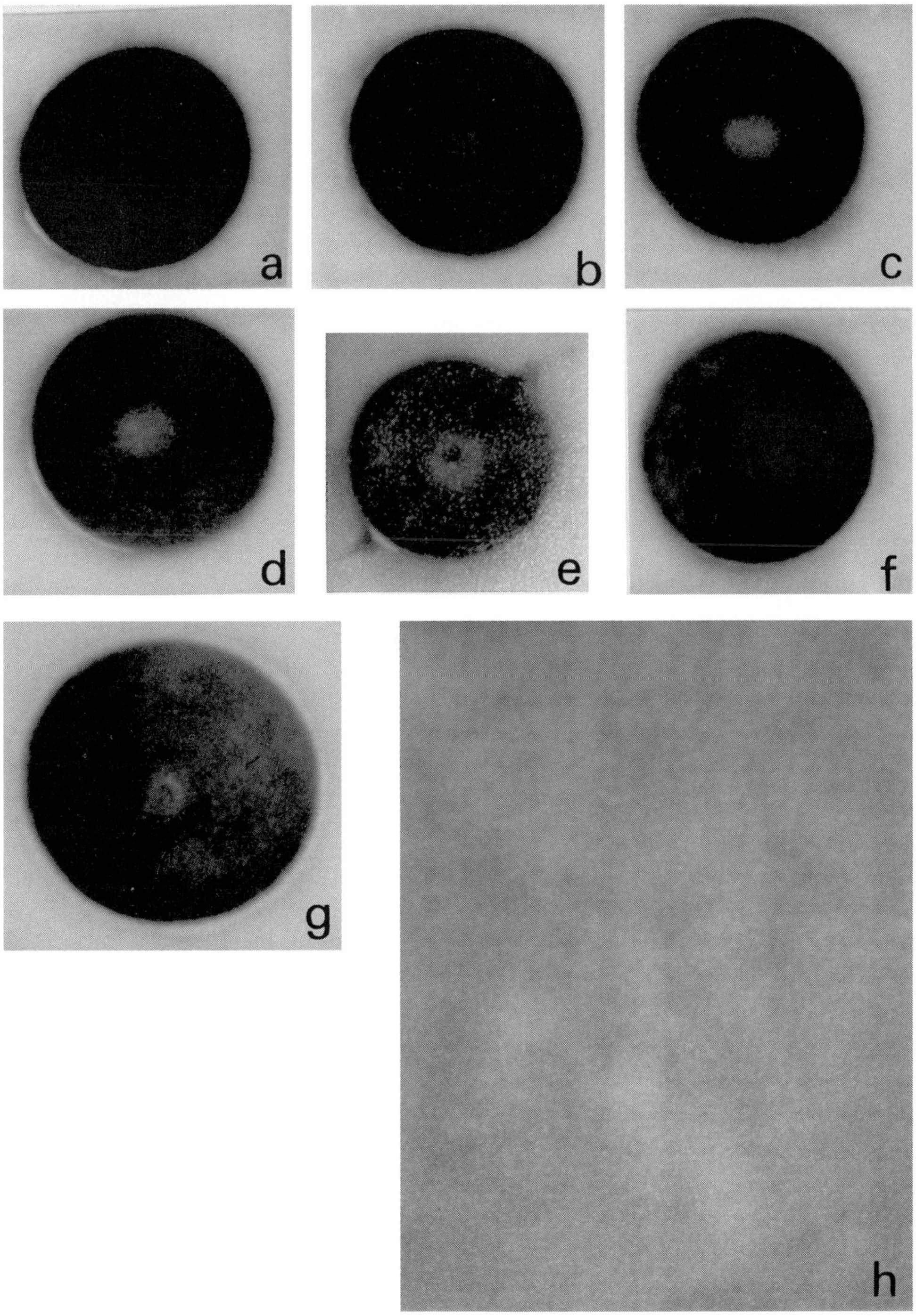

Figure 3. Fe-P solubilizing activity of carboxylic acids, root exudates, and its release portion. (a) control, (b) 5 mM citric acid, (c) 100 mM citric acid, (d) 100 mM malic acid, (e) 100 mM piscidic acid, (f) 0.5 M succinic acid, (g) root exudates collected from Manak roots after low-P treatment (2 μM) for 30 days, (h) the roots of Manak treated with low-P for 30 days were placed on the filter paper containing $FePO_4$ solution. After 24 h, the paper was stained with molybdenum-ascorbic acid. Intensive white spots indicate where the root tips were placed.

Table 3. Change in the release rates of citric and piscidic acids depending on change in internal P concentration

Genotype	P treatment[a] (μM P)	Period (days)	Citric acid (μmol g^{-1} root dry wt 24 h^{-1})	Piscidic acid	Shoot (mg P g^{-1} dry wt)	Root
ICPH 8	L	15	0.058 (0.03)	1.63 (0.10)	1.80 (0.16)	3.81 (0.42)
	L-L	15-7	0.069 (0.01)	4.30 (1.24)	1.92 (0.15)	4.69 (0.91)
	L-H	15-7	0.041 (0.01)	2.45 (0.05)	6.58 (0.31)	9.54 (0.76)
ICPL 87	L	15	0.045 (0.03)	1.42 (0.11)	1.90 (0.22)	6.37 (0.80)
	L-L	15-7	0.048 (0.02)	2.90 (0.67)	1.78 (0.01)	6.14 (0.69)
	L-H	15-7	0.029 (0.01)	2.26 (0.47)	6.65 (1.12)	9.78 (0.61)
Manak	L	15	0.085 (0.04)	2.12 (0.07)	1.89 (0.40)	5.38 (1.02)
	L-L	15-7	0.048 (0.00)	5.54 (0.17)	1.49 (0.23)	6.83 (0.56)
	L-H	15-7	0.057 (0.01)	4.72 (0.23)	5.74 (0.46)	8.89 (0.31)
ICPL 88039	L	15	0.063 (0.01)	1.63 (0.10)	1.82 (0.22)	6.07 (0.92)
	L-L	15-7	0.049 (0.00)	5.23 (0.30)	1.74 (0.04)	6.56 (0.78)
	L-H	15-7	0.026 (0.01)	2.89 (0.50)	7.20 (0.47)	7.85 (0.37)

[a] Seedlings pre-exposed to 2 μM P for 15 days (L, 15) were transferred to the 2 (L-L) or 80 μM P (L-H) and then treated for more 7 days (15-7). Values are means (n= 2–3) with standard error in parentheses.

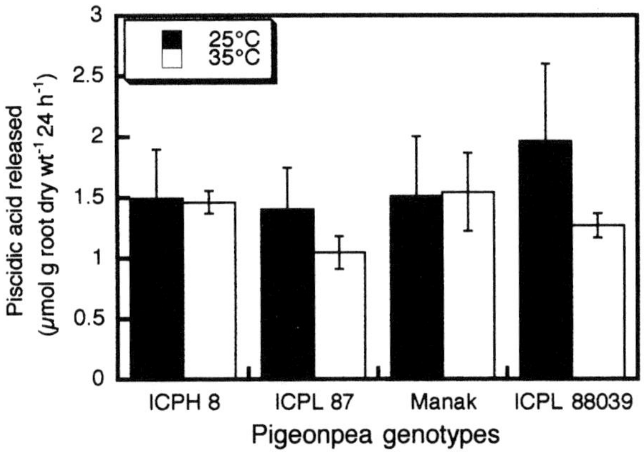

Figure 4. Effect of temperature on piscidic acid release from roots. After 15 days at low-P treatment, seedlings were transferred to container containing 1 mM $CaCl_2$ solution. Root exudates were collected and put into two water baths that were continuously maintained at 25 °C or 35 °C. The columns and vertical bars represent the means and the standard errors from two replications.

roots between treatments or among genotypes (data not shown). The release of piscidic acid from roots increased remarkably in all the genotypes after continuous treatment with the same low P concentration (2 μM) for 7 days (L-L) following low-P pretreatment for 15 days (L), but the P concentrations in shoots and roots remained similar despite this increase in piscidic acid exudation (Table 3). The P concentrations in shoots and roots increased remarkably with increasing external P concentration, and the release rate of citric acid decreased in all the genotypes. On the other hand, the release of piscidic acid did not decrease but was rather enhanced as compared with that of 15 days at low-P treatment. Two temperature treatments (25 °C and 35 °C) had no significant effect on the release of piscidic acid by genotypes although the release from ICPL 87 and ICPL 88039 decreased at 35 °C (Figure 4).

Among the four organic acids, citric acid was most able to dissolve insoluble P in the Alfisol (Figure 5). Succinic acid could hardly dissolve insoluble P even at a high concentration. Piscidic acid had higher solubilizing ability than malic acid. At least 100 μM of any organic acid was required to release 1.6 μM P of insoluble P in the Alfisol into the soil solution.

Discussion

Ae et al. (1990) first reported that pigeonpea exuded piscidic acid from roots under low available P conditions. This is a unique carboxylic acid detected mainly in pigeonpea and there are no reports of its occurrence in other plants. Piscidic acid was identified in root exudates of all pigeonpea genotypes tested in this study, irrespective of external P levels (Table 2). Pis-

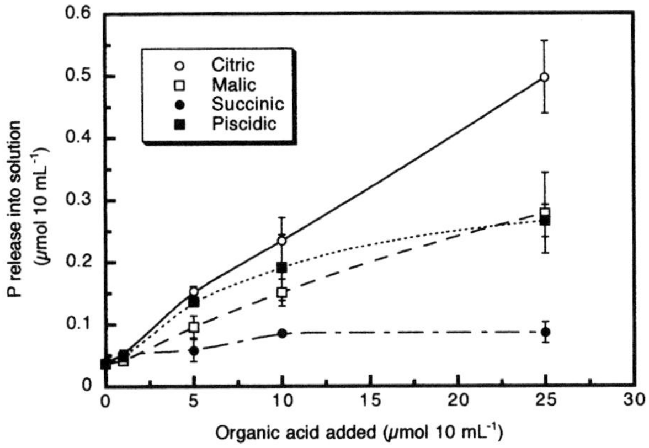

Figure 5. Extraction efficiency of insoluble P from an Alfisol by four organic acids. The P concentration in solution was assessed by extracting the Alfisol with each organic acid solution (pH 4.5) for 60 min. The values represented are the means and standard errors from six replications.

cidic acid was also observed in root exudates of four other pigeonpea genotypes in addition to those shown in Table 2 (data not shown). The release of piscidic acid was enhanced by low-P conditions (Table 2) and increased with extended severe low-P treatment period (Table 3). These findings suggest that piscidic acid release from roots is a general characteristic of pigeonpea, and that its release is induced more by P deficiency. Citric acid release also was induced by low-P treatment, but the amount released was approximately 1/5–1/100 that of piscidic acid (Tables 2 and 3). Citric acid was better able to dissolve Fe-P than piscidic acid (Figures 3 and 5). However, citric acid seemed to have a lesser role in solubilizing Fe-P by pigeonpea root exudates (Figure 3g, h) than piscidic acid because of it is only released in very small amounts (Tables 2 and 3).

The release of citric and piscidic acids depended on internal P concentration, especially that in shoots (Table 3). In addition, piscidic acid release increased without further decreasing in the P concentrations in shoots and roots. This result suggests that piscidic acid release from the roots might depend on low-P treatment period and plant age besides internal P concentration (Otani and Ae, 1997). The release of piscidic acid was enhanced regardless of the significant increase in P concentration of shoots after high-P treatment following low-P pretreatment (Table 3). This suggests that inducible piscidic acid synthesis may be involved in its release from pigeonpea roots as shown in the release pattern of citric acid from roots of *Cassia tora* L. in response to Al toxicity where a significant secretion of citric acid was induced by 6 h without Al after 3-h pulse with Al (Ma et al., 1997). The mechanisms of synthesis and release of piscidic acid from pigeonpea roots need further investigation. The release of piscidic acid in all the genotypes tested was hardly affected by the comparatively high temperature of 35 °C, a temperature to be expected in soils in tropical areas (Figure 4). Therefore, piscidic acid exudation was thought to be a specific response to P deficiency in shoots.

Genotypic variability in the release rate of piscidic acid from roots was not found among pigeonpea genotypes, but the releases from Manak and ICPL 88039 were a little higher than those from the other two genotypes under low-P treatment throughout this study (Tables 2 and 3, Figure 4). Jones (1998) pointed out that organic acid release rates calculated based on total root mass are greatly underestimated, and that weight ratio between treatments can severely bias the results. The Fe-P solubilizing substances were localized to root-tips (Figure 3h), corresponding with the findings of Hoffland et al. (1989) who observed rhizosphere acidification of the root-tips of rape under P-stressed conditions. This study had no alternative but to express the exudates in terms of total root mass due to the difficulty of collecting root-tips. Small difference in the release rate of piscidic acid among genotypes in low-P treatment may be attributed to calculation based on total root dry mass. Phosphorus deficiency significantly increases the leakage of the plasma membrane because of the decrease in the level phospholipids (Ratnayake et al., 1978). Therefore, the approximately 2- or 3-fold increase in piscidic and citric acids in the low-P treatment shown in Table 2 might not indicate an active release implicated in the P acquisition, but a secondary root response.

Pigeonpea could utilize sparingly soluble P sources, irrespective of genotype (Figure 1a, b). This is thought to be efficient utilization of available P quantified by the Bray II method (Table 1). The greater dry matter production and P uptake of Manak at P_0 and Fe-P application than those of the other genotypes were presumed to be due to its superior ability in Fe-P utilization (Figure 1a, b). The dry weight and P uptake of ICPL 88039 were considerably lower than Manak under P_0 application, although the amount of piscidic acid released from the roots of ICPL 88039 was comparable to that of Manak. This result suggests that Manak might have another mechanism (s) related to P acquisition and utilization from low-P availability soils in addition to piscidic acid release.

Solubilization of P compounds in the soil by organic acids is achieved by the replacement between organic acids and P on the ligand exchange surface and by formation of complexes between organic acids and metal ions such as Fe, Al and Ca. Complex formation depends on the number and position of carboxyl (-COOH) and hydroxyl (-OH) functional groups in the organic acids (Hue et al., 1986). Solubilization of P by several organic acids from Alfisol where the Fe-bound P dominates corresponded to the stability constant for Al (log K_{Al}; 12.26 for citric acid, 6.21 for tartaric acid, 6.00 for malic acid, and 4.62 for succinic acid, cited in the data of Hue et al., 1986) (Figure 5). This also was confirmed by the assay method for Fe-P solubilizing activity (Figure 3). The stability constant for Fe^{3+} is comparable to that of Al^{3+} for citric and malic acids (Cline et al., 1982). The stability constant of Fe-piscidic acid is predicted to be similar to that of Fe-tartaric acid because carboxyl and hydroxyl groups of the tartarate portion of piscidic acid are involved in chelation of Fe^{3+} (Ae et al., 1990).

It remains unknown whether the amount of organic acid released from the roots is sufficient to dissolve sparingly soluble P in soils, thereby enabling to survive and alleviate P deficiency. The total amount of citric and piscidic acids released from pigeonpea roots was in the range of 0.1–0.7 μmol plant^{-1} 24 h^{-1} depending on P stress status, and its amount was considerably lower than the amounts of citric (25–175 μmol plant^{-1} 24 h^{-1}) and malic (25–200 μmol plant^{-1} 24 h^{-1}) acids released from P-deficient *Lipinus albus* (Johnson et al., 1996). In addition, the amount was not sufficient to extract insoluble P in Alfisols (Figure 5) even if the root exudates collected from Manak could solubilize the Fe-P on filter paper (Figure 3g). The concentration of organic acids at the root surface is predicted by a simulation model based on data of organic acid efflux in bulk solution (Jones et al., 1996a; Pellet et al., 1997; Ryan et al., 1995). Sorption of organic acids to a soil's solid phase (Hue, 1991) and degradation by micro-organisms (Jones et al., 1996b) are important factors to be considered when estimating the concentration of organic acids at the root surface. Figure 5 indicates that the concentration of organic acids at the pigeonpea root surface would need to be more than 100 μM to mobilize P in Alfisols. The real concentration of organic acids at the root surface will need to be measured when a crop is growing.

In conclusion, we could not explain the difference in the plant growth and the P uptake among pigeonpea genotypes grown in an Alfisol through the citric and piscidic acids released from their roots. However, there is need of a similar research using a large number pigeonpea genotypes. The filter paper technique demonstrated in Figure 3 might be useful as a rapid and simple screening system for identifying those genotypes effective in extracting P by root exudates from a large number of genotypes. An increase in P available in an Alfisol was observed after cultivation of pigeonpea without application of P (Figure 2). This finding indicates the possibility that pigeonpea may be a leguminous crop effective in enhancing soil fertility.

Acknowledgements

We are grateful to all the supporting staff of the Project funded by the Government of Japan for their help with the experiments.

References

Adu-Gyamfi J J, Ishikawa S, Nakamura T and Nakano H 2000 Genotypic variability and physiological characteristics of crop plants adapted to low-nutrient environments. *In* Food Security in Nutrient-Stressed Environments: Exploiting Plants' Genetic Capabilities, Proceedings of an International Workshop 27–30 September 1999, Patancheru, India. Developments in Plant and Soil Sciences volume 94. Ed. J J Adu-Gyamfi. Kluwer Academic Publishers, Dordrecht, The Netherlands. (in press).

Ae N, Arihara J, Okada K, Yoshihara T and Johansen C 1990 Phosphorus uptake by pigeonpea and its role in cropping systems of the Indian subcontinent. Science 248, 477–480.

Cline G R, Powell, P E, Szaniszlo P J and Reid C P P 1982 Comparison of the abilities of hydroxamic, synthetic and other natural organic acids to chelate iron and other ions in nutrient solution. Soil Sci. Soc. Am. J. 46, 1158–1164.

Fujita K, Chaudhary M I, Adu-Gyamfi J J and Yoshizawa D 1995 Dinitrogen fixation and growth responses to phosphorus and aluminum application in pigeon pea (*Cajanus cajan* L.). Soil Sci. Plant Nutr. 41, 729–735.

Gardner W K, Barber D A and Parberry D G 1983 The acquisition of phosphorus by *Lupinus albus* L. III. The probable mechanism by which phosphorus movement in the soil/root interface in enhanced. Plant Soil 70, 107–124.

Hoffland E 1992 Quantitative evaluation of the role of organic acid exudation in the mobilization of rock phosphate by rape. Plant Soil 140, 279–289.

Hoffland E, Findenegg G R and Nelemans J A 1989 Solubilization of rock phosphate by rape. II. Local root exudation of organic acids as a response to P-starvation. Plant Soil 113, 161–165.

Hue N V 1991 Effects of organic acids/anions on P sorption and phytoavailability in soils with different mineralogies. Soil Sci. 152, 463–471.

Hue N V, Craddock G R and Adams F 1986 Effect of organic acids on aluminum toxicity in subsoils. Soil Sci. Soc. Am. J. 50, 28–34.

Johnson J F, Vance C P and Allan D L 1996 Phosphorus deficiency in *Lupinus albus*. Altered lateral root development and enhanced expression of phospho*enol*pyruvate carboxylase Plant Physiol. 112, 31–41.

Jones D L 1998 Organic acids in the rhizosphere – a critical review. Plant Soil 205, 25–44.

Jones D L and Darrah P R 1994 Role of root derived acids in the mobilization of nutrient from the rhizosphere. Plant Soil 166, 247–257.

Jones D L, Darrah P R and Kochian L V 1996a Critical evaluation of organic acid mediated iron dissolution in the rhizosphere and its potential role in root iron uptake. Plant Soil 180, 57–66.

Jones D L, Prabowo A M and Kochian L V 1996b Kinetics of malate transport and decomposition in acid soils and isolated bacterial populations: The effect of microorganisms on root exudation of malate under Al stress. Plant Soil 182, 239–247.

Ma J F, Zheng S J and Matsumoto H 1997 Specific secretion of citric acid induced by Al stress in *Cassia tora* L. Plant Cell Physiol. 38, 1019–1025.

Murphy J and Riley J P 1962 A modified single solution method for the determination of phosphate in natural waters. Anal. Chem. Acta 27, 31–36.

Ohwaki Y and Hirata H 1992 Difference in carboxylic acid exudation among P-starved leguminous crops in relation to carboxylic acid contents in plant tissues and phospholipid level in roots. Soil Sci. Plant Nutr. 38, 235–243.

Otani T and Ae N 1997 The exudation of organic acids by pigeonpea roots for solubilizing iron- and aluminum-bound phosphorus. *In* Plant Nutrition-for Sustainable Food Production and Environment. Eds T Ando, K Fujita, T Mae, H Matsumoto, S Mori and J Sekiya. pp. 325–326. Kluwer Academic Publishers, Dordrecht, The Netherlands.

Pellet D M, Paperinik L A, Jones D L, Darrah P R, Grunes D L and Kochian L V 1997 Involvement of multiple aluminium exclusion mechanisms in aluminium tolerance in wheat. Plant Soil 192, 63–68.

Ratnayake M, Leonald R T and Menge J A 1978 Root exudation in relation to supply of phosphorus and its possible relevance to mycorrhizal formation. New Phytol. 81, 543–552.

Rengel Z 1999 Physiological mechanisms underlying differential nutrient efficiency of crop genotypes. *In* Mineral Nutrition of Crops: Mechanisms and Implications. Ed. Z Rengel. pp. 227–265. The Haworth Press, New York.

Ryan P R, Delhaize E and Randall P J 1995 Malate efflux root apices and tolerance to Aluminium are highly correlated in wheat. Aust. J. Plant Physiol. 22, 531–536.

Sekiya K 1983 Fractionation of soil inorganic phosphorus. *In* Methods of Soil Analysis. Ed. Dojo Yobun Bunsekihou. pp. 235–238. Yokendo, Tokyo, Japan (in Japanese).

Subbarao G V, Ae N and Otani T 1997a Genetic variation in acquisition, and utilization of phosphorus from iron-bound phosphorus in pigeonpea. Soil Sci. Plant Nutr. 43, 511–519.

Subbarao G V, Ae N and Otani T 1997b Genetypic variation in iron- and aluminum-phosphate solubilizing activity of pigeonpea root exudates under P deficient conditions. Soil Sci. Plant Nutr. 43, 295–305.

Toshima H, Saito M and Yoshihara T 1999 Total synthesis of (+)-(2*S*, 3*R*)-piscidic acid via catalytic asymmetric dihydroxylation of a trisubstituted olefin. Biosci. Biotechnol. Biochem. 63, 964–967.

Yoshihara T, Ichihara A and Sakamura S 1971 The synthesis of dimethyl esters of *dl-O, O*′-dimethylfukiic acid and *dl-O, O*′-dimethylepifukiic acid. Agr. Biol. Chem. 35, 1822–1824.

Zheng S J, Ma J F and Matsumoto H 1998 High aluminum resistance in buckwheat. I. Al-induced specific secretion of oxalic acid from root tips. Plant Physiol. 117, 745–751.

J.J. Adu-Gyamfi (Ed.), *Food security in nutrient-stressed environments: exploiting plants' genetic capabilities*: 123–131, 2002.

Root cell-wall properties are proposed to contribute to phosphorus (P) mobilization by groundnut and pigeonpea

N. Ae[1] & R.F. Shen
Division of Soil Sciences, National Institute of Agro-Environmental Sciences, Tsukuba, Ibaraki, 305-8604, Japan.
[1]*Corresponding author**

Key words: cell wall, groundnut, iron-bound phosphorus, pigeonpea, root epidermis

Abstract

Groundnuts showed a superior ability to take up phosphorus (P) from two soils of extremely low fertility, where sorghum and soybean died of P deficiency. This ability could not be attributed to differences in root development, to P uptake parameters such as C_{min}, or to the excretion of root exudates capable of solubilizing iron- (Fe-P) and aluminum-bound P (Al-P), the sparingly soluble P forms in soils. A new P solubilizing mechanism (called 'contact reaction') which occurs at the interface between root surface and soil particles, is therefore proposed. Isolated cell walls from groundnut roots solubilized more P from P-fixing minerals than those from sorghum and soybean roots. The P-solubilizing activity of groundnut root cell-walls might therefore be related to the superior growth of this crop under P-deficient conditions. The P-solubilizing active sites in groundnut root cell walls were located at the root surface and could act as chelating agent with Fe(III). This P-solubilizing active component in the cell walls could be extracted by NaOH, but not by HCl, and was identified as a small molecule through column chromatography with Sephadex LH-20. The P-solubilizing ability of pigeonpea root cell-walls was examined and found to be as high as that of groundnut. As pigeonpea plants excrete significant amount of root exudates with Fe-P solubilizing ability only after they flower, the P-solubilizing ability of root cell-walls may partially explain the high P efficiency of this species before it flowers.

Introduction

Fertilization of groundnut with phosphorus (P) has rarely shown positive growth responses (Hartzog and Adams, 1988; Walker et al., 1979; Walker et al., 1982). Unlike other crop species as maize and sorghum, groundnut can make use of the residue of fertilizers applied to preceding crops in rotation; and it can take advantage of minerals that are not normally available to other crops (Hallock, 1962). Chapman and Carter (1976) concluded that the fertilizer management of groundnut was unusual. In previous experiments, it was found that groundnut could survive on low-P soils. The major P fractions in these soils were iron-bound P (Fe-P) and aluminum-bound P (Al-P), both of which are sparingly soluble (Otani and Ae, 1996a). The P status of Ishigaki (Acrisol, Okinawa, Japan) and Nishinasuno (Andosol, Tochigi, Japan) soils, that had extremely low P availability (Table 1) and superior P uptake by groundnut (cv. Chibahandachi) compared to sorghum and soybean are shown (Table 2) for a reference (Ae et al., 1996). We could not explain this ability by the following factors: 1, C_{min} (Otani et al., 1996); 2, root length (Otani and Ae, 1996b), and 3, root exudates (Otani et al., 1996). Tanaka et al. (1995) examined root exudates of groundnut, and detected formic acid which had no chelating ability with Fe(III), but could reduce Fe(III) to Fe (II) resulting in an increase in the solubility of Fe-P. However, Otani et al. (1996) found that the overall Fe(III) reduction activity increased under Fe-deficient conditions and decreased under P-deficient conditions and that there were hardly any differences in the reduction activity among pigeonpea, groundnut and soybean even though these three crops are very different in their ability to survive in low-P soils. The exudation of the formic acid could not explain the vigorous growth of groundnut on low-P Andosol

* FAX No: +81-298-38-8271. E-mail: noriae@niaes.affrc.go.jp

This paper previously appeared in *Plant and Soil* **245**, 2002, pp. 95–103.

whose major P fraction was Al-P. The exudation of formic acid by groundnut roots might not be so important in releasing P from Fe-P as thought earlier. Also, it was not possible to detect considerable amount of root exudates that had $FePO_4$ solubilizing ability in groundnut (Otani et al., 1996). Thus another explanation for the superior P uptake by the groundnut roots was needed.

The driving force that moves P from soil to the plant root surface is diffusion. The strategy groundnuts use to survive in extremely low P surroundings must be related to the interface between root and soil particles. Nagarajah et al. (1970) reported that mucilage, consisting of partially methylated polygalacturonic acid, formed a complex with Fe(III) that resulted in the desorption of P from minerals. Gessa et al. (1997) cited that the three-dimensional arrangement of carboxyl groups of D-galacturonic acid in the pectic network of plant cell walls had a high affinity for Fe(III) and that this network may possess the ability to solubilize Fe-P. These findings suggest that roots could have the ability to release P from Fe-P and Al-P. Ae et al. (1996) found that groundnut roots had high P-solubilizing ability for $FePO_4$, and $AlPO_4$, based on their contact reaction, but that this ability was not derived from polygalacturonic acid in pectic substances. Many questions remain open: is this contact reaction applicable to crop species other than groundnut? What reactions are involved in the P-releasing procedure? And, what is the nature of the chemical compounds with this P-solubilizing activity that are located at the groundnut root surface? In this paper, the authors try to answer these questions on the basis of the proposed contact reaction theory.

Materials and methods

Cell wall preparation and CEC measurement

Pigeonpea (*Cajanus cajan* L. cv. ICPL 87), soybean (*Glycine max* (L.) Merrill cv. Tachinagaha), sorghum (*Sorghum bicolor* (L.) Moench cv. Hybrid-sorgo), groundnut (*Arachis hypogaea* L. cv. Chibahandachi) and maize (*Zea mays* L. cv. DK250) were grown for 2 months on sand to which nutrient solution was applied twice a week. Fresh roots were collected from pots and 'crude cell walls' and 'purified cell walls' were prepared according to the method described by Ae et al. (1996). Thirty g fresh roots were cut into 5-mm long pieces and ground with 250 mL distilled water in a blender (Polytron PT300) at 27 000 rpm for 5 min. The suspension was filtered through Whatman No. 6 filter paper and roots were resuspended in 250 mL of 0.5 *M* HCl for 30 min to remove minerals on the root surface. The suspension was again filtered and washed three times thoroughly with water, rinsed with 200 mL ethanol, and dried at 50 °C to obtain 'crude cell walls'. To prepare 'pure cell walls', the root suspension collected after treatment with 0.5 *M* HCl was filtered and washed three times with water. The washed roots were resuspended in 250 mL of 0.1% sodium deoxycholate for 1 h to remove intercellular components and plasma membranes. The suspension was filtered again and washed with water, rinsed with 200 mL of ethanol and dried at 50 °C. Root samples were collected 19, 35, 54, 78 and 109 days after sowing (DAS) to investigate the effect of growth stage on their P-solubilizing ability.

Cation exchange capacity (CEC) was measured as described by Ida et al. (1958). Root powder (50 mg) was suspended in 40 mL of 0.5 *M* HCl and shaken occasionally for 1.5 h. The suspended sample was filtered using Whatman glass microfibre (GF/C) and washed with until Cl^{-1} was no longer detected. The sample was then transferred to 50 mL aliquots of 0.1 *M* calcium (Ca) acetate with pH that varied 4 to 7 (adjusted with HCl). The root suspension was shaken occasionally for 1.5 h, filtered and washed with water. Roots were again put into 50 mL of 0.005 *M* HCl and shaken for 1.5 h and filtered through GF/C. An atomic absorption spectrophotometer (Hitachi 170-50) was used to determine the Ca concentration in the filtered solution and CEC was estimated.

Solubilizing activity of cell walls

Forty mg of the cell-wall preparation was added to a centrifuge tube containing either 2.0 ml of 2.0 *M* acetate buffer at pH 5.0 or 1/7 *M* barbital-acetate buffer (pH 3.8 to 6.7) and 2.0 mL of mineral suspended solution (10 mg of mineral mL^{-1} water) were applied to this tube. After shaking the tube reciprocally (10 cm, 60 rpm) for 2 h, it was centrifuged at 4000 × *g* for 15 min. The resulting supernatant was passed through a 0.2 μm filter and the P concentration of the filtrate was determined (Murphy and Riley, 1962). The P-solubilizing activity of the root cell walls was obtained by subtracting the P contents of the control (with the mineral suspension only) and the blank (with cell walls only) from the P content in the sample (cell walls with the mineral suspension). There were discrepancies in the activities but this could possibly be because different batches of cell walls and/or different Fe-P sources were used in the experiments, but the trend was always the same regardless of the batch and Fe-P source.

Preparation of P fixed minerals

Goethite was prepared according to Atkinson et al. (1967). Allophane was collected from clay fraction of an Andosol in Kanuma, Tochigi prefecture Japan. Fe_2O_3, a common contaminant on the surface of Allophane was removed as described by Jackson et al. (1986). Silica-alumina (Si/Al) gel was prepared by the method of Nanzyo and Watanabe (1981). Hematite was purchased from Wako Chemicals, Japan. Phosphorus (P) as 0.5 *M* NaH_2PO_4 was adsorbed onto these 4 minerals at pH 5.0 using acetate buffer, and then the minerals were washed thoroughly with water. Rock minerals, Variscite with Wavelite (location: Avant, Arkansas, USA), Strengite with Dufrenite and Rockbridgite (location: Indian Mountain, Alabama, USA), were ground thoroughly before use. $FePO_4$ and $AlPO_4$ (Wako Chemicals, Japan) were also used as scarcely soluble P-fixing minerals. These P-fixing minerals were used in suspension (10 mg mL^{-1}). Most P solubilization experiments were performed using $FePO_4$ that was well washed with water to remove soluble components before use.

Effect of 0.5 M NaOH treatment on groundnut roots

Roots of 50-day groundnut grown in solution culture (Arnon, 1936) were put into a 0.5 *M* NaOH solution for various periods, 30 seconds, 5 min, 20 min, and 1 h. Four groundnut plants treated with NaOH were thoroughly washed with water and of these, two were put into water again for observation, the roots of the other two plants were cut off, and used to prepare 'pure cell walls'. Subsequently, the effect of soaking time in NaOH on the ability of their cell wall preparation to solubilize P from $FePO_4$ was determined.

Reaction between groundnut root cell wall and chelating chemicals

To a test tube containing 40 mg of pure root cell-walls (Ae et al., 1996) from 2-month- old groundnut plants 5mL of 0.1 *M* $FeCl_3$ were added and the tube was gently shaken (30 times rolling per min.) for 2 h. Fe-treated cell walls were washed with 5 mL aliquots of water 10 times. Washed root cell-walls were air-dried on filter paper. Fe-treated root cell-walls were soaked in 5 mL aliquots of various concentration of cheating chemicals (1/200 m*M* to 1/20 *M*) in 1 *M* acetate buffer at pH 5.0 and the suspensions shaken for 2 h (30 times rolling per min). The mixtures were passed through 0.65 μm membrane filters, and their Fe contents determined using an atomic absorption spectrophotometer (Hitachi A-50, Japan).

Extraction of P-solubilizing active site

Root cell-walls of four crops, i.e., groundnut, maize, pigeonpea and soybean, were used to extract P-solubilizing active site using. 1.0 *M* NaOH and 0.5 *M* HCl as extractant. 1.5 mL of either NaOH or HCl were added to 40 mg of cell walls suspension in a glass mortar and ground for 1 min, then another 8.5 mL of extractant were added to make the total volume of extractant 10 mL. The mixtures of cell walls and extractant were put into a water bath at 80 °C for 2 h. Then, mixtures were filtered through a glass filter and washed with water. This extraction procedure was repeated twice. After neutralizing the supernatant with HCl or NaOH and drying the residue, it was suspended in water and its P-solubilizing ability was determined. To purify P-solubilizing active substances, the neutralized supernatants were dried and then dissolved in 50% ethanol. Samples were applied to a Sephadex LH-20 column chromatography (2.5 cm × 50 cm) and eluted 50% ethanol. The absorbance of each 3 mL by fraction was monitored at 280 nm and its P-solubilizing activity measured.

Results

Superior ability of groundnut to take up P from a low P-soil

The P availability of the Ishigaki soil was extremely low. The major P source was Fe-P, with a little Al-P, and negligible Ca-P. In contrast, Al-P was the major inorganic-P fraction in Nishinasuno soil (Table 1). In these soils the P uptake of groundnut was much higher than that of maize, sorghum, and soybean that died from lack of P (Table 2). This ability of groundnut to grow and survive in such condition indicated that this species might possess the ability to make use of Fe-P and Al-P.

Table 1. Phosphorus (P) status of soils used in the pot experiment

Soil	pH(H_2O)	Total-P (mg P kg^{-1})	Available P (mg P kg^{-1})		Inorganic P (mg P kg^{-1})			Organic P (mg P kg^{-1})		
			Truog	Bray 2	Ca-P	Al-P	Fe-P	Ca-P	Al-P	Fe-P
Ishigaki	4.5	109	0.9	0.9	0	20	58	0	4	7
Nishinasuno	5.1	829	0.2	1.6	2	90	75	1	111	439

Table 2. Phosphorus (P) uptake by groundnut, soybean, maize and sorghum from Ishigaki and Nishinasuno soils (pot experiment: nutrient solution without P was applied every 3 days)

Soil	pH (H_2O)	P uptake (mg P pot^{-1})			
		Groundnut	Soybean	Sorghum	Maize
Ishigaki	4.5	2.0±0.22	0.0±0.05[a]	0.0±0.01[a]	0.0±0.01[a]
Nishinasuno	5.1	2.1±0.48	0.1±0.13[a]	1.2±0.01	0.1±0.15[a]

Mean ± SE.
[a]Dead within 2 months after sowing.

Solubilization by root cell walls from P-fixed minerals

Cell walls isolated from groundnut roots could solubilize more P from all the minerals tested except Variscite than cell walls from sorghum and soybean roots (Table 3). Groundnut root cell walls could solubilize relatively more P from Hematite-P than from other P-fixing mineral (see P solubility % in Table 3). The cell walls of all crops, could solubilize relatively more P from P-fixing Fe minerals than from P-fixing Al minerals.

Effect of pH on CEC and on P-solubilizing ability from $FePO_4$

Groundnut cell wall showed a higher P solubilizing ability from Fe-P than soybean, sorghum, or maize irrespective of the variations in pH (Figure 1). For all crops the P- solubilizing ability decreased and CEC increased with increasing of pH. Soybean had the highest CEC, followed by groundnut and the cereals.

Effect of growth stage on P-solubilizing ability of groundnut roots

The root cell walls of groundnut showed high P-solubilizing ability even at 19 DAS and this ability increased gradually with the highest values at the pod-filling stage (Figure 2).

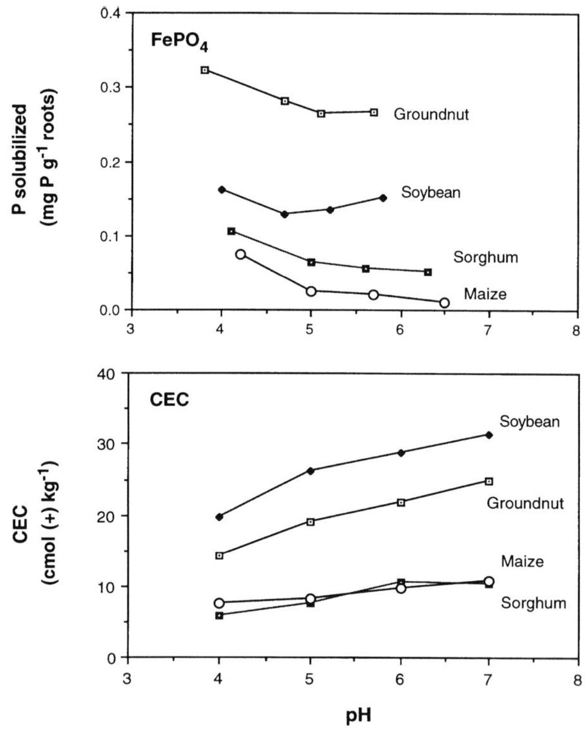

Figure 1. Relationship between pH and phosphorus (P) solubilizing-ability of plant root cell-walls from $FePO_4$ and on cation exchange capacity (CEC) in four different crop species. SE values were negligible compared to symbols in this figure.

Table 3. Effects of 'crude root cell-wall' preparations of various plants on P solubility from minerals containing sparingly soluble P

P form	P content	Released P by root cell-walls (μg P solubilized g^{-1} cell-wall)			P solubility (%)[a]		
Minerals	(mg P g^{-1})	Sorghum	Soybean	Groundnut	Sorghum	Soybean	Groundnut
Fe-P							
$FePO_4$	164.63	273.4	247.1	382.2	0.36	0.33	0.50
Strengite	32.29	46.7	47.1	81.6	0.28	0.29	0.50
Goethite-P	2.95	26.4	32.8	58.1	1.79	2.22	3.94
Hematite-P	0.29	2.6	7.8	12.1	1.78	5.35	8.33
Al-P							
$AlPO_4$	241.09	103.4	123.5	201.1	0.12	0.14	0.23
Variscite	49.81	112.1	140.2	135.6	0.45	0.56	0.54
Si/Al gel-P	160.76	123.6	126.4	229.9	0.15	0.16	0.29
Allophane-P	87.81	94.8	117.8	232.7	0.22	0.27	0.53

[a]Calculated from the ratio of solubilized-P to added mineral-P.

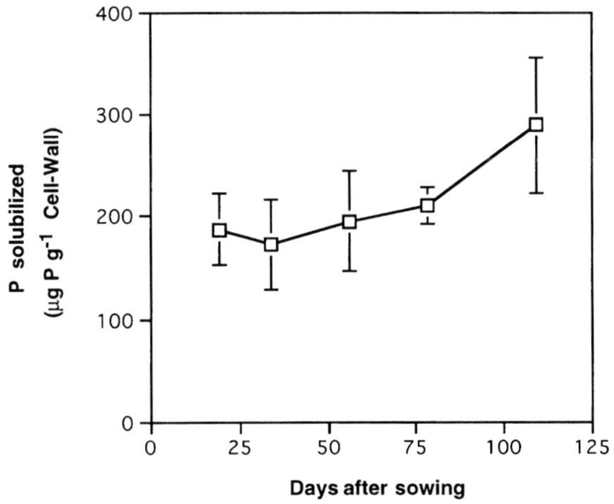

Figure 2. Effect of plant growth stage on phosphorus (P) solubilizing-activity of groundnut root cell-wall.

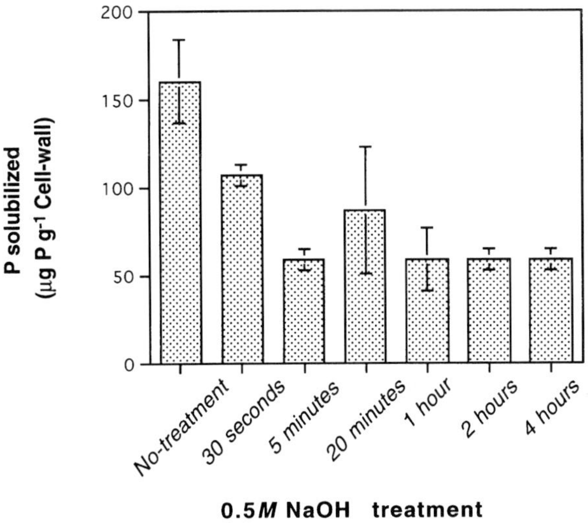

Figure 3. Effect of pre-treatment in 0.5 *M* NaOH for varying times on ability to solubilize P from $FePO_4$ by groundnut root cell-walls. (Roots were incubated in 0.5 *M* NaOH for 30 sec, 5 min, 20 min, and 1 h. Then they were thoroughly washed with water and cut off for pure cell-wall preparation and their P-solubilizing ability was measured).

Effects of soaking of roots in NaOH solution on P solubilization

It is expected that P-solubilizing ability of cell walls is located at the surface of roots, otherwise no *in vitro* P solubilization should occur. To test this, roots were soaked in 0.5 *M* NaOH for different times. Short exposure to NaOH was assumed to affect only outer layers of the roots, whereas longer exposure to NaOH should affect the entire roots, including the cell walls inside the root cylinder. Exposure of roots to NaOH solution for 30 s decreased their solubilizing activity by 30% compared to untreated cell walls (Figure 3). Soaking for 5 min or longer decreased solubilizing activity by additional 30%. When roots were soaked in NaOH for 5 min, the solution turned yellowish in color; groundnut plants wilted when re-transferred to water, but recovered the next day. In contrast, plants soaked for more than 20 min did not recover and died.

Table 4. Phosphorus (P) solubilizing ability of root cell-wall as digested by HCl and NaOH (μg P g^{-1} Cell-wall)

Digestion	Groundnut	Soybean	Maize	Pigeonpea
'Pure cell wall'	135±20.0	78±8.7	29±4.6	117±8.2
0.5 *M* NaOH				
Residue	55±2.9	18±7.5	– nil –	30±0.8
Supernatant	500±109.0	629±17.3	281±48.0	580±6.4
1.0 *M* HCl				
Residue	122±6.8	61±8.2	15±3.7	69±1.0
Supernatant	– nil –	– nil –	– nil –	– nil –

(Mean±SE).

Reaction mode of P solubilization by groundnut root cell-walls

The reaction between chelating chemicals and Fe-fixing groundnut root cell-walls was studied. Citric acid, EDTA and triethanolamine were used as chelating chemicals. Log Ks of the stability constant for Fe(III) with these chelating agents are as follows, 24 in citric acid, and EDTA and about 2 in triethanolamine; i.e., EDTA and citric acid are much more higher stable with Fe(III) than triethanolamine. Fe(III) could be released from Fe-bound cell walls when the concentrations of citric and EDTA exceeded 10^{-3} *M* (Figure 4). Triethanolamine did not release any Fe(III) from Fe-bound cell walls even when its concentration was higher than 0.1 *M*.

Extraction of P-solubilizing active compounds

The $FePO_4$ solubilizing ability of pure cell walls from pigeonpea roots was nearly the same as that of groundnut (Table 4). An attempt was made to extract the P-solubilizing active compounds from the cell walls of groundnut, pigeonpea, soybean and maize. By hydrolysis of cell walls with NaOH, the P-solubilizing active site was shifted to supernatant, but HCl was not able to extract the P-solubilizing component. The P-solubilizing activity of supernatants with NaOH was higher than that of untreated cell walls from all 4 crop species. After hydrolysis with NaOH solution at 80 °C, the supernatant was applied to a column for chromatography with Sephadex LH-20. Each 3 mL eluent from this column was monitored at 280 nm and its P-solubilizing activity measured. Some P-solubilizing activity was found in one major peak and one minor peak. The minor peak was in the higher molecular weight (HMW-) fraction, and the major one in the lower molecular weight (LMW-) faction (data not shown). After 24-h hydrolysis with NaOH at 80 °C, the P-solubilizing activity was not detected in the HMW-fraction but appeared in the LMW-fraction (Figure 5).

Discussion

Hallock (1962) reported that groundnut could take advantage of minerals that are not available to other crops. Our results confirmed that, in Ishigaki soil, in which Fe-P is predominant P form, groundnut could grow better than other crop species (Ae et al., 1996). These results suggested that groundnut has a superior ability to take up Fe-P than the other crops tested. In order to explain this ability of groundnut, its root exudates were checked. However, no significant amount of root exudates that could solubilize Fe-P were found (Otani et al., 1996). Tanaka et al. (1995) reported that formic acid was secreted from groundnut roots and it might react to reduce Fe(III) to Fe(II), resulting in a release of P from Fe-P in a soil. However, Otani et al. (1996) found that the overall Fe(III) reduction activity increased under Fe-deficient conditions and decreased under P-deficient conditions, and that the reduction activity was almost the same in pigeonpea, groundnut, and soybean, even though these crops differ widely in their ability to survive in low-P soils. Therefore, the exudation of formic acid by groundnut roots might not be important in releasing P from Fe-P. Apparently the 'root exudate' theory failed to explain this genotype's superior P-uptake ability.

Soybean and sorghum died from lack of P in the Ishigaki soil where groundnut survived and set seeds (Ae and Otani, 1997). This suggests that the P concentration in soil solution was almost nil, so P by solubilization must occur at the interface between the groundnut root surface and soil particles. Attention was focused on the root surface, i.e., the epidermal cell wall in which the contact reaction could take place. Groundnut root cell walls had higher solubilizing abilities than those of sorghum and soybean (Table 3). Hamatite and goethite are common minerals in acid soils, and P-fixing minerals are common products when P-fertilizer is applied to the soil. Strengite, the amorphous end product of the reaction between Fe and P, is very stable due to its low solubility product (10^{-32}) (Sample et al., 1980). Despite this, groundnut root cell-walls solubilized P from strengite more efficiently than those of soybean or sorghum. This higher

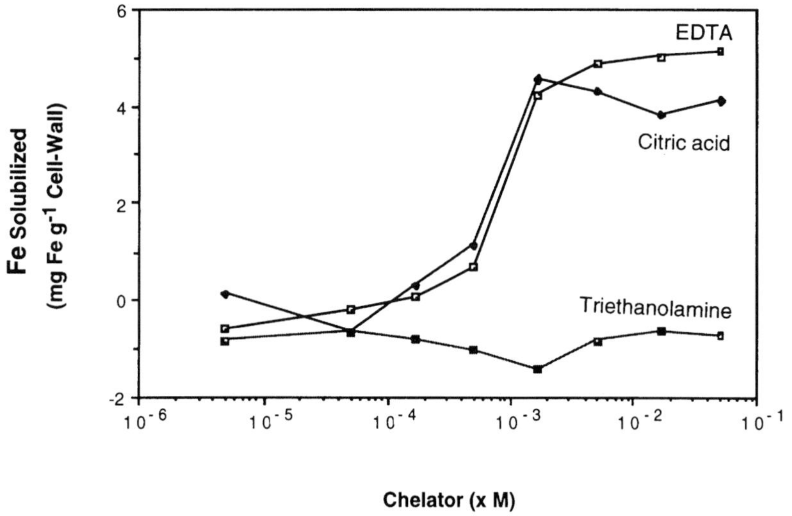

Figure 4. Effect of three chelating compounds on solubilization of Fe(III) from Fe-saturated groundnut root cell-walls (5 mL of 0.1 *M* $FeCl_3$ was added to a test tube containing 40 mg of pure root cell-walls and gently shaken for 2 h. Fe-treated cell-walls were washed 10 times with 5 mL water. Fe-treated root cell-walls were soaked in 5 mL of various concentration of chelating chemicals (1/200 m*M* to 1/20 *M*) in 1 *M* acetate buffer at pH 5.0 and this solution was shaken for 2 h. The solution was passed through a 0.65 μm membrane filter and Fe contents determined by atomic absorption spectrophotometry). SE values were negligible compared to symbols in this figure.

solubilizing activity may explain groundnut's superior P uptake compared to soybean and sorghum.

As mentioned before, mucilage, consisting of partially methylated polygalacturonic acid, was reported to form a complex with Fe(III) resulting in desorption of P from the surface of iron-oxide minerals (Nagarajah et al., 1970). These polyanionic pectic substances may be related to the chelating ability for Fe(III). Therefore, it is reasonable to assume that higher CEC value of cell walls may reflect their higher P-solubilizing activity. The relationship between CEC and the solubilizing ability for Fe-P were examined at various pH levels using $FePO_4$ instead of P-fixed by minerals (Figure 1). The CEC increased with the increasing pH in all crop species. In contrast, P-solubilizing activity decreased with increasing of pH. The root cell walls of soybean showed a higher CEC than those of groundnut, but the latter showed the highest P solubility. These results support the conclusion that P solubility may not rely on the anions derived from pectic substances, i.e., on polygalacturonic acid.

This solubilizing ability of groundnut root cell-walls increased gradually with the growth-stage and was very stable (Figure 2). Thus, the proposed contact reaction hypothesis, i.e., P-solubilization by root cell walls, seems to have an advantage over the root exudates theory for groundnut, because the P-solubilizing activity was evident at the young seedling stage and was not destroyed by soil microorganisms. In general, P absorption by a plant is related to root surface-area including root hairs (Ito and Barber, 1983). The contact-reaction hypothesis seems to support the fact that with root elongation, new active site for P-solubilization becomes available.

The reaction site responsible for P-solubilization has to be on the root surface, otherwise, P-fixed by minerals will not be in contact with the reaction site. Figure 3 showed that 30 s exposure to 0.5 *M* NaOH reduced P solubilization by 30%, but that groundnut plants recovered when transferred to water. Soaking for more than 20 min resulted in plant death. This suggested that at least 30% of the total root-solubilizing ability was located on the surface of the root cell-walls and supports this contact reaction hypothesis. A microscopic observation reported by Uheda et al. (1997) showed that mucilage was surrounded by the root surface and that soil particles were embedded in this thin mucilage layer. This means that soil particles would be tightly attached to the root surface (Iijima, personal communication). Groundnut roots lack root hairs except a small number of rosette-type root hairs around lateral root tips (Meisner and Karnox, 1991). Yarbrough (1949) pointed out that the surface cell layer of groundnut roots shed continuously. Also, he suggested that at least for the absorption of mineral ions, cells of the roots' outer zone provided continuous renewal of cells to substitute for the lack of root

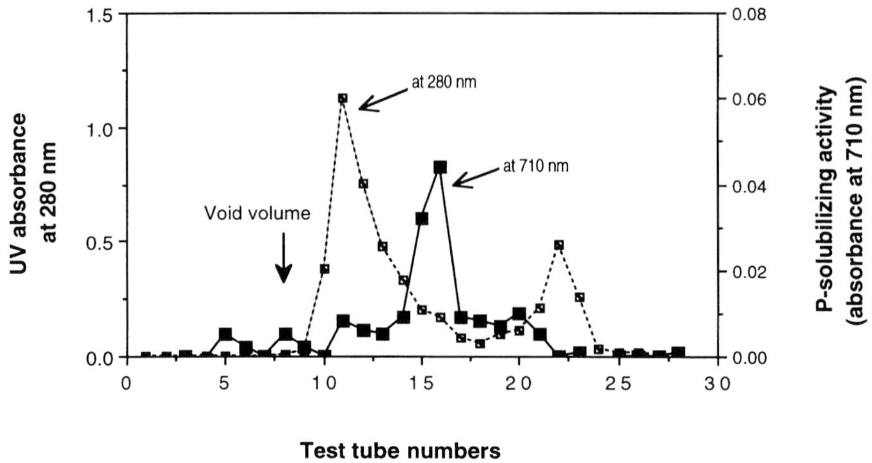

Figure 5. Elution pattern of the extracted cell-wall components with NaOH (left vertical axis UV absorbance at 280 nm; right P-solubilizing activity expressed as absorbance at 710 nm).

hairs. The mode of absorption of water and nutrients by groundnut is assumed to differ from that of other plants. Though the physiological function of special trait has not well understood, this trait could help the root maintain sufficient active substances to continuously chelate with the Fe(III) and then release P for plants uptake. Groundnut plant start shedding root cells 7 days after germination (Uheda et al., 1997), so this effect could also support the contact reaction hypothesis.

The mechanism of P-solubilizing ability in groundnut root cell-walls was examined using three chelating chemicals (Figure 4). Iron would be released from Fe-bound cell walls into the solution of high concentrations of citric acid or EDTA, but not of triethanolamine. This implies that the reaction of the P solubilization is related to its chelating effect on the root surface and that it act like EDTA.

To identify chemical compounds related to P solubilization, the active site had to be extracted from the cell walls. Alkaline hydrolysis was a useful method. After NaOH treatment, P-solubilizing ability increased by a factor of 5–10 compared to that original root cell-walls of groundnut, soybean, maize and pigeonpea (Table 4). This suggests that the P-solubilizing ability is not only located on root surface of the cell wall, but also on the inner root cell walls, and also that the P-solubilizing ability commonly exists in root cell walls irrespective of whether plants are monocotyledonous or dicotyledonous. However, its significance was evident only in groundnut, and/or the reason for this may related not only to groundnuts higher solubilizing ability, but also to the continuous shedding of its root surface cells as described earlier.

Extracted P-solubilizing sites following NaOH hydrolysis were applied to column chromatography with LH-20 (Figure 5). The P-solubilizing active fractions appeared 2 peaks, minor one was in the void fraction, i.e., the HMW-fraction, and a major one in the LMW-fraction. With longer hydrolysis with NaOH at 80 °C, the peak of the P-solubilizing component in the HMW-fraction disappeared, and the 2nd peak in the LMW-fraction increased. This phenomenon might confirm that P-solubilization was not related to pectic substance, since after hydrolysis the smaller molecules still had P-solubilizing activity. In addition, the characteristics of the P-solubilizing ability of the original root cell-walls and the cell-wall components extracted by NaOH were compared and found similar. Both of them are inhibited by the addition of Fe(III) but not by the addition of Ca(II) or Mg(II) (data not shown).

Interestingly, the root cell-walls of pigeonpea had also a high P-solubilizing ability. Pigeonpea also has superior ability to take up P from sparingly soluble P sources in low-P soils. This was attributed to its root exudation of organic acids, especially piscidic acid (Ae et al., 1990), malonic acid, etc. (Otani and Ae, 1997; Otani et al., 1996). But the P-solubilizing ability due to root exudation may be controlled by the plant growth stage because not much root exudation occurs before flowering (Otani and Ae, 1997; Subbarao et al., 1997). Therefore root exudation could not explain P uptake by pigeonpea before the flowering stage. However, this new finding that root cell-walls of pigeonpea showed as high P-solubilizing activity as do those of groundnut may enable us to elucidate P uptake in young pigeonpea plants.

In summary, though the P-solubilizing active substances remain to be elucidated, the proposed P-solubilizing mechanism (contact reaction) could explain at least partially the superior P uptake by some crop like groundnut and pigeonpea. To better understand the role of the root cell walls in plant nutrition, it is necessary to test more crops to see if other crops have similar strategies to acquire P from low P soils.

References

Ae N, Arihara J, Okada K, Yoshihara T and Johansen C 1990 Phosphorus uptake by pigeonpea and its role in cropping systems of Indian subcontinent. Science 248, 477–480.

Ae N, Otani T, Makino T and Tazawa J 1996 Role of cell wall of groundnut roots in solubilizing sparingly soluble phosphorus in soil. Plant Soil 186, 197–204.

Arnon D I 1936 Microelements in culture-solution experiments with higher plants. Am. J. Bot. 25, 322–325.

Atkinson R J, Posner A M and Quirk J P 1967 Adsorption of potential-determining ions at the ferric oxide-aqueous electrolyte interface. J. Phys. Chemi. 71, 550–558.

Chapman S R and Carter L P 1976 Crop Production: Principles and Practices. W H Freeman and Company, San Francisco, USA.

Gessa C, Deiana S, Premoli A and Ciurli A 1997 Redox activity of caffeic acid towards iron (III) complexed in a polygalacturonate network. Plant Soil 190, 289–299.

Hallock D L 1962 Effect of lime and rate of fertilizer application on yield and seed size of jumbo runner peanut. Agron. J. 54, 428–430.

Hartzog D L and Adams J F 1988 Relationship between soil-test P and K and yield response of runner peanuts to fertilizer. Commun. Soil Sci. Plant Anal. 19, 1645–1653.

Ida T, Hori S and Okuda A 1958 Studies on cation absorption by plant roots. T. Cation exchange capacity and cation absorption selectivity of roots. Jpn. J. Soil Sci. Plant Nutr. 29, 259–262. *(In Japanese).*

Itoh S and Barber S A 1983 Phosphorus uptake by six plant species as related to root hairs. Agron J. 75, 457–461.

Jackson M L, Lim C H and Zelazny L W 1986 Oxides, hydroxides, and aluminosilicates. *In* Methods of Soil Analysis. Part 1. Physical and Mineralogical Methods. Ed. A Klute. pp 101–150. American Society of Agronomy, Madison, WI, USA.

Meisner C A and Karnox K J (1991) Root hair occurrence and variation with environment. Agron. J. 83, 814–818.

Murphy J and Riley J P 1962 A modified single solution method for the determination of phosphate in natural water. Anal. Chem. Acta 27, 31–36.

Nagarajah S, Posner A M and Quirk J P 1970 Competitive adsorption of phosphate with polygalacturonate and other organic anions on kaolinate and oxide surface. Nature 228, 83–84.

Nanzyo M and Watanabe Y 1981 Material balance associated with phosphate sorption by amorphous clay, silica-alumina gel and synthetic goethite. Soil Sci. Plant Nutr. 27, 329–337.

Otani T and Ae N 1996a Phosphorus (P) uptake mechanisms of crop grown in soils with low P status. I. Screening crops for efficient P uptake. Soil Sci. Plant Nutr. 42, 155–163.

Otani T and Ae N 1996b Sensitivity of phosphorus uptake to changes in root length and soil volume. Agron. J. 88, 371–375.

Otani T and Ae N 1997 The exudation of organic acids by pigeonpea roots for solubilizing iron- and aluminum-bound phosphorus. *In* Plant Nutrition – for sustainable food and environment. Eds T Ando et al. pp. 325–326. Kluwer Academic Publishers, Dordorecht, the Netherlands.

Otani T, Ae N and Tanaka H 1996 (P) uptake mechanisms of crop grown in soils with low P status. II. Significance of organic acids in root exudates of pigeonpea. Soil Sci. Plant Nutr. 42, 533–560.

Sample E C Soper R J and Racz G J 1980 Reactions of phosphate fertilizers in soils. *In* The Role of Phosphorus in Agriculture. Eds. F e Khasawneh, E C Sample and E J Kamprath, pp 263–310. American Society of Agronomy, Madison, WI. USA.

Subbarao G V, Ae N and Otani T 1997 Genetic variation in iron-aluminum-phosphate solubilizing activity of pigeonpea root exudates under P deficient conditions. Soil Sci. Plant Nutr. 43, 295–305.

Tanaka H, Adachi N and Isoi T 1995 Secretion of formate in phosphorus-deficient nutrient solution from peanut seedlings (*Arachis hypogaea* L.). Soil Sci. Plant Nutr. 41, 389–392.

Uheda E, Akasaka Y and Daimon H 1997 Morphological aspects of the shedding of surface layers from peanut roots. Can. J. Bot. 75, 607–611.

Walker M E, Flowers R A, Henning R J, Keisling T C and Mullinix B G 1979 Response of early bunch peanuts to calcium and potassium fertilization. Peanut Sci. 6, 119–123.

Walker M E, Gaines T P and Henning R J 1982 Foliar fertilization effects on yield, quality, nutrient uptake, and vegetative methods for soil testing, Peanut Sci. 9, 53–57.

Yabrough J A 1949 *Arachis hypogaea*. The seedling, its cotyledons, hypocotyl and roots. Am. J. Bot. 36, 758–722.

J.J. Adu-Gyamfi (Ed.), *Food security in nutrient-stressed environments: exploiting plants' genetic capabilities:* 133–143, 2002.

Role of soil microorganisms in improving P nutrition of plants

P. Gyaneshwar[1,4], G. Naresh Kumar[2], L. J. Parekh[2] & P. S. Poole[3]
[1]*Soil and Water Sciences Division, International Rice Research Institute, MCPO BOX 3127, 1271, Makati City, Philippines (Present address: Department of Plant and Microbial Biology, 111 Koshland Hall, UC Berkeley, Berkeley CA 94720, USA).* [2]*Department of Biochemistry, M. S. University of Baroda, Vadodara, 390 002 India.* [3]*School of Animal and Microbial Sciences, University of Reading, Whiteknights, P. O. Box 228, Reading RG6 6AJ, UK.* [4]*Corresponding author**

Key words: aluminum, iron, organic phosphates, phosphate solubilizing micro-organisms, rhizosphere

Abstract

Phosphorus (P) is one of the major plant growth-limiting nutrients although it is abundant in soils in both inorganic and organic forms. Phosphate solubilizing micro-organisms (PSMs) are ubiquitous in soils and could play an important role in supplying P to plants in a more environmentally friendly and sustainable manner. Although solubilization of P compounds by microbes is very common under laboratory conditions, results in the field have been highly variable. This variability has hampered the large-scale use of PSMs in agriculture. Many reasons have been suggested for this variability, but none of them have been extensively investigated. In spite of the importance of PSMs in agriculture, the detailed biochemical and molecular mechanisms of P solubilization are not known. Recent work in our laboratory has shown that the conditions employed to isolate PSMs do not reflect soil conditions and that PSMs capable of effectively releasing P from soil are not so highly abundant as was suggested in earlier studies. These studies have also indicated that the mineral phosphate solubilizing (mps) ability of microbes could be linked to specific genes, and that these genes are present even in non P solubilizing bacteria. Understanding the genetic basis of P solubilization could help in transforming more rhizosphere-competent bacteria into PSMs. Further research should also focus on the microbial solubilization of iron (Fe) and aluminum (Al) phosphates, as well as mobilization of the organic phosphate reserves present in the soils.

Introduction

The Green Revolution in agriculture has been one of the most successful human achievements in this century. This revolution resulted in global food security and played an important role in transforming developing countries, such as India, from being food-deficient to having a food surplus. However, the yield increases in the food productions are being offset with a constant increase in human population and now the world food security is again being threatened. The world today needs a second Green Revolution in order for food production to increase by 50% in the next 20 years to sustain the increase in population (Leisinger, 1999; Vasil, 1998).

Chemical fertilizers have played a significant role in the Green Revolution but excessive use of them has led to reduction in soil fertility and to environmental degradation. Moreover, the use of chemical fertilizers is reaching the theoretical maximum use beyond which there will be no further increase in yields (Ahmed, 1995). Nitrogen (N) is the most limiting nutrient for crop yields, and N fertilizers are an expensive input in agriculture costing more than US$45 billion per year globally (Kush and Bennett, 1992).

After N, phosphorus (P) is the major plant growth-limiting nutrient despite being abundant in soils in both inorganic and organic forms. However, many soils throughout the world are P-deficient because the free phosphorus concentration (the form available to plants) even in fertile soils is generally not higher than 10 μM even at pH 6.5 where it is most soluble

* E-mail: gyani@nature.berkeley.edu

This paper previously appeared in *Plant and Soil* **245**, 2002, pp. 83–93.

(Arnou, 1953). On an average, most mineral nutrients in soil solution are present in millimolar amounts, however, phosphorus is present only in micromolar or lesser quantities (Ozanne, 1980). These low levels of P are due to high reactivity of soluble P with Calcium (Ca), iron (Fe) or aluminum (Al) that lead to P precipitation. Inorganic P in acidic soils is associated with Al and Fe compounds (Sharpley et al., 1984) whereas calcium phosphates are the predominant form of inorganic phosphates in calcareous soils. Organic P may also make up a large fraction of soluble P, as much as 50% in soils with high organic matter content (Barber, 1984). Phytate, a hexaphosphate salt of inositol, is the major form of P in organic matter contributing between 50 and 80% of the total organic P (Alexander, 1977). Although micro-organisms are known to produce phytases, that can hydrolyze phytate, phytate tends to accumulate in virgin soils because it is rendered insoluble as a result of forming complex molecules with Fe, Al and Ca (Alexander, 1977). Phospholipids and nucleic acids form a pool of labile P in soil that is easily available to most of the organisms present there (Molla and Chowdary, 1984).

To circumvent the problem of P deficiency, chemical fertilizers are added to the soils. The production of chemical phosphatic fertilizers is a highly energy-intensive process requiring energy worth US$4 billion per annum in order to meet the global need (Goldstein et al., 1993). The situation is further compounded by the fact that almost 75–90% of added P fertilizer is precipitated by Fe, Al and Ca complexes present in the soils (Stevenson, 1986; Vig and Dev, 1984).

Need for biofertilizers in plant P nutrition

Theoretical estimates have suggested that the accumulated P in agricultural soils is sufficient to sustain maximum crop yields worldwide for about 100 years (Goldstein et al., 1993). Phosphorus biofertilizers in the form of micro-organisms can help in increasing the availability of accumulated phosphates for plant growth by solubilization (Goldstein, 1986; Kucey et al., 1989; Richardson, 1994; Subba Rao, 1982; Tandon, 1987). In addition, the micro-organisms involved in P solubilization as well as better scavenging of soluble P (P biofertilizers) can enhance plant growth by increasing the efficiency of biological nitrogen fixation, enhancing the availability of other trace elements such as Fe, zinc (Zn), etc. and by production of plant growth promoting substances (Kucey et al., 1989). Most of the impact with the use of micro-organisms as biofertilizers has been directed towards understanding of the biological nitrogen fixation that occurs in the symbiotic system of legume and bacteria of the Rhizobiaceae family. In contrast, the fundamental work on microbial P solubilization has been substantially less, though it is known that P is the most limiting factor for N_2 fixation by *Rhizobium*-legume symbiosis.

Importance of P in symbiotic biological nitrogen fixation

Major research on biofertlizers has concentrated on understanding and improving N_2 fixation in the unique symbiosis between legumes and *Rhizobia* that results in the formation of a specialized structure called nodule, in which the bacterial symbiont fixes N in return for reduced carbon (C) from the plant host (Denarie et al., 1996; Long, 1996; Mylona et al., 1995; Pawlowski and Bisseling, 1992). However, it is known that every aspect of the process of formation of the N_2 fixing nodule is limited by the availability of P (MacDermott, 1999). Legumes like Alfalfa (Al-Niemi et al., 1997; Deng et al., 1998), clover, Common bean, cow pea (Cassman et al., 1981) and Pigeon pea (Itoh, 1987) show a high positive response to P supplementation. The N_2 fixing potential of aquatic legumes like *Sesbania rostrata* (Ventura and Ladha, 1997), an important constituent of the green-manure technology for rice (Ladha et al., 1992), is also limited by P.

There are only a few reports of P solubilization by *Rhizobium* (Chabot et al., 1993, 1996, Halder et al., 1991, 1992). A P-solubilizing *Rhizobium leguminosarum* has been shown to increase the growth of maize and lettuce (Chabot et al., 1996). In *Rhizobium*, P solubilization was shown to be the result of the producing 2-ketogluconic acid, which was abolished by addition of NaOH, indicating that pH reduction of the system was responsible for the P-solubilizing ability of this bacterium. However, the detailed biochemical and molecular mechanisms of P solubilization by *Rhizobium* are not known.

Nature of P biofertilizers

Many plants have shown to have benefited from the association with micro-organisms under P-deficient conditions. This association could result either in better uptake of the available P, or rendering unavailable

P-sources accessible to the plant. The arbuscular mycorrhizae (AM) belong to the former category and the later category includes various bacteria and fungi isolated for their ability to solubilize insoluble mineral phosphate complexes especially those of calcium phosphate complexes.

Mycorrhizae

AM fungi are known to be ubiquitous in agricultural soils and are believed to enhance P nutrition of plants by scavenging the available P due to the large surface area of their hyphae, and by their high affinity P uptake mechanisms (Hayman, 1974, 1983; Mosse, 1980; Sanders and Tinker, 1973). There are also reports of organic acid production by AM (Lapeyrie, 1988; Paul and Sundara Rao, 1971) that could solubilize the insoluble mineral phosphates. It has also been suggested that there could be further effects on the availability of Fe phosphates (Bolan et al., 1987; Cress et al., 1984), but so far no alternative to the original mechanism of Sanders and Tinker (1973) has been accepted.

Production of organic acids by AM would certainly affect the availability of acid-labile insoluble phosphate and the whole issue of AM mediated increase in available P needs re-examination. Ectomycorrhizal fungi have been shown to possess P solubilizing activity (Lapeyrie et al., 1991). They are capable of utilizing P from inositol phosphates and possess phosphatase activity, that could further affect their ability to release P from soil organic matter (Antibus et al., 1991; Koide and Schreiner, 1992). However, the use of AM as phosphate biofertilizer is hindered by the inability to culture them *in vitro*, since they are obligate symbionts. In addition, AM infection also seems to depend on the P status of the plant (Abbott et al., 1984). It is known that the AM fungi are not able to colonize plant roots strongly under P sufficient conditions (Amijee et al., 1989; Jasper et al., 1979; Koide and Schreiner, 1992). In certain cases, the growth rates of plants were reduced by AM coloniation in the presence of available P (Buwalda and Goh, 1982; Peng et al., 1993; Son and Smith, 1995).

Phosphate-solubilizing micro-organisms (PSMs)

The involvement of micro-organisms in solubilization of inorganic phosphates was known as early as 1903 (Kucey et al., 1989). Since then, there have been extensive studies on the solubilization of mineral phosphates by micro-organisms. These have been reviewed by Goldstein (1986), Kucey et al. (1989), Subba Rao (1982) and Tandon (1987). Phosphate-solubilizing micro-organisms (PSMs) are ubiquitous, and their numbers vary from soil to soil. In soil, P-solubilizing bacteria constitute 1–50% and fungi 0.5%–0.1% of the total respective population. In general, P solubilizing bacteria generally out-number P solubilizing fungi by 2–150 fold (Banik and Dey, 1982; Kucey, 1983; Kucey et al., 1989). The majority of the PSMs solubilize Ca–P complexes and only a few can solubilize Fe–P and Al–P (Banik and Dey, 1983; Kucey et al., 1989). Hence, these PSMs could be effective in calcareous soils in which CaP complexes are present, but not in other soils such as alfisols in which phosphates are complexed with Fe and Al ions. However, these PSMs may be effective even in these soils when supplemented with rock phosphate.

Most P-solubilizing bacteria were isolated from the rhizosphere of various plants and are known to be metabolically more active than those isolated from sources other than rhizosphere (Baya et al., 1981; Katznelson and Bose, 1959). The P-solubilizing ability in bacteria was lost upon repeated subculturing but no such loss has been observed in the case of P-solubilizing fungi (Kucey, 1983; Sperber, 1958). In general, fungal isolates exhibit greater P-solubilizing ability than bacteria in both liquid and solid media (Banik & Dey, 1982; Gaur et al., 1973; Kucey, 1983).

The P-solubilizing ability of PSMs also depends on the nature of N source used in the media, with greater solubilization in the presence of ammonium salts than when nitrate is used as N source. This has been attributed to the extrusion of protons to compensate for ammonium uptake, leading to a lowering of extracellular pH (Roos and Luckner, 1984). In some cases, however, ammonium can lead to decrease in P solubilization (Reyes et al., 1999). In addition, other media components were also found to affect the P solubilization ability (Cunningham and Kuiack, 1992).

Mechanisms of P solubilization

The ability to solubilize Ca–P complexes has been attributed to the ability of the PSMs to reduce the pH of their surroundings, either by the release of organic acids or protons. The organic acids secreted can either directly dissolve the mineral phosphate as a result of anion exchange of PO_4^{2-} by acid anion or can chelate both Fe and Al ions associated with phosphate (Bajpai and Sundara Rao, 1971; Bardiya and Gaur, 1972; Katznelson and Bose, 1959; Moghimi et al., 1978; Sperber, 1957). The PSMs produce organic acids such as acetate, lactate, oxalate, tartarate, succinate, citrate, gluconate, ketogluconate, glycolate, etc. (Banik and Dey, 1982; Cunningham and Kuiack, 1992; Duff et al., 1969; Goldstein, 1986; Gyaneshwar et al., 1998a; Kim et al., 1998, 1999; Louw and Webley, 1959; Sperber, 1958; Taha et al., 1969). In certain cases, P solubilization has been found to be inducible by P starvation (Goldstein and Liu, 1987; Gyaneshwar et al., 1999).

Many micro-organisms can reduce the pH of the medium and solubilize P from rock phosphate (Gyaneshwar et al., 1998a; Nahas, 1996). The P-solubilizing activity of *Rhizobium* was abolished by addition of NaOH indicating that P solubilizing activity of this strain was entirely due to its ability to reduce the pH of the media (Halder and Chakrabarty, 1993). However, acidification does not seem to be the only mechanism of solubilization, as the ability to reduce the pH in some cases did not correlate with the ability to solubilize mineral phosphates (Subba Rao, 1982). The chelating property of the organic acids is also important as it has been shown that the addition of 0.05 M EDTA to the media had the same solubilizing effect as inoculation with *Penicillium bilaii* (Kucey, 1988).

Molecular genetic studies on P solubilization

Molecular genetic approaches have been used to understand the genetic basis of gluconic acid production from P solubilizing bacteria. *Escherichia coli* is known to possess a gene for apo glucose dehydrogenase (enzyme responsible for conversion of glucose to gluconate), but not for the essential cofactor pyrroloquinoline quinone (PQQ) (Matsushita et al., 1997). The genes involved in PQQ biosynthesis from P-solubilizing *Erwinia herbicola* were cloned by using its genomic DNA library to select *E. coli* transformants for a mps phenotype. Using this approach, two genes were obtained from *E. herbicola*, a gene encoding PQQ synthase (Goldstein and Liu, 1987) and a second gene that is a putative PQQ transporter (Goldstein, 1995; Liu et al., 1992). Another gene putatively involved in PQQ transport was similarly cloned (Babu-khan et al., 1995). This approach led to the isolation of two genes similar to the *pqqD* and *pqqE* genes from P-solubilizing *Rahnella aquatilis* (Kim et al., 1998). Genes conferring mps phenotype have been cloned in *E. coli* from a non-P solubilizing *Synechocystis* PCC 6803 (Gyaneshwar et al., 1998b). The fact that mps trait can be linked to particular genes can pave the way for generating novel PSMs by incorporating the mps genes in rhizosphere competent bacteria like *Rhizobium* and *Pseudomonas*. Additionally, a high-affinity P transporter from mycorrhizal fungi, *Glomus vermiformis*, has been cloned and its expression was found to be localized to the external hyphae of the fungus (Harrison and Van Buuren, 1995). Over-expression of these high affinity transporters could allow the development of better P scavenging AM strains.

P assimilation in microbes

Some non P-solubilizing bacteria can also be important as P biofertilizers as they can take up the sparingly soluble P through their high affinity transporters and this P can become available to plants through mineralization as the bacteria die. P uptake and metabolism has been extensively studied in the model bacterium *Escherichia coli* (Torriani-Gorini et al., 1987; Wanner, 1994, 1996) and similar systems have been found in other bacteria. In *E. coli,* P stress can affect the expression of over 400 proteins, and this global affect is mediated by a two component regulatory system PhoR and PhoB, in which PhoR is the sensor and PhoB is the cognate positive regulator. Under P-starvation conditions, PhoR phosphorylates PhoB, that in turn binds specific DNA sequences called PHO box (Willsky and Malamy, 1976; Wanner and Chang, 1987; Wanner, 1996). Many genes that are regulated by P starvation may not be involved in P acquisition or assimilation, e.g. *gltB* and *gltD* (which code for glutamate synthase) and *glpB* (coding for anaerobic glycerol-3-phosphate dehydrogenase) (Metcalf et al., 1990). Recently, it has been shown that P starvation resulted in induction of glucose dehydrogenase, an enzyme involved in oxidative glucose metabolism in *Enterobacter asburiae*, a related member of enterobacteraciae family (Gyaneshwar et al., 1999).

In *Rhizobium*, P limitation can result in higher P transport rates and the induction of alkaline phosphatase (Al-Niemi et al., 1997). *Rhziobium* has a functional homologue of PhoB (Wanner, 1996), but a PhoR counterpart has not yet been detected in *Rhizobium* (Bardin and Finan, 1998). As in *E coli*, PHO box sequences are present upstream of the genes regulated by P starvation in *Rhizobium* (McDermott, 1999).

Plant-PSM inoculations

Studies involving plants inoculated with PSMs showed growth enhancement and increased P contents but large variations were found in PSMs effectiveness (Kucey et al., 1989; Subbarao, 1982; Tandon, 1987). *Penicillium bilaii* and *Bacillus megatherium* are considered the most effective PSMs according to field experiments (Asea et al., 1988; Kucey, 1988). *Bacillus megatherium* has been shown to release P from organic phosphates, but does not solubilize mineral phosphates. The PSM-plant inoculations resulted in 10–15% increases in crop yields in 10 out of 37 experiments (Tandon, 1987). These studies also demonstrated an increase in P uptake by plants. The results of PSM-plant inoculations, however, may not conclusively show a direct role for the PSMs in supplementing soil phosphates for plants because: 1. No increase was found in about 70% of the experiments; 2. the increase in crop yeilds were not compared with crop yields with addition of superphosphates; 3. the plant growth promoting activity of PSMs, other than P solubilization, has not been determined; and 4. enhancement in P uptake mediated by AM fungi as a consequence of PSM inoculation was not considered in interpreting the results.

Doubts have also been raised on the ability of PSMs to liberate P under soil conditions (Brown, 1972; Tinker, 1980). Enhancement of growth and crop yields following superphosphate supplementation, despite the abundance of PSMs in the rhizosphere and bulk soils, raises important issues on the variations in the effectiveness of PSM-plant inoculations. Why is P limited for plants when PSMs are abundant in soils? Could inoculation with selected PSMs overcome the problems encountered by the native PSMs?

Many reasons were proposed to account for the variations in the effectiveness of PSM inoculations on plant growth enhancement and crop yields (Kucey et al., 1989): 1. Survival and colonization of inoculated PSMs in the rhizosphere; 2. competition with native micro-organisms; 3. nature and properties of soils and plant varieties; 4. insufficient nutrients in the rhizosphere to produce enough organic acids to solubilize soil phosphates; and 5. inability of PSMs to solubilize soil phosphates.

It has been shown that at least two PSMs showing P solubilization in laboratory conditions could not release P from alkaline vertisols even when supplemented with other nutrients (Gyaneshwar et al., 1998a). Similarly, no soluble P was liberated when the non-sterilized bulk and rhizosphere soil samples were supplemented with C and N (Table 1). By demonstrating that the ability of the PSMs to liberate P from mineral phosphates was significantly decreased under buffered conditions, it was concluded that the inability of PSMs to release P from soils was possibly due to the high buffering capacity of the alkaline soils, coupled with PSMs inability to secrete high concentrations of organic acids. These observations cast doubts on the earlier estimates of PSMs in the rhizosphere which suggested PSMs to be present in the order of 10^8 g^{-1} soil (Kucey et al., 1989). This could be explained as partly due to the use of inappropriate screening media. Most of the PSMs have been screened based on their

Table 1. Effect of carbon and nitrogen supplementation on growth and P solubilization by indigenous PSMs

Time (Days)	Total number of culturable micro-organisms (cfu g^{-1} soil)	pH	Solution P (mM)
0	2×10^5	7.5	ND
1	3.5×10^5	7.2	ND
2	6.3×10^5	6.8	ND
3	1.3×10^6	6.7	ND
4	2.7×10^6	6.4	ND

Results are average of three independent observations in three different soil samples. Ten mM glucose and 1 mM ammonium chloride was added to the soil (1 g ml^{-1} water) and the culturable counts were determined on different days by serial dilution and plating on luria agar plates. The soil was centrifuged at different time points and the supernatant was used for the estimation of pH and P (Gyaneshwar et al., 1998a). ND = Not detectable.

Table 2. Recovery of phosphate-solubilizing micro-organisms from rhizosphere soils[a] by conventional and modified screening media

Sample	Number of PSMs (Showing zone of pH reduction)	
	Conventional (Unbuffered)	Modified (Buffered)
1.	8×10^3	Nil
2.	1×10^6	Nil
3.	3×10^5	Nil
4.	5×10^5	Nil
5.	2×10^4	2×10^2

[a] Various agricultural soils were suspended in sterile normal saline and serial dilutions were plated on minimal media with rock phosphate as P source and methyl red as pH indicator. PSMs were defined as the microbes showing zone of pH reduction (Gyaneshwar et al., 1999).

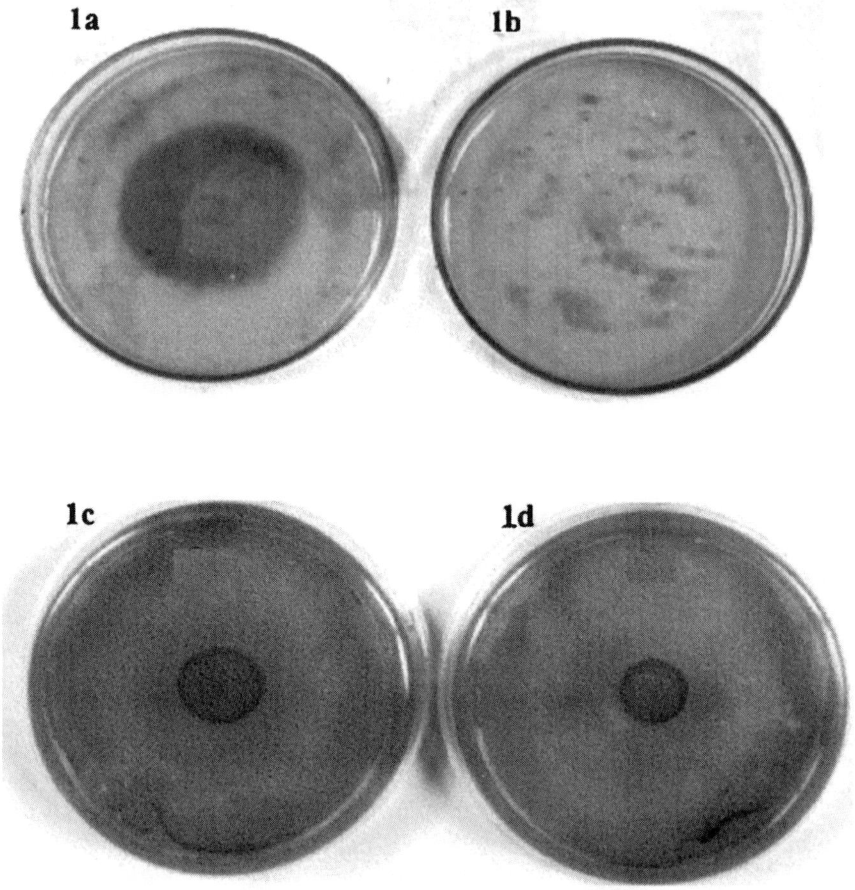

Figure 1. Effect of buffering on solubilization of rock phosphate (RP) by *Citrobacter koseri* (isolated using unbuffered screening medium) and by *Enterobacter asburiae* (isolated using a modified screening medium). The zone of pH reduction and P solubilization could be seen by change of color of pH indicator dye methyl red. *C. koseri* could solubilize RP only in unbuffered medium (a) but was not able to do so in medium buffered with 100 mM Tris–HCl pH 8.0 (b). In contrast, *E. asburiae* could solubilize RP in both unbuffered (c) and buffered (d) medium.

ability to solubilize Ca–P complexes in vitro (Louw and Webley, 1958; Sperber, 1958). The Ca–P complexes can be solubilized by reduction in pH (Ae et al., 1991) and the screening medium often does not have a buffering component. However, the soils with Ca–P a major P source also have high buffering capacities (Ae et al., 1991). The low buffering capacity of the screening media would lead to the isolation and designation of any bacteria that can lower the pH of the medium as PSM do, but the bacteria probably will not be able to do this in soil because the soil is buffered. Therefore, it will be better to isolate PSMs in conditions that resemble those of soil. For example, addition of buffer in the screening medium proved that the PSMs able to secrete large amount of acid were significantly few in numbers (Table 2). The incorporation of a buffer allowed the isolation of PSMs capable of solubilizing P from rock phosphate under buffered media conditions in contrast to those isolated under unbuffered screening media (Figure 1). Addition of methyl red indicator dye allowed the isolation of PSMs that showed large zone of pH reduction (Figure 1). In contrast to the PSMs isolated using unbuffered medium, these PSMs were also able to release P and Fe from alkaline soils (Gyaneshwar et al., 1999).

Although inoculations with PSMs have not been very effective, joint inoculation of PSMs with AM and N_2 fixing bacteria have been more successful. Co-inoculation of *Azospirillum, Rhizobium* and *Azotobacter* with PSMs showed synergistic effect on plant growth and crop yields (Barea et al., 1975; Kundu and Gaur, 1980, 1984). Mycorrhizae, when used along with PSMs, showed either positive or negative influence or no effect on plant yields (Lee and Bagyaraj,

1986; Toro et al., 1996, 1997). Recent experiments of AM inoculation along with PSMs supplemented with rock phosphate or degraded organic matter have suggested mycorrhizosphere interactions that benefit both plant growth and biogeochemical cycling (Toro et al., 1997; Vassilev et al., 1998).

PSMs can also increase the growth of plants by mechanisms other than P solubilization, e.g. production of phytohormones such as Indole-acetic acid (Arshad and Frankenbergar Jr., 1998; Brown, 1972; Datta et al., 1982; Leinhos and Nacek, 1994). This has raised doubts about the actual contribution of P solubilization in plant growth enhancement by PSMs. Indeed, it has been observed in some cases that the growth enhancement was not due to an increase in P uptake (De Freitas et al., 1997). Nevertheless, plant growth enhancement is an important and desired characteristic, irrespective of the mechanism by which it is achieved. However, care should be taken in interpreting the results of increase in growth enhancement by PSMs.

Survival of inoculants in the fields

The competitiveness of a P solubilizing microorganism in natural environments will depend upon its ability to survive and multiply in soil. However, understanding of this part of the use of PSMs is the most limiting factor and it is difficult, if not impossible, to predict the behavior and efficacy of the inoculated PSM in a particular location. In general, the population sizes of the introduced microbe decline rapidly upon the introduction in soils (Ho and Ko, 1985). The survival of the inoculant strain depends upon various factors such as soil composition (Bashan et al., 1995; Heijnen et al., 1988), temperature and the presence of the recombinant plasmids (Van Veen et al., 1997). The biotic factors that affect the survival of the inoculated microbes include competition, predation and root growth that provides the substrates to the microbes. The abiotic factors include texture, pH, temperature, moisture content and substrate availability in the soils (Gray, 1975; Paul and Clark, 1988; Van Elsas et al., 1991). The biotic factors play a very important role in the survival of the inoculated strains as the decline observed in non-sterile soils can often be abolished in sterile soils (Heijnen et al., 1988; Heijnen and Van Veen, 1991). Additionally, an increase in the population of the introduced microbe can also be observed (Postma et al., 1988). Availability of the utilizable substrates often limits the success of the microbial inoculants. In most of the natural environment, nutrients are present in small amounts (Gottschal et al., 1992; Gray, 1975; Hatorri and Hatorri, 1976; Paul and Clark, 1988). The effectiveness of the introduced microbes also depends upon their physiological status. In a recent study, the survival of inoculant strains of various bacteria were shown to be not dependent on the availability of C or on selective predation but only on the initial inoculum density (Jjemba and Alexander, 1999).

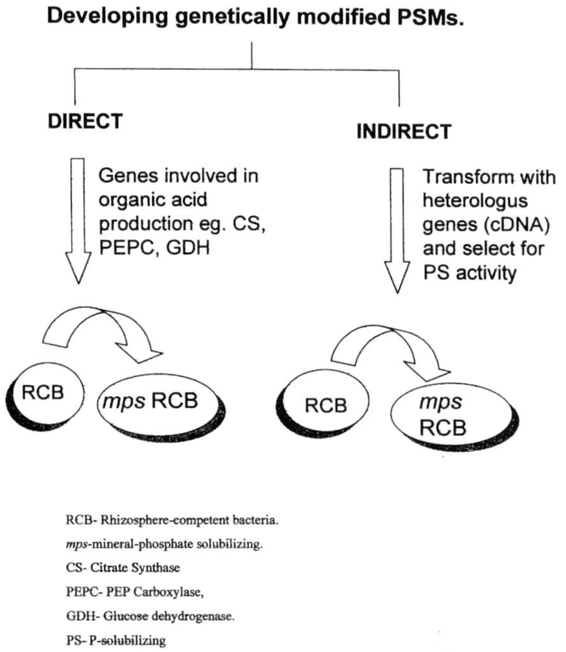

Figure 2. Development of genetically modified phosphate-solubilizing micro-organisms. PSMs can be generated in a direct way by incorporation of genes known to be involved in organic acid production like citrate synthase (CS), PEP carboxylase (PEPC) and glucose dehydrogenase (GDH). Alternatively, rhizosphere competent bacteria can be transformed with genomic DNA/cDNA library, and transformants secreting high amounts of acids can be isolated using the modified screening medium (Indirect).

Use of genetic engineering in developing better and effective PSMs

Mineral phosphates in general are solubilized by organic acids, either by reduction in the pH, or chelating the cat-ions associated with P (Goldstein, 1995; Kucey et al., 1989). Better understanding of the genetic basis of the secretion of organic acids could pave the way for transferring the mps ability to various bacteria that are competent to colonize a particular rhizosphere. The rhizosphere competence is a major determinant for the success of the inoculant, because in the rhizosphere C, sources are available to the microbes to produce organic acid. Additionally, the solubilized P can be better utilized by the plant before it gets precipitated again. The secretion of gluconic acid seems to be the major mechanism of P-solubilization by gram negative bacteria (Duff et al., 1959; Goldstein and Liu, 1987; Goldstein et al., 1993; Goldstein 1995; Kim et al., 1998). Gluconic acid is produced by the oxidative metabolism of glucose by glucose dehydrogenase (GDH) which requires pyrroloquinoline quinone (PQQ) as a co-factor. As mentioned earlier, genes involved in biosynthesis/transport of PQQ have been cloned from various bacteria. Since rhizophere-competent bacteria like *Rhizobium* posses apo-GDH, it would be interesting to transfer the genes involved in PQQ biosynthesis to *Rhizobium* to make a PSM (Figure 2). A P solubilizing *Rhizobium* would also be helpful in increasing N_2 fixation. *Pseudomonas* spp., the other important group of rhizosphere competent bacteria can form gluconic acid through the oxidative glucose metabolism. Over-expression of PQQ biosynthesis and GDH genes could also make them better PSMs.

Another approach is to screen the mps genes directly in the desired bacteria by over/under expression of genes and selection for transformants with mps ability (Figure 2). Such an approach has been used to obtain mps genes from *Synechosystis* PCC 6803 in *E. coli* (Gyaneshwar, 1998b). However, it remains to be seen if this will also be effective in other bacteria.

Genetic engineering could also help in increasing the survival of the inoculant strain by incorporating the ability to utilize certain nutrients better than the rest of the microbial population (Glick and Bashan, 1997). Use of ACC, a precursor of ethylene normally present in the root exudates, can give selective advantages of the introduced microbe that can utilize it as a nitrogen source (Glick et al., 1995, 1995; Jacobson et al., 1994). Also, genes for utilization of salicylate were transferred to a growth promoting bacteria, and the recombinant bacterium was able to survive and enhance plant growth better than the wild type (Colbert et al., 1993). The presence of such organic pollutants as herbicides and pesticides may also facilitate the proliferation of the bacteria that have the ability to metabolize them (Brazil et al., 1995).

Conclusions and future prospects

Soil micro-organisms have enormous potential in providing soil phosphates for plant growth. Their potential is limited by lack of PSMs which could release phosphates from soils, by limited knowledge on the factors governing the association of the PSMs in the rhizosphere, and by the nature and amounts of root exudates. AM fungi together with PSMs could be much more effective in supplementing soil P. Understanding AM-plant symbiosis, developing AM fungi that could be cultured *in vitro*, and developing P-solubilizing AM will help realize their potential as phosphate biofertilizers. The development of better screening procedures and understanding of the genetic basis of P solubilization and rhizospheric competence will help in developing novel PSMs that will be better suited to survive and solubilize P in soil conditions.

Acknowledgment

The authors wish to thank J.K. Ladha and E.K. James for critical reading of the manuscript and T.R. Dermott for sharing reprints and in-press articles.

References

Al-Niemi T S, Kahn M L and McDermott T R 1997 P metabolism in the *Rhizobium tropici*-bean symbiosis. Plant Physiol. 113, 1233–1242.

Ae N, Arihara J and Okada K 1991 Phosphorus response of chick pea and evaluation of phosphorus availability in Indian alfisols and vertisols. *In* Phosphorus Nutrition of Grain Legumes. Eds. C Johansen, KK Lee and KL Sharawat. pp. 33–41. ICRISAT, India.

Abbott L K, Robson A D and De Bore G 1984 The effect of phosphorus on the formation of hyphae in soil by the vesicular-arbuscular mycorrhizal fungus *Glomus fasciculatum*. New Phytol. 97, 437–446.

Ahmed S 1995 Agriculture-Fertilizer Interface in Asia-Issues of Growth and Sustainability. Oxford and IBH Publ. Co. New Delhi.

Alexander M 1977 Introduction to soil microbiology. Wiley New York.

Amijee F, Tinker P B and Stribley D P 1989 Effects of phosphorus on the morphology of vesicular-arbuscular mycorrhizal root system of leek (*Allium porrum* L). Plant Soil 119, 334–336.

Antibus R K, Sinsabaugh R L and Linkins A E 1991 Phosphatase activities and phosphorus uptake from inositiol phosphates by ectomycorrhizal fungi. Can. J. Bot. 70, 794–801.

Arnou, D I 1953 *In* Soil and Fertilizer Phosphorus in Crop Nutrition (IV). Ed. WH Pierre. Noramn, AG. Acad. Press NY.

Arshad M and Frankenberger W T Jr. 1998 Plant growth-regulating substances in the rhizosphere: Microbial production and functions. Adv. Agron. 62, 46–151.

Asea, P E A, Kucey R M N and Stewart J W B 1988 Inorganic phosphate solubilization by two Penicillium species in solution culture and in soil. Soil Biol. Biochem. 20, 459–464.

Babu-Khan S, Yeo T C, Martin W L, Duron M R, Rogers R D and Goldstein A H 1995 Cloning of a mineral phsophate-solubilizing gene from *Pseudomonas cepacia*. Appl. Environ. Microbiol. 61, 972–978.

Bajpai P D and Sundra Rao W V B 1971 Phosphate solubilizing bacteria. Part 2. Extracellular production of organic acids by selected bacteria solubilizing insoluble phosphate. Soil Sci. Plant Nutr. 17, 44–45.

Banik S and Dey B K 1982 Available phosphate content of an alluvial soil as influenced by inoculation of some isolated phosphate solubilizing bacteria. Plant Soil 69, 353–364.

Banik S and Dey B K 1983 Phosphate solubilizing potentiality of micro-organisms capable of utilizing aluminum phosphate as sole phosphate source. Zentrabl. Backteriol. Prasitenkd. Infektionskr. Hyg. II 138, 17–23.

Barber S A 1984 Soil nutrient bioavailability. John Wily, New York, USA.

Bardiya M C and Gaur A C 1972 Rock phosphate dissolution by bacteria. Indian J. Microbiol. 12, 269–271.

Barea J M, Azcon R and Hyman D S 1975 Possible synergistic interactions between endogone and phosphate-solubilizing bacteria in low phosphate soils. *In* Endomycorrhizas. Eds. FE Sanders, B Mooose and PB Tinkler pp. 409–417. Academic Press, London.

Bashan Y, Puente M E, Rodriquea M N, Toledo G, Holguin G, Ferrera-Cerrato R and Pedrin S 1995 Survival of *Azospirillum brasilense* in the bulk soil and rhizosphere of 23 soil types. Appl. Environ. Microbiol. 61, 1938–1945.

Baya A M, Robert S B and Ramos C A 1981 Vitamin production in relation to phosphate solubilization by soil bacteria. Soil Biol. Biochem. 13, 527–532.

Bolan N S, Robson A D and Barrow N I 1987. Effect of vesicular arbuscular mycorrhiza on avalibility of iron phosphates to plants. Plant Soil 99, 401–410.

Brazil G M, Kenefick L, Callanan M, Haro A, De Lorenzo V, Dowling D N and O'Gara 1995 Construction of a rhizosphere pseudomonad with potential to degrade polychlorinated biphenyls and detection of *bph* gene expression in the rhizosphere. Appl. Environ. Microbiol. 61, 1946–1952.

Brentnall B A 1991 Phosphate fertilizers: Economic aspects. *In* Ullman's Encyclopedia of Industrial Chemistry, Vol. A19. VCH Verlagsgesellschaft, New York. 425 pp.

Brown M E 1972. Plant growth substances produced by microorganisms of soil and rhizosphere. J Appl. Bacteriol. 35, 443–451.

Buwalda J G and Goh K M 1982 Host-fungus competition for carbon as a cause of growth depressions in vesicular-arbuscular mycorrhizal rye grass. Soil Biol. Biochem. 14, 103–106.

Cassman K G, Whiteny A S and Fox R L 1981 Phosphorus requirements of soybean and cow pea as affected by mode of N nutrition. Agron. J. 73, 17–22.

Chabot R, Antoun H and Cecas P 1993 Stimulation de la croissancedu mais et de la laitue romaine par des microorganismes dissolvant le phosphore inorganique. Can. J. Microbiol. 39, 941–947.

Chabot R, Antoun H and Cesas M P 1996 Growth promotion of maize and lettuce by phosphate solubilizing *Rhizobium leguminosarum* biovar *phaseoli*. Plant Soil 184, 311–321.

Colbert S F, Hendson M, Ferri M and Schroth M N 1993 Enhanced grwoth and activity of a biocontrol bacterium genetically engineered to utilize salicylate. Appl. Environ. Microbiol. 59, 2071–2076.

Cress W A, Johnson G V and Barton L L 1984 The role of endomycorrhizal fungi in iron uptake by *Hilaria jamesii*. J. Plant Nutr. 40, 547–555.

Cunningham J E and Kuiack C 1992 Production of citric and oxalic acids and solubilization of calcium phosphates by *Penicillium bilaii*. Appl. Environ. Microbiol. 58, 1451–1458.

Datta M, Banik S and Gupta R K 1982 Studies on a phytohormone producing phosphate solubilizing *Bacillus firmus* in augmenting paddy yield in acid soils of Nagaland. Plant Soil 69, 365–373.

Deng S, Kahn M L and McDermott T R 1998 Characterization and transposon mutagenesis of a non-specific acid phosphatase cloned from *Rhizobium meliloti*. Arch. Microbiol. 170, 18–26.

De Freitas J R, Banerjee M R and Germida J J 1997 Phosphate solubilizing rhizobacteria enhance the growth and yield but not the phosphorus uptake of canola. Biol. Fert. Soil 24, 358–364.

Denarie J, Debelle F and Prome J C 1996 *Rhizobium* lipochitooligosaccharide nodulation factors: Signaling molecules mediating recognition and morphogenesis. Ann. Rev. Biochem. 65, 503–535.

Duff R B, Webley D M and Scott R O 1969 Solubilization of minerals and related materials by 2-ketogluconic acid producing bacteria. Soil Sci. 95, 105–114.

Gaur A C, Medan M and Ostwal K P 1973 Solubilization of phosphate ores by native microflora of rock phosphates. Indian J. Expt. Biol. 11, 427–429.

Glick B R 1995 The enhancement of plant growth by free-living bacteria. Can J. Microbiol. 41, 109–177.

Glick B R and Bashan Y 1997 Genetic manipulation of plant growth-promoting bacteria to enhance biocontrol of phytopathogens. Biotechnol. Adv. 15, 353–378.

Golstein A H 1986 Bacterial phosphate solubilization: Historical perspective and future prospects. Am. J. Alt. Agric. 1, 57–65.

Goldstein A H 1995 Recent progress in understanding the molecular genetics and biochemistry of calcium phosphate solubilization by gram negative bacteria. Biol. Agric. Hort. 12, 185–193.

Goldstein A H and Liu S T 1987 Molecular cloning and regulation of a mineral phosphate solubilizing gene from *Erwinia herbicola*. Bio/Technol. 5, 72–74.

Goldstein A H, Rogers R D and Mead G 1993 Mining by microbe. Bio/Technol. 11, 1250–1254.

Gottschal J C 1992 Substrate capturing and growth in various ecosystems. J. Appl. Bacteriol. Symp. Suppl. 73, 39S–48S.

Gray T R G 1975 Survival of vegetative microbes in soil. Symp. Soc. Gen. Micorbiol. 26, 327–364.

Gyaneshwar P, Naresh Kumar G and Parekh L J 1998a Effect of buffering on the phosphate-solubilizing ability of micro-organisms. World J. Microbiol. Biotechnol.14, 669–673.

Gyaneshwar P, Naresh Kumar G and Parekh L J 1998b Cloning of mineral phosphate solubilizing genes from *Synechocystis* PCC 6803. Curr Sci. 74, 1097–1099.

Gyaneshwar P, Parekh L J, Archana G, Poole P S, Collins M D, Hutson R A and Naresh Kumar G 1999 Involvement of a phosphate starvation inducible glucose dehydrogenase in soil phosphate solubilization by *Enterobacter asburiae*. FEMS Microbiol. Lett. 171, 223–229.

Halder A K, Misra A K and Chakrabarty P K 1991 Solubilization of inorganic phosphates by *Bradyrhizobium*. Indian J. Exp. Biol. 29, 28–31.

Halder A K, Banerjee A, Misra A K and Chakrabarty P K 1992 Role of ammonium and nitrate on release of soluble phosphate from hydroxyapatite by *Rhizobium* and *Bradyrhizobium*. J. Basic Microbiol. 32, 325–330.

Halder A K and Chakrabarty P K 1993 Solubilization of inorganic phosphates by *Rhizobium*. Folia Microbiol. 38, 325–330.

Harrison M J and Van Buuren 1995 A phosphate transporter from mycorrhizal fungus *Glomus versiforme*. Nature 378, 626–629.

Hattori T and Hattori R 1976 The physical environment in soil microbiology: An attempt to extend principles of microbiology to soil micro-organisms. CRC Crit. Rev. Microbiol. 4, 423–461.

Hayman D S 1974 Plant growth responses to vesicular-arbuscular mycorrhiza VI. Effects of light and temperature. New Phytol. 73, 71–80.

Hayman D S 1983 The physiology of vesicular-arbuscular endomycorrhizal symbiosis. Can. J. Bot. 61, 944–963.

Heijnen C E and Van Veen J A 1991 A determination of protective microhabitats for bacteria introduced into soil. FEMS Microbiol. Ecol. 85, 73–80.

Heijnen C E, Van Elsas J D, Kuikman P J and Van Veen J A 1988 Dynamics of *Rhizobium leguminosarum* bv. *trifolii* introduced into soil: The effect of betonite clay on predation by protozoa. Soil Biol. Biochem. 20, 483–488.

Ho W C and Ko W H 1985 Soil microbiostasis: Effects of environmental and edaphic factors. Soil Biol. Biochem. 17, 167–170.

Itoh H S 1987 Characteristics of phosphorus uptake of chick pea in comparison with pigeon pea, soybean and maize. Soil Sci. Plant Nutr. 33, 417–422.

Jacobson C B, Pasternak J J and Glick B R 1994 Partial purification and characterization of ACC deaminase from the plant growth promoting rhizobacterium *Pseudomonas putida* GR 12–2. Can J. Microbiol. 40, 1019–1025.

Jasper D A, Robson A D and Abbott L K 1979 Phosphorus and the formation of vesicular-arbuscular mycorrhizae. Soil Biol. Biochem. 11, 501–505.

Jjemba P K and Alexander M 1999 Possible determinants of rhizosphere competence of bacteria. Soil Biol. Biochem. 31, 623–632.

Katznelson H and Bose B 1959 Metabolic activity and phosphate dissolving capability of bacterial isolates from wheat roots in the rhizosphere and non rhizosphere soil. Can. J. Microbiol. 5, 79–85.

Kim K Y, McDonald G A and Jordan D 1999 Solubilization of hydroxyapatite by *Enterobacter agglomerans* and cloned *E. coli* in culture medium. Biol. Fert. Soil 24, 347–352.

Kim K Y, Jordan D and McDonald G A 1998a *Enterobacter agglomerans*, Phosphate solubilizing bacteria and microbial activity in soil: Effect of carbon sources. Soil Biol Biochem. 30, 995–1003.

Kim K Y, Jordan D and Krishnan H B 1998b Expression of genes from *Rahnella auqatilis* that are necessary for mineral phosphate solubilization in *Escherichia coli.* FEMS Microbiol. Lett. 159, 121–127.

Koide T R and Schreiner P R 1992 Regulation of Vesicular-arbuscular mycorrhizal symbiosis. Ann. Rev. Plant Physiol. Plant Mol. Biol. 43, 557–581.

Kucey R M N 1983 Phosphate solubilizing bacteria and fungi in various cultivated and virgin Albreta soils. Can J. Soil Sci. 63, 671–678.

Kucey R M N 1987 Increased phosphorus uptake by wheat and field beans inoculated with a phosphorus solubilizing *Penicillium billai* strain and with Vesicular Arbuscular Mycorrhizal fungi. Appl. Environ. Microbiol. 53, 2699–2703.

Kucey R M N 1988 Effect of *Penicillium billai* on the solubility and uptake of P and micronutrients from soil by wheat. Can. J. Soil Sci. 68, 261–270.

Kucey R M N, Jenzen H H and Leggett M E 1989 Microbially mediated increases in plant available phosphorus. Adv. Agron. 42, 199–228.

Kundu B S and Gaur A C 1980 Establishment of nitrogen fixing and phosphate solubilizing bacteria in the rhizosphere and their effect on the nutrient uptake of the wheat crop. Plant Soil 57, 223–230.

Kundu B S and Gaur A C 1984 Rice responses to inoculation with nitrogen fixing and P-solubilizing micro-organisms. Plant Soil 79, 227–234.

Kush G S and Bennet J (eds) 1992 Nodulation and Nitrogen Fixation in Rice: Potential and Prospects. International Rice Research Institute, Manila Philippines.

Ladha J K, Parrek R P and Beker M 1992 Stem-nodulating legume-rhizobium symbiosis and its agronomic use in low land rice. Adv. Soil Sci. 20, 148–192.

Lapeyrie F 1988 Oxalate synthesis from soil bicarbonate by fungus *Paxillus involutus.* Plant Soil 110, 3–8.

Lapeyrie F, Rangers J and Vairelles D 1991 Phosphate-solubilizing activity of ectomycorrhizal fungi *in vitro*. Can J. Bot. 69, 342–346.

Lee A and Bagyaraj A J 1986 Effect of inoculation with vesicular-arbuscular mycorrhizal fungi and either phosphate rock dissolving bacteria or *thiobacilli* on dry matter production and uptake of phosphorus by tomato plants. N. Z. J. Agric. Res. 29, 525–531.

Leinhos V and Nacek O 1994 Biosynthesis of auxins by phosphate solubilizing rhizobacteria from wheat (*Triticum aestivum*) and rye (*Secale cereale*). Microbiol. Res. 149, 31–35.

Leisinger K M 1999 Biotechnology and food security. Curr. Sci. 76, 488–500.

Liu S T, Lee L Y, Tai C-H, Hung- Y-S Chang H, Wolform R, Rogers R and Goldstein A H 1992 Cloning of an *Erwinia herbicola* gene necessary for gluconic acid production and enhanced mineral phosphate solubilization in *E. coli* HB 101: Nucleotide sequence and probable involvement in biosynthesis of the coenzyme Pyrroloquinoline quinone. J. Bacteriol. 174, 5814–5819.

Long S R 1996 *Rhizobium* symbiosis: Nod factors in perspective. Plant Cell 8, 1885–1896.

Louw H A and Webley D M 1958 A plate method for estimating the number of phosphate dissolving and acid producing bacteria. Nature 182, 1317–1318.

Louw H A and Webley D M 1959 A study of a soil bacteria dissolving certain mineral phosphate fertilizers and related compounds. J. Appl. Bacteriol. 22, 227–233.

Matsushita K, Arents J C, Bader R, Yamada M, Adachi O and Postma P W 1997 *Escherichia coli* is unable to produce pyrroloquinolinequinone (PQQ). Microbiol. 143, 3149–56.

McDermott T R 1999 Phosphorus assimilation and regulation in Rhizobia. *In* Nitrogen Fixation in Prokaryotes: Molecular and Cellular Biology. Ed. EW Triplett. Horizon Scientific Press USA (in press).

Metcalf W W, Steel P M and Wanner B L 1990 Identification of phosphate starvation inducible genes in *Escherichia coli* K 12 by DNA sequence analysis of *psi:lacZ* (Mudl) transcriptional fusions. J. Bacteriol. 172, 3191–3200.

Moghimi A, Tate M E and Oades J M 1978 Characterization of rhizospheric products especially 2-ketogluconic acid. Soil Biol. Biochem. 10, 283–287.

Molla M A Z and Chowdary A A 1984 Microbial mineralization of organic phosphate in soil. Plant Soil 78, 393–399.

Moose B 1980 Vesicular-arbuscular mycorrhiza research for tropical agriculture. Research Bulletin 194, Hawaii Institute of Tropical Agriculture and Human Resources, Honolulu, HI, USA; University of Hawaii.

Mylona P, Pawlowski K and Bisseling T 1995 Symbiotic Nitrogen fixation. Plant Cell 7, 869–885.

Nahas E 1996 Factors determining rock phosphate solubilization by micro-organisms. World J. Microbiol. Biotechnol. 112, 567–572.

Ozanne P G 1980 Phosphate nutrition of plants: A general treatise. *In* The Role of Phosphorus in Agriculture. Eds. FE Khasawneh, EC Sample and EJ Kamprath. Soil Sci. Soc. Am. Madison WI.

Patrick W H Jr., Mikkelsen D S and Wells B R 1985 Plant nutrient behavior in flooded soils. *In* Fertilizer Technology and Use. 3rd edn. Ed. OP Engelstad. SSA, Madison WI.

Paul E A and Clark F E 1988 Soil Microbiology and Biochemistry. Acad. Press Inc. San Diego. Calif.

Paul N B and Sundara Rao W V B 1971 Phosphate dissolving bacteria and VAM fungi in the rhizosphere of some cultivated legumes. Plant Soil 35, 127–132.

Pawlowski K and Bisseling T 1992. Rhizobial and actinorhizal symbiosis: What are the shared features. Plant Cell 4, 1899–1913.

Peng S, Eissentat D M, Graham J H, Williams K and Hodge N C 1993 Growth depressions in mycorrhizal citrus at high phosphorus supply: Analysis of carbon costs. Plant Physiol. 101, 1063–1071.

Postma J, Van Elsas J D, Govaert J M and Van Veen J A 1988 Dynamics of *Rhizobium leguminosarum* bv. *trifolii* introduced into soil as determined by immunofluorescence and selective plating techniques. FEMS Microbiol. Ecol. 53, 251–260.

Reyes I, Bernier L, Simard R R and Antoun H (1999) Effect of nitrogen source on the solubilization of different inorganic phosphates by an isolate of *Penicillium rugulosum* and two UV induced mutants. FEMS Microbiol. Ecol. 28, 281–290.

Richardson A E 1994 Soil micro-organisms and phosphate availability. *In* Soil Biota Management in Sustainable Agriculture. Eds. CE Pankhurst, BM Doube, VVSR Gupts and PR Grace pp. 50–62. CSIRO, Melbourne, Australia.

Roos W and Luckner M 1984 Relationships between proton extrusion and fluxes of ammonium ions and organic acid in *Penicillium cyclopium*. J. Gen. Microbiol. 130, 1007–1014.

Sah R N and Millelsen D S 1986 Transformation of inorganic phosphorus during the flooding and draining cycle of soil. Soil Sci. Soc. Am. J. 50, 62–67.

Sanders F E and Tinker P B 1973 Phosphate flow into mycorrhizal roots. Pest. Sci. 4, 385–395.

Sperber J I 1957 Solubilization of mineral phosphates. Nature 180, 994–995.

Sperber J I 1958 The incidence of apatite-solubilizing organisms in the rhizosphere and soil. Aust. J. Aric. Res. 9, 778–781.

Son C L and Smith S E 1995. Mycorrhizal growth responses: interaction between photon irridiance and phosphorus nutrition. New Phytol. 108, 305–314.

Stevenson F J 1986 Cycles of Soil Carbon, Nitrogen, Phosphorus, Sulfur Micronutrients. Wiley New York.

Subba Rao N S 1982 *In* Advances in Agricultural Microbiology. Ed. NS Subba Rao. pp. 229–305. Oxford and IBH Publ. Co.

Taha S M, Mahmoud S A Z, El Damtay A H and El-Hafez 1969 Activity of phosphate dissolving bacteria in Eyptian soils. Plant Soil 31, 149–160.

Tandon H L S 1987 Phosphorus Research and Production in India. Fertilizer Development and Consultation Organization New Delhi. 160 pp.

Thomson B D, Robson A D and Abbott L K 1986 Effects of phosphorus on the formation of mycorrhizas by *Gigasopra calospora* and *Glomus fusciculatum* in relation to root carbohydrates. New Phytol. 103, 751–756.

Tinker P B 1980 *In* The Role of Phosphorus in Agriculture. Eds. FE Khasawneh, EC Sample and EJ Kamprath. Soil Sci. Soc. Am. Madison WI.

Toro M, Azcon R and Herrera R 1996 Effects on yield and nutrition of mycorhizal and nodulated *Pueraria phaseoloides* exerted by P-solubilizing bacteria. Biol. Fert. Soil 21, 23–29.

Toro M, Azcon R and Barea J 1997 Improvement of arbuscular mycorrhiza development by inoculation of soil with phosphate-solubilizing rhizobacteria to improve rock phosphate bioavailability (32P) and nutrient cycling. Appl. Environ. Microbiol. 63, 4408–4412.

Torriani-Gorini A and Ludtket D 1985 The *pho* regulon in *E. coli* K 12. *In* The Molecular Biology of Bacterial Growth. Eds. M Schaechter, FC Neidhardt, J Ingrahm and NO Kjeldegaard. pp. 224–243. Jones and Barlett Inc, Boston.

Van Veen J A, Leonard S, Van Overbeek L S and Van Elsas J D 1997 Fate and activity of micro-organisms introduced into soil. Microbiol. Molecular Biol. Rev. 61, 121–135.

Van Elsas J D, Van Overbeek L S and Fouchier R 1991 A specific marker, *pat*, for studying the fate of introduced bacteria and their DNA in soil using a combination of detection techniques. Plant Soil 138, 49–60.

Vasil I K 1998 Biotechnology and food security for 21st century: A real world perspective. Nature Biotechnol. 16, 399–400.

Vassielv M, Azcon R, Barea J M and Vassilev N 1998 Application of an encapsulated fungus in solubilization of inorganic phosphate. J. Biotechnol. 63, 67–72.

Ventura W and Ladha J K 1997. Sesbania phosphorus requirements when used as biofertilizers for long term rice cultivation. Soil Sci. Soc. Am. J. 61, 1240–1244.

Vig A C and Dev G 1984 Phosphorus adsorption characteristics of some acid and alkaline soils. J. Indian Soc. Soil Sci. 32, 235–239.

Wanner B L 1994 Multiple controls of the *Escherichia coli pho* regulon by the Pi sensor Pho R, the catabolic regulatory protein Cre C and acetyl phosphate. *In* Phosphate in Micro-organisms: Cellular and Molecular Biology. Eds. A Torriani-Gorrini, E Yagil and S Silver. pp. 13–21. ASM Press, Washington.

Wanner B L 1996 Phosphorus assimilation and control of phosphate regulon. *In Escherichia coli* and *Salmonella:* Cellular and Molecular Biology. Eds. FC Neidhart, R Curtiss, JL Ingraham, M Schaechter and HE Umbarger. pp. 1357–1381. ASM Press. Washington.

Wanner B L and Chang B D 1987 The *pho* BR operon in *E. coli* K12. J. Bacteriol. 169, 5569–5574.

Willsky G R and Malamy M H 1976 Control of Synthesis of alkaline phosphatase and the phosphate binding protein in *Escherichia coli*. J. Bacteriol. 127, 595–609.

J.J. Adu-Gyamfi (Ed.), Food security in nutrient-stressed environments: exploiting plants' genetic capabilities: 145–154, 2002.

Phosphorus use efficiency as related to sources of P fertilizers, rainfall, soil, crop management, and genotypes in the West African semi-arid tropics

A. Bationo[1] & K. Anand Kumar[2]
[1]*Tropical Soil Biology and Fertility Institute of CIAT, c/o ICRAF, P.O. Box 30677, Nairobi, Kenya*
[2]*International Crops Research Institute for the Semi-Arid Tropics (ICRISAT), BP 12404, Niamey, Niger*

Key words: cowpea, pearl millet, P-use efficiency, rainfall, soil, and crop management, West Africa

Abstract

The rainfall of agricultural areas of the West African semi-arid tropics varies from 300 to 1200 mm. Although in absolute terms rainfall is low only in the northern half of the desert margins, the high inter-annual variability associated with eratic distribution of rainfall both in location and during the growing season, constitute major limitations for agricultural production. Continuous and intensive cropping without restoration of the soil fertility has depleted the nutrient base of most of the soils. For many cropping systems in the region, nutrient balances are negative, indicating soil mining. Among soil fertility factors, phosphorus (P) deficiency is a major constraint to crop production. Phosphorus use efficiency (PUE) as defined as yield increase per kg fertilizer P added, is related to P sources, environmental factors, soil, and crop management. In addition to water-soluble P-fertilizers, the effect of rock phosphate (PR) sources from Niger (Parc-W PR and Tahoua PR), Mali (Tilemsi PR), and Burkina Faso (Kodjari PR) and modified partially acidulated phosphate rocks (PAPR) on P-use efficiency is reported. PAPR improved the PUE of all PR sources. Among the four PR sources in the region, Tahoua PR (TPR) recorded the highest PUE as compared to Kodjari (KPR) or Parc-W (PRW). It is concluded that rainfall received in September at grain filling and maturation stage was best correlated to PUE. There are large differences in the PUEs of different pearl millet cultivars, and values varied from 25 to 77 kg grain (kg P ha)$^{-1}$. Hill placement of 4 kg P ha^{-1} at sowing time improved PUE more than the present recommendation of 13 kg ha^{-1} broadcast, and also improved the efficiency of PAPR. The rotation of cereals and cowpea, and soil amendment with crop residue application drastically increased PUE in the region.

Introduction

The West African semi-arid tropics (SAT) is home to the world's poorest people, 90% of whom live in villages and depend for their livelihood on subsistence agriculture. In this zone, the length of the crop growing season ranges from 75 to 150 days. Recurrent droughts, soils of poor native fertility and low water-holding capacity, wind erosion, and surface crusting are the main abiotic constraints to crop production.

In traditional agricultural systems, when crop yields declined to unacceptable levels, over-cropped land was left fallow until soil fertility was built up, and new land was opened for cultivation. Increasing population pressure is decreasing the availability of land and is leading to shorter fallows relative to the duration of cropping. As a result, shifting cultivation is losing its effectiveness and soil fertility is rapidly declining in many areas. The present farming systems are unsustainable without external inputs of nutrients, will continue to be low in productivity, and have long-term destructive potential to the environment. In such systems, plant nutrient balances are negative (Stoorvogel and Smaling, 1990).

Among soil fertility factors, phosphorus (P) deficiency is a major constraint to crop production and response to nitrogen (N) is substantial only when both moisture and P are not limiting (Traoré, 1974). Although lack of water limits crop production in the drier zones in the Sahel, all available evidence indicates

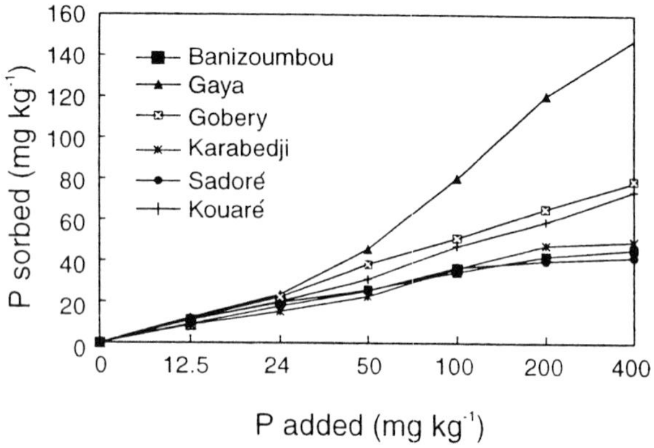

Figure 1. Phophorus (P) sorption isotherms of soils samples from six benchmark sites in Niger and Burkina Faso, West Africa.

that inherent low fertility (mainly P) is a more serious problem (Breman and de Wit, 1983; van Keulen and Breman, 1990).

For many years, research has been undertaken to assess the extent of soil-P deficiency, to estimate P requirements of major crops, and to evaluate the agronomic potential of various phosphate fertilizers including phosphate rock (PR) from local deposits (Bationo et al., 1987, 1990; Goldsworthy, 1967a,b; Pichot and Roche, 1972; Thibaut et al., 1980).

In a survey of the fertility status of representative sites, Manu et al. (1991) found that the total P in these soils ranged from 25 to 349 mg kg^{-1} with a mean of 109 mg kg^{-1}. Available P determined by Bray P1 was also generally low, ranging from 1 to 30 mg kg^{-1} with an average of 6 mg kg^{-1}. However, 77% of samples had available P values of less than 8 mg kg^{-1} determined to be the critical P-level required to obtain 90% of the maximum pearl millet yield in the sandy soils of Niger (Bationo et al., 1989a).

The method of Fox and Kamprath (1970) was used to study the P-sorption characteristics of soils from representative sites, and selected adsorption isotherms are presented in Figure 1. Sorption data were fitted to the Langmuir equation (Langmuir, 1918) and P adsorption maxima were calculated. From these representative sites, Manu et al. (1991) found that the values of maximum P sorbed ranged from 27 to 253 mg kg^{-1} with a mean of 94 mg kg^{-1}. Soils of this region can be considered as having relatively low P-sorption capacities compared to the clay-rich Utisols and Oxisols found in humid tropical regions (Sanchez and Uehera, 1980). As a consequence of the low P-retention capacity of these soils, relatively small quantities of P-fertilizers needed to be applied to obtain optimum crop growth.

Phosphorus use efficiency (PUE) in this paper is calculated by dividing the difference in yield between P-treatments and control by the rate of P applied. In addition to such water-soluble P sources as single superphosphate (SSP) and triple superphosphate (TSP), phosphate rocks (PRs) indigenous to this region, e.g., Tahoua PR (TPR) and Parc-W PR (PRW) from Niger, and Kodjari PR (KPR) from Burkina Faso were evaluated in crop production field trials on the main soil types. In this region, the use of water-soluble imported P-fertilizers is severely limited because of their high cost. The direct application of PR indigenous to the region may be an economical alternative to the use of more expensive imported P-fertilizers. Some PRs may not be suitable for direct application because of their low chemical reactivity (Hammond et al., 1986). Partial acidulation of PR (PAPR) represents a technology that improves the agronomic effectiveness of an indigenous PR at a lower cost than would be required to manufacture conventional, fully acidulated fertilizers from the same rock (Bationo et al., 1990; Chien and Hammond, 1978; Hammond et al., 1986).

In this paper, after a brief review of the PUE of crops as affected by sources of P fertilizers, the effects of rainfall, soil, and crop management on PUE are discussed.

Materials and methods

Effect of P sources and rainfall on PUE

In 1982, a benchmark field trial was initiated on the sandy Sahelian soils of the ICRISAT Sahelian Center campus at Sadoré, Niger to evaluate the agronomic efficiency of different sources of P-fertilizers. The sources of P-fertilizers in the trial were Parc W PR, partially acidulated Parc W at 50% (Parc W PAPR 50), TSP, and SSP. P was applied at 0, 4.8, 8.8, 13 and 17.6 kg ha^{-1} to a test crop of pearl millet cultivar CIVT.

In 1996, field trials were conducted at Sadoré, Gobery, and Gaya to evaluate the agronomic effectiveness of KPR and TPR compared to SSP.

Effect of soil and crop management on PUE

In a researcher-managed trial at Karabedji, Niger hill placement of small quantities of fertilizers (TPR and KPR) were evaluated on pearl millet and cowpea.

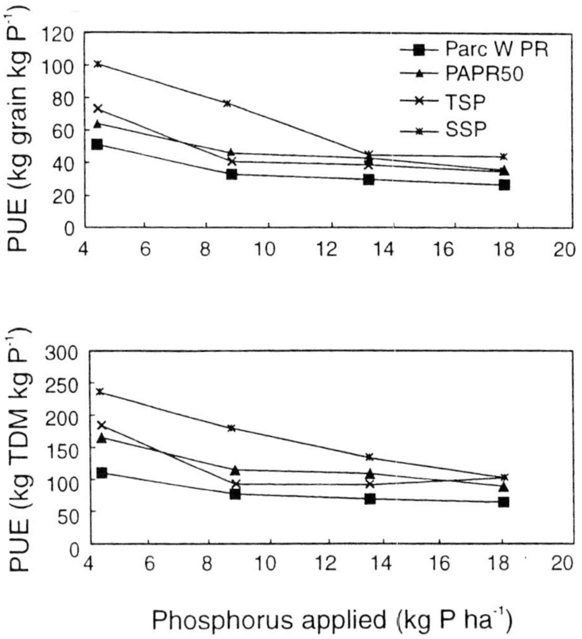

Figure 2. Relationship between different phophorus (P) sources and application rates and PUE for pearl millet grain and total dry matter yields, rainy season, Sadoré, Niger, Average of 6 years' data (1982–1987).

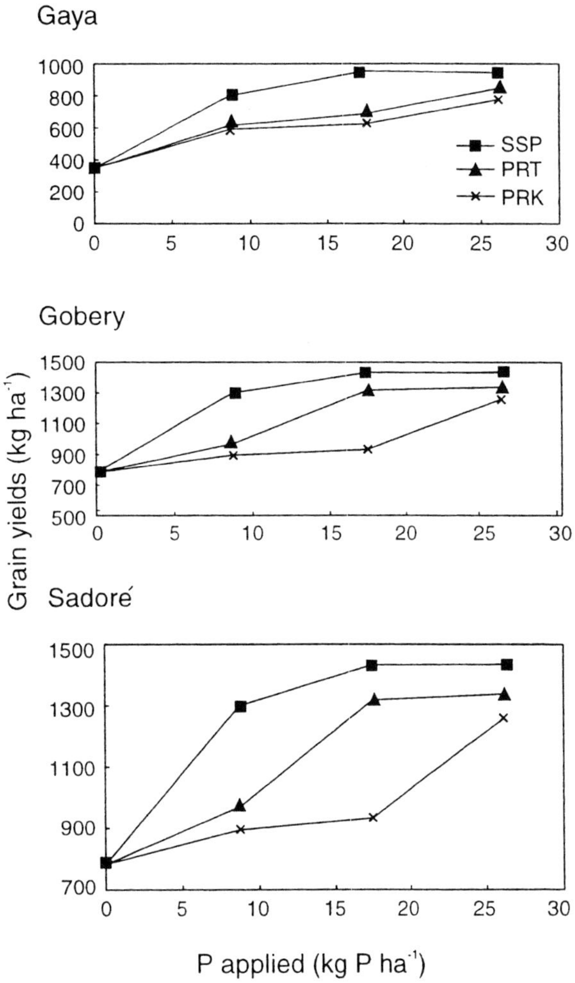

Figure 3. Relationship between P sources and application rates on pearl millet grain yields in three agroecological zones of Niger, rainy season 1986.

In 1998, data were collected in an experiment to evaluate the effect of N application, crop residues, ridging, and rotation of pearl millet with cowpea on PUE.

From 1992 to 1995, an experiment was conducted to study the effect of crop rotation of pearl millet and cowpea on PUE at the ICRISAT Sahelian Center. P as SSP was applied at 0, 6.5 and 13 kg ha^{-1}.

Results and discussion

PUE as related to different sources of P fertilizers and rainfall

For the benchmark experiment conducted during 1982–1987 SSP outperformed the other sources and its superiority to sulphur-free TSP indicates that with continuous cultivation, sulphur deficiency develops (Frisen, 1991). For both pearl millet grain and total dry matter yields, the relative agronomic effectiveness of TSP was almost similar to PAPR50 indicating that partial acidulation of PRW at 50% can significantly increase its effectiveness (Figure 2). SSP had the highest PUE values at all rates of P application. Increasing the rate of SSP application resulted in a decrease in PUE. For pearl millet grain, application of 4.4 kg P ha^{-1} resulted in a PUE of 100 kg grain kg P^{-1}, but the PUE decreased to 45 kg grain kg P^{-1} when 13 kg P ha^{-1} were applied. The difference between the P sources is better resolved at lower application rates of P fertilizers than at higher application rates. The difference between PRW and SSP at 4.4 kg P ha^{-1} was 50 kg grain kg P^{-1} while at 17.5 kg P ha^{-1}, this difference was reduced to only 27 kg grain kg P^{-1} (Figure 2).

From the agronomic evaluation of P sources in different agro-ecological zones of Niger, the response of pearl millet to different sources of P-fertilizers indicates that TPR outperformed KPR in agronomic effectiveness (Figure 3 and Table 1). These results are in agreement with the fact that the molar PO_4/CO_4 ratio is 23.0 for KPR and 4.88 for TPR, and that

Table 1. Relative agronomic effectiveness measured as yield (kg ha^{-1}) for pearl millet and cowpea as compared to SSP (%) of Tahoua phosphate rock (TPR) and Kodjari phosphate rock (KPR) in three agroecological zones of Niger

Yields	Sadoré		Goberi		Gaya	
	TPR	KPR	TPR	KRP	TPR	KPR
Pearl millet grain	63	32	76	41	80	57
Pearl millet total biomass	65	35	60	40	68	63
Cowpea fodder	43	28	73	51	42	42

TPR also has a higher solubility in neutral ammonium citrate (NAC).

Mokwunye (1995) found that the level of isomorphic substitution of carbonate for phosphate within the lattice of the apatite crystal influences the solubility of the apatite in the rock, and therefore controls the amount of P that is released when PR is applied to soils. Chien (1977) found that the solubility of PR in NAC was directly related to the level of carbonate substitution. As a result of the higher value of Tilemsi PR in NAC, and the high substitution of carbonate for phosphate, Bationo et al., (1997) found that application of Tilemsi PR can result in net returns and value/cost ratios similar to those from the use of recommended cotton or cereal complex imported fertilizers.

The PUE at Gobery was 31 kg grain kg P^{-1} for TPR, but decreased to 9 kg grain kg P^{-1} with KPR application at 17 kg P ha^{-1}. As soils in Gaya and Gobery are more acidic and receive more rain than the Sadoré site, the agronomic effectiveness is higher at those sites. The ability of the soil to provide H^+ ions is essential to ensure the effectiveness of PR application to crops (Chien, 1977; Khasawneh and Doll, 1978). Therefore, acidic soils with a high pH buffering capacity provide an ideal environment for PR dissolution. Results presented for upland rice by Bado et al. (1995) indicate that the PUE of the unreactive KPR on an acidic (pH in H_2O=5) soil is similar to the PUE of water-soluble TSP (Mahaman et al., 1998).

The agronomic effectiveness of the leguminous cowpea was no better than that of a cereal pearl millet crop (Table 1). This is in contradiction to other reports where legumes are better able to solubilize PR than cereals through their capacity for rhizosphere acidulation (Aguilar and van Diest, 1981; Hedley et al., 1982; Kirk and Nye, 1986) and exudation of organic acids (Ohwaki and Hirata, 1992).

The increase in soil pH resulting from flooding rice fields is expected to depress the dissolution of PR. Enhanced performance of PR in flooded systems has been reported (Hammond et al., 1986). Kirk and Nye (1986) explain the enhanced PR performance in flooded soils by arguing that rice roots will acidify surrounding soil, and that dissolved organic matter may chelate calcium (Ca) and P. In the irrigated system the PUE of PR often was higher than that of TSP (INRAN, 1988).

Using data of PUE at 13 kg P ha^{-1} and details from the experimental results presented in Figure 2 that were collected over 10 years, it was found that the rainfall received in September at grain-filling and maturation stages was best correlated to PUE (Figure 4). From the results presented in Figure 4, it could be concluded that PUE of SSP was most affected by the amount of September rainfall due to the higher biomass produced by crops in that treatment than by those that received applications of PR or PAPR50. The predictions indicate that SSP, combined with 40 mm rainfall in September will result in a PUE of 118 kg dry matter kg P^{-1} while for 100 mm rainfall the PUE will increase to 160 kg dry matter kg P^{-1}.

In the West African SAT, both water and nutrients limit crop production, but from multilocational water-balance studies in Niger, it was shown that an important outcome of fertilizer use is an increase in water-use efficiency (WUE) (Breman and de Wit, 1983; van Keulen and Breman, 1990). In long-term experiments, WUE for grain yield increased dramatically from 5.4 kg mm^{-1} ha^{-1} without the use of fertilizers to 14.4 kg mm^{-1} ha^{-1} with the use of fertilizers. Increased root growth due to P application is associated with greater rooting depth and deeper extraction of moisture during dry spells (Payne et al., 1995). Early vigor and enhanced growth due to P application results in more complete ground cover early in the season, which reduces the proportion of water lost through evaporation to some extent, thus facilitating effective and efficient use of rainfall.

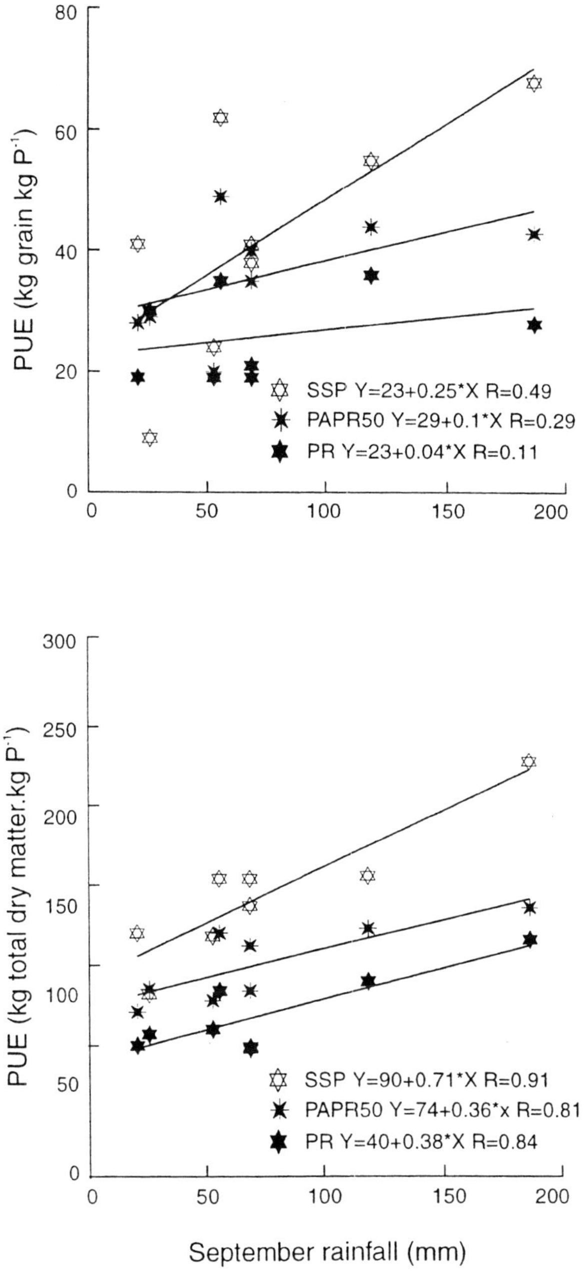

Figure 4. Relationship between September rainfall and PUE for pearl millet grain and total dry matter yields, Sadoré, Niger, rainy seasons 1982–1993.

Although the application of fertilizers improves WUE, the efficiency of fertilizer depends on the amount of rainfall received by the crop. For N, Bationo et al. (1989b) developed a model relating grain yield of pearl millet to mid-season rainfall (45 days, from mid-July to end of August). This model predicts that response to N in dry years will be limited, with little

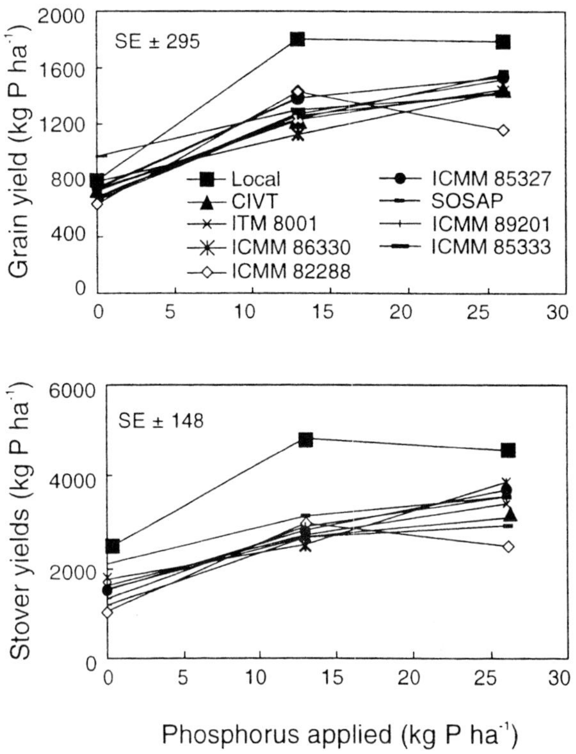

Figure 5. Relationship between phosphorus applied and grain and stover yields for nine pearl millet cultivars, Sadoré, Niger, rainy season 1991–1993.

benefit to the farmers from investment in N fertilizers.

Relationship between crop and soil management on PUE

Over a period of 3 years, nine pearl millet cultivars were evaluated to determine their PUE. For both grain and stover, there are very large differences among the nine cultivars for their response to the application of P fertilizer (Figure 5). PUE at 13 kg P ha^{-1} varied among the nine cultivars from 25 kg grain kg P^{-1} for variety ICMV IS 85333 to 77 kg grain kg P^{-1} for local variety Haini-Kirei. The local cultivar matures 3 weeks later and has a very dense root system. Figure 6 shows that the PUE of different genotypes was significantly correlated with grain yield at 13 kg P ha^{-1}, and explained 77% of the total variation in this relationship. This significant relationship indicates that PUE cultivars can be first identified using their grain yield performance at 13 kg P ha^{-1}. The relationship between the PUE of the different genotypes with grain yield in the absence of P application was not significant, and only

Table 2. Effect of different sources and placement of P on pearl millet yield and phosphorus use efficiency (PUE kg grain kg P^{-1}) Karabedji, Niger rainy season 1998

P source[a] and method of application	Grain		TDM[d]	
	Yield (kg ha^{-1})	PUE	Yield (kg ha^{-1})	PUE
Control	281		1726	
SSP broadcast[b]	535	23	3726	154
SSP broadcast+SSP HP[c]	743	27	5563	226
SSP HP	611	83	3774	514
15-15-15 broadcast	660	29	4226	192
15-15-15 broadcast+15-15-15 HP	1493	71	7677	350
15-15-15 HP	690	102	4767	760
PRT broadcast	690	31	4135	185
PRT broadcast+SSP HP	663	22	4365	155
PRT broadcast+15-15-15 HP	806	31	5061	196
PRK broadcast	465	14	3302	121
PRK broadcast+SSP HP	747	27	5052	196
PRK broadcast+15-15-15 HP	806	31	5010	193
SE	84		194	

[a]SSP, single superphosphate; 15-15-15, compound fertilizer containing 15% N, 15% P_2O_5, 15% K_2O; TPR, Tahoua phosphate rock; KPR, Kodjari phosphate rock.
[b]Broadcast at 4 kg P ha^{-1}.
[c]HP, hill placed at 4 kg P ha^{-1}.
[d]TDM, total dry matter.

Table 3. Effect of different sources of phosphorus (P) and their placement on cowpea yield and phosphorus use efficiency (PUE), Karabedji, 1998 rainy season

P source and method of application	Grain		Fodder	
	Yield (kg ha^{-1})	PUE (kg grain kg P^{-1})	Yield (kg ha^{-1})	PUE (kg grain kg P^{-1})
Control	505		1213	
SSP broadcast[b]	1073	44	2120	70
SSP broadcast+SSP HP[c]	1544	61	3139	113
SSP HP	1050	136	2021	452
15-15-15 broadcast	1165	51	2381	90
15-15-15 broadcast+15-15-15 HP	2383	110	3637	142
15-15-15 HP	1197	173	2562	337
PRT broadcast	986	37	2220	77
PRT broadcast+SSP HP	1165	68	3127	113
PRT broadcast+15-15-15 HP	1724	72	3163	115
PRK broadcast	920	32	1791	44
PRK broadcast+SSP HP	1268	45	2588	81
PRK broadcast+15-15-15 HP	1440	55	2792	93
SE	±164		±313	

[a]SSP, single superphosphate; 15-15-15 = compound fertilizer containing 15% N, 15% P_2O_5, 15% K_2O; TPR, Tahoua phosphate rock; KPR, Kodjari phosphate rock.
[b]Broadcast at 13 kg P ha^{-1}.
[c]HP, hill placed at 4 kg P ha^{-1}.

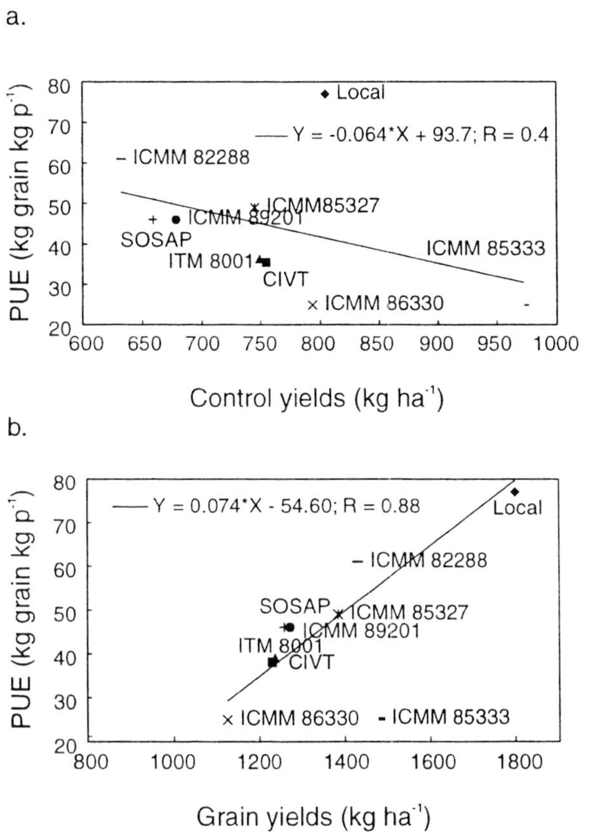

Figure 6. Relationship between PUE at an application rate of 13 kg P ha^{-1} and grain yield of (a) unfertilized, and (b) fertilized pearl millet.

explained 15% of the total variation. This observation shows that a cultivar with a high PUE coefficient will not necessarily perform better under low-P conditions than one with a low PUE coefficient. This also implies that genotypes selected for high grain yield under low-P situations will not necessarily be P-use efficient. There is ample evidence that indicates marked differences exist between species and genotypes for P uptake (Caradus, 1980; Föhse et al., 1988; McLachlam, 1976; Nielsen and Schjorring, 1983; Spencer et al., 1980).

For the researcher-managed on-farm trials conducted to study interaction between hill placement of small quantities of P fertilizers on the efficiency of water-soluble (SSP, 15-15-15) and phosphate rock (PRT and PRW), results are presented in Tables 2 and 3. They indicate that hill placement increases the agronomic effectiveness of both water-soluble and PR sources for pearl millet and cowpea. Compared to the control, the pearl millet grain yield increased from 281 for the control to 1493 kg ha^{-1}, for the 15-15-15 broadcast plus 15-15-15 hill-placed treatments; the application of 15-15-15 alone resulted in yields of 661 kg ha^{-1}. The PUE results in Table 3 indicate that hill placement of 4 kg P ha^{-1} with broadcast PRK can improve the PUE of the unreactive PRK. For cowpea fodder, PUE increased from 44 kg kg P^{-1} with the addition of KPR only, to 93 when KPR is broadcast with hill placement of 15-15-15. Whereas PUE efficiency was 14 for pearl millet grain yield with KPR broadcast, it increased to 31 kg grain kg P^{-1} when additional hill placement for 15-15-15 was applied (Table 2). Although hill placement of 4 kg P ha^{-1} alone gave higher PUE values than 3 kg P ha^{-1} broadcast, this treatment will result in a net negative P balance. With the association of hill placement and low-cost PR sources the net balance of P will be positive and soil mining will be avoided. For most of cases, 15-15-15 hill placement efficiency is higher than SSP hill placement. This is due in part to germination failures probably because of deleterious pH and salt effects on the seedling. The most effective NPK placement is also likely due to a stimulation of early root growth by the ammonium component (Marschner et al., 1986), and an enhanced availability of P in the immediate seedling environment. Over the past few years, on-station research at ICRISAT's Sadoré location has focussed on the placement of small quantities of P fertilizers at sowing in order to develop an optimum farmer-affordable P application recommendation. Compared to the control, pearl millet grain yield increased between 60 and 70% when 5 kg P ha^{-1} was hill placed, and by 100% when 13 kg P ha^{-1} was broadcast. PUE on total dry matter (TDM) and grain yield indicate that PUEs at 3, 5 and 7 kg P ha^{-1} hill application were higher than broadcasting 13 kg P ha^{-1}. For example, in 1995, for total dry matter (TDM) the PUE for 13 kg P ha^{-1} was 159 kg TDM kg P^{-1} as compared to 402 kg TDM kg P^{-1} with the application of 3 kg P ha^{-1} hill-placed. This is due in part to the placement of P where the soil is humid compared to the surface broadcast where some fertilizers will remain in the dry zone of the soil (Muhelhig-Versen et al., 1997).

In long-term soil management trials, application of N, crop residues and ridging, and rotation of pearl millet with cowpea were evaluated to determine their effect on PUE. The results show that the productivity of sandy soils can be dramatically increased with the adoption of improved crop and soil management technologies. Whereas the absolute control recorded

Table 4. Effect of mineral fertilizers, crop residue (CR), nitrogen (N), and crop rotation on pearl millet yield (kg ha^{-1}) and phosphorus use efficiency (PUE kg grain kg P^{-1}), Sadoré, Niger, rainy season 1998

Treatment	Without Cr without N[a]				Without Cr with N				With Cr without N				With Cr with N			
	TDM		Grain		TDM		Grain		TDM		Grain		TDM		Grain	
	Yield	PUE	Yield	PUE	Yield	PUE	Yield	PUE	Yield	PUE	Yield	PUE	Yield	PUE	Yield	PUE
Control	889		33		2037		58		995		61		1471		98	
13 kg P ha^{-1}	2704	140	633	46	4339	177	1030	75	4404	185	726	51	240	4594	1212	86
13 kg P ha^{-1} +ridge	2675	137	448	32	4057	155	946	68	3685	210	785	56	4530	235	1146	81
13 kg P ha^{-1} +rotation	5306	340	1255	94	6294	327	1441	106	5392	338	1475	109	6124	358	1675	121
13 kg P ha^{-1} +ridge+rotation	5223	333	1391	104	5818	291	1581	117	6249	404	1702	126	7551	468	1829	133
SE	±407		±407		±407		±407		±407		±407		407		407	

[a] CR, crop residue; N, nitrogen; TDM, total dry matter; P, phosphorus.

33 kg ha^{-1} of grain, 1829 kg ha^{-1} was obtained when P, N, and crop residues were applied to plots that were ridged and followed a leguminous cowpea crop grown in the previous season (Table 4). Results indicate that for grain yield, PUE will increase from 46 with P application alone to 133 when P is combined with N and crop residue applications, and the crop is sown on ridges in a rotation.

In a study on the long-term effect of different cropping systems on PUE it was found that rotation of pearl millet with cowpea could significantly increase both pearl millet and cowpea production (Figure 7). For pearl millet, total dry matter, PUE increased from 149 kg ha^{-1} in the continuous cultivation to 252 kg ha^{-1} in a rotation. For cowpea fodder, PUE increased from 40 kg ha^{-1} in continuous cultivation to 65 kg ha^{-1} in rotation. In a long-term field trial to study the effect of crop residue application on PUE, the PUE was 67 kg kg P^{-1} when P-fertilizers alone were applied, its value doubled when P-fertilizers were combined with crop residues (Bationo et al., 1985).

Conclusion

In the West African SAT, lack of volcanic rejuvenation has caused the region to undergo several cycles of weathering erosion, leaving soils poor in nutrients. Both total and available P values are very low, and P deficiency is a major constraint to crop production. With their sandy texture, these soils have low P-retention capacity. The PUE is highly variable and depends on P sources, rainfall, soil, and crop management. In the West African SAT there has been little research on understanding the factors affecting P uptake such as the ability of plants to: (1) solubilize soil P through pH changes and the release of chelating agents and phosphates enzymes; (2) explore a large soil volume; and (3) absorb P from low soil solution P concentration.

Genotypic improvement can come through an increased capacity of plants to extract P from the soil, or through decreased internal P requirement per unit dry matter produced. The opportunities for increased efficiency of P utilization through cultivar improvement include selection for treatments that favour strong plant demand such as late maturity, increased rootlet activity, and increased P-solubilization capacity.

The available and total P values are very low in the region. With those extremely low values of total P, it can be questionable to select cultivars adapted to low-P conditions, as one cannot mine what is not there. Direct application of indigenous PR can be an economic alternative to the use of more expensive imported water-soluble P fertilizers.

The effectiveness of mycorrhizae in utilizing soil-P has been well documented (Daft, 1991; Lee and Wani, 1991; Silberbush and Barber, 1983). An important future research opportunity is the selection of plant genotypes that are conducive to colonization by efficient vesicular-arbuscular mycorrhizal (VAM) associations for better utilization of P from PR.

Previous agronomic research has already identified a significant number of technologies to enhance PUE, but future research needs to screen technologies under

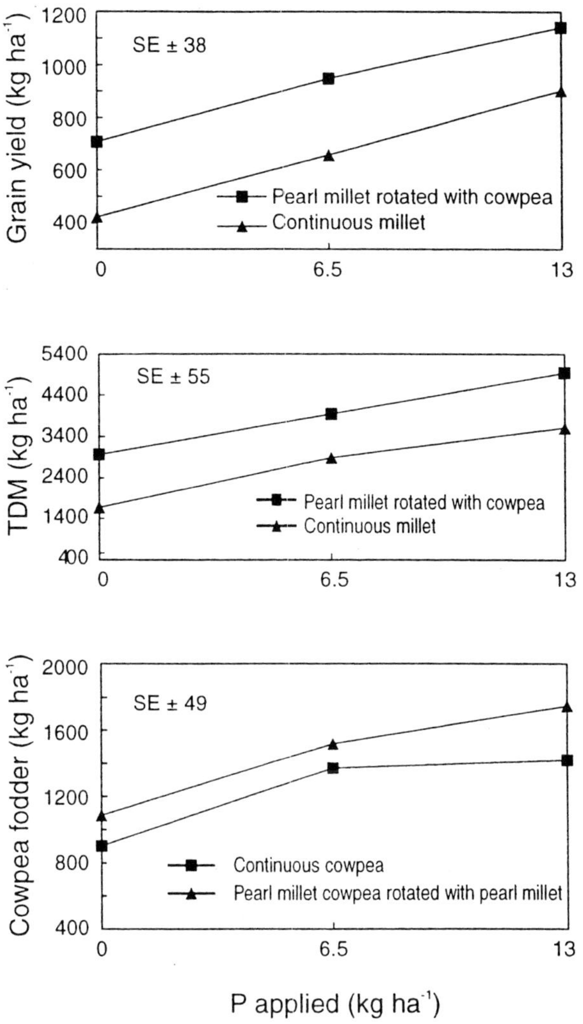

Figure 7. Effects of P and cropping systems on pearl millet grain, total dry matter (TDM) and cowpea fodder yields, Sadoré, Niger, rainy seasons 1992–1995.

farmer management in order to recommend those with the highest economic returns.

References

Aguilar S A and van Diest 1981 Rock phosphate mobilization induced by the alkaline uptake pattern of legumes utilizing symbiotically fixed N. Plant Soil 61, 27–42.

Bado V B 1998 Efficacite agronomique des phosphates naturels du Burkina Faso sur le riz pluvial en sol ferralitique. Cahiers Agric. 7(3), 236–238.

Bationo A, Chien S H and Mokwunye A U 1987 Chemical characteristics and agronomic values of some phosphate rocks in West Africa. *In* Food Grain Production in Semi-Arid Africa. Eds J M Menyonga, T Bezuneh and A Youndeowei. pp 399–407. Coordination Office OAU/SAFGRAD Essex, UK.

Bationo A, Mokwunye A U and Christianson C B 1989a Gestion de la fertilité des sols sableux au Niger: un aperçu de quelques résultats de recherche de l'IFDC/ICRISAT. *In* Les Actes du Séminaire National sur l'Aménagement des Sols, la Conservation de l'Eau, et la Fertilisation. pp 177–194. Tahoua, Niger.

Bationo A, Christianson C B and Baethgen W E 1989b Plant density and nitrogen fertilizer effects on pearl millet production in Niger. Agron. J. 82(2), 290–295.

Bationo A, Chien S H, Christianson C B, Henao J and Mokwunye A U 1990 A 3-year evaluation of two unacidulated and partially acidulated phosphate rocks indigenous to Niger. Soil Sci. Soc. Am. J. 54, 1772–1777.

Bationo A and Mokwunye A U 1991 Alleviating soil fertility constraints to increased crop production in West Africa. The experience in the Sahel. Fertil. Res. 29, 98–115.

Bationo A, Buerkert A, Sedogo M P, Christianson B C and Mokwunye A U 1995 A critical review of crop residue uses a soil amendment in the West African semi-arid tropics. *In* Livestock and Sustainable Nutrient Cycling in Mixed Farming Systems of Sub-Saharan Africa, Proceedings of an International Conference. Eds J M Powell, S Fernandez-Rivera, T O Williams and C Renard. pp 305–322. International Livestock Centre for Africa (ILCA) Addis Ababa, Ethiopia.

Bationo A, Ayuk E D, Ballo and Kone M 1997 Agronomic and economic evaluation of Tilemsi phosphate rock in different agro-ecological zones of Mali. Nutr. Cycling Agroecosyst. 48, 179–189.

Breman H and de Wit C T 1983 Rangeland productivity and exploitation in the Sahel. Science 221, 1341–1347.

Caradus J R 1980 Distinguishing between grass and legume species for efficiency of phosphorus use. N.Z. J. Agric. Res. 23, 75–81.

Chien S H 1977 Thermodynamic considerations on the solubility of phosphate rock. Soil Sci. 123, 117–121.

Chien S H and Hammond L L 1978 A simple chemical method for evaluating the agronomic potential of granulated phosphate rock. Soil Sci. Soc. Am. J. 42, 615–617.

Daft M J 1991 Influences of genotypes, rock phosphate and plant densities on mycorrhizal development and the growth responses of five different crops. Agric. Ecosyst. Environ. 35, 151–169.

Föhse D N, Claasen, and Jungk A 1988 Phosphorus efficiency of plants. I. External and internal P requirement and P uptake efficiency of different plant species. Plant Soil 110, 101–109.

Fox R L and Kamprath E J 1970 Phosphate sorption isotherm for evaluating the requirements of soils. Soils Sci. Soc. Am. Proc. 31, 902–907.

Friesen D 1991 Fate and efficiency of sulfur fertilizer applied to food crops in West Africa. *In* Alleviating Soil Fertility Constraints to Increased Crop Production in West Africa. Ed. A U Mokwunye. pp 59–68. Kluwer Academic Publishers. Dordrecht, The Netherlands.

Goldsworthy P R 1967a Responses of cereals to fertilizers in Northern Nigeria I. Sorghum. Exp. Agric. 3, 29–40.

Goldsworthy P R 1967b Responses of cereals to fertilizers in Northern Nigeria II. Maize. Exp. Agric. 3, 263–268.

Hammond L L, Chien S H and Mokwunye A U 1986 Agronomic value of unacidulated and partially acidulated phosphate rocks indigenous to the tropics. Adv. Agron. 40, 89–140.

Hedley M J, Nye P H and Bolan S N 1982 Plant-induced changes in the rhizosphere of rape (*Brassica napus* var. Emerald) seedings. II: Origin of the pH change. New Phytol. 91, 31–44.

INRAN 1988 Départment des recherches écologiques. Bilan de cinq années de recherches Institut National des Recherches Agronomiques du Niger, Niamey, Niger (1982–1987). 80 pp.

Khasawneh F E and Doll E C 1978 The use of phosphorus rock for direct application to soils. Adv. Agron. 30, 155–206.

Kirk G J D and Nye P H 1986 A simple model for predicting the rate of dissolution of sparingly soluble calcium phosphate in soil: The basic model. J. Soil Sci. 37, 529–540.

Langmuir J 1918 The adsorption of gases on plain surface of glass, mica and platinum. J. Am. Chem. Soc. 40, 1361–1403.

Lee K K and Wani S P 1991 Possibilities for manipulating mycorrhizal associations in crops. *In* Phosphorus Nutrition of Grain Legumes in the Semi-Arid Tropics. Eds C Johansen, K K Lee and K L Sahrawat. pp 107–166. International Crops Research Institute for the Semi-Arid Tropics, Patancheru, Andhra Pradesh, India.

Mahamane I, Bationo A, Seyni F and Hamidou Z 1997 Acquis récents des recherches sur les phosphates naturels du Niger. *In* Soil Fertility Management in West African Land Use Systems: Proceedings of a Regional Workshop Niamey, Niger, 4–8 March 1997. Eds G Renard, A Neef, K Becker and M von Oppen. pp 73–78. University of Hohenheim, ICRISAT, INRAN, Margraf Verlag, Weikersheim, Germany.

Manu A, Bationo A and Geiger S C 1991 Fertility status of selected millet producing soils of West Africa with emphasis on phosphorus. Soil Sci. 152, 315–320.

Marschner H, Römheld V, Horst W J and Martin P 1986 Root-induced changes in the rhizosphere: Importance for the mineral nutrition of plants. Z. Planzenernahr. Bodenk. 149, 441–456.

McClellan G H and Northolt A J G 1986 Phosphate deposits of tropical sub-Saharan Africa. *In* Management of Nitrogen and Phosphorus Fertilizers in Sub-Saharan Africa. Eds A U Mokwunye and P L G Vlek. pp 173–223. Martinus Nijhoff, Dordrecht, The Netherlands.

McLachlan K D 1976 Comparative phosphorus response in plants to a range of available phosphorus situations. Aust. J. Agric. Res. 27, 323–341.

Muelhig-Versen B A, Buerkert A, Bationo A and Maschner H 1997 Crop residue and phosphorus management in millet based cropping systems on sandy soils of Sahel. *In* Soil Fertility Management in West African Land Use Systems: Proceedings of a Regional Workshop. Niamey, Niger, 4–8 March 1997. Eds G Renard, A Neef, K Becker and M von Oppen. pp 31–40. University of Hohenheim, ICRISAT, INRAN, Margraf Verlag, Weikersheim, Germany.

Nielsen N E and Schjorring J K 1983 Efficiency and kinetics of phosphorus uptake from soil by various barley genotypes. Plant Soil 72, 225–230.

Ohwaki Y and Hirata H 1992 Differences in carboxilic acid exudation among P-starved leguminous crops in relation to carboxylic acid contents in plant tissues and phospholipid levels in roots. Soil Sci. Plant Nutr. 38, 235–243.

Payne W A, Hossner L R, Onken A B and Wendt C W 1995 Nitrogen and phosphorus uptake in pearl millet and its relation to nutrient- and water-use efficiency. Agron. J. 87, 425–431.

Pichot J and Roche P 1972 Le phosphate dans les sols tropicaux. Agron. Trop. 27, 939–965.

Sanchez P A and Uehara G 1980 Management considerations for acid soils with high phosphorus fixation capacity. *In* The Role of Phosphorus in Agriculture. Ed. F E Khasawneh. pp 471–509. Am. Soc. Agron., Madison, WI, USA.

Silberbush M and Barber S A 1983 Sensitivity of simulated phosphorus uptake to parameters used by a mechanistic-mathematical model. Plant Soil 74, 93–100.

Spencer K, Govaars A G and Hely F W 1980 Early phosphorus nutrition of eight forms of two clover species. *Trifolium ambiguam* and *T. repens*. N.Z. J. Agric. Res. 23, 457–475.

Stoorvogel J J, Smaling E M A 1990 Assessment of soil nutrient depletion in sub-Saharan Africa: 1983–2000. Report 28, DLO Winand Staring Centre for Integrated Land, Soil and Water Research (SC-DLO), Wageningen, The Netherlands. 173 pp.

Thibaut F, Traoré M F C and Pichot J 1980 L'utilisation agricole des phosphates naturels de Tilemsi (Mali). Synthèse des résultats de la recherche agronomique sur les cultures vivrières et oléagineuses. Agron. Trop. 35, 240–249 (in French).

Traoré F 1974 Etude de la fumure azotée intensive des céréales et du role specific de la matière organique dans la fertilité des sols du Mali. Agron. Trop. 29, 567–586 (in French).

van Keulen H and Breman H 1990 Agricultural development in the West African Sahel region: a cure against land hunger? Agric. Ecosyst. Environ. 32, 177–197.

J.J. Adu-Gyamfi (Ed.), Food security in nutrient-stressed environments: exploiting plants' genetic capabilities: 155–163, 2002.

Modelling root-induced solubilization of nutrients

G. J. D. Kirk
Soil and Water Sciences Division, International Rice Research Institute (IRRI), MCPO Box 3127, 1271 Makati City, Philippines

Key words: diffusion, iron oxidation, organic acid, pH change, rhizosphere, rice

Abstract

Theory for coupled diffusion processes in soil is briefly described and three examples of its application to understand root-induced solubilization of nutrients given. The examples are: (1) solubilization of P through root-induced pH changes in the rhizosphere of rice plants growing in flooded soil; (2) solubilization of P through excretion of organic chelating agents from rice roots growing in aerobic soil; and (3) the effects of root geometry on P solubilization, particularly cylindrical versus planar geometry and the effect of excretion of a solubilizing agent being localized along the root axis. The theory is tested by comparing measured concentration profiles of P near roots with the predictions of the theory made using independently measured parameter values. In the examples given, the agreement between the observed and predicted concentration profiles is very good, indicating that the theory is sound and the processes involved well understood.

Introduction

Roots interact with the soil around them in various complicated ways, and identifying the processes that are important in regulating nutrient and toxin dynamics, and, from these, processes that might be manipulated by plant breeding or agronomy, is a complicated matter. Mathematical modelling can help. In fact it may be essential where several interacting rate processes must be considered and plant physiological and soil processes linked together.

Early models of nutrient uptake by roots considered the roots to behave as cylindrical sinks to which nutrients moved by mass flow and diffusion (Tinker and Nye, 1999). Root demand for the nutrient could vary, but the roots did not otherwise influence the soil. These models worked reasonably well for the more soluble nutrient ions, such as NO_3^-. But for the less soluble nutrients, such as P, it was evident that the roots played a much more active role and were in some cases able to bring about an increase in the concentration of the nutrient in the soil solution. Nye (1983, 1984) developed theory to account for such solubilization processes. In this paper I briefly describe this theory and then illustrate its application with three examples.

This paper previously appeared in *Plant and Soil* **245**, 2002, pp. 49–57.

Theory for coupled diffusion processes in soil

Consider the diffusion of two interacting solutes X and Y in soil near a planar surface that simultaneously absorbs X and releases Y. The following two simultaneous second order differential equations describe the system (after Nye, 1983)

$$\frac{\partial}{\partial t}([X_L] - \lambda[Y_L]) = (1 - \lambda\nu)\left(\frac{\partial}{\partial x}D_X\frac{\partial[X_L]}{\partial x}\right) \quad (1)$$

$$\frac{\partial}{\partial t}([Y_L] - \nu[X_L]) = (1 - \lambda\nu)\left(\frac{\partial}{\partial x}D_Y\frac{\partial[Y_L]}{\partial x}\right), \quad (2)$$

where: $[X_L]$, $[Y_L]$ = concentration of X and Y in the soil solution; λ, ν = interaction coefficients $(\partial[X_L]/\partial[Y_L])_{[X]}$, $(\partial[Y_L]/\partial[X_L])_{[Y]}$; and D_X, D_Y = diffusion coefficients of X and Y in the soil.

Equations (1) and (2) can be solved subject to appropriate boundary and initial conditions to develop models of root-induced solubilization processes. In general the boundary conditions are such that the equations must be solved numerically rather than analytically, but well-developed methods are available for this. In the examples that follow I compare measured concentration profiles of P near roots with those predicted with the equations using parameter values measured independently of the concentration profiles. This comparison provides a rigorous test of how well the theory accounts for the processes involved.

Solubilization through pH changes

Saleque and Kirk (1995) and Kirk and Saleque (1995) applied the above theory to the solubilization of P by rice plants growing in chemically reduced, flooded soil as a result of root-induced acidification. To measure changes in soil P fractions and other root-induced changes in the soil, Saleque and Kirk grew rice in cylinders of a P-deficient flooded Ultisol separated into upper and lower cells by a fine nylon mesh that the roots could not penetrate. The roots formed a planar layer above the mesh, and the soil below could be sectioned parallel to the root-plane to obtain concentration profiles. In both P-fertilized and -unfertilized soil, the quantity of readily plant-available P was negligible in comparison with the quantity of P extracted by the plants. Ninety percent of the P taken up was drawn from acid-soluble pools associated with ferrous carbonates and hydroxides (Figure 1). There were also small zones of accumulation of P in an alkali-soluble pool which coincided with zones of $Fe(OH)_3$ accumulation near the roots. The zone of P depletion coincided with a zone of acidification in which the pH fell from near 6 in the soil bulk to less than 4 near the roots. The acidification was caused by (1) H^+ export from the roots to balance excess intake of cations over anions under NH_4^+ nutrition, and (2) H^+ generation in Fe^{2+} oxidation by root-released O_2: $4Fe^{2+} + O_2 + 10H_2O = 4Fe(OH)_3 + 8H^+$. Independent in vitro measurements confirmed that H^+ addition increased the P concentration in the soil solution and the quantity of P that could be desorbed per gram of soil (Figure 2).

To model this system it is necessary to account for the diffusion of H^+ away from the roots, its reaction with the soil in solubilizing P, and the diffusion of P back to the absorbing roots. Equations (1) and (2) can be used. Since we are concerned with changes in pH it is convenient to express the P concentrations in solution as a logarithmic variable, pP ($= -\log[P_L]$). In many soils pP and pH are linearly related to the concentrations of labile P and acid in the whole soil over wide concentration ranges. Also, because P occurs only in low concentrations in the soil solution, the effect of changes in pP on pH are small and therefore $\nu \approx 0$. Thus Eqs. (1) and (2) reduce to

$$\frac{\partial}{\partial t}(\mathrm{pP} - \lambda \mathrm{pH}) = \left(\frac{\partial}{\partial x} D_{\mathrm{P}} \frac{\partial \mathrm{pP}}{\partial x}\right) \quad (3)$$

$$\frac{\partial}{\partial t}\mathrm{pH} = \left(\frac{\partial}{\partial x} D_{\mathrm{HS}} \frac{\partial \mathrm{pH}}{\partial x}\right), \quad (4)$$

where: λ is the H^+–P interaction coefficient ($= (\partial \mathrm{pP}/\partial \mathrm{pH})_{[P]}$), and D_P and D_{HS} are the soil P and soil acidity diffusion coefficients.

The values of λ, D_P and D_{HS} are discussed below.

It is implicit in Eq. (4) that all the H^+ produced in the system is released across the root surface. In the flooded soil—rice system, H^+ is both released across the root surface and produced at a distance from it in Fe^{2+} oxidation. But if Fe^{2+} oxidation is fast, as it generally will be, the spread of the oxidation zone will be small and as a first approximation it can be assumed that all the H^+ is released across the root–soil boundary.

The H^+–P interaction coefficient λ is evaluated from the data in Figure 2 as follows. It can be shown that for a given addition of H^+ to the soil, the quantity of P that must be removed to leave the concentration in solution unchanged (i.e., $-(\Delta[P]/\Delta[HS])_{[P_L]}$) is equal to $\lambda\, b_p/b_{HS}$ where b_P and b_{HS} are the pP and pH buffer powers: $-(\partial[P]/\partial \mathrm{pP})_{[HS]}$ and $-(\partial[HS]/\partial \mathrm{pH})_{[P]}$. Therefore $\lambda = (\Delta[P]/\Delta[HS] \times (b_{HS}/b_P)$ and λ at a given $[P_L]$ is found by comparing the values of [P] between different pHs at that $[P_L]$ in Figure 2, and dividing the decrease in [P] per unit decrease in pH by b_P.

Figure 3 shows the observed P depletion and pH profiles and those predicted with the model using λ, b_P and b_{HS} values obtained from Figure 2 and the other parameter values measured independently. The agreement between the observed and predicted depletion profiles is very good. Since the model makes no arbitrary assumptions and uses only independently measured parameter values, and since the predictions are very sensitive to the parameter values, this indicates that the processes involved are well understood.

The model shows that, under the conditions of the root-plane experiments, solubilization by acidification accounted for at least 80% of the P taken up in both the P-unfertilized and -fertilized soil, but less than 50% of the P solubilized diffused to the roots and was taken up. The amount of solubilized P taken up for a given release of H^+ in a particular soil depends on the relative rates of diffusion of acidity away from the roots

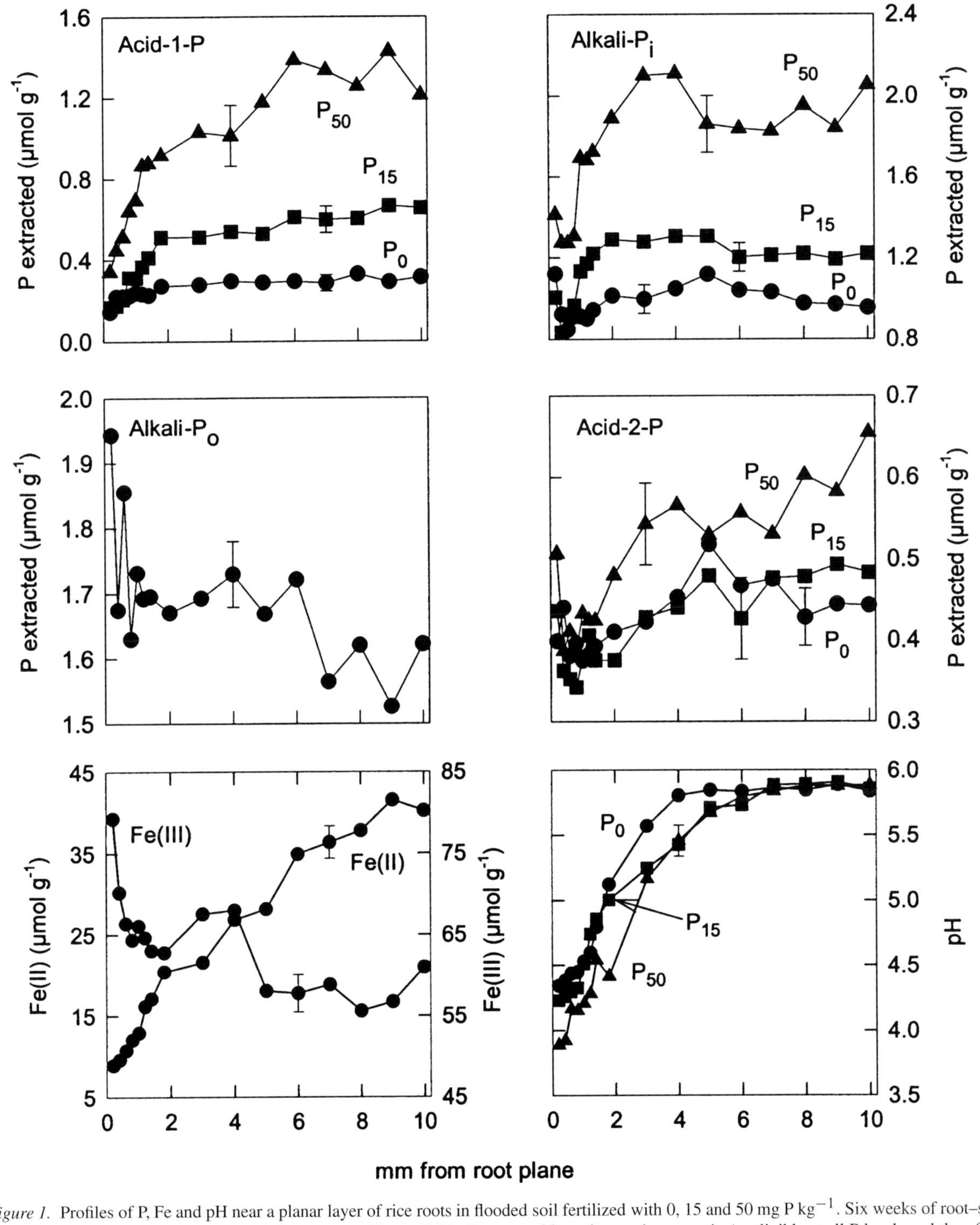

Figure 1. Profiles of P, Fe and pH near a planar layer of rice roots in flooded soil fertilized with 0, 15 and 50 mg P kg^{-1}. Six weeks of root–soil contact. The P pools measured sequentially were: readily available P extracted by anion-exchange resin (negligible at all P levels and therefore not shown); readily acid-soluble P (Acid-1-P), extracted by anion-exchange resin + H^+-form cation-exchange resin; alkali-soluble inorganic P (Alkali-P_i), by 0.1 M NaOH; alkali-soluble organic P (Alkali-P_o), by digesting the previous extract and subtracting the alkali-soluble inorganic P; the more recalcitrant acid-soluble P (Acid-2-P), by 1 M HCl + 1 M H_2SO_4. The data points in the P and pH profiles are each the mean of the three replicate runs for each P level. In the Fe profiles the differences between P levels were not significant, so each point is the mean of nine values from the three replicate runs for each P level. Bars on lines are $LSD_{0.05}$.

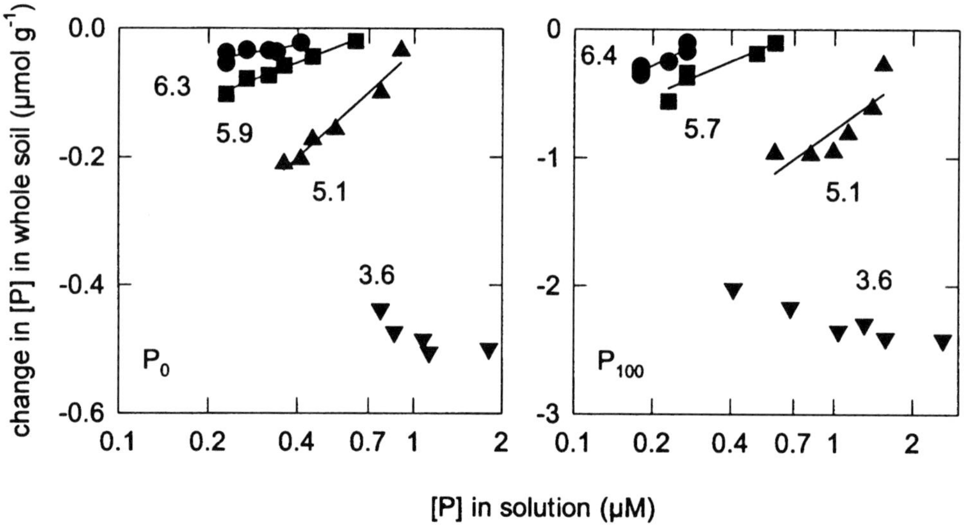

Figure 2. Effect of pH on desorption of P from the experimental soil in Figure 1, obtained by shaking the soil with varying quantities of HCO_3^-–form anion exchange resin and H^+–form cation exchange resin. Plots on left are for unfertilized soil and plots on right for soil with 100 mg P kg^{-1}. Numbers next to data are pHs in the equilibrium soil suspensions. Lines are linear regressions of [P] against pP (i.e., –log[P in solution]); the linear relation breaks down at low pH. Note *x*-axis is logarithmic.

and P towards them. The uptake per unit H^+ release is roughly proportional to

$$1/\left(1+\sqrt{D_{HS}/D_P}\right)$$

(Kirk and Saleque, 1995, Equation 13). When $D_{HS} \ll D_P$, this term approaches 1 and most of the P solubilized will be taken up because the acidification operates close to the roots. But when $D_{HS} \gg D_P$, the term approaches 0 and little of the solubilized P is taken up.

The values of D_P and D_{HS} in Eqs. (3) and (4) are given by:

$$D_P = 2.303\theta f D_{LP}[P_L]/b_P$$

$$D_{HS} = 2.303\theta f\,(D_{LH}[H_3O_L^+] + D_{LC}[HCO_3^-{}_L])\;/b_{HS}$$

where: D_{LP}, D_{LH}, D_{LC} = diffusion coefficients of P, H_3O^+ and HCO_3^- in free solution; $\theta\ f$ = volumetric water content × diffusion impedance factor; and $[H_3O^+{}_L]$, $[HCO_3^-{}_L]$ = concentration of H_3O^+ and HCO_3^- in the soil solution.

The expression for D_{HS} allows for the fact that H^+ is transferred through the soil not only by the diffusion of H_3O^+ away from the zone of acidification but also by the diffusion of HCO_3^- towards it and the simultaneous removal of H_2CO_3 as CO_2. In the pH range in which H_3O^+ and HCO_3^- concentrations are both low — roughly pH 4.5–5.5 depending on the CO_2 pressure — the value of D_{HS} passes through a minimum. In this pH range, a flux of acid through the soil will therefore cause a large pH change. In flooded soils, because $\theta\,f$ and the CO_2 pressure are large, D_{HS} tends to be large and pH changes small. However in acid soils with flooded steady-state pHs below about 6.5, low D_{HS} values are expected with pH changes near roots of 1–2 units, as in the example here.

Table 1. Rates of organic anion excretion and concentrations in roots of rice (cv Azucena) grown in nutrient solution. Rates and concentrations were measured 17 days after planting. Data are means of two replicates

	Rate of excretion ($pmol\ g^{-1}$ root FW s^{-1})		Concentration in root ($\mu mol\ g^{-1}$ root FW)	
	+P	−P	+P	−P
Citrate	33.6	66.4	0.27	0.24
Malate	0	0	1.42	0.68

Solubilization through chelation by organic anions

A similar root-plane system to that in the first example was used to study P solubilization by rice growing in aerobic soil (Hedley et al., 1994). We found that the bulk of the P absorbed by rice in a highly weathered

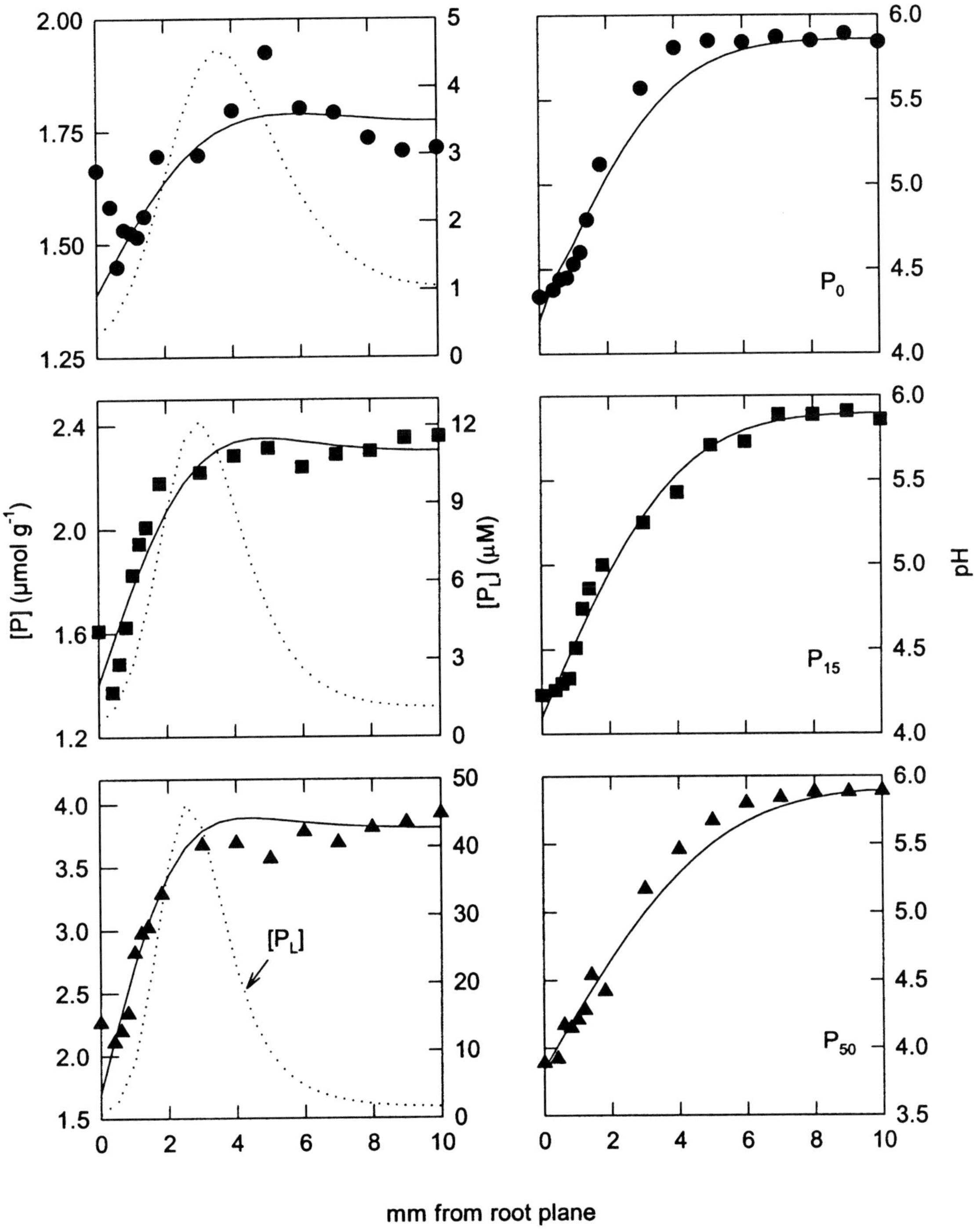

Figure 3. Observed (points) and calculated (lines) profiles of soil P ([P]) and pH in the experiments in Figure 1, and calculated (dotted lines) profiles of P in solution ($[P_L]$). The [P] values at each distance are the sum of the Acid-1-P, Alkali-P_i and Acid-2-P concentrations in Figure 1. The parameter values for the model predictions were measured independently (see text).

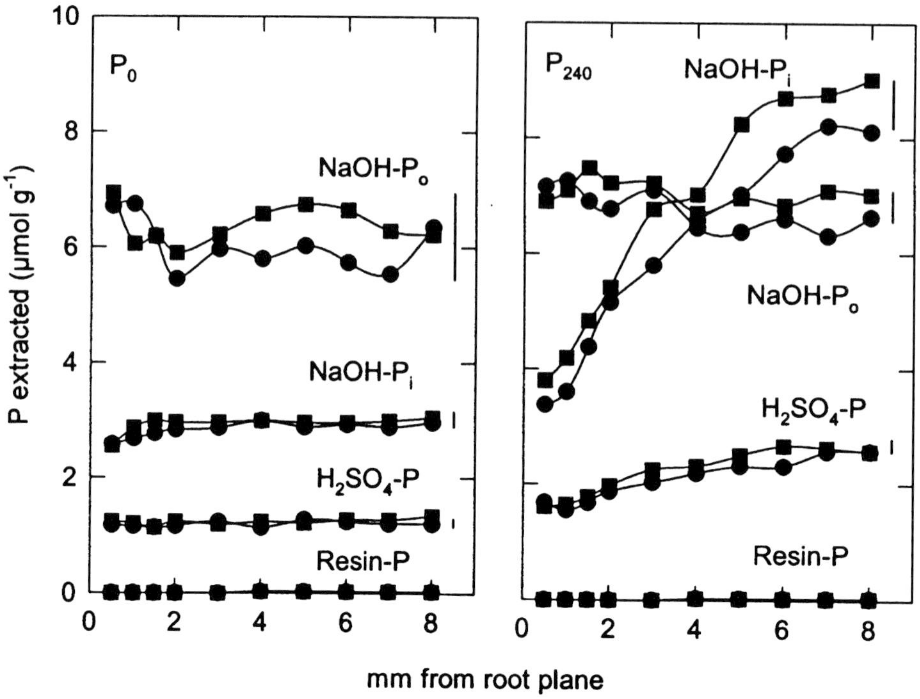

Figure 4. Profiles of P near a planar layer of rice roots in aerobic soil fertilized with 0 and 240 mg P kg^{-1}. Six weeks of root–soil contact. The P pools measured sequentially were: readily-available P (Resin-P), extracted by anion-exchange resin; alkali-soluble inorganic P (NaOH-P_i), by 0.1 M NaOH; alkali-soluble organic P (NaOH-P_o), by digesting the previous extract and subtracting the alkali-soluble inorganic P; acid-soluble P (H_2SO_4-P), by 0.5 M H_2SO_4. Squares and circles indicate runs with different rice cultivars but otherwise identical. Bars are $LSD_{0.05}$.

P-deficient aerobic soil was solubilized from alkali-soluble inorganic forms (Figure 4). The solubilization could not be explained by root-induced pH changes, nor by increased mineralization of organic P because there was no net depletion of organic P in spite of increased phosphatase activity near the roots. Subsequently a substantial release of citrate from rice roots growing in nutrient solution (Kirk et al., 1999a) and in soil (Kirk et al., 1999b) was found. Citrate excretion by plants growing in nutrient solutions tended to increase under P stress and was greater than malate excretion even though concentrations within the roots were lower (Table 1), indicating that the synthesis or excretion of citrate or both were stimulated under P deficiency. We postulated that the stimulated excretion of citrate was responsible for solubilizing soil P.

We obtained an estimate of the expected change in the concentration of P in the soil solution as the plants remove P from the soil, and the effect of citrate on this change, from P desorption isotherms with different citrate additions (Figure 5). The curves in Figure 5 are complicated because they reflect the combined influences of the original soil P sorption isotherm and the isotherm for the soil modified by reaction with citrate. However a citrate–P interaction coefficient, λ, can be derived from the curves in Figure 5 in the same way that the H^+–P interaction coefficient was derived from Figure 2, as shown by Kirk (1999).

In applying Eqs. (1) and (2) to this system it is necessary to allow for the decomposition of the citrate in the soil. If the decomposition follows first order kinetics with rate constant k, and if $\nu \approx 0$ as for solubilization by pH changes, then

$$\frac{\partial}{\partial t}([P_L] - \lambda[C_L]) = \left(\frac{\partial}{\partial x} D_P \frac{\partial[P_L]}{\partial x}\right) \qquad (5)$$

$$\frac{\partial}{\partial t}([C_L] = \left(\frac{\partial}{\partial x} D_C \frac{\partial[C_L]}{\partial x}\right) - \theta k[C_L], \qquad (6)$$

where $[C_L]$ = concentration of citrate in the soil solution, $\lambda = (\partial[P_L]/\partial[C_L])_{[P]}$, $D_P = \theta\ f\ D_{LP}/b_P$, $D_C = \theta\ f\ D_{LC}/b_C$, $b_P = -(\partial[P]/\partial[P_L])_{[C]}$ and $b_C = -(\partial[C]/\partial[C_L])_{[P]}$. An estimate of the rate constant k is obtained by following the change in $[C_L]$ in soil incubated with known amounts of citrate.

Figure 6 compares the P depletion profiles from Hedley et al.'s (1994) experiment with those calculated

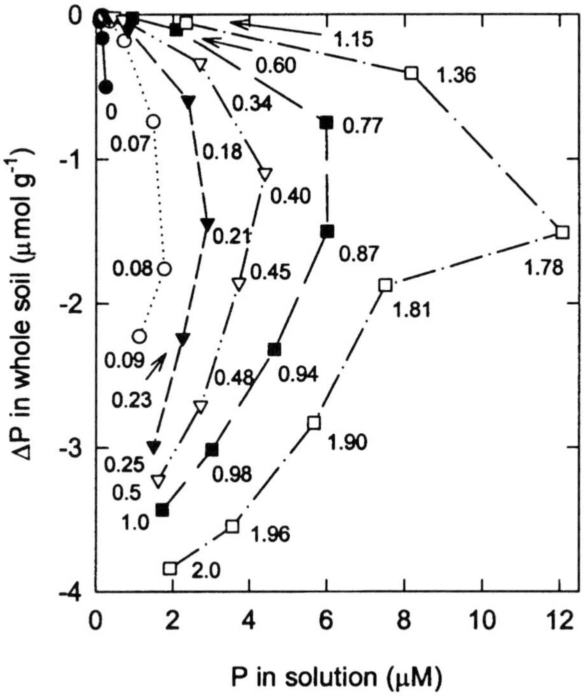

Figure 5. The effect of citrate on the desorption of P from the P_{240} soil used in Figure 4. Common symbols indicate common initial citrate concentrations in solution (● = 0, o = 0.1, ▼ = 0.25, ▽ = 0.5, ■ = 1.0 and □ = 2.0 mM). Numbers on curves are final citrate concentrations in solution (mM). Each point is the mean of two replicates.

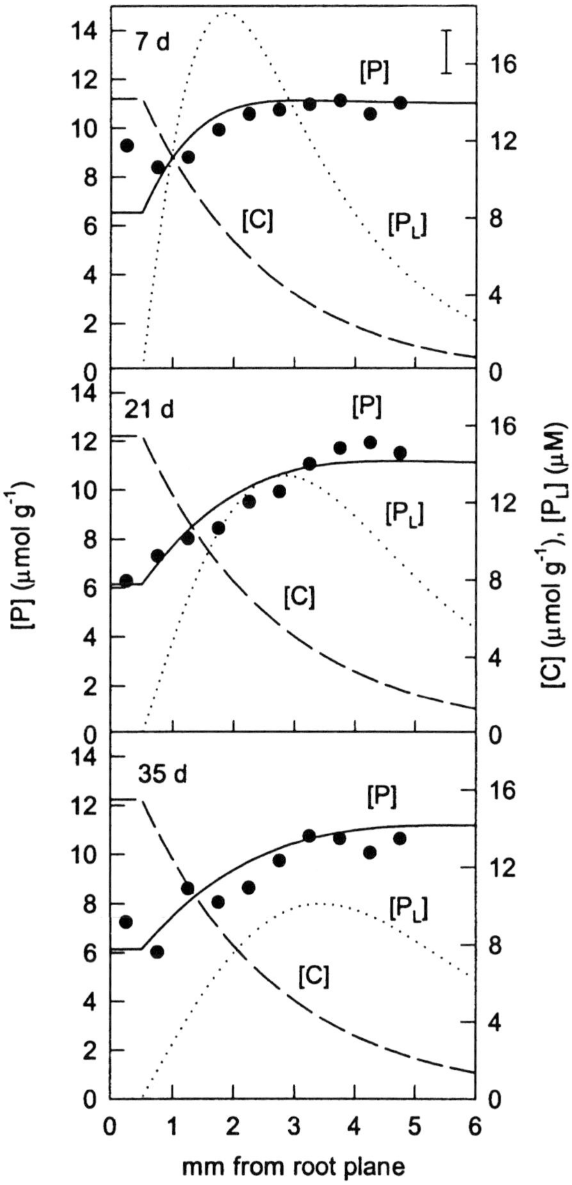

Figure 6. Comparison of observed (points) and predicted (solid lines) concentration profiles of P in the whole soil ([P]) from the experiment with rice in aerobic soil in Figure 4, and the corresponding predicted profiles of citrate in the whole soil and P in solution. P_{240} soil and the indicated three periods of root–soil contact. The [P] values at each distance are the sum of [NaOH-P_i] and [H_2SO_4-P] values. Vertical bar in top figure indicates $LSD_{0.05}$ for the measured [P] values. The parameter values for the model predictions were measured independently (see text).

with Eqs. (5) and (6), and the calculated profiles of citrate in the whole soil and P in solution. The flux of citrate across the roots was estimated from the results of Kirk et al. (1999a,b). It will be seen that the agreement between the observed and predicted depletion profiles at the indicated three times of root–soil contact is very good.

Excretion of organic anions may solubilize P by changing the soil pH, by displacing P from adsorption sites, or by chelating metal ions that would otherwise immobilise P or that form soluble organic anion–metal–P complexes. Changes in pH may be important where the anion is excreted in large quantities in association with protons. But in general the plant's inorganic cation–anion balance has a much larger effect on rhizosphere pH. Displacement of adsorbed P will only be important where the amounts of organic anion adsorbed are comparable to the amounts of P desorbed. Unless very large amounts are excreted, this would require that the anion be strongly adsorbed on the soil; it would therefore diffuse only very slowly and the effect would be confined to a narrow zone near the root. From the diffusion coefficients of P and citrate in the present example, it is evident that chelation is the more important mechanism, and is likely to be in other highly weathered soils containing large amounts of metal oxides.

The model is very sensitive to the rate constant for citrate decomposition, k. As k decreases, more citrate accumulates in the soil, and the extent of solubilization and consequent depletion of P increase. It may be that rates of decomposition in rhizosphere soil are somewhat less than those in non-rhizosphere soil as a result of microbial growth kinetics or interactions between different root exudates or both. The model shows that any process that increases the longevity of citrate in the rhizosphere will have a large effect on P solubilization and uptake.

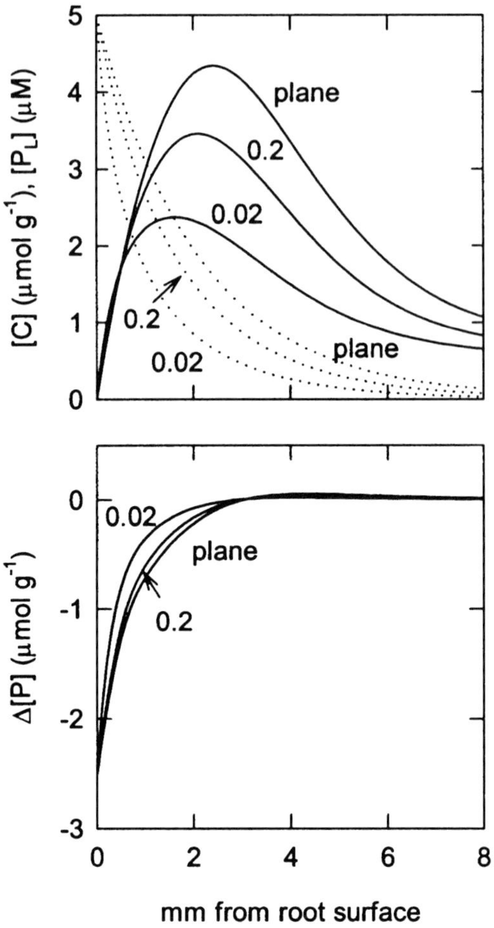

Figure 7. Concentration profiles of P and citrate in a hollow cylinder compared with a linear system. Above: P in solution (solid lines) and citrate in whole soil (dotted lines); below: P in whole soil. Numbers on curves are root radii in cm. Time = 21 days, other parameter values as in Figure 6. Root hairs extended 0.5 mm from the root plane into the soil and P depletion across the root hair zone was taken to be uniform.

Geometry effects

The experiments in the first two examples were designed so that roots formed a planar layer next to a semi-infinite block of soil so that concentration profiles of P and other reactants near the roots could be measured. This system differs from the natural root–soil interface in that (1) the geometry of the natural system is cylindrical rather than planar; (2) the zone that the root may influence is not semi-infinite but is bounded by the zones of influence of neighbouring roots, and neighbouring roots may benefit from P solubilized by each other; (3) excretion of solubilizing agents is often localized along the root axis rather than distributed uniformly.

For the cylindrical geometry appropriate for an individual root, $\partial/\partial x(\partial/\partial x)$ in Eqs. (1) and (2) is replaced by $1/r\partial/\partial r(r\partial/\partial r)$. The effect of neighbouring roots is allowed for by assigning each root a cylindrical zone of influence of radius $x = 1/\sqrt{\pi L_V}$ where L_V is the root length density.

Figure 7 shows the predicted concentration profiles for a cylinder excreting citrate compared to a plane with, for illustration, a constant change in $[C_L]$ at the root surface imposed instantaneously at $t = 0$. Note that at sufficiently large root radii the cylindrical system approaches the planar. Figure 7 shows that the effect of citrate on the concentration of P in solution at a given distance from the root surface is less marked for a cylinder than a plane. However, when there is no solubilization, uptake per unit surface area will be greater for a cylinder than a plane. The net effect appears to be that release of the solubilizing agent is less effective in cylindrical geometry. Thus the ratios of P uptake to citrate release for the results in Figure 7 were 0.010 and 0.019 for cylindrical geometry with root radius = 0.02 and 0.2 cm, and 0.025 for planar geometry.

The effect of citrate excretion being localized along the root axis is shown in Figure 8. The figure shows that localization can greatly amplify the effect of excretion. For a given release of citrate per unit total root length, as the fraction of the root length excreting citrate decreased from 1 to 0.1, P uptake from the part of the root excreting citrate decreased by only 13%; total P uptake by the whole root length increased by 11%. Thus by localizing the excretion a given P uptake could be attained by a much shorter length of root.

In the example in Fig. 8 the inter-root distance was sufficiently large that neighbouring roots did not influence each other. For more dense roots there may be complicated interactions between root density and the

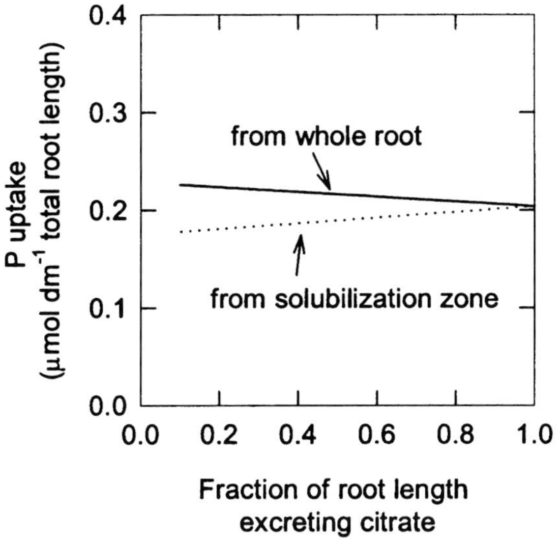

Figure 8. Effect of localization of citrate excretion along the root length. The release of citrate per unit total root length is constant. Root radius = 0.2 mm, time = 21 days, other parameter values as in Figure 6.

relative positions of localization along the lengths of neighbouring roots. In principle, the model could be adapted for such situations.

Conclusions

The good agreement between the observed and predicted results in the examples given shows that it is possible to model root-induced solubilization processes mechanistically. When a model has been corroborated in this way, it can be used to decide which processes should be the targets of attempts to manipulate solubilization, either through plant breeding or agronomy, and which would not be worth investing in.

References

Hedley M J, Kirk G J D and Santos, M B 1994 Phosphorus efficiency and the forms of soil phosphorus utilized by upland rice cultivars. Plant Soil 158, 53–62.

Kirk G J D 1999 A model of phosphate solubilization by organic anion excretion from plant roots. Eur. J. Soil Sci. 50, 369–378.

Kirk G J D and Saleque M A 1995 Solubilization of phosphate by rice plants growing in reduced soil: prediction of the amount solubilized and the resultant increase in uptake. Eur. J. Soil Sci. 46, 247–255.

Kirk G J D, Santos E E and Findenegg G R 1999a Phosphate solubilization by organic anion excretion from rice (*Oryza sativa* L.) growing in aerobic soil. Plant Soil 211, 11–18.

Kirk G J D, Santos E E and Santos M B 1999b Phosphate solubilization by organic anion excretion from rice growing in aerobic soil: rates of excretion and decomposition, effects on rhizosphere pH, and effects on phosphate solubility and uptake. New Phytol. 142, 185–200.

Nye P H 1983 The diffusion of two interacting solutes in soil. J. Soil Sci. 34, 677–691.

Nye P H 1984 On estimating the uptake of nutrients solubilized near roots or other surfaces. J. Soil Sci. 35, 439–446.

Saleque M A and Kirk G J D 1995 Root-induced solubilization of phosphate in the rhizosphere of lowland rice. New Phytol. 129, 325–336.

Tinker P B and Nye P H 2000 Solute movement in the rhizosphere. Oxford University Press, New York.

J.J. Adu-Gyamfi (Ed.), Food security in nutrient-stressed environments: exploiting plants' genetic capabilities: 165–176, 2002.

Scope for genetic manipulation of mineral acquisition in chickpea

Md. Yusuf Ali[1], L. Krishnamurthy[2], N.P. Saxena[2], O.P. Rupela[2], Jagdish Kumar[2] & C. Johansen[3,4]
[1]*Bangladesh Agricultural Research Institute (BARI), Rajshahi 6000, Bangladesh;* [2]*International Crops Research Institute for the Semi-Arid Tropics (ICRISAT), Patancheru 502324, A.P., India* [3]*Apartment 2B, Palmdale, Plot 6, Road 104, Gulshan-2, Dhaka 1212, Bangladesh.* [4]*Corresponding author**

Key words: chickpea, *Cicer arietinum* L., genetic variation, nutrients, root exudates, roots

Abstract

Nutrient acquisition in chickpea needs to be efficient, because it is mainly grown as a post-rainy season, rainfed crop, and generally on soils inferior in physical characteristics and poor in fertility. Nutrient deficiencies have been reported to cause yield losses of varying magnitude in chickpea, e.g., 22–50% due to iron (Fe), around 10% due to sub-optimal nodulation and hence nitrogen (N) deficiency, 29–45% due to phosphorus (P), up to 100% due to boron (B), and 16-30% due to sulphur (S). Yield losses due to salinity are equally large but are difficult to estimate because of its heterogeneous occurrence. In chickpea, genotypic differences in morpho-physiological (including root size) and functional (exudates) root traits, and in nodulation capacity for increased nitrogen fixation have been identified. Genotypic differences in response to application of Fe, B and zinc (Zn) have also been found among chickpea genotypes. A drought tolerant chickpea genotype ICC 4958, which has a relatively large root system, acquired more P than other genotypes during the vegetative period in a pot experiment at ICRISAT. The recent thrust on identifying QTLs for root size should facilitate progress in incorporating useful root traits through marker assisted selection in desirable agronomic backgrounds. Selection for nodulation capacity in released cultivars has resulted in high nodulating chickpea genotypes that produced 10% higher yield than the control varieties. Information on targeted crop improvement for higher nutrient-use efficiency for P, S, Zn, B and Fe is not readily available. Methods to screen for tolerance to salinity are available, but sufficiently high levels of tolerance have not yet been found in germplasm or wild relatives of chickpea to warrant breeding for salinity tolerance. Use of alternative approaches, such as mutation to generate genetic diversity or introgression of alien genes from other crops (transgenic) are thus required, and these remain long-term objectives.

Introduction

Chickpea is an important food legume crop in many countries of the world. It is a source of high quality protein in food and feed. Since it is grown mostly as a rainfed, postrainy season crop (in winter in the subtropics and in spring in Mediterranean and temperate climates), it is often subjected to terminal drought stress. Being a legume, and having secondary status to cereals, it is often grown on land less preferred for cultivation of cereals, where soils are generally marginal in their soil physico-chemical characteristics. Thus, availability and interaction between soil moisture and nutrients are important considerations in chickpea management. The crop is efficient in biological nitrogen fixation (Peoples and Crasswell, 1992) and in accessing native soil P not utilized by crops such as cereals (Ae et al., 1991). These factors confer on it the characteristics for greater productivity under adverse environmental conditions and sustainability of the production systems in which it is grown. Mineral nutrient deficiencies commonly observed in major chickpea producing areas are: nitrogen (N) (due to sub-optimal nitrogen fixation), phosphorus (P), sulfur (S), iron (Fe), zinc (Zn) and boron (B). Salinity is another abiotic stress of economic importance: globally nearly 323 million ha of agricultural land is affected by saline or sodic soil conditions (Brinkman,

* E-mail: Ljohanse@citechco.net

This paper previously appeared in *Plant and Soil* **245**, 2002, pp. 123–134.

1980). These soils occur mainly in arid and semi-arid regions where evaporation exceeds precipitation, coinciding with regions otherwise suitable for chickpea cultivation.

In this paper, recent findings on the chickpea root system, the nutrient-acquiring organ, are reviewed. The extent of genetic variation in chickpea is explored with respect to response to the major mineral limitations, and the scope for favorable genetic manipulation assessed, in terms of alternative management options. The scope for using emerging technologies in molecular genetics to enhance mineral acquisition of chickpea is considered.

The chickpea root system

The key to improving the ability to acquire often scarce resources of both water and nutrients lies in a plant's root system. Important factors are the ability of the root system to explore the available soil volume, to release substances which may enhance availability of nutrients and to host symbioses that favor nutrient acquisition (e.g. nitrogen fixation, mycorrhizae). Over recent years there have been several studies indicating genetic variation in key root traits. These studies are reviewed here in the context of recent studies of the chickpea root system in relation to nutrient and water acquisition.

Rooting depth

Rooting depth indicates the extent of soil that crop roots can exploit for nutrients and water. There are large variations in maximum rooting depth reported for chickpea (28 to 270 cm) (Krishnamurthy et al., 1996), mostly governed by growth duration (Aujla and Cheema, 1985; Sheldrake and Saxena, 1979; Subramanian et al., 1980) and the soil environment. Rooting depth is limited by soil bulk density, texture and moisture. In Vertisols at ICRISAT, Andhra Pradesh, India, maximum rooting depth is found to be 120–135 cm (Krishnamurthy et al., 1996; Sheldrake and Saxena, 1979). In clayey soils, very low or very high levels of soil moisture do not permit root penetration. When dry, these soils tend to harden by shrinking and impede root penetration. When wet, these soils expand but soil pores are filled with water resulting in anaerobic conditions and thus impeded root respiration (Hodgson and MacLeod, 1989; Okada et al., 1991), thereby limiting root penetration (Krishnamurthy et al., 1998). Limitation due to oxygen deficiency may not allow the intrinsic genetic potential for rooting depth to express in such soils. A classic example was provided by Gregory (1996). The average maximum rooting depth of lupin (*Lupinus* spp.), pea (*Pisum sativum* L.) and wheat (*Triticum aestivum*) on three deep sands (xeric psamments) in Western Australia was 190 cm, 65 cm and 113 cm, respectively, with significant differences between genotypes and species but not between sites (Hamblin and Hamblin, 1985). By contrast, when planted in a duplex soil in Western Australia, both lupin (*Lupinus angustifolius*) and wheat were able to extend roots only to 80 cm depth due to physical impediments to growth in both the sand and clay layers and restriction in depth of soil following re-wetting by rain (Dracup et al., 1993; Gregory and Eastham, 1996).

Genetic differences have been reported for effective rooting depth (75 to 150 cm), as estimated by soil moisture depletion, in spring-sown chickpeas in a Calcic Luvisol in Syria (Silim and Saxena, 1993). Wide varietal variation in chickpea was demonstrated in effective rooting depth in an alluvial sandy loam under nonirrigated conditions, and was positively correlated with water uptake (Nagarajrao et al., 1980).

Root distribution

Chickpea is generally grown on residual soil moisture and is exposed to varying degrees of terminal drought stress depending upon the soils, rainfall and evaporative demand. However, in several regions, chickpeas are grown with irrigation. Therefore, root proliferation is expected to vary across the soil profile influencing the timing and extent of nutrient and water uptake between irrigated and nonirrigated environments. In order to better understand this variation a field experiment was conducted in the 1998-99 postrainy season on a Vertisol (fine montmorillonitic isohyperthermic typic pallustert) having 230 mm available water to a soil depth of 150 cm at ICRISAT, Patancheru, India (17°30′ N, 78°16′ E, altitude 549 m) (Ali, 2000). Root distribution of chickpea genotypes ICC 4958 (a prolific rooting accession), Annigeri (a landrace well adapted to peninsular India) and ICCV 10 (a widely adapted variety) was compared with and without irrigation. Roots were extracted by the monolith method (Heeraman and Juma, 1993) and root length density (RLD) and dry weights measured.

The root distribution patterns of Annigeri and ICCV 10 were similar and thus only results for ICCV 10, for comparison with ICC 4958, are presented (Fig-

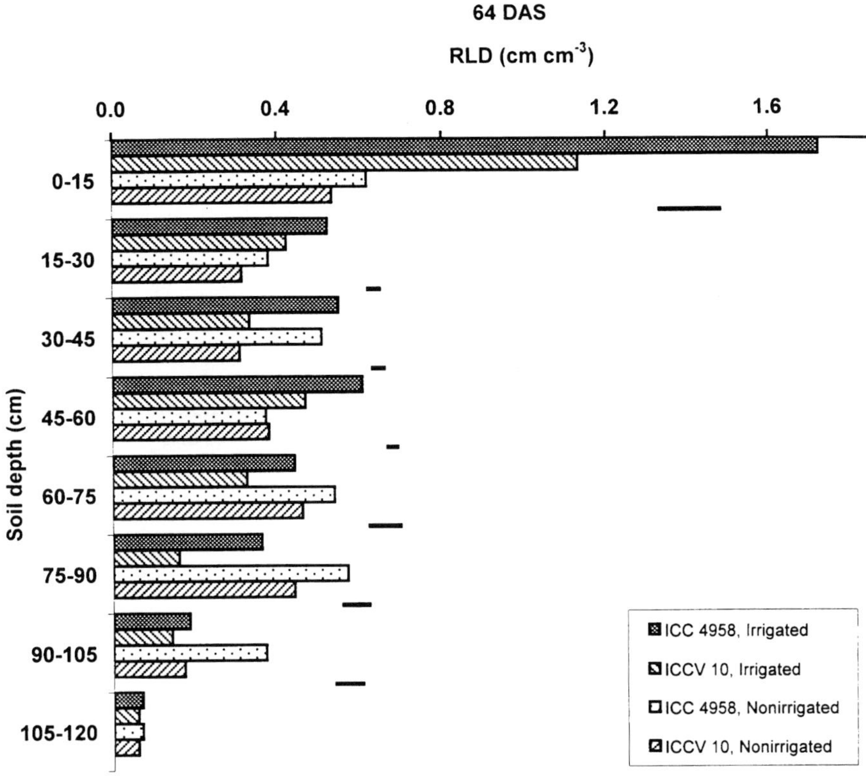

Figure 1. Root length density (RLD cm cm^{-3}) of two chickpea genotypes at 64 days after sowing under nonirrigated and optimum irrigated conditions in a Vertisol during the postrainy season 1998-99. (Values were means of 3 replicates. Horizontal bars denote SE of mean for irrigation × genotype interaction)

ure 1). The time of this observation, 64 days after sowing, coincides with the mid-pod fill stage of the crop in nonirrigated and early pod fill stage in the irrigated conditions. Three major points are apparent in this figure. A significant increase in RLD with irrigation occurred in the 0–15 cm soil layer in both the genotypes. However, at depth there were more roots without irrigation, indicating the negative interaction between irrigation and rooting depth. There is an overall superiority of ICC 4958 over ICCV 10 in total root length per unit area of soil. These observations support the contention that genetic improvements to even well adapted genotypes are possible to increase root proliferation and thereby increase root surface area available for nutrient uptake from soil. This study also shows the advantage of irrigation in increasing root proliferation for nutrient uptake, as most nutrients are distributed in the surface soil layer.

Root activity

Uptake of nutrients such as P (Ae et al., 1991) and Fe (Ohwaki and Sugahara, 1997) is enhanced by root exudates. Chickpea produces more exudates compared to other legumes or cereals (Ae et al., 1991). A field study was conducted in a Vertisol at ICRISAT during the 1992–93 postrainy season in order to measure root respiration, as an estimate of root activity, over soil depth and time. The following chickpea genotypes/selections were grown under rainfed conditions (except for a post-sowing irrigation to ensure even crop establishment): Annigeri, ICC 4958, ICCV 94912, ICCV 94913 and ICCV 94916. The rate of respiration, measured after an hour of incubation of freshly extracted roots, did not differ among the genotypes/selections and therefore the means of genotypes are presented in Figure 2. The rate of respiration was the highest in the seedling stage (15 days after sowing) and this rate declined with the age of the crop. With the advance in crop growth and drying up of the surface soil layers the higher root activities moved to moist

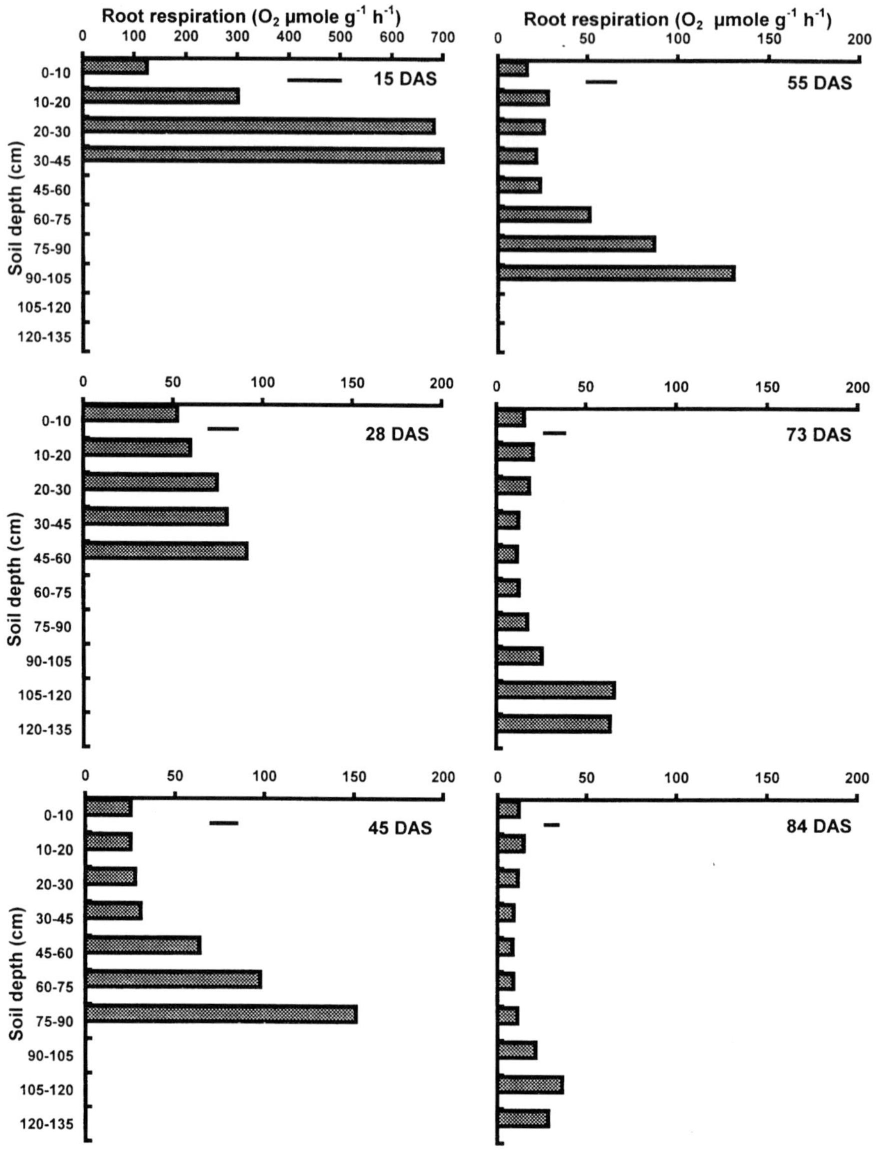

Figure 2. Rate of respiration of extracted roots as measured from various soil layers under receding soil moisture conditions of a Vertisol field, in 1992-93 season. (In this study, two genotypes, ICC 4958 and Annigeri, and three selections made from progenies of these genotypes were grown. The means of 3 replications and 5 genotypes are presented as the genotypic variation was nonsignificant. Horizontal bars denote SE of mean for irrigation × genotype interaction).

deeper soil layers over time. Therefore total root activity at deeper, moist soil layers is directly dependent on the amount of roots in those zones.

Molecular markers to select for root traits

Root studies in the field are laborious, fraught with errors and are limited in their interpretation because of normally variable soil conditions. Concerted efforts are being made to map quantitative trait loci (QTLs) for root traits in chickpea by several groups of researchers. Currently at ICRISAT, root volume and length of over 140 recombinant inbred lines of an interspecific *Cicer* cross are being measured. Molecular markers have been identified for these recombinant inbred lines (RILs). Studies on the segregation of molecular markers and root traits are expected to help tag root traits. This procedure should make it feasible to incorporate desirable root traits through marker assisted selection in disease resistant and high yielding genetic backgrounds of chickpea.

Scope for genetically enhancing mineral acquisition

Against the background of the accumulated knowledge of chickpea root behavior, the scope for enhancing acquisition of major mineral nutrient limitations to chickpea productivity is examined. Although chickpea may be considered efficient in this regard, in comparison with most other cultivated crop species, there is scope for further improving this efficiency to increase and stabilize yields of the crop in stress environments.

Nitrogen

Like other legumes chickpea can access soil mineral-N through its roots and atmospheric N through fixation in root nodules formed in symbiosis with rhizobia. Chickpea can potentially fix about 80% of its N-needs through N fixed by nodules (Peoples and Crasswell, 1992) under optimum growth conditions. Under these conditions, biomass yield and N_2-fixation traits correlated well in agronomic studies (Herridge et al., 1998). Also, several authors have reported a significant correlation in biomass yield and N_2-fixation traits, when a large number of legume cultivars (including those of chickpea) are screened (Rupela and Dart, 1982). Apparently, this led to the conclusion that breeding for high yield will result in high N_2-fixation as well. Therefore, in chickpea breeding programmes, the N_2-fixation trait is generally not considered in progeny selection (Bahl et al., 1990; Saxena and Singh, 1987). However, large plant-to-plant differences in N_2-fixation traits within particular cultivars of chickpea and pigeonpea (perhaps the only two legumes studied in this regard) speak differently (Rupela, 1994). Nodulation was assessed in two chickpea cultivars, G 130 (ICC 4948) and K 850 (ICC 5003). For each cultivar, plants were rated for nodulation on a '1' (lowest nodulation) to '5' (highest nodulation) visual rating scale. G 130, a selection from a landrace, has low nodulation capacity and K 850, a bred line, has high nodulation capacity. In G 130, about 70% plants had low nodulation (rating '1'), about 30% plants had medium nodulation (rating '2' and '3') and only 1% plants had high nodulation (rating '4' and '5'). In K 850, on the other hand, almost 70% plants had high nodulation (rating '4' and '5'), about 30% had medium nodulation (rating around '3') and only 1% plants had low nodulation (rating '1'). Plants of contrasting nodulation capacities within each of the two cultivars were selected and advanced, by pure line selection. In initial verification at ICRISAT (Rupela, 1994) the contrasting pairs (HN: high nitrogen fixation, LN: low nitrogen fixation) within each cultivar and their parents (ICC 4948, ICC 5003) remained consistent for nodulation characteristics. In further studies, these were evaluated in seven experiments at five different locations in India (Dudeja et al., 1997) and one location each in Bangladesh (for two years), Nepal and Pakistan (Khanam et al., 1997). Relative differences for nodule number and nodule mass between the HN and LN selections within a cultivar were consistent across locations and years. The HN selections generally yielded higher (range 4–41% in ICC 5003 HN and 4–106% in ICC 4948 HN) than the LN selections of the same cultivar. However, the differences were significant ($P<0.05$) only in two of the seven experiments in India, one of the two experiments in Bangladesh and in one experiment in Pakistan.

Another study examined a maximum of 353 plants derived from 86 entries including advanced breeding lines, recently released cultivars and germplasm accessions. Seed from plants selected for HN and LN at two soil-N levels (high = suppressive for BNF, low = control) were evaluated and advanced for 2–3 years following pure line selection. In the 1995/96 postrainy season, 84 entries involving HN and LN selections from 17 parent lines, each having 1–10 selections were evaluated in the field for N_2-fixation traits and yield at two soil N levels (Rupela, 1997). Three of the 17 advanced breeding lines (ICCV 91016, ICCV 91019 and ICCV 91026) had HN selections that were significantly ($P< 0.05$) better than parents. Some selections had superiority over parents in nodule mass at low soil N and others had superiority at high soil N. This superiority, however, was not necessarily associated with superiority in total dry matter yield. Some selections were, however, superior both for nodule mass and yield, suggesting the scope of simultaneous selection for these traits.

In another field study, the HN selection of ICC 4948 had greater (13%) overall N-yield in a chickpea-sorghum cropping system (Rupela et al., 1995). In this experiment, the soil health (measured as microbial biomass carbon and N) in the plots of the HN selection was superior (by 1.8 to 2.0 times) to those of the LN selection.

The studies confirmed that high N_2-fixing selections can be developed from advanced breeding chickpea lines having disease tolerance/resistance. It also demonstrated the scope of developing stress tolerant lines (high mineral-N tolerant symbiosis in this case) by pure line selection without resorting to a crossing program. The occurrence of large plant to plant differences in nodulation (including non-nodulation) within land races and advanced breeding lines (Rupela, 1992) was considered to be due to the ability of chickpea to effectively use both soil-N (through roots) and aerial-N (due to symbiotic N_2-fixation). As a result, both the HN and LN type plants within a line remain undetected from above ground traits, and are only detectable when nodulation or N_2-fixation are measured.

Phosphorus

Phosphorus is normally the most limiting nutrient for growth of leguminous crops in tropical and subtropical regions. The reports on the response of chickpea to applied P are highly inconsistent. Response of chickpea to P in the field can be 0–45% (Ae et al., 1991; Dadhich and Mali, 1991; Johansen and Sahrawat, 1991; Ninje, 1991), depending upon the agroclimatic environment. Phosphorus limitation is particularly severe in soils with high contents of Fe, Ca or Al oxides where P is strongly bound and is thus less available for uptake by crop plants (Ae et al., 1991). Chickpea is capable of P extraction from otherwise unavailable forms of P, particularly Ca-bound P, by acidifying alkaline calcareous Vertisols (Ae et al., 1991; Marschner and Romheld, 1983). Gas chromatographic analysis of the root exudates from the different crops showed that citric acid was the major component exuded and that chickpea exuded the most (Ae et al., 1991). The greater acid secretion by chickpea roots would imply a greater volume of rhizosphere acidification in chickpea than in the other crop species. This ability makes chickpea efficient in mining P from soils and converting it into organic forms, thus increasing its significance in recycling of soil P in cropping systems.

The availability of soil moisture plays a key role in P uptake. Drilling of P fertilizer below the soil surface compared to surface broadcasting and mixing was found to be superior for chickpea (Tandon, 1987). Later work demonstrated deep banding of P to be advantageous in chickpea as it enables more P uptake (Arihara et al., 1991).

Table 1. Root dry weight (g pot^{-1}) of five chickpea genotypes, as affected by levels of P fertilizer application, at 40 and 73 days after sowing[a]

Genotype	P levels (kg ha^{-1})				
	0	20	40	60	100
40 days after sowing					
Annigeri	3.2abc	4.7ab	5.3ab	4.2bc	5.9a
ICC4958	4.6a	3.9bc	6.2a	7.0a	6.2a
ICCV10	2.5bc	2.8c	4.00bc	4.7b	4.9ab
ICCV94916-4	3.7ab	6.2a	5.0cb	7.4a	5.9a
ICC5680	2.0c	2.6c	3.0c	2.8c	3.6b
73 days after sowing					
Annigeri	11a	37a	28a	29ab	28a
ICC4958	22a	29ab	24a	24ab	20a
ICCV10	14a	25ab	27a	21b	26a
ICCV94916-4	16a	31ab	32a	34a	27a
ICC5680	12a	12a	23a	20b	23a

[a]Three plants were grown in 25-cm diameter, 120-cm tall pots during 1998/99 postrainy season at ICRISAT. Values are means of three replicates.
[b]In a column, means followed by a common letters are not significantly different at the 5% level – according to Duncan's multiple range test.

In an attempt to quantify differences between chickpea genotypes in their ability to acquire P, a pot culture study was conducted at ICRISAT to measure the effect of P application on extent and timing of root proliferation in different chickpea genotypes (Annigeri, ICC 4958, ICCV 94916-4, ICCV 10 and ICC 5680). Top soil (0–15 cm) from a P-deficient Vertisol (1 ppm Olsen's P) was uniformly packed into 120 cm tall polyvinyl chloride cylinders and different rates of P (0, 20, 40, 60, 100 kg P ha^{-1}, applied as $CaHPO_4.2H_2O$) were mixed in the top 30 cm of the soil column. The 3 plants grown in each cylinder were harvested at 20, 40 and 73 days after sowing (DAS) and the roots separated from the soil.

The root dry weight harvested at 20 DAS showed no difference due to P application, possibly because of reliance on P derived from seed to this stage. However, at 40 DAS there was a significant increase in root dry

Table 2. Root:shoot ratio of five chickpea genotypes, as affected by levels of P fertilizer application, at 40 and 73 days after sowing[a]

40 days after sowing

Genotype	P levels (kg ha^{-1})					
	0	20	40	60	100	Mean
Annigeri	0.71	0.64	0.59	0.55	0.54	0.60
ICC 4958	0.73	0.52	0.45	0.50	0.41	0.52
ICCV 10	0.84	0.59	0.57	0.63	0.56	0.64
ICCV 94916-4	0.68	0.60	0.50	0.55	0.46	0.56
ICC 5680	0.60	0.49	0.48	0.52	0.44	0.51
Mean	0.71	0.57	0.52	0.55	0.48	
SE (±) for comparison of means of:						
Genotype	0.019***					
P level	0.019***					
Genotype × P level	0.043^{NS}					

73 days after sowing

Genotype	P levels (kg P ha^{-1})					
	0	20	40	60	100	Mean
Annigeri	0.59	0.80	0.65	0.64	0.52	0.64
ICC 4958	0.59	0.60	0.38	0.38	0.29	0.45
ICCV 10	0.74	0.65	0.62	0.63	0.45	0.62
ICCV 94916-4	0.54	0.59	0.57	0.47	0.36	0.51
ICC 5680	0.53	0.60	0.65	0.58	0.51	0.57
Mean	0.60	0.65	0.57	0.54	0.43	
SE ± for comparison of means of:						
Genotype	0.042**					
'P' level	0.042**					
Genotype × 'P' level	0.094^{NS}					

[a]Three plants were grown in each 25-cm diameter, 120-cm tall pots during 1998–99 postrainy season at ICRISAT. Values are means of three replicates.

weight up to 60 kg P ha^{-1} and at 73 DAS up to 20 kg P ha^{-1} (Table 1). Increased P application encouraged early shoot growth as well as root growth but a decrease in root:shoot ratio with P application indicated that a greater proportion of dry matter was diverted to the shoot (Table 2). Root:shoot ratio decreases with plant age (Krishnamurthy et al., 1996; Table 2) which implies that the bulk of plant P requirements would be acquired at early growth stages, when the top soil containing a higher P concentration is likely to be moist. The data of Tables 1 and 2 illustrate the strategy of chickpea in investing in a greater partitioning of dry matter to roots under P-limiting conditions.

At 40 and 73 DAS, there were significant genotypic differences in P response, with ICC 4958 and ICCV 94916-4 being least responsive with most root production at low P levels (Table 1). These two genotypes have been selected for more prolific root growth under drought stress. It therefore appears that they retain this characteristic under P stress, and thus have a comparative advantage over other genotypes in exploiting a greater soil volume and potentially extracting more P from the soil when soil P is in limiting supply. There is indeed a correlation between this root vigor and total P uptake by the plants at 40 and 73 DAS under conditions of low P supply, with more P uptake in the vigorous rooting ICC 4958 and ICCV 94916-4 and least in the poorly rooting ICC 5860 (Table 3). ICC 5680 was included in this study because of its poor rooting ability under drought conditions, and it also showed least root development among the genotypes at low soil P levels. Thus differences in root proliferation should reflect in differences in P acquisition ability. Exploration of a larger soil volume should permit greater acidification of that volume and consequent increased availability of P, in alkaline soil, for uptake by roots. It therefore seems feasible that P nutrition of chickpea can be improved by genetic improvement of root system proliferation.

Sulfur

Crops in general require as much S as they need P. Sulfur deficiency has been reported from over 70 countries worldwide including nearly 175 million hectares of cropped area in the Indian subcontinent (Tandon, 1991). Chickpea responds to S fertilization across locations in India with yield increases ranging from 16 to 30% in environments with a mean yield of 1851 kg ha^{-1} without S application (Tandon, 1991). Although genotypic differences in response to applied S have been documented for wheat and rapeseed mustard, we can find no reports of such genotypic comparisons for chickpea. In view of the fact that S availability generally tends to increase with soil depth, deeper rooting capability in chickpea may assist its acquisition of S, and deeper rooting genotypes would therefore have an advantage.

Zinc

Zinc deficient soils are common in both tropical and temperate climates (Graham et al., 1992; Sillanpaa and Vlek, 1985; Welch et al., 1991). In India, Zn deficiency is recognized as one of the most common micronutrient deficiencies, and about 50% of the Indian soils have been classified as Zn deficient (Katyal, 1985). Most arable land in Pakistan has inadequate levels of available soil Zn, and Zn deficiency is a common nutritional disorder in alkaline and calcareous soils (Kausar et al., 1979; Khattak and Perveen, 1986; Rashid et al., 1987). Chickpea, particularly, has been shown to be more sensitive to Zn deficiency (Ahlawat, 1990; Khan et al., 1998a; Sakal et al., 1988) than wheat and other cereals (Tiwari and Pathak, 1982). Therefore, chickpea production can be adversely affected by Zn deficiency as chickpea is mainly grown on Zn-deficient soils, although there are few recorded field responses of chickpea to Zn application.

Wide genotypic variation is available for sensitivity to Zn deficiency in chickpea (Khan et al., 1998b). Less increase in root:shoot ratio and transport of proportionally more Zn to the shoot (70%) under conditions of Zn deficiency were the salient characteristics of less sensitive genotypes. More studies are needed to understand the mechanisms underlying Zn efficiency in chickpea. However, it is feasible that production of more Zn-chelating exudates depends upon the extent of root system. Therefore, improving the root system should also improve Zn acquisition ability of chickpea plants. Inexpensive alleviation of Zn deficiency by fertilizer application is possible once the problem diagnosed. However, it is prudent to have genotypes efficient in Zn acquisition. The fact that breeding lines and newly released genotypes show improved Zn efficiency (Khan et al. 1998b) indicates an unconscious selection for this trait already. More conscious efforts are needed to select Zn-efficient genotypes of chickpea. Genetic transformation, by transferring rye chromosomes 1R and 7R into wheat, has increased the Zn-efficiency of wheat (Cakmak et al., 1997). Similar approaches could be tried for chickpea using genetic variation for Zn-efficiency from within the germplasm collection or wild species of *Cicer* or related legumes.

Table 3. Phosphorus uptake by whole plants (mg $plant^{-1}$) of five chickpea genotypes at 0 (P_0) and 100 (P_{100}) kg ha^{-1} levels of P fertilizer application, at 40 and 73 days after sowing (DAS)[a]

Genotype	40 DAS		73 DAS	
	P_0	P_{100}	P_0	P_{100}
Annigeri	4.4	12.8	12.0	47.0
ICC 4958	6.4	13.4	25.7	46.8
ICCV 10	2.4	11.0	17.0	55.4
ICCV 94916-4	4.3	13.4	18.3	51.3
ICC 5680	4.7	9.7	15.8	38.4
SE (genotype × P level)		0.62		1.77

[a]Three plants were grown in each of 25-cm diameter, 120-cm tall pots during 1998–99 postrainy season at ICRISAT. Values are means of three replicates.

Boron

Boron deficiency is a major cause of crop yield loss in China, India, Nepal and Bangladesh (Anantawiroon et al., 1997). Some of the major chickpea growing areas with soils diagnosed to be B deficient are found in the inner Terai region of Nepal, adjacent regions of Bihar state of India and in northern Bangladesh. The yield losses due to B deficiency can be as high as 100% in chickpea (Srivastava et al. 1997). Boron deficiency was noted to be more severe in soils containing a high sand fraction and low organic matter (Phetchawee and Ratanarat, 1989).

There has been little quantification of the extent of genetic variation in response of chickpea to B, except from the observation in Nepal that introduced (exotic) lines that are much more susceptible to the problem than local landraces, such as Dhanush. However, the grain yields of these landraces are also poor when compared with the yield potential of chickpea in other regions (Srivastava et al., 1997). A wide range of genetic variation in abilty to yield at low B supply was found in wheat (*Triticum aestivum*) (Anantawiroon et al., 1997) and an acceptable screening method (Rerkasem and Loneragan, 1994) has been developed and used. The backcross reciprocal monosomic method (Snape and Law, 1980) has been successfully employed to compare a particular monosomic chromosome from one variety with the same monosomic chromosome from another variety (Chantachume et al., 1997; Law et al., 1983) to identify the chromosome responsible for B efficiency. Similar efforts are needed to develop a screening method for selecting B-efficient genotypes in chickpea. It will be possible to choose an appropriate genetic enhancement method depending on the type and extent of variation available in chickpea.

Iron

Iron deficiency can occur in chickpea when it is grown in soils of high pH (8.0). Yield losses due to Fe-deficiency in susceptible chickpea genotypes can be in the range 22–50% (Saxena et al., 1994). Literature on the effects of alkaline soil conditions (high pH, calcareous soils), appearance of Fe-chlorosis, yield losses due to deficiency, genotypic differences in Fe-deficiency and its inheritance, screening methods, and criteria for selection have been reviewed by Saxena et al. (1994). Good progress has been made in understanding the physiological basis of development of Fe-deficiency. Present evidence suggests that the large genotypic differences in Fe response found may be related to differences in both root (Malewar et al., 1982) and shoot (Ohwaki and Sugahara, 1993) traits. Further work is warranted to develop this information and to use it in identifying tolerant sources in germplasm and in segregating populations derived from these sources. It does not seem necessary to develop a specific breeding program to produce Fe-efficient genotypes because the inefficient genotypes can easily be identified by the distinctive Fe deficiency symptoms and rejected in a breeding population. If necessary, these populations could be grown under conditions that favor induction of Fe-deficiency under natural conditions (wet, alkaline, clayey soils).

Salinity

A large potentially arable area, nearly 323 million ha on a global basis, is estimated to be saline or sodic (Brinkman, 1980). In some countries, e.g. Pakistan, saline land areas are substantial and cover nearly 40% of the cultivated area (Mohammad, 1978). Excessive and injudicious application of irrigation water, among various factors, have contributed to recent increases in saline land area. Legumes, in general, have been found to be more sensitive to saline soil conditions than other crop species (Maas and Hoffman, 1977). Among legumes, cool season food legumes – chickpea, lentil and faba bean – are more sensitive to soil salinity (Lauchli, 1984; Maas and Hoffman, 1977).

Responses of cool season food legumes (including chickpea), to salinity, methods to screen for tolerance to saline conditions, inheritance of the trait, and genotypic differences in tolerance, have been reviewed (Saxena et al., 1993, 1994). The levels of tolerance to soil salinity are indeed low in chickpea as seen in a 50% reduction in shoot biomass at an electrical conductivity (EC) of 5–6 dS m^{-1}. At this relatively low level of tolerance, L 550, a Kabuli chickpea variety from India was found to be most tolerant, and this has been confirmed in many experiments grown in pots and field (Johansen et al., 1990; Saxena, 1987). However, it is disappointing to find that the wild relatives of chickpea are even more sensitive to salinity than the cultivated species (Johansen et al., 1990).

Some progress has been made in generating understanding of mechanisms of salinity tolerance, e.g., Na^+ accumulation in the tops in chickpea (Lauter and Munns, 1986). It is likely that other mechanisms, e.g., selective retention of Na^+ in the roots and stems of soybean (Jacoby, 1964; Salim, 1987), may also be present in the germplasm and are worth investigating. Many studies have shown that *in vitro* selection of cells for high salinity levels does not lead to development of salt tolerant plants regenerated from those single cells (Rowland et al., 1989; Watad et al., 1991). This is because expression of tolerance at the cellular level does not necessarily translate into tolerance at the whole plant level. It does not seem that empirical methods of breeding for yield are likely to be effective in selecting tolerant plants until sources of greater salinity tolerance are found in cultivated accessions of chickpea or its wild relatives. A systematic evaluation of genotypes that have evolved in naturally saline soils (e.g. as in Iran or Iraq) would be a first step in this regard. Research efforts should also be directed towards characterization of the physiological mechanisms or traits conferring salinity tolerance to the point of identifying genes controlling them. Then enhancement of salinity tolerance in chickpea by transgenic methods may become feasible.

In the short term, it appears that use of high yielding varieties developed for rainfed cultivation should increase production on saline lands as well, because salinity occurs heterogeneously across a field. These varieties would contribute more from non-saline or less saline patches, compared to other varieties. Increasing production from the non-saline patches should therefore lead to increased productivity. Some even question whether increasing salinity tolerance, with mechanisms such as Na/K discrimination, would indeed lead to increased productivity of these species in dry saline soils. Evidence in barley and wheat (Richards, 1993) suggests that it may not be so.

Conclusion

With a shrinking area of quality arable land for agriculture and its increasing allocation to staple cereal or high value crops, it is unlikely that in future chickpea will be grown on lands which are better endowed with nutrients than on which it is grown at present. Mineral nutrient deficiency will therefore continue to be a major and increasing constraint to chickpea production. It is encouraging that in recent years substantial progress has been made in generating information and knowledge on the crop. Also conventional and recently-developed tools are available to impart increased nutrient acquisition ability for important nutrient elements. It seems, therefore, realistic to expect some recovery of losses in yield due to nutrient deficiency through genetic improvement for increased nutrient acquisition. Mechanisms of efficient uptake of nutrients, mediated through either physical root size, microbial symbioses or surface chemical characteristics of roots (exudates) are now better understood. Tools of modern molecular biology, for identification, mapping and tagging of the genes are now being investigated to develop flanking markers for QTLs indicating the large root trait in chickpea. Once this is done on a set of RILs, laborious root measurements on a large scale need not be necessary and marker-assisted selection for improvement of root size, and hopefully also for root exudates could be carried out in the near future. Outputs of transgenic research in other crops, in due course of time, could be extended for enhancement of nutrient uptake mechanisms and salinity tolerance in chickpea. Increased emphasis on better understanding of mechanisms of nutrient uptake and their genetic control are required for chickpea if the crop is to maintain production under seemingly inevitable decreasing soil fertility conditions.

References

Ae N, Arihara J and Okada K 1991 Phosphorus response of chickpea and evaluation of phosphorus availability in Indian Alfisols and Vertisols. *In* Phosphorus Nutrition of Grain Legumes in the Semi-Arid Tropics. Eds C Johansen, K K Lee and K L Sahrawat. pp. 33–41. International Crops Research Institute for the Semi Arid Tropics, Patancheru, Andhra Pradesh, India.

Ahlawat I P S 1990 Diagnosis and alleviation of mineral nutrient constraints in chickpea. *In* Chickpea in the Nineties. Eds H A van Rheenen and M C Saxena. pp. 93–99. International Crops Research Institute for the Semi Arid Tropics, Patancheru, Andhra Pradesh, India.

Ali M Y 2000 Influence of phosphorus fertilizer and soil moisture regimes on root system development, growth dynamics and yield of chickpea. Ph.D. Thesis, Bangabhandu Sheikh Mujibur Rahman Agricultural University, Salna, Gazipur, Bangladesh.

Anantawiroon P, Subedi K D and Rerkasem B 1997 Screening wheat for boron efficiency. *In* Boron in Soils and Plants. Eds R W Bell and B Rerkasem. pp. 101–104. Kluwer Academic Publishers, Dordrecht, The Netherlands.

Arihara J, Ae N and Okada K 1991 Improving responses of chickpea and pigeonpea to phosphorus application in Vertisols and Alfisols. *In* Phosphorus Nutrition of Grain Legumes in the Semi-Arid Tropics. Eds C Johansen, K K Lee and K L Sahrawat. pp. 157–165. International Crops Research Institute for the Semi Arid Tropics, Patancheru, Andhra Pradesh, India.

Aujla T S and Cheema S S 1985 Water use, root density and yield of chickpea *Cicer arietinum* as affected by levels of soil-moisture storage with and without postsowing irrigation. Indian J. Ecol. 12, 91–98.

Bahl P N, Salimath P M, Malik B A, Dahiya B S, Deshmukh R B and Rangasamy P 1990 New approaches in chickpea breeding and prospects in the nineties. *In* Chickpea in the Nineties. Proceedings of the Second International Workshop on Chickpea Improvement, 4–8 Dec 1989, ICRISAT Asia Center, Patancheru, Andhra Pradesh, India. Eds H A van Rheenen and M. C. Saxena. pp. 225–231. International Crops Research Institute for the Semi-Arid Tropics, Patancheru, Andhra Pradesh, India.

Brinkman R 1980 Saline and sodic soils. *In* Problem Soils: Their Reclamation and Management, Eds K J Beek, W A Blokhuis, P M Driessen, N van Breemen, R Brinkman and L J Pons. International Soil Reference and Information Centre, Wageningen, The Netherlands.

Cakmak I, Derici R, Torun B, Tolay I, Braun H J and Schlegel R 1997 Role of rye chromosomes in improvement of Zn efficiency in wheat and triticale. Plant Soil 196, 249–253.

Chantachume Y, Shepherd K W, Paull J G and Rathjen A J 1997 Chromosomal variations for boron tolerance in wheat (*Triticum aestivum*). *In* Boron in Soils and Plants. Eds R W Bell and B Rerkasem. pp. 105–109. Kluwer Academic Publishers, Dordrecht, The Netherlands.

Dadhich S C and Mali A L 1991 Effect of herbicide and phosphorus on crop-weed competition in chickpea (*Cicer arietinum*). Indian J. Agron. 36, 283–285.

Dracup M, Gregory P J and Belford R K 1993 Restricted growth of lupin and wheat roots in the sandy A horizon of a yellow duplex soil. Aust. J. Agric. Res. 44, 1273–1290.

Dudeja S S, Potdukhe S R, Namdeo S L, Datar V V, Kumar V, Tilak K V B R, Khurana A L and Rupela O P 1997 Multilocational evaluation of some selected chickpea nodulation variants in India. *In* Extending Nitrogen Fixation Research to Farmers' Fields: Proceedings of an International Workshop on Managing Legume Nitrogen in the Cropping Systems of Asia, 20–24 Aug. 1996, ICRISAT Asia Center, India. Eds O P Rupela, C Johansen and D F Herridge. pp. 261–276. International Crops Research Institute for the Semi-Arid Tropics, Patancheru, Andhra Pradesh, India.

Graham R D, Ascher J S and Hynes S C 1992 Selecting Zn efficient cereal genotypes for soils of low Zn status. Plant Soil 146, 241–250.

Gregory P J 1996 Spatial distribution of root systems and root activities. *In* Dynamics of Roots and Nitrogen in Cropping Systems of the Semi-Arid Tropics. Eds O Ito, C Johansen, J J Adu-Gyamfi, K Katayama, J V D K Kumar Rao and T J Rego. pp. 129–143. JIRCAS Agriculture Series No. 3. Japan International Research Center for Agricultural Sciences, Tsukuba, Japan.

Gregory P J and Eastham J 1996 Growth of shoots and roots and interception of radiation by wheat and lupin crops on a shallow duplex soil in response to time of sowing. Aust. J. Agric. Res. 47, 427–447.

Hamblin A P and Hamblin J 1985 Root characteristics of some temperate legume species and varieties on deep, free-draining Entisols. Aust. J. Agric. Res. 36, 63–72.

Herridge D F, Marcellos H, Felton W L, Turner G L and Peoples M B 1998 Chickpea in wheat-based cropping systems of northern New South Wales III. Prediction of N_2 fixation and N balance using soil. Aust. J. Agric. Res. 49, 409–418.

Hodgson A S and MacLeod D A 1989 Oxygen flux, air-filled porosity and bulk density as indices of vertisol structure. Soil Sci. Soc. Am. J. 53, 540–543.

Jacoby B 1964 Function of bean roots and stems in sodium retention. Plant Physiology 39, 445–449.

Johansen C and Sahrawat K L 1991 Strategies for maximizing the efficiency of phosphorus utilization in cropping systems involving chickpea and pigeonpea. *In* Phosphorus Nutrition of Grain Legumes in the Semi-Arid Tropics. Eds C Johansen, K K Lee and K L Sahrawat. pp. 227–241. International Crops Research Institute for the Semi-Arid Tropics, Patancheru, Andhra Pradesh, India

Johansen C, Saxena N P, Chauhan Y S, Subbarao G V, Pundir R P S, Kumar Rao J V D K and Jana M K 1990 Genotypic variation in salinity response of chickpea and pigeonpea. *In* Proceedings of the International Congress of Plant Physiology, New Delhi, India. February 1988. Eds S K Sinha, P V Sane, S C Bhargava and P K Agrawal. Vol. 2. pp. 977–983. Society for Plant Physiology and Biochemistry, New Delhi, India.

Katyal J C 1985 Research achievements of all India coordinated scheme of micronutrients in soils and plants. Fert. News 30, 67–80.

Kausar M A, Alam S M, Sharif M and Pervaiz M 1979 Micronutrient status of Pakistan soils. Pak. J. Sci. Ind. Res. 22, 156–161.

Khan H R, McDonald G K and Rengel Z 1998a Assessment of the Zn status of chickpea by plant analysis. Plant Soil 198, 1–9.

Khan H R, McDonald G K and Rengel Z 1998b Chickpea genotypes differ in their sensitivity to Zn deficiency. Plant Soil 198, 11–18.

Khanam D, Bhuiyan M A H, Hossain A K M, Ali S, Mushtaq N, Bhattarai S and Maskey 1997 Evaluation of selected chickpea nodulation variants in Bangladesh, Nepal and Pakistan. *In* Extending Nitrogen Fixation Research to Farmers' Fields: Proceedings of an International Workshop on Managing Legume Nitrogen in the Cropping Systems of Asia, 20–24 Aug 1996, ICRISAT Asia Center. Eds O P Rupela, C Johansen and D F Herridge. pp. 277–287. International Crops Research Institute for the Semi-Arid Tropics, Patancheru, Andhra Pradesh, India.

Khattak J K and Perveen S 1986 Trace element status of Pakistan soils. *In* Proc. of XII International Forum on Soil Taxonomy and Agrotechnology Transfer, Lahore, Pakistan. pp. 119–124. Soil Survey of Pakistan and Soil Management Services, USA.

Krishnamurthy L, Ito O and Johansen C 1996 Genotypic differences in root growth dynamics and its implications for drought resistance in chickpea. *In* Dynamics of Roots and Nitrogen in Cropping Systems of the Semi-Arid Tropics. Eds O Ito, C Johansen, J J Adu-Gyamfi, K Katayama, J V D K Kumar Rao and T J Rego. pp. 235–250. JIRCAS Agriculture Series No. 3. Japan International Research Center for Agricultural Sciences, Tsukuba, Japan.

Krishnamurthy L, Ito O, Johansen C and Saxena N P 1998 Length to weight ratio of chickpea roots under progressively receding soil moisture conditions in a Vertisol. Field Crops Res. 58, 177–185.

Lauchli A 1984 Salt exclusion: An adaptation of legumes for crops and pastures under saline conditions. *In* Salinity Tolerance in Plants: Strategies for Crop Improvement. Ed. R C Staples. pp. 171–187. John Wiley, New York, USA

Lauter D J and Munns D N 1986 Salt resistance of chickpea genotypes in solutions salinized with NaCl or Na_2SO_4. Plant Soil 95, 271–279.

Law C N, Snape J W and Worland A J 1983 Quantitative genetic studies in wheat. *In* Proc. Sixth International Wheat Genetic Symposium. Ed. S Sakamoto. pp. 539–547. Kyoto University, Japan.

Maas E V and Hoffman G J 1977 Crop salt tolerance – current assessment J. Irrigation and Drainage Division, Proc. Am. Soc. Civil Engineers 103, 115–134.

Malewar G V, Jadhav D K and Ghonsikar C P 1982 Screening techniques and sources of tolerance to salinity and mineral nutrient imbalances in cool season food legumes. International Chickpea Newsletter 6, 13–14.

Marschner H and Römheld V 1983 *In-vitro* measurement of root-induced pH changes at the soil-root interface: Effect of plant species and nitrogen source. Zeitschript für Pflanzenphysiologie 111, 241–251.

Mohammad 1978 *In* Membrane biophysics and salt tolerance in plants. Eds R H Qureshi, S Mohammad and A Aslam. pp. 47–64. Pakistan University of Agriculture Press, Faislabad, Pakistan.

Nagarajrao Y, Mallick S and Gurcharan Singh 1980 Moisture depletion and root growth of different varieties of chickpea under rainfed conditions. Indian J. Agron. 25, 289–293.

Ninje P M 1991 Irrigation and fertilizer requirement of chickpea (*Cicer arietinum*) genotypes in deep Vertisol. Indian J. Agron. 36, 111–114.

Okada K, Ae N and Arihara J 1991 Soil aeration status of alfisols and vertisols as a limiting factor for growth of pigeonpea. *In* Phosphorus Nutrition of Grain Legumes in the Semi-Arid tropics. Eds C Johansen, K K Lee and K L Sahrawat. pp. 139–155. International Crops Research Institute for the Semi-Arid Tropics, Patancheru, Andhra Pradesh, India.

Ohwaki Y and Sugahara K 1993 Genotypical differences in responses to iron deficiency between sensitive and resistant cultivars of chickpea (*Cicer arietinum*). Plant Soil 155/156, 473–476.

Ohwaki Y and Sugahara K 1997 Active extrusion of protons and exudation of carboxylic acids in response to iron deficiency by roots of chickpea (*Cicer arietinum*). Plant Soil 189, 49–55.

Peoples M and Crasswell E T 1992 Biological nitrogen fixation: investments, expectations and actual contributions to agriculture. Plant Soil 141, 13–39.

Phetchawee S and Ratanarat S 1989 Surveys on boron deficiency of groundnut in farmers fields in central and eastern Thailand in 1987. *In* Proceedings of the 7th Thailand National Groundnut Meeting, Pattaya, Chonburi, 16–18 Mar. 1988. Ed. A Patanothai. pp. 373–376. Khon Kaen University, Thailand.

Rashid A, Ullah R and Salim M 1987 Soil conditions and crop factors inducing Zn deficiency in plants. *In* Proceedings of the National Seminar on Micronutrients in Soils and Crops, Peshawar, Pakistan. pp. 94–117. North West Frontier Province Agriculture University, Peshawar, Pakistan.

Rerkasem B and Loneragan J F 1994 Boron deficiency in two wheat genotypes in a warm, subtropical region. Agron. J. 86, 887–890.

Richards R A 1983 Should selection for yield in saline regions be made on saline or non-saline soils? Euphytica 32, 431–438.

Rowland G G, McHughen A and McOnie C 1989 Field performance at saline-affected sites of a somaclonal variant of McGregor flax selected for salt tolerance *in vitro*. Can. J. Plant Sci. 69, 49–60.

Rupela O P 1992 Natural occurrence and salient characters of nonnodulating chickpea plants. Crop Sci. 32, 349–352.

Rupela O P 1994 Screening for intracultivaral variability of nodulation in chickpea and pigeonpea. *In* Linking Biological Nitrogen Fixation Research in Asia: Report of a meeting of the Asia Working Group on Biological Nitrogen Fixation in Legumes, 6–8 Dec. 1993, ICRISAT Asia Center. Eds O P Rupela, J V D K Kumar Rao, S P Wani and C Johansen. pp. 75–83. International Crops Research Institute for the Semi-Arid Tropics, Patancheru, Andhra Pradesh, India.

Rupela O P 1997 Field evaluation of some recently developed selections for high nodulation and value of nodulation variants of chickpea. *In* Extending Nitrogen Fixation Research to Farmers' Fields. Proceeding of an International Workshop on Managing Legume Nitrogen Fixation in Cropping Systems of Asia, 20–24 Aug. 1996, ICRISAT Asia Center. Eds O P Rupela, C Johansen and D F Herridge. pp. 305–317. International Crops Research Institute for the Semi-Arid Tropics, Patancheru, Andhra Pradesh, India.

Rupela O P and Dart P J 1982 Screening for nodulation characteristics in chickpea and subsequent generation of seeds. *In* Biological Nitrogen Fixation Techonology for Tropical Agriculture. Proceedings of a workshop, 9–13 Mar. 1981, Cali, Colombia. Eds P H Graham and S C Harris. pp. 57-61. Centro International de Agricultura Tropicale (CIAT), Cali, Colombia.

Rupela O P, Wani S P, Danso S K A and Johansen C 1995 Effect of high nodulating selection of chickpea cultivar ICC 4948 on yield and soil properties of a chickpea-sorghum cropping system. J. Soil Biol. Ecol. 15, 127–134.

Sakal R, Sinha R B and Singh A P 1988 Effect of boron application on blackgram and chickpea production in calcareous soil. Fertiliser News 33, 27–30.

Salim M 1987 Screening techniques and sources of tolerance to salinity and mineral nutrient imbalances in cool season food legumes. FABIS Newsletter 18, 11–13.

Saxena M C and Singh K B (Eds) 1987 The Chickpea. CAB International and The International Center for Agricultural Research in the Dry Areas. Wallingford, Oxon, UK. 409 pp.

Saxena N P 1987 Problems and prospects to screen and breed for tolerance to soil salinity: A case study with chickpea. *In* Adaptation of Chickpea and Pigeonpea to Abiotic Stresses. Proceedings of the Consultants' Workshop, 19–21 December 1984. Eds N P Saxena and C Johansen. pp. 115–120. International Crops Research Institute for the Semi-Arid Tropics, Patancheru, Andhra Pradesh, India.

Saxena N P, Johansen C, Saxena M C and Silim S N 1993 Selection for drought and salinity tolerance in cool-season food legumes. *In* Breeding for Stress Tolerance in Cool Season Food Legumes. Eds K B Singh and M C Saxena. pp. 245–270. John Wiley and Sons, Chichester, UK.

Saxena N P, Saxena M C, Ruckenbauer P, Rana R S, El-Fouly M M and Shabana R 1994 Screening techniques and sources of tolerance to salinity and mineral nutrient imbalances in cool season food legumes. *In* Expanding the Production and Use of Cool Season Food Legumes. Eds F J Muehlbauer and W J Kaiser. pp. 457–471. Kluwer Academic Publishers, Dordrecht, The Netherlands.

Sheldrake A R and Saxena N P 1979 Growth and development of chickpea under progressive moisture stress. *In* Stress Physiology in Crop Plants. Eds H Mussell and R C Staples. pp. 466–483. John Wiley and Sons, Chichester, UK.

Silim S N and Saxena M C 1993 Adaptation of spring-sown chickpea to the Mediterranean basin. I. Response to moisture supply. Field Crops Res. 34, 121–136.

Sillanpaa M and Vlek P L G 1985 Micronutrients and the agroecology of tropical and Mediterranean regions. Fert. Res. 7, 151–167.

Snape J W and Law C N 1980 The detection of homologous chromosome in wheat using backcross reciprocal monosomic lines. Heredity 45, 187–200.

Srivastava S P, Yadav C R, Rego T J, Johansen C and Saxena N P 1997 Diagnosis and alleviation of boron deficiency causing flower and pod abortion in chickpea (*Cicer arietinum* L.) in Nepal. *In* Boron in Soils and Plants. Eds R W Bell and B Rerkasem. pp. 95–99. Kluwer Academic Publishers, Dordrecht, The Netherlands.

Subramanian T L, Sree Ramulu U S, RajaRajan A, Ramanathan S and Natarajan C P 1980 A study on rooting pattern of bengal gram (*Cicer arietinum* L.) (CO.1) using tracer tecniques. Madras Agric. J. 67, 51–53.

Tandon H L S 1987 Phosphorus Research and Agricultural Production in India. Fertilizer Development and Consultation Organization, New Delhi, India. 160 pp.

Tandon H L S 1991 Sulphur Research and Agricultural Production in India. Fertilizer Development and Consultation Organization, New Delhi, India. 140 pp.

Tiwari K N and Pathak A N 1982 Studies of the Zn requirements of different crops. Exp. Agric. 18, 393–398.

Watad A A, Swartzberg D, Bressan R A, Izhar S and Hasegawa P M 1991 Stability of salt tolerance at the cell level after regeneration of plants from a salt tolerant tobacco cell line. Physiol. Plant. 83, 307–313.

Welch R M, Allaway W H, House W A and Kabota J 1991 Geographical distribution of trace element problems. *In* Micronutrients in Agriculture. 2nd edn. Eds J J Mortvedt, F R Cox, L M Shuman and R M Welch. pp. 31–57. Soil Sci. Soc. Am., Madison, Wisconsin, USA.

J.J. Adu-Gyamfi (Ed.), Food security in nutrient-stressed environments: exploiting plants' genetic capabilities: 177–182, 2002.

Aluminum tolerance in nodulated N_2-fixing legumes species native to two contrasting savanna sites

M-L Izaguirre-Mayoral[1,4], S. Flores[2] & T. Oropeza[3]

[1]*Instituto Venezolano de Investigaciones Cientificas (IVIC). Laboratorio de Biotecnologia y Virologia Vegetal, Centro de Microbiologia y Biologia Celular Apdo. 21827, Caracas 1020-A, Venezuela.* [2]*Centro de Ecología, IVIC.* [3]*Universidad Experimental Simon Rodriguez, Caracas, Venezuela.* [4]*Corresponding author**

Received 10 May 2000. Accepted in revised form 3 November 2000

Key words: aluminium, aluminium accumulators, calcium, iron, nitrogen fixation

Abstract

Eight native legume species were analyzed for their tolerance to high soil aluminium (Al) levels in terms of the Al, iron (Fe) and calcium (Ca) contents in different plant organs and spatial distribution in two contrasting savanna sites in Venezuela. The less fertile soils in site 1 (S-1) are characterized by a pH of 4 and contents of Al 8.4, Fe 0.9 and Ca 0.2 cmol kg^{-1}. The relatively more fertile soils in site 2 (S-2) have a pH of 5.4, with Al 1.3, Fe 0.4 and Ca 2.2 cmol kg^{-1}. Examples of nodulated *Chamaecrista flexiuosa, Clitoria guianensis, Galactia jussieuana, Stylosanthes elegans* and *Zornia curvata* were harvested at S-1. Nodulated examples of the same species plus examples of *Centrosema pubescens, Chamaecrista tetraphilla* and *Phaseolus gracillis* were harvested at S-2. All S-1 species were categorized as Al-tolerant. In turn, *C. pubescens, C. tetraphilla* and *P. gracillis* were classified as non-tolerant to soil Al contents above 1.3 cmol kg^{-1}. With the exception of *S. elegans*, all S-1 species accumulate Al in underground organs, mainly in secondary roots, and were considered root-Al accumulators. The accumulation of Al in aboveground organs of *S. elegans* allowed the classification of this species as leaf-Al accumulator. Over all the species, the Al contents in underground and aboveground organs were 11 and 8-fold higher in S-1 than in S-2 species, respectively. However, the accumulation of Al in roots of S-1 species did not inhibit their symbiotic performance as indicated by the absence of significant differences in the relative abundance of ureides between S-1 and S-2 species. This situation could be partially explained by the exclusion of Al from nodule tissues. In S-1 and S2 species, Al contents in nodules were below detectable values. Significant differences in Ca content were also encountered between S-1 and S-2 species, with the highest values in S-2 species. Except for *S. elegans*, there was a significant positive correlation between Al and Fe contents in the different organs of S-1 and S-2 species, as well as a significant negative correlation between Al and Ca contents in S-1 species.

Introduction

Aluminium toxicity is considered to be the main factor limiting plant growth and, thus, crop cultivation on acid soils. Acid soils comprise large areas of the world's agricultural regions, particularly in tropics, and therefore soil acidity is an important factor limiting food production in developing countries. In Venezuela, the strong increase in population and the concomitant utilization of the northern fertile areas for housing expansion has forced the less suitable savanna regions to be developed for agricultural production. These neotropical savannas are characterized by a marked seasonality in water availability, their soils are usually acid, well drained, with low levels of available nutrients and high Al saturation (Izaguirre-Mayoral et al., 1992). However, in spite of such deleterious conditions, a wide range of types of native vegetation are capable to grow and withstand the high soil Al content without visible effects on biomass production (Medina and Bilbao, 1991). The physiological and biochemical mechanisms underlying the plant tolerance to Al are,

* FAX No. 0058-5041382. E-mail: mizaguir@pasteur.ivic.ve

This paper previously appeared in *Plant and Soil* **245**, 2002, pp. 163–168.

however, poorly understood and a matter of controversy. Previous data support the hypothesis that part of the mechanism for Al tolerance in plants is based on the metabolism-dependent exclusion of Al from the sensitive meristems (Rengel, 1996).

In nodulated commercial legumes, high soil Al levels causes a significant reduction in the rates of symbiotic N_2 fixation and nodule mass (Chong et al., 1987) and affects the growth and survival of rhizobial strains (Flis et al., 1993). Nevertheless, in savanna soils containing high levels of Al, the diversity of native legumes is larger than that of grasses and a large number of nodulated legume species are able to fix N_2 at high rates (Izaguirre-Mayoral et al., 1995; Sicardi De Mallorca and Izaguirre-Mayoral, 1994a, b;). These observations are probably due to an intrinsic tolerance of nodulated native legumes to soil acid conditions and high levels of exchangeable Al. The tolerance to Al was previously demonstrated in a legume species native of a tropical cloud forest (Izaguirre-Mayoral and Flores, 1995). For this species, the accumulation of Al in roots and its partial exclusion from nodule tissues were proposed as the main strategy underlying symbiotic efficiency in seedlings and adult plants. The present investigation was, therefore, undertaken to identify Al tolerant-nodulated legume species that are native of two contrasting sites of a neotropical savanna. Within the savanna context, in site 1 (S-1) are located the less fertile soils due to the high Al content and low pH. In contrast, the relatively more fertile soils in site 2 (S-2) are currently used for agricultural purposes. The categorization of native legumes as non-tolerant or tolerant to Al was based on the spatial distribution of the species. The possible strategies underlying the Al tolerance in the native legumes were analyzed taking into consideration the Al, Ca and Fe contents as well as the Al/Ca ratio in different plant organs.

Materials and methods

Plant and soil sampling were carried out, during the rainy season, in two contrasting soil sites of a *Trachypogon* savanna located in La Iguana Research Station of the Simon Rodriguez University, Guarico State, Venezuela (8° 25′ N, 65° 24′ W). For soil analyses, triplicate samples were collected to a depth of 20 cm, at random, in representative areas of site 1 (S-1) and site 2 (S-2). The chemical properties and texture of the soils were determined as in Izaguirre-Mayoral et al. (1992), except for the Al content that was extracted and quantified following the methodology described by Hargrove and Thomas (1981). The chemical analyses of S-1 soils indicated the following composition: pH 4.0, N 0.3 mg g^{-1}, available P 0.1 $\mu g\ g^{-1}$, Mg 0.8 cmol kg^{-1}, K 0.02 cmol kg^{-1}, Ca 0.2 cmol kg^{-1}, Mn 0.04 cmol kg^{-1}, Fe 0.9 cmol kg^{-1} and Al 8.4 cmol kg^{-1}. Chemical properties of S-2 soils were: pH 5.4, N 0.7 mg g^{-1}, available P 0.2 $\mu g\ g^{-1}$, Mg 1.0 cmol kg^{-1}, K 0.06 cmol kg^{-1}, Ca 2.2 cmol kg^{-1}, Mn 0.1 cmol kg^{-1}, Fe 0.4 cmol kg^{-1} and Al 1.3 cmol kg^{-1}. S-2 soils were also characterized by an 84% sand, 6% clay and 10% silt. The texture of S-1 soils was, in contrast, constituted by 42% sand, 36% clay and 20% silt.

Plant sampling at S-1 included the species: *Galactia jussieuana, Chamaecrista flexuosa, Clitoria guianensis, Zornia curvata* and *Stylosanthes elegans*. Individuals of the same species plus examples of *Centrosema pubescens, Chamaecrista tetraphilla* and *Phaseolus gracillis* were harvested at S-2. At least 12 individuals of each species were sampled by collecting their entire aboveground biomass and underground organs contained in a soil volume of approximately 0.03 m^3. All plants were nodulated and at the vegetative stage of growth.

In order to identify which organs potentially accumulate Al, plants were divided into shoots, leaves, xylopodium, nodules and lateral roots. Plant samples were dried to constant weight in an oven at 60 °C and ground to a fine mesh. Dried subsamples of 500 mg were predigested overnight with 2 ml of concentrated nitric acid. Thereafter, 5 ml of the mixture were digested for 4 h at 200 °C and diluted to a final volume of 50 ml with desionized water. Total AL, Fe and Ca content were determined with atomic absorption espectrophotometry according to Izaguirre-Mayoral et al. (1992). The relative abundance of ureides (RAU), considered as an index of the symbiotic efficiency in tropical nodulated legumes, was analyzed following the protocol described in Sicardi De Mallorca and Izaguirre-Mayoral (1994a).

For statistical analyses, data in percentage were transformed to degrees by angular transformation. Statistical differences for individual plant parameters were analyzed using one-way Analysis of Variance. Statistical differences between data for each variable were determined using Least Significant Differences (LSD). Test comparison between means was carried out using the Student' *t* test.

Table 1. Al and Ca content and the Al/Ca ratio in different plant organs of legume species native of Venezuelan savannas sites S-1 and S-2

Species	Al ($g\ kg^{-1}$ dry wt)		Ca ($g\ kg^{-1}$ dry wt)		Al/ca ratio	
	S-1	S-2	S-1	S-2	S-1	S-2
Chamaecrista flexuosa						
Leaves	6.0c[a]	1.0b (S)	45.5a	96.5a (S)	0.1c	0.01c (S)
Shoots	5.2c	1.2b (S)	47.3a	79.3b (S)	0.1c	0.01c (S)
Xylopodium	15.8b	1.9a (S)	2.1b	43.7c (S)	7.6b	0.04b (S)
Secondary roots	52.5a	2.8a (S)	1.6b	42.0c (S)	33.3a	0.07a (S)
Clitoria guianesis						
Leaves	5.8c	1.1b (S)	59.2a	96.7a (S)	0.1c	0.01b (S)
Shoots	4.8c	1.7b (S)	42.2a	80.3b (S)	0.1c	0.02b (S)
Xylopodium	11.1b	3.0a (S)	3.4b	42.3c (S)	3.3b	0.07a (S)
Secondary roots	58.2a	3.1a (S)	1.5c	42.1c (S)	38.9a	0.07a (S)
Galactia jussieuana						
Leaves	21.4c	1.0c (S)	52.9a	88.0a (5)	0.5c	0.01b (S)
Shoots	16.1c	1.9b (S)	43.4b	66.7b (S)	0.4c	0.03b (S)
Xylopodium	40.5b	3.0a (S)	3.7c	45.6c (S)	10.9b	0.07a (S)
Secondary roots	54.8a	3.1a (S)	1.9d	50.3c (S)	28.8a	0.06a (S)
Stylosanthes elegans						
Leaves	27.7a	2.7a (S)	50.7a	92.9a (S)	0.5a	0.03a (S)
Shoots	18.9b	2.4a (S)	30.9a	73.5b (S)	0.6a	0.03a (S)
Xylopodium	1.7c	1.0b (NS)	35.2b	52.4b (S)	0.05b	0.02a (NS)
Secondary roots	1.6c	1.5b (NS)	37.1b	63.3b (S)	0.04b	0.02a (NS)
Zornia curvata						
Leaves	4.8c	1.8c (S)	55.6a	70.1a (S)	0.09c	0.03b (S)
Shoots	3.8c	1.2d (S)	41.2b	59.3b (S)	0.09c	0.02b (S)
Xylopodium	15.1b	2.7b (S)	1.3c	22.4c (S)	11.7b	0.12a (S)
Secondary roots	22.6a	3.2a (S)	1.1c	22.1c (S)	20.2a	0.14a (S)
Centrocema pubescens						
Leaves		1.3c		44.5a		0.03b
Shoots		1.4c		53.4a		0.03b
Xylopodium		4.6b		21.9c		0.20a
Secondary roots		5.3a		35.8b		0.15a
Chamaecrista tetraphylla						
Leaves		0.7b		70.5a		0.01b
Shoots		0.7b		74.0a		0.01b
Xylopodium		2.3a		55.5b		0.04a
Secondary roots		2.7a		54.4b		0.05a
Phaseolus gracillis						
Leaves		2.1b		12.1b		0.2a
Shoots		1.2c		10.8b		0.1b
Xylopodium		4.0a		74.4a		0.06c
Secondary roots		4.4a		68.2a		0.06c

[a]For individual plant species, number in columns followed by the same letter are not statistically different at $p<0.05$. (S), difference between means significant at $p<0.05$. (NS), difference between means not significant at $p<0.05$.

Table 2. Iron (Fe) content in different plant organs of legume species natives of Venezuelan savanna sites S-1 and S-2

Species	Fe (g kg^{-1} dry wt)	
	S-1	S-2
Chamaecrista flexuosa		
Leaves	3.1c[a]	0.8b (S)
Shoots	3.4c	0.7b (S)
Xylopodium	6.9b	1.2ab (S)
Secondary roots	8.5a	1.7a (S)
Clitoria guianesis		
Leaves	2.7c	0.9c (S)
Shoots	2.6c	1.1c (S)
Xylopodium	5.3b	3.6b (S)
Secondary roots	11.3a	5.5a (S)
Galactia jussieuana		
Leaves	4.1c	0.4bc (S)
Shoots	3.1cd	0.9b (S)
Xylopodium	8.7b	1.5a (S)
Secondary roots	12.5a	1.6a (S)
Stylosanthes elegans		
Leaves	1.4a	1.3a (NS)
Shoots	1.1a	0.8a (NS)
Xylopodium	1.2a	1.1a (NS)
Secondary roots	1.2a	1.3a (NS)
Zornia curvata		
Leaves	2.3b	0.6b (S)
Shoots	2.1b	0.9b (S)
Xylopodium	5.7a	1.2a (S)
Secondary roots	6.1a	1.4a (S)
Centrocema pubescens		
Leaves		0.9b
Shoots		0.9b
Xylopodium		0.8b
Secondary roots		1.6a
Chamaecrista tetraphylla		
Leaves		0.7b
Shoots		0.7b
Xylopodium		2.5a
Secondary roots		2.7a
Phaseolus gracillis		
Leaves		0.4c
Shoots		0.9b
Xylopodium		1.4a
Secondary roots		1.5a

[a] For individual plant species, number in columns followed by the same letter are not statistically different at $p<0.05$. (S), difference between means significant at $p<0.05$. (NS), difference between means not significant at $p<0.05$.

Results

From the five legume species harvested from S-1, *C. flexuosa, C. guianensis, G. jussieuana* and *Z. curvata* were found to accumulate Al in underground organs, mainly in the secondary roots (Table 1). *S. elegans* was the only species with significantly higher Al content in above than in its underground organs. At S-2, all species, with the exception of *S. elegans*, had significantly higher Al content in their underground than in aboveground organs, with no significant differences between xylopodia and lateral roots. Over all the species, the Al contents in underground organs were 11-fold higher in S-1 than in S-2 species. Concomitantly, an 8-fold difference in Al content in aboveground organs was observed between S-1 and S-2 species. For all S-1 and S-2 species, the Al content in nodules was below detectable levels.

Calcium accumulation followed an opposite pattern to that displayed by Al. As shown in Table 1, the highest significant Ca content was detected in leaves and shoots of all legume species, regardless of the collection site. The inverse pattern of Al and Ca accumulation was reflected in a significantly negative correlation between Al and Ca contents in different plant organs (r^2=−0.665, $P<0.05$) of S-1 species. In contrast to results from S-1 species, the negative relationship between Al and Ca content was not significant in S-2 legumes. Significant differences in Ca content were also encountered between S-1 and S-2 species, with the highest values in S-1 species. Except for *S. elegans*, all legume species displayed significantly higher Al/Ca ratios in underground organs.

Except for *S. elegans*, Fe concentrations in all S-1 and S-2 species were always significantly higher in underground than in aboveground organs, the highest values being detected in secondary roots of S-1 species (Table 2). With the exception of *S. elegans*, positive correlation values of r^2=0.936 for S-1 species and r^2=0.462 for S-2 species at $p<0.05$ between Al and Fe concentrations were established for all analyzed organs.

The analysis of RAU (Table 3) did not reveal significant differences in the efficiency of the symbiotic process between individual legume species native of S-1 and S-2.

Discussion

According to local information, the Al levels in S-1 hamper the growth of commercial Al-sensitive crops.

Table 3. Relative abundance of ureides (RAU) in legume species native of Venezuelan savanna sites S-1 and S-2

Species	(RAU %)	
	S-1	S-2
Chamaecrista flexuosa	38.5b[a]	44.7b (NS)
Clitoria guianesis	37.3b	40.6b (NS)
Galactia jussieuana	30.1b	38.3b (NS)
Stylosanthes elegans	30.2b	35.2b (NS)
Zornia curvata	35.3b	43.9b (NS)
Centrocema pubescens		40.6b
Chamaecrista tetraphylla		36.3b
Phaseolus gracillis		69.6a

[a]Numbers in column followed by the same letter are not statistically different at $p<0.05$. (NS), difference between means not significant at $p<0.05$.

This site has been considered unsuitable for agriculture unless the soils are subjected to pH amendments. Nevertheless, under the deleterious conditions of S-1 soils, there are a number of legume species capable of growing there normally with no visual effects of Al toxicity, such as reduced number of nodules per plant (not shown) or limited symbiotic performance. The high Al concentrations in underground organ's lack of effect on the symbiotic efficiency of S-1 species was ascertained by the absence of significant differences between RAU values of native legumes from both sites. The latter is probably due to the observed exclusion of Al from nodule tissues in S-1 and S-2 species. This point requires further investigations.

From present data, we might postulate that all legume species growing in S-1 are Al-tolerant. The total absence of *C. pubescens, C. tetraphilla* and *P. gracillis* in S-1 was, on the other hand, considered as an indicator of the non-tolerance of those species to levels of Al above 1.3 cmol kg^{-1} of soil. The average Al content in leaves of all legumes collected at S-1 were 3.5-fold higher and from S-2 2.4-fold lower than that reported for legumes from the Cerrado regions of Brazil (Geoghegan and Sprent, 1996), but were similar to values encountered in plants from various deciduous forest sites of Africa (Högberg, 1986).

The capacity of S-1 legume species to withstand the high levels of Al in S-1 soils could result from the establishment of two different strategies for accumulation of toxic ions in their underground or aboveground organs. The first strategy consists on the accumulation of Al in underground organs probably by an effective Al-chelating mechanism located in endodermal cells (Cuenca et al., 1991) that decreases the rate of Al transport to leaves (root-Al accumulators). The second strategy allows the xylem movement of Al from roots toward aerial sinks and its accumulation mainly in leaves (leaf-Al accumulators) (Cuenca and Herrera, 1987). The localization, distribution and chemical speciation of Al in leaves of Al-accumulating species has been previously reported (Haridasan et al., 1986).

The first strategy for Al-tolerance is carried out by *C. flexuosa, C. guianensism* and *Z. curvata.* In these species, the ratio of Al concentration in underground to the aboveground organs ranks around 7. The second strategy for Al-tolerance was exclusively observed in *S. elegans* and seems to be based on the xylem translocation of Al from roots to aerial organs and its accumulation in leaves. The abscission of leaves enriched in Al might be the mechanism underlying the Al tolerance of this species. Field observations of individual plants of *S. elegans* revealed a rapid turnover of their leaves.

An analysis of results obtained in *G. jussieuana* indicate that this species might represent an intermediate between both strategies. Individual plants of this species accumulate only 2.7-fold higher Al content in underground than in aboveground organs. It could be postulated that the root system of *G. jussieuana* is not so efficient in hindering the transport of Al to aboveground organs as that in other root-Al accumulator species. However, this species is the most widely spread in savanna areas and can successfully explore soils with pH below 4.0.

Legume three species categorized as root-Al accumulators were also characterized by a significantly accumulation of Fe in secondary roots. It has been postulated that similarities in the ionic radii of Al and Fe allow both ions the use of the Fe^{3+} phytosiderophore transport system to gain access to endodermal cells (Kochian, 1995). The significant positive correlation between Fe and Al contents detected in different plant organs and the preferential accumulation of Fe in underground organs of S-1 and S-2 species cannot be, however, explained at present. There is no available information on the requirement of Fe by nodulated native legumes. It is only known that cultivated legumes dependent on the symbiotic process for their N economy have a higher demand for Fe than that of non-nodulation C_3 species (Marschner, 1995). Concomitantly, the significantly lower Ca concentration in S-1 when compared to that in S-2 species could be, on the other hand, attributed to the lower Ca content in S-1 soils or to an impairment in the S-1 species Ca uptake mechanisms elicited by Al (Rengel, 1996). Present results question the possibility that Ca might alleviate Al toxicity in S-1 species (Keltjens and Dijkstra, 1991; Shen et al., 1993). However, Cuevas (1995) has postulated the importance of Ca availability for effective nodulation in low pH soils.

Based on the Al/Ca ratio values in lateral roots, the authors propose the range of Al-tolerance within root-Al accumulators S-1 species as follows: *C. guianensis*>*C. flexuosa*>*G. jussieuana*>*Z. curvata*. The high degree of tolerance of *S. elegans* to high soil Al levels was ascertained by the important value index of this species in S-1 (data not shown). It is also suggested that the Al/Ca ratio should be considered as an indicator of the capacity of different plant organs to tolerate high levels of Al, rather than as an index of toxicity. This suggestion is based on: (1) the significant negative correlation between Al and Ca contents in different plant organs of S-1 species, and (2) the low significant values of Al/Ca rations in all S-2 plants and in the organs of S-1 plants with low Al concentrations.

The type of strategy used by Al-tolerant plants seems to be genetically fixed, since the pattern of Al accumulation displayed by each species was maintained, regardless of the soil-Al content. Genetic control of Al tolerance has been demonstrated in several commercial crops (Gallego and Benito, 1997; Spehar and Galwey, 1996). Finally, the identified nodulated Al-tolerant species might constitute a valuable source of genes for future breeding programs and could be considered as important components for the sustainable exploitation of savanna regions.

Acknowledgements

The authors would like to thank Dr Elvira Cuevas (IVIC) and two referees for their enlightening comments and very helpful suggestions on the manuscript.

References

Chong K, Wynne J C, Elkan G H and Schneeweis T J 1987 Effects of soil acidity and aluminium content on *Rhizobium* inoculation. Trop. Agric. 64, 97–104.

Cuenca G and Herrera R 1987 Ecophysiology of aluminium in terrestrial plants, growing in acid and aluminium-rich tropical soils. Ann. Sco. Zool. Belg. 117, 57–73.

Cuenca G, Herrera R and Merida T 1991 Distribution of aluminium in accumulator plants by X-ray microanalysis in *Richeria grandis* Vahl leaves from a cloud forest in Venezuela. Plant Cell Environ. 14, 437–441.

Cuevas E 1995 Biology of the belowground system of tropical dry forests. *In* Seasonally Dry Tropical Forests. Eds. SH Bullock, HA Mooney and E Medina. pp. 362–383. Cambridge University Press, England.

Flis S E, Glenn A R and Dilworth M J 1993 The interaction between aluminium and root nodule bacteria. Soil Biol. Biochem. 25, 403–417.

Gallego F J and Benito C 1997 Genetic control of aluminium tolerance in rye (*Secale cereale*, L.). Theor. Appl. Genet. 95, 393–399.

Geoghegan I E and Sprent J 1996 Aluminum and nutrient concentrations in species native of central Brazil. Commun. Soil Sci. Plant Anal. 27, 2925–2934.

Hargrove W L and Thomas G W 1981 Extraction of aluminum from aluminum-organic matter complexes. Soil Sci. Soc. Am. J. 45, 151–153.

Haridasan M, Paviani T I and Schiavini 1 1986 Localization of aluminium in the leaves of some aluminium-accumulating species. Plant Soil 94, 435–437.

Högberg P 1986 Soil nutrient availability root symbiosis and tree species composition in tropical Africa: A review. J. Trop. Ecol. 2, 359–372.

Izaguirre-Mayoral M L, Carballo O, Flores S, Mallorca M S and Oropeza T 1992 Quantitative analysis of the symbiotic N_2 fixation, non-structural carbohydrates and chlorophyll content in sixteen native legume species collected in different savanna sites. Symbiosis 12, 293–312.

Izaguirre-Mayoral M L and Flores S 1995 Symbiotic nitrogen fixation in *Dioclea guianensis* Benth., a shade-tolerant and aluminum-accumulator legume species native of tropical cloud forests. Symbiosis 19, 65–83.

Izaguirre-Mayoral M L, Vivas A I and Oropeza T 1995 New insights into the symbiotic performance of native tropical legumes: I. Analysis of the response of thirty-seven native legume species to artificial shade in a neotropical savanna. Symbiosis 19, 111–129.

Keltjens W G and Dijkstra W J 1991 The role of magnesium and calcium in alleviating aluminium toxicity in wheat plants. Dev. Plant Soil Sci. 45, 763–768.

Kochian L V 1995 Cellular mechanisms of aluminum toxicity and resistance in plants. Ann. Rev. Plant Physiol. Plant Mol. Biol. 46, 237–260.

Marschner H 1995. Adaptation of plants to adverse chemical soil conditions. *In* Mineral Nutrition in Higher Plants. 2nd edn. pp 596–679. Academic Press, London.

Medina E and Bilbao B 1991 Significance of nutrient relations and symbiosis for the competitive interaction between grasses and legumes in tropical savannas. *In* Modern Ecology: Basic and Applied Aspects. Eds. G Esser and D Overdieck. pp. 295–319. Elsevier Science Publ, Amsterdam.

Rengel Z (1996) Tansley Review No 89. Uptake of aluminium by plant cells. New Phytol 134, 389–406.

Shen Z, Wang J and Guan H 1993 Effect of aluminium and calcium on growth of wheat seedlings and germination of seeds. J. Plant Nutr. 16, 2135–2148.

Sicardi De Mallorca M and Izaguirre-Mayoral M L 1994a A comparative evaluation of the symbiotic N_2-fixation and physiological performance of 36 native legume species collected in a tropical savanna during the rainy and dry seasons. Symbiosis 16, 225–247.

Sicardi De Mallorca M and Izaguirre-Mayoral M L 1994b Seasonal dynamic, host range and symbiotic efficiency of native rhizobial populations in three soil horizons of four contrasting savanna sites. Symbiosis 17, 43–63.

Spehar C R and Galwey N W 1996 Diallel analysis for aluminium tolerance in tropical soybean (*Glycine max* (L.) Merrill). Theor. Appl. Genet. 92, 267–272.

J.J. Adu-Gyamfi (Ed.), *Food security in nutrient-stressed environments: exploiting plants' genetic capabilities*: 183–187, 2002.

Effect of rock phosphate and superphosphate fertilizer on the productivity of maize var. Bisma

D. R. Lukiwati
Faculty of Animal Agriculture, Diponegoro University, Campus Tembalang, Semarang, Indonesia, 50275
E-mail: dwiretno_lukiwati@yahoo.com

Key words: crude protein, dry matter, maize grain, rock phosphate, superphosphate, *Zea mays*

Abstract

Superphosphate (SP) fertilizers have been used widely to improved food crop production in Indonesia. However, the high cost of SP is now focusing attention on rock phosphate (RP) fertilizer as a natural phosphorus (P) source and because of its relatively low cost. A field experiment was conducted to evaluate the effects of RP and SP-36 fertilizers and interaction between treatments on the productivity of maize on a Latosolic soil (low available Bray-II extractable P). A split-plot design with three blocks as replicates was used, with sources of P fertilizer as main plots and four levels of P fertilizer as subplots. Results showed that RP and SP-36 fertilizers increased maize grain yield, dry matter (DM) stover yield, and crude protein (CP) contents of maize stover over those of the control that did not receive P fertilizer. Superphosphate-36 application resulted in higher maize grain yield, DM yield, and CP contents of maize stover than crops treated with RP fertilizer at the same rate. The highest P-use efficiency for grain and stover was obtained with RP and SP-36 fertilizers that provided 66 kg P ha^{-1}.

Introduction

In Indonesia corn or maize (*Zea mays* L) is an important crop as a source of food and livestock feed. Maize stover is used as ruminant livestock feed. However, most of the land in Indonesia used for food crops production has non-productive Latosolic soils characterized by low phosphorus (P) and low pH.

An insufficiency of P in the medium from which the crop draws its mineral nutrition results in retarded growth, reduced development, vegetative mishaps and, often severely reduced grain yield. In cereals the problems associated with P deficiency mainly appear during the active growth stages that are most demanding of this element. Phosphorus deficiency can be corrected by applying P fertilizer to the soil at or before sowing (Grundon, 1987). According to Coates et al. (1990), the application of P fertilizer during periods of active growth increased forage crop yield and quality. The extent to which added P fertilizer increases the supply of P to plants growing in a specific soil depends on the type of fertilizer and the characteristics of that soil (Helyar and Godden, 1976).

Superphosphate fertilizers have been widely used to improve food crop production on non-productive land in Indonesia. However, the high cost of SP is now focusing attention on RP fertilizer. Rock phosphate is a natural P sources, it is relatively cheaper than SP, and it is locally available. There are RP deposits in Cirebon (West Java), and in Gresik (East Java). According to Kerridge and Ratcliff (1982), RP can also be used as a maintenance fertilizer for pasture species on acid soils in humid areas.

Rock phosphate contains P in a form that is not readily soluble in water but soluble in neutral ammonium citrate solution making it slowly available to plants. Water-insoluble P-fertilizers do not all behave in the same way as SP. Finely ground RP is an apatite mineral and when added to acid soils, the solubilization of RP is increased so that below pH 5.5 it begins to make a significant contribution, and on really acid soils it can be a useful source of phosphate. The agronomic effectiveness of RP depends on the chemical and mineralogical nature of the rock, the pH, and other characteristics of the soil, and the crop (Young et al., 1985).

Table 1. Maize var. Bisma grain yield, and P-use efficiency with rock phosphate (RP) and SP-36 fertilizers[a]

Level of P fertilizer (kg P ha^{-1})	RP		SP-36	
	Grain yield (kg ha^{-1})	P-use efficiency (kg grain kg^{-1} P)	Grain yield (kg ha^{-1})	P-use efficiency (kg grain kg^{-1} P)
0	4028 d	0	4085 d	0
66	5199 c	79	6285 b	95
132	5885 bc	44	7342 a	55
198	6257 bc	31	7599 a	38

[a]Means followed by the same letters are not significantly different at DMRT 5%.

Superphosphate-36 contains 36% P_2O_5, with the P in a water-soluble form. Water-soluble phosphate when added to acid soils will be precipatated as iron (Fe) or aluminium (Al) phosphates and tends to be tightly absorbed by mineral colloids in the soil. Soils with strong adsorbtion properties, such as acid clays, will often need very high rates of added phosphate to significantly raise their phosphate-supplying power. Both types of P fertilizer (RP and SP) are directly applied to the soil (Young et al., 1985).

The objective of the reported research was to study the effect of RP and SP-36 fertilizers and the interaction between treatments on the maize productivity in Latosolic soil.

Materials and methods

A field experiment of split-plot design with three block as replicates was conducted during 3 months on three blocks of Latosolic soil with the following pH values: 4.73 (block I), 4.48 (block II), and 4.26 (block III). Available P (Bray-II extractable) was 2.59 (block I), 2.15 (block II), and 1.72 ppm (block III). Soil nitrogen (N) contents (%) were; 0.029 (block I), 0.020 (block II) and 0.019 (block III). Thus, the soil was acid, low in P and of low-medium N content. The experiment was conducted on 240 m^2 divided into 24 plots. Each plot size was 3.5×2.5 m, or 8.75 m^2. Four levels of P fertilizer from two sources of P were established as treatments.

The main plot consisted of P fertilizers from two sources; RP containing 27% P_2O_5, and SP-36 containing 36% P_2O_5. Four levels of P fertilizer (0, 66, 132 and 198 kg P ha^{-1}) were used as sub-plots. Standard fertilizers, i.e., urea (46% N) at 100 kg N ha^{-1}, and KCl (60% K_2O) at 83 kg K ha^{-1}, were applied to each plot.

Maize seed var. Bisma was dibbled into small holes made with a wooden stick at the rate of two seeds per hole, spaced 70×40 cm. Each plot contained 60 plants or 68 571 plants ha^{-1}. Fertilizer-P (RP and SP-36) and KCl were dibbled close to planting holes at sowing. Shallow sub-surface banding (5 cm beside and below the seeds) can significantly increase P fertilizer use efficiency over broadcast fertilizer application (Randall and Hoeft, 1988). One-third of the urea was applied at the time of germination a week after sowing. The remaining two-thirds of the urea was applied at the time of flowering, or at 4 weeks after sowing dibbled close to the maize plants.

The maize was harvested 3 months after sowing and the grain sun-dried for 3 days. After harvesting the maize grain, the stover was cut close to the ground and measured for dry matter (DM) yield and crude protein (CP) content (N content in %×6.25) Each replicate was the average measurement from four plants/two holes. To measure DM yield (DM content in %×fresh stover yield) the harvested stover material (leaves and stalk) was chopped, sub-sampled, and oven-dried to constant weight at 70 °C for 48 h. These samples were finely ground and analyzed for N content by Kjeldahl digestion and titration (Islam et al., 1992). The N content of maize stover was assessed as the total leaf and stalk N, and as DM yield.

The analyses of variance on grain yield, DM yield, and CP content were made using the general linear model procedure of SAS. Significant differences among the treatments were calculation using Duncan's Multiple Range Test (DMRT).

Results

The application of RP and SP-36 fertilizers resulted in the production of maize grain yields (Table 1), and DM yields of maize stover (Table 2) significantly

Table 2. Dry matter yield of maize stover and P-use efficiency with rock phosphate (RP) and SP-36 fertilizers[a]

Level of P fertilizer (kg P ha^{-1})	RP		SP-36	
	Stover (kg ha^{-1})	P-use efficiency (kg grain kg^{-1} P)	Grain yield (kg ha^{-1})	P-use efficiency (kg grain kg^{-1} P)
0	3604 d	0	3613 d	0
66	4429 cd	67	6506 b	99
132	5255 bc	40	8226 a	62
198	5934 b	30	9016 a	45

[a]Means followed by the same letters are not significantly different at DMRT 5%.

Table 3. Crude protein content of maize stover with rock phosphate (RP) and SP-36 fertilizers[a]

Level of P fertilizer (kg ha^{-1})	RP (%)	SP-36 (%)
0	5.82 ab	5.92 ab
66	5.98 ab	6.09 ab
132	6.14 a	6.22 a
198	6.26 a	6.38 a

[a]Means followed by the same letters are not significantly different at DMRT 5%.

higher than those from control plants that did not receive fertilizer. However, maize grain yields and DM yields of maize stover were not significantly affected by increasing levels of SP-36 fertilizer from 132 to 198 kg P ha^{-1}, or RP fertilizer from 66 to 198 kg P ha^{-1}.

Application of different levels of RP and SP-36 fertilizers (0, 66, 132 and 198 kg P ha^{-1}) were able to increase the CP content of maize stover. However, RP application was not significantly different from SP-36 application at the same level of P. The CP stover contents of were not significantly different from the control (Table 3).

The highest grain yields, DM stover yields, and CP stover content of were obtained with RP and SP-36 fertilizers applied at 198 kg P ha^{-1}. However, the highest P-use efficiency for grain (grain yield kg/P applied kg) and DM (DM yield kg/P applied kg) was obtained with application of 66 kg P ha^{-1}, 79 kg grain/kg P and 67 kg DM/kg P after RP application, and 95 kg grain/kg P and 99 kg DM/kg P after SP-36 application. Maize grain yield increased 83% and DM yields 68% more after RP application than after SP-36 application.

Superphosphate-36 produced grain yields (Table 1) and DM yields (Table 2) significantly (P<0.05) higher than RP at the same level of P. Superphosphate-36 gave P-use efficiency for grain and DM higher than RP at the same level of P fertilizer. Increasing levels of RP and SP-36 fertilizers decreased P-use efficiency for both grain and stover DM.

Discussion

Phosphorus fertilization (RP and SP-36) improved maize grain yield (Table 1), DM stover yields (Table 2), and CP contents of maize stover (Table 3). These results indicate that the growth of maize could be markedly improved by the application of P fertilizers to soils deficient in P. According to Lukiwati et al. (1997), increased P fertilization resulted in improved plant growth. The maize grain yield and the DM yield of maize stover increased significantly more with RP and SP-36 fertilizers than the control (without P fertilizer). There was no response if levels of P fertilizer were increased from 132 to 198 kg P ha^{-1} for SP-36, and from 66 to 198 kg P ha^{-1} for RP. Yield responses to P generally decline or have no effect as the rate of P application is increased (Jones, 1990). The highest maize grain yield, DM stover yield, and CP content of maize stover were obtained with SP-36 and RP inputs of 198 kg P ha^{-1}, the highest level studied. But, these yields were not significantly different from those obtained with 132 kg P ha^{-1}(SP-36) and with 66–132 kg P ha^{-1} (RP). The highest P-use efficiency for grain and DM was obtained also with RP and SP-36 applied at 66 kg P ha^{-1}. The result confirm that levels of 66 kg P ha^{-1} will ensure that maize plants are adequately supplied with P, and that fertilizer at this rate is capable of increasing maize grain and DM stover yields of maize.

The crude protein content of maize stover was not significantly increased by application of RP or SP-36 fertilizers. There was no response if levels of P fertilizers increased from 66 to 198 kg P ha^{-1}. Crude protein content of maize stover with RP was not significantly different to SP-36 at the same level of P fertilizer. Result showed that CP contents of maize stover were generally unresponsive to increased levels of P or to different sources of P fertilizer. This suggests that the CP content of maize stover is a relatively poor indicator of the P status of soil and plant. According to Coates et al. (1990), the quality of crops could be improved with P fertilization during periods of active growth.

Superphosphate-36 produced maize grain yields and DM yields of maize stover significantly higher than RP at the same level of P application. Maize gave more response to SP-36 than to RP. Similar reports have been presented by Hewitt (1978) and Kerridge and Ratcliff (1982), with difference species of *Gramineae* mainly *Urochloa mosambiensis* (sabi grass) and *Panicum maximum*.

The price of SP-36 is six times more than that of RP. According to Terman (1976), P sources of low solubility such as RP frequently give lower limiting yields, especially of short-season crops. It is suggested this was due to slow release of P from RP. With time, the relative efficiency of RP might be expected to increase. A field trial could be designed to allow measurement of the residual effects of P and the maintenance requirements for this element for at least 5 years (Grundy et al., 1981). At the end of this period, a more acaurate assessment of the relative effectiveness of RP and SP will be possible.

The reputation of maize as a plant able to grow in acid soil low in P may be due to mycorrhizal interactions. Soil pH lower than 5.5 did not affect the P availability of SP-36. The extent of native mycorrhizal fungi (data not presented) could increased P availability from fertilizers in acid soil of low-P content (Lukiwati et al., 1997). It is also assumed that P deficiency induces the secretion of organic acids by roots and this enhances the availability of P by releasing P ions from poorly soluble inorganic P compounds (Luo et al., 1997).

Based on these results, RP and SP-36 fertilizers applied at 66 kg P ha^{-1} improved productivity and gave the highest P-use efficiency for grain and DM in the Latosolic soil use for this experiment. The response obtained to P fertilizers is a function of many factors including the form of P fertilizer applied. Yield response to P fertilization and P-use efficiency generally declines as the rate of P application is increased. Phosphorus sources of low solubility such as RP frequently give lower limiting yields, especially in annual crops.

Alternatively, maize yields in the Latosolic soil could improved by the combination of RP and a small amount of SP-36. The decision is an economic one, based on the relative agronomic effectiveness and the relative prices of RP and to SP-36. The next field trial could be designed to study the effect of some levels of combination of RP and SP-36 on maize productivities in the Latosolic soil.

Acknowledgements

Thanks to Ms Miggie Handayani and Ms Ida Susilowati in whose crop the experiment was conducted.

References

Coates D B, Kerridge P C, Miller C P and Winter W H 1990 Phosphorus and beef production in Northern Australia. 7. The effect of phosphorus on the composition, yield and quality of legume-based pasture and their relation to animal production. Trop. Grassld. 24(3), 209–220.

Grundon N J 1987 Hungry Crops: A Guide to Nutrient Deficiencies in Field Crops. Department of Primary Industries, Brisbane, Australia. 242 pp.

Grundy M J, Bell L C, Asher C J and Evenson J P 1981 Mineral nutrient requirements for pasture establishment on bauxite mined land at Weipa, North Queensland. Trop. Grassld. 15(3), 163–176.

Helyar K R and Godden D P 1976 Soil phosphate as a capital asset. *In* Review in Rural Science III. Proc. of a Symposium Prospects for Improving Efficiency of Phosphorus Utilization. Ed. G J Blair. pp 23–30. University of New England, Armidale, New South Wales, Australia.

Hewitt B R 1978 Role of rock phosphate as a plant nutrient. *In* Annual Report – Pasture Improvement Project. pp 84–85. Fac. of Agric., Khon Kaen University, Thailand.

Islam A K M S, Kerven G and Oweczkin J 1992 Methods of Plant Analysis. ACIAR 8904 IBSRAM QC. Department of Agriculture, The University of Queensland, Australia. 22 pp.

Kerridge P C and Ratcliff D 1982 Comparative growth of four tropical pasture legumes and guinea grass with different phosphorus sources. Trop. Grassld. 16(1), 33–40.

Lukiwati D R, Hardjosoewignjo S, Fakuara Y, Anas I, Wiradarya T R and Rambe A 1997 Improvement of phosphorus uptake of forage legumes by rock phosphate fertilization and vesicular-arbuscular mycorrhiza inoculation. *In* Plant Nutrition for Sustainable Food Production and Environment. Proceedings of the XIII International Plant Nutrition Colloquim, 13–19 September 1997, Tokyo Japan. Developments in Plant and Soil Sciences, Vol. 78. Eds T Ando, K Fujita, T Mae, H Matsumoto, S Mori and J Sekiya. pp 945–946. Kluwer Academic Publishers, Dordrecht, The Netherlands.

Luo H M, Osaki M and Tadano T 1997 Mechanisms of differential tolerance of crop plants to high aluminium and low phosphorus growth conditions. *In* Plant Nutrition for Sustainable Food Production and Environment. Proceedings of the XIII International Plant Nutrition Colloquim, 13–19 September 1997, Tokyo Japan. Developments in Plant and Soil Sciences, Vol. 78. Eds T Ando, K

Fujita, T Mae, H Matsumoto, S Mori and J Sekiya. pp 473–474. Kluwer Academic Publishers, Dordrecht, The Netherlands.

Randall G W and Hoeft R G 1988 Placement methods for improved efficiency of P and K fertilizers: A Review. J. Prod. Agric. 1, 70–78.

Terman G L 1976 Past, present, and prospective phosphorus fertilizers and their evaluation. *In* Reviews in Rural Science III. Proceedings of a Symposium Prospects for Improving Efficiency of Phosphorus Utilization. Ed. G J Blair. pp. 91–95. University of New England, Armidale, New South Wales, Australia.

Young R D, Weatfall D G and Colliver G W 1985 Production, marketing, and use of phosphorus fertilizers. *In* Fertilizer Technology and Use. Third edition. Ed. O P Engelstad. pp. 323–376. Soil Sci. Soc. of Am., Madison, WI, USA.

J.J. Adu-Gyamfi (Ed.), *Food security in nutrient-stressed environments: exploiting plants' genetic capabilities*: 189–191, 2002.

Candidate mechanisms

J. J. Adu-Gyamfi
International Crops Research Institute for the Semi-Arid Tropics (ICRISAT), PMB 3491, Kano, Nigeria

Introduction

Understanding nutrient efficient mechanisms is paramount for designing a simple, fast and effect screening technique that would be a cornerstone of future attempts to breed nutrient-efficient genotypes. It is well known that efficient genotypes grow and yield well on nutrient-deficient soils by exploring specific physiological mechanisms that allow them to gain access to sufficient quantities of nutrients and/or to more effectively utilize them (Rengel, 1999).

The papers presented in this Session covered a diverse range of tropics on the candidate mechanisms of the adaptation of crop plants to low-nitrogen (N) and low-phosphorus (P) environments. We have attempted to group the presentations and extracted and highlighted the major points that were made. Three broad categories of mechanisms that exist in plants to increase the availability, uptake and utilization of N and P are:

1. Increase N and P uptake;
2. More effectively utilize nutrients particularly N and P;
3. Use the aluminium (Al) tolerance exhibited by native legume species.

However, reliable and well-targeted screening procedures using the appropriate parameters is a prerequisite to breeding for genotypes with enhanced tolerance to low-nutrient environments. Nakano et al. (2002) did a good job in screening 100 genotypes each of sorghum and pigeonpea on an Alfisol and a Vertisol and evaluating the morphological and ecophysiological characteristics associated with the adaptation to low-N and -P soils under field conditions. The genotypes were selected from a large collection of germplasm held in the ICRISAT genebank in Patancheru, India. Differences in shoot, grain, N, and P yields were reported among the genotypes. Dry matter was shown to be positively correlated to leaf size, leaf weight, and leaf/stem ratio but not to specific leaf weight. The conclusion that hybrids and improved varieties performed better than the local lines in low-N and -P soils should be quoted with caution.

There are many controversies on the performance of local landraces and farmer-selected local varieties compared to hybrids and newly improved varieties in low-nutrient soils. Biotic and abiotic factors are known to influence the adaptation of crop plants to low-nutrient environments. The data presented were based on single treatments at a single location. Future screening methods should take into consideration the interaction of nutrients, water, the chemical and physical characteristics of the soil type, temperature and photoperiod, and the production system.

During the discussion it was suggested that future screening and selection of genotypes adapted to low-N and -P conditions for specific traits should be done in the target environment where the crop is widely cultivated before any physiological study is undertaken under controlled conditions.

Mechanisms to increase N and P availability and uptake

The mechanisms highlighted were mainly: (1) the secretion or exudation of chemical compounds into the rhizosphere, (2) the association with microorganisms, and (3) the alteration of the geometry or architecture of the root system.

Ae et al. (1990) demonstrated that root exudates from pigeonpea contained piscidic acid that is effective in solubilizing Fe-P and releasing P from Alfisols. This finding, published in *Science* was based on a single genotype of pigeonpea. Therefore Ishikawa et al. (2002) investigated the genotypic variation in the P-solubilizing activity of piscidic acid and other carboxylic anions from the root exudates of pigeonpea. They found that all the pigeonpea genotypes tested exuded piscidic acid from their roots and confirmed in vitro that piscidic acid can solubilize sparingly soluble P in Alfisols. However, they suggested the need to investigate more effective P-solubilizing mechanisms than piscidic acid exudation to explain genotypic variation in low-P availability of pigeonpea.

Another important part of the work by Ishikawa et al. (2002) was the development of a rapid, simple, and easy technique for testing the solubilizing activity of root exudates. This method uses a simple filter paper and allows the screening of large samples of root exudates. Screening large numbers of crop genotypes for their ability to solubilize P has been hampered by lack of simple and relatively cheap techniques. This method, if improved, will help in-depth research into the genetic basis of qualitative and quantitative differences in root exudates from crop plants.

A new concept of 'contact reaction' which focuses on the interaction between root surfaces and soil particles was presented by Ae and Shen (2002) as a possible mechanism for the adaptation of pigeonpea and groundnut to low-P soils. It was shown that the better adaptation of groundnut and pigeonpea than sorghum and soybean to P-deficient soils might be explained by the higher solubilizing ability of their root cell walls. The author argued that one of the weakness of the 'root exudate' theory is microbial degradation. It is shown that phenolic compounds (including piscidic acid) disappeared within 3–4 days after application because of microbial decomposition. Another weakness of the 'root exudate' theory is the high amount of chelating chemicals in root exudates required to release of P. Participants expressed interest in the new concept.

Kirk (2002) applied the theory of coupled diffusion processes in soil to quantify root-induced solubilization of nutrients. The fact that measured concentration profiles of P near roots coincided with the predictions of the model based on independently measured parameters, stressed the validity of the model and the importance of root exuded citric acid for P acquisition by roots in tropical soils. The conclusion that it is possible to model root-induced solubilization processes mechanistically would help in understanding and identifying the processes that are relevant in regulating nutrient and toxic dynamics. The author also attempted to use a mechanistic model to explain the alteration of geometry or architecture of the root system to the adaptation to low-P and -P soils. This is a major challenge to crop physiologists.

Nakamura et al. (2002) demonstrated that higher specific-N absorption rate per root mass (SRA) and per root length (SLRA) by hybrids than a local variety of sorghum in low-N conditions accounted for the better adaptation of hybrids to low-N conditions than the local variety.

It is well known that the association between roots and mycorrhizae allows plants to improve their access to soil-P pools. Gyaneshwar et al. (2002) reported the importance of phosphate solubilizing microorganisms (PSMs). The authors presented data on the successful isolation of P-solubilizing *Enterobacter asburiae* that could release P from the soil. This finding was made under laboratory conditions, and there is a need to verify the results in the field. During discussions, some participants however, expressed the importance of isolating indigenous bacteria in low-P soils to solubilize P rather than using genetically engineered microorganisms.

Mechanism to more effectively utilize N and P

A study by Adu-Gyamfi et al. (2002) clearly demonstrates large significant differences in P utilization efficiency among pigeonpea and sorghum genotypes. An important trait observed in their study is the rapid and efficient translocation and remobilization of P from leaves and stem reserves to support crop growth and grain filling in P-stressed conditions. A ^{32}P study showed that in low-P conditions, P-efficient genotypes translocated more P from roots to the flag leaves. It was shown that the contribution from leaf and stem P offered powerful resources for grain filling for sorghum, and that hybrids and improved varieties were more efficient in the translocation of P from roots to leaves, accounting for their relatively high adaptability to low-P conditions. The work by Fujita et al. (2002) showed that at low-P availability, there is increased partitioning of photo-assimilates to roots and decreased partitioning to the growing leaves. The synthesis of hemicellulose and cellulose in leaves is suppressed causing an inhibition in leaf area expansion. It was therefore concluded that the variability among crop plants in the translocation of photo-assimilates to leaves and the other growing parts at low-P availability provides scope for genetic manipulation of the trait to improve the adaptation of crops to low-P soils.

The long-term experiments by Bationo et al. (2002) showed that the rotation of cereals and legumes help increased the P use efficiency (PUE) of crops. It was clearly shown that PUE of pearl millet increased from 46 kg grain kg fertilizer P^{-1} in continuous rotation to 76 kg grain kg fertilizer P^{-1} when rotated with cowpea. In the West African Sahel region, hill placement of small quantities of P fertilizer (4 kg

P ha^{-1}) was shown to increase PUE more than the recommended rate of 13 kg P broadcast ha^{-1}.

Bationo et al. (2002) showed that partially acidulated rock phosphate (PR) improved PUE by crops, and that rainfall and soil acidity were the main determinants of the agronomic efficiency of different PR sources. For maize although superphosphate application gave a higher yield than RP, it is more economical for farmers to use RP (Lukiwati, 2002).

Mechanisms of aluminium tolerance by native legume species

The paper by Izaguirre-Mayoral et al. (2002) identified legumes species tolerant to aluminium (Al) toxicity. These native legume species can be considered promising germplasm sources for Al-tolerance and could be used as sources of genes for future breeding programs. Aluminium tolerance in transgenic plants by alteration of citrate synthesis has been reported earlier by de la Fuente et al. (1987).

References

Ae N and Shen R-F 2002 Root cell-wall properties are proposed to contribute to phosphorus (P) mobilization by groundnut and pigeonpea. Plant Soil 245, 95–103.

Ae N, Arihara J, Okada K, Yoshihara T and Johansen C 1990 Phosphorus uptake by pigeonpea and its role in cropping systems of the Indian subcontinent. Science 248, 477–480.

Adu-Gyamfi J J, Ishikawa S, Nakamura T and Nakano H 2002 Genotypic variability and physiological characteristics of crop plants adapted to low-nutrient environments. *In* Food Security in Nutrient-Stressed Environments: Exploiting Plants' Genetic Capabilities, Proceedings of an International Workshop 27–30 September 1999, Patancheru, India. Developments in Plant and Soil Sciences volume 95. Ed. J J Adu-Gyamfi. pp 67–79.Kluwer Academic Publishers, Dordrecht, The Netherlands.

Bationo A, Henao J and Anand Kumar K 2002 Phosphorus use efficiency as related to sources of P fertilizers, rainfall, soil, crop management, and genotypes in the West African semi-arid tropics. *In* Food Security in Nutrient-Stressed Environments: Exploiting Plants' Genetic Capabilities, Proceedings of an International Workshop 27–30 September 1999, Patancheru, India. Developments in Plant and Soil Sciences volume 95. Ed. J J Adu-Gyamfi. pp 145–154. Kluwer Academic Publishers, Dordrecht, The Netherlands.

de la Fuente J M, Ramirez-Rodriguez V, Cabrera-Ponce J L and Herrera-Estrella L 1997 Aluminium tolerance in transgenic plants by alteration of citrate synthesis. Science 276, 1566–1568.

Fujita K, Chaudhary M I, Fujita A, Kai Y, Takayanagi M, El Sherry H and Adu-Gyamfi J J 2002 Photosynthesized carbon translocation and distribution of crops adapted to low-nutrient environments. *In* Food Security in Nutrient-Stressed Environments: Exploiting Plants' Genetic Capabilities, Proceedings of an International Workshop 27–30 September 1999, Patancheru, India. Developments in Plant and Soil Sciences volume 95. Ed. J J Adu-Gyamfi. pp 91–101. Kluwer Academic Publishers, Dordrecht, The Netherlands.

Gyaneshwar P, Naresh Kumar G, Parekh L J and Poole P S 2002 Role of soil microorganisms in improving P nutrition of plants. Plant Soil 245, 83–93.

Ishikawa S, Adu-Gyamfi J J, Nakamura T, Yoshihara T and Wagatsuma T 2002 Genotypic variability in phosphorus solubilizing activity of root exudates by pigeonpea grown in low-nutrient environments. Plant Soil 245, 71–81.

Izaguirre-Mayoral M-L, Flores S and Oropeza T 2002 Aluminium tolerance in nodulated N_2-fixing legumes species native to two contrasting savanna sites. Plant Soil 245, 163–168.

Kirk G J D 2002 Modelling root-induced solubilization of nutrients. Plant Soil 245, 49–57.

Lukiwati D R 2002 Effect of rock phosphate and superphosphate fertilizer on the productivity of maize var. Bisma. *In* Food Security in Nutrient-Stressed Environments: Exploiting Plants' Genetic Capabilities, Proceedings of an International Workshop 27–30 September 1999, Patancheru, India. Developments in Plant and Soil Sciences volume 95. Ed. J J Adu-Gyamfi. pp 183–187. Kluwer Academic Publishers, Dordrecht, The Netherlands.

Nakamura T, Adu-Gyamfi J J, Yamamoto A, Ishikawa S, Nakano H and Ito O 2002 Varietal differences in root growth as related to nitrogen uptake by sorghum plants in low-nitrogen environment. Plant Soil 245, 17–24.

Nakano H, Adu-Gyamfi J J and Nakamura T 2002 Genetic adaptation of crop plants to low-nutrient environments: Morphological and ecophysiological characteristics of adaptation. *In* Food Security in Nutrient-Stressed Environments: Exploiting Plants' Genetic Capabilities, Proceedings of an International Workshop 27–30 September 1999, Patancheru, India. Developments in Plant and Soil Sciences volume 95. Ed. J J Adu-Gyamfi. pp 59–66. Kluwer Academic Publishers, Dordrecht, The Netherlands.

Rengel Z 1999 Physiological mechanisms underlying differential nutrient efficiency of crop genotypes. *In* Mineral Nutrition of Crops: Mechanisms and Implications. Ed. Z Rengel. pp 227–265. The Haworth Press, New York, USA.

J.J. Adu-Gyamfi (Ed.), *Food security in nutrient-stressed environments: exploiting plants' genetic capabilities*: 193–200, 2002.

Assessing germplasm collections for traits useful in plant nutrition

P. J. Bramel

Genetic Resources and Enhancement Program, International Crops Research Institute for the Semi-Arid Tropics (ICRISAT), Patancheru 502 324, Andhra Pradesh, India

Key words: accessions, empirical selection, genepools, germplasm collections nutrient use efficiencies

Abstract

In the past, empirical selection by breeders has been successful at changing such regulatory processes as the pattern of partitioning and the timing of development. It has done little to increase the efficiency of the major metabolic or assimilatory processes, such as photosynthesis or nutrient use efficiency. A number of possible agents of change to develop crop varieties that produce greater yield per unit of nutrient are available. A combination of enhanced use of limited nutrients and better extraction of available nutrients from the soil would likely result in the best possible option for crops grown under low-nutrient conditions. The unpredictable nature of such conditions will require plasticity to efficiently use a sustainable portion of the available nutrients, irrespective of the source. The primary agent of change will be a source of genetic variation for the trait of interest. The two main options are intra-specific or inter-specific genetic variation. Intra-specific variation is more easily exploited by the plant breeder, but is less likely to exist, except in crops where the evolution has included significant production under very diverse marginal conditions. A more likely source of significant variation will be differences among species that have evolved under distinctly different environments, such as the wild relatives of the crop. The objective of this review is to explore the various agents of change required for enhanced nutrient use efficiency. A number of studies have been undertaken in sorghum to assess the value of intra-specific and inter-specific variation as improved sources of drought tolerance and nutrient use efficiency (NUE). A number of accessions were identified based upon an assessment of the various components of these two traits within very diverse germplasm. The key to progress in breeding nutrient use efficient cultivars will be in the enhanced utilization of genetic resources. Intra-specific strategies will concentrate on the use of landraces or farmer varieties and wild relatives in the primary genepool. The use of these sources could be direct, but could involve backcrossing to minimize the impact of undesirable alleles. The secondary and tertiary genepools might serve as very important sources of unique alleles. One option for the future will be the use of gene mapping and isolation. Once isolated and sequenced the gene can be directly introduced into the crop cultivars with transformation. This same approach could be used in unrelated species where mechanisms are carefully evaluated using comparative approaches.

Introduction

Evans (1993) stated "Empirical selection is an extremely powerful agent of change but selection by design may yet prove to be even more powerful when our understanding of the physiology of crop yield is more comprehensive than it is at the present". This empirical selection had been successful in changing such regulatory processes as patterns of partitioning and the timing of development. It had done very little to increase the efficiency of the major metabolic or assimilatory processes, e.g., photosynthesis or nutrient use efficiency (NUE). These processes will be more difficult to 'improve' since there are multiple trade-offs for which an adaptive role may not be directly related to yield under specific environmental constraints that is an integrated product of a great variety of processes. The key to altering these processes with plant breeding will be the genetic variant that can be found with the targeted assessment of genetic resources. This source of variation will impact the potential genetic gain to be made for the process and, consequently, yield under conditions with reduced external inputs.

Agents of change

There are a number of agents of change for the development of new varieties that produce greater yield per unit of nutrient, or greater yield from a higher rate of extraction of the available nutrients from the soil. A combination of these agents would likely be the best possible option for crops grown under low-nutrient conditions. The unpredictable nature of low-nutrient conditions will require plasticity to efficiently use a sustainable portion of the available nutrients, irrespective of the source.

Genetic variation

The most important agent of change will be genetic variation. The two main options are intra-specific or inter-specific genetic variation. Intra-specific variation within the wild relatives of a crop or the cultivated landraces will be more useful in the short-term to the plant breeder than inter-specific variation. Significant variation is least likely to exist in the crop genepool, except in those crops which have evolved under diverse marginal conditions. The more likely source of variation will be due to differences among species that have evolved in distinctly different environments. The impact of natural selection would result in the greatest differences in the efficiency of metabolic processes in this comparison.

Initially, the main target for genetic enhancement of NUE will be ex situ collections. The future collection of potential sources of improvement will need to be targeted to varieties, wild relatives, or related species of the crop of interest. This targeted collection should focus on environments or agronomic conditions that would have favored natural selection for increased efficiency or uptake. The collection should also account for the impact of farmer selection for specific adaptation. Such farmer selection would likely increase the possibility of increased NUE or uptake associated with a higher productivity phenotype. During the collection process, information from farmers should be solicited on the specific adaptation of the variety to production constraints such as nutrient or drought stresses. The specific unique response of the farmers' variety should be documented to increase its value for evaluation.

The process of genetic resources conservation could impact its potential use as a source of this specific adaptation. The consequences of the sampling and regeneration of a collected accession for ex situ conservation could effect the subsequent value of that accession. This is a consequence of both genetic drift from the random loss of alleles due to sampling, and possibly as a result of changes in adaptation from regeneration in environments different from those in which the accession evolved. This effect is more likely to be of greater consequence in the wild species. Thus, it may be very critical to maintain these species in situ and sample directly for evaluation, thus avoiding any undesirable changes in adaptation that may influence the utility of a specific collection.

Exploitation of genetic resources

The targeted evaluation of genetic resources should depend upon a careful consideration of evolutionary relationships and processes in the primary, secondary, and tertiary genepools. The process of domestication according to Evans (1993) involves three phases. These include the initial domestication from the wild relative and associated changes due to cultivation. Domestication occurs through reproductive isolation in some crops or as a consequence of the intense selection of only a few desirable types. Ultimately, the process of domestication results in only a sample of alleles or loci from the progenitor being incorporated into crop plants. This effect results in greater value for the wild relatives for genetic enhancement of such factors as metabolic processes. The second phase is the adaptation of the new crop to environments outside the range of its wild relatives. This phase would result in selection for shifts in the geographical adaptation that could include nutrient- or drought-stressed conditions. The final phase involves changes brought about by intense artificial selection for broader adaptation. One assumption from this process is that variation for adaptation to different nutrient conditions would have a higher probability of being fixed in accessions selected for production under these conditions, rather than result from a random mutation. Thus, environmental information on the specific conditions from which a genotype was collected would be very beneficial in the identification of probable sources of improvement.

The consequences of the shift in selection from specific adaptation to broader adaptation, usually with the addition of external inputs, would reduce the probability of finding significant variation among improved breeding lines. Therefore, the primary target of evaluation should be landraces or wild species. In wheat, Hetrick et al. (1992, 1993) evaluated the impact of evolutionary history and breeding history on adaptation to low-phosphorus (P) environments. Various ancestral species and landraces were evaluated for mycorrhizal dependence and degree of root colonization under very low soil-P levels. Hetrick et al. (1992) concluded that primitive wheat cultivars, ancestors, and landraces benefit more from mycorrhizal symbiosis than modern cultivars. This could be due to the impact of cultivars development in this century at locations where inorganic-P fertilizers are applied. Varieties selected under such conditions had a reduced frequency of the genes that foster mycorrhizal associations. No relationship was found between the degree of root colonization and growth response, but there was a significant correlation between root fibrousness and mycorrhizal dependence. The disadvantage of these changes was more pronounced in test conditions where P levels were low.

The genetic consequence of introgression by landraces or wild species is related not only to the impact of increased genetic variance for the trait of interest, but also to the linkage associations between the desirable and undesirable alleles or loci that are introduced from the donor parent and contribute to linkage drag. One of the consequences of use of genetic resources with a high percentage of undesirable alleles is the negative effect this has on the mean and variance of the elite, adapted genepool. Such alleles may have been nearly fixed during previous natural and artificial selection into an optimum arrangement. The introduction of the desirable variation from this exotic source would be accompanied by small segments of DNA that contain variants for these undesirable genetic changes. The impact of this introduction will be to slow overall progress due to the unpredicted impact on yield from these undesirable changes. This linkage drag has been shown to significantly limit the use of very divergent germplasm, especially the wild relatives or primitive landraces of crops, that are likely to be the main source of genetic variation for these metabolic processes.

The ICRISAT sorghum collection is very large (Table 1). ICRISAT maintains 36 292 accessions of cultivated sorghum of which 75% are landraces. The collection of wild relatives includes 461 accessions, 72% from the primary, 9% from the secondary, and 15% from the tertiary genepools. Such collection is very large to evaluate completely so a sampling strategy will need to be used. There are two options, one would be to initially focus on the core collection that has been identified to represent the diversity of the total collection. This subset would sample the variation available in the specific trait. This evaluation could be used to further target the collection to accessions closely related to those found useful. It would also allow intra-specific variation to be assessed for diversity in the trait of interest. The lack of variation in the metabolic process would indicate where effort would need to be concentrated on wild species. A secondary, more-targeted assessment could be made of accessions that were previously collected from closely related environmental conditions. A focus only on the geographical origin of the germplasm would be limited by the available information, but would be the best approach if possible. The best strategy is likely to be a combination of the two opinions, i.e., to survey the range of response, initially intra-specific, and then to sample species-level diversity. An initial screen of species-level variation could be followed by a more targeted assessment of variation within related and unrelated species.

Selection environment

The second agent of change will be the appropriate selection environment for both evaluations of genetic resources and selection within breeding populations. This environment should be homogeneous in its nature and repeatable over time. An environment that uniformly applies selection pressure for the trait of interest will increase the heritability for that trait under those specific conditions. Repeatability over time is critical to reduce the impacts of genotype × environment interactions, and again to increase the realized heritability or genetic gain from selection. This environment must allow maximum expression of the direct trait of interest, such as the metabolic process per se. The trait of interests' relationship to yield under variable field conditions need not be optimized in this selection environment since its indirect effect is well understood.

The independence of NUE and nitrogen (N) responsiveness would indicate that selection for germplasm could initially be based on NUE at very low N levels. This would enhance the differences between genotypes for their use at nutrient-stressed levels. Grain yield's potential and adaptation influences N responsiveness, so this would be evaluated later when breeding or testcrossing had increased the yield level. The use of a preliminary screen to select sorghum germplasm with adaptation to low-N conditions was reported in Reddy Podduturi (1990). The selection environment had less than 2 ppm N due to the continuous cropping of sorghum in the field for the past

Table 1. Taxonomic classification of sorghum collection held at the International Crop Research Institute for the Semi-Arid Tropics (ICRISAT), Patancheru, India

	Section	Species/subspecies of *Sorghum*	Race	Number of assessions
Primary genepool				
$2n$=20	Sorghum	*S. bicolor bicolor*	5 races; 15 intermediates	36 292
		S. bicolor arundinaceum	*arundinaceum*	33
			aethiopicum	16
			verticilliflorum	102
			virgatum	18
		S. bicolor drummondii	*drummondii*	157
Secondary genepool				
$2n$=(40,20)		*S. halepense*	*halepense*	21
			controversum	4
			miliaceum	6
			almum	6
		S. propinquum		2
Tertiary genepool				
$2n$=(10,20,30,40)	Chaetosorghum	*S. macrospermum*		2
	Heterosorghum	*S. laxiflorum*		14
	Parasorghum	*S. australiense*		3
		S. brevicallosum		2
		S. dimidiatum		2
		S. matarankense		3
		S. nitidum		2
		S. purpureosericeum		7
		S. timorense		17
		S. versicolor		6
	Stiposorghum			
		S. angustum		10
		S. bulbosum		
		S. ecarinatum		1
		S. extans		1
		S. interjectum		2
		S. intrans		2
		S. plumosum		3
		S. stipoideum		2
Unclassified				17
Total wild				461
Total cultivated				36 292

5 years. The selection criteria included measures of N uptake, translocation, and efficient use. The accessions screened included adapted breeding lines, adapted old varieties, exotic landraces, exotic breeding lines, wild relatives of sorghum, and pearl millet. The results of the screening identified three groups of genotypes that differed in their overall NUE per day. The low NUE genotypes are listed in Table 2. In this group, there is a strong relationship between rates of uptake and dry matter accumulation. Thus the increased growth rate within the group is a product of the increased N uptake rate. Six of the 12 genotypes in this group are adapted inbred lines. The moderate NUE genotypes were also characterized by the same relationship between growth and N uptake rates (Table 3). This second group did contain a few exceptional accessions, such

Table 2. Sorghum genotypes identified by Reddy Podduturi (1990) as having low nutrient-use efficiencies (NUEs)

Genotype	NE1[a]-seedling stage (g TDW/g TN[b])	NE1[a]-flowering stage (g TDW/g TN[b])	Dry matter accumulated (g TDW day^{-1})	Nitrogen uptake (g TN day^{-1})	NE1[a] (g TDW$(g\ TN)^{-1}$ day^{-1})
784	37.6	103.3	1.67	0.010	1.25
Wheatland	41.9	106.9	1.69	0.011	1.24
SC35-6	47.6	125.2	2.32	0.013	1.50
N94	44.6	105.2	2.45	0.018	1.15
SC33-6	45.3	119.2	2.90	0.018	1.41
N91	46.4	115.0	2.96	0.019	1.36
IA28	45.2	119.0	2.97	0.021	1.42
DR144	45.8	120.8	2.97	0.020	1.46
Tx 2752	45.0	112.3	3.02	0.021	1.44
KS18	39.9	97.3	3.12	0.025	1.46
1247-2	44.2	124.2	3.23	0.021	1.49
1300	43.9	108.6	3.40	0.025	1.26

[a]NE1, total above-ground dry matter per unit plant nitrogen i.e., g TDW/g TN.
[b]TDW, total dry weight; TN, total nitrogen.

Table 3. Sorghum genotypes identified by Reddy Podduturi (1990) as having moderate nutrient-use efficiencies (NUEs)

Genotype	NE1[a]-seedling stage (g TDW/g TN[b])	NE1[a]-flowering stage (g TDW/g TN[b])	Dry matter accumulated (g TDW day^{-1})	Nitrogen uptake (g TN day^{-1})	NE1[a] (g TDW$(g\ TN)^{-1}$ day^{-1})
SC599-6	41.3	137.6	2.21	0.011	1.85
N32	46.2	122.7	2.35	0.013	1.61
IS1347	43.7	131.2	2.36	0.012	1.72
KS9	41.6	116.7	2.44	0.015	1.58
SC642	37.3	131.3	2.95	0.018	1.88
SC103	46.3	135.0	2.96	0.017	1.72
Martin	48.5	104.5	3.02	0.018	1.72
SC630	44.5	108.6	3.28	0.020	1.72
Dorado	46.7	129.2	3.34	0.019	1.69
1276	38.6	121.2	3.49	0.023	1.68
Atlas	40.1	128.1	3.66	0.024	1.62
IR204	43.2	113.6	3.85	0.022	1.65
M35-6	48.4	130.2	3.91	0.024	1.61
DR14	40.8	128.8	4.44	0.025	1.85
S. verticiliflorum	35.0	134.5	5.01	0.037	1.68

[a]NE1, total above-ground dry matter per unit plant nitrogen, i.e., g TDW/g TN.
[b]TDW, total dry weight; TN, total nitrogen.

as the wild relative *Sorghum verticiliflorum*, which had very high N uptake and growth rates. The third group was characterized by high to very high NUE per day (Table 4). In this group there were a number of exceptions to the relationship between N uptake and growth rates. These were genotypes with low growth and N uptake rates, but high NUE (i.e., MLS, a pearl millet accession). There were also genotypes with high growth and N uptake rates but a lower NUE (Tx 7000, KS56, San Chi San, and Ell_{IS}). The group also contained genotypes with high growth rates, low N uptake rates but high NUEs (SC566 and SC279). The exceptional accession was China 17 that had a high growth rate, a low N uptake rate, but the highest NUE. These exceptional genotypes should then be characterized further for their use in the enhancement of NUE in sorghum in relation to both adaptation to low-N conditions and their response to increased N levels.

Table 4. Sorghum genotypes identified by Reddy Podduturi (1990) as having high nutrient-use efficiencies (NUEs)

Genotype	NE1[a]-seedling stage (g TDW/g TN[b])	NE1[a]-flowering stage (g TDW/g TN[b])	Dry matter accumulated (g TDW day^{-1})	Nitrogen uptake (g TN day^{-1})	NE1[a] (g TDW(g TN)$^{-1}$ day^{-1})
ICTP 8203	42.9	134.4	1.68	0.007	2.22
Tx 7078	43.9	119.8	2.22	0.011	2.03
DR50	47.1	152.1	2.79	0.010	2.16
Tx 623	43.8	142.1	2.82	0.014	2.13
SC303	39.3	132.8	2.92	0.015	2.39
Tx 430	46.3	124.6	2.99	0.015	2.05
SC414-12	47.2	106.8	2.82	0.018	2.12
SC566	43.3	124.6	3.29	0.017	2.26
SC279	48.2	129.2	3.24	0.015	2.64
Tx 7000	41.8	117.5	4.12	0.023	2.08
KS56	37.3	111.9	4.08	0.024	2.15
San Chi San	55.5	137.6	4.21	0.020	2.28
MLS[b]	37.7	116.7	1.80	0.005	3.07
China-17	48.8	154.3	3.32	0.012	3.73
Ellis	45.3	151.8	6.41	0.032	2.75

[a]NE1, total above-ground dry matter per unit plant nitrogen, i.e., g TDW/g TN.
[b]TDW, total dry weight; TN, total nitrogen.

Selection criteria

The third agent of genetic change is a selection criterion that is rapid, low-cost, and repeatable. This is very critical in the evaluation of genetic resources where the criteria of direct selection will need to be well-targeted for use in unrelated or unadapted genetic material. The genetic enhancement of a trait that is quantitatively inherited, such as processes governing NUE, will depend upon an increase in the additive genetic variance for that trait. This will result from the introduction of new alleles or loci, or increased frequency of alleles or loci that are at very low frequency in the adapted breeding populations. This objective implies that evaluation for the genetic source per se will not be predictive, since it is the complementarity of its genes to those already in the breeding populations that will influence its use. An alternative could be to evaluate germplasm for a more specific targeted process that is known to be a constraint to yield in the present cultivars. Thus, a clearly targeted selection criteria, more simply inherited than yield, will need to be used for the evaluation of donor parents so that it will enhance the potential gain for yield by introducing genes for changes in a process vital to yield.

Youngquist et al. (1992) compared the use of total above-ground plant nitrogen content (PN), grain nitrogen content (GN), grain dry weight (GDW), total above ground dry weight (TDW), total above-ground dry matter per unit plant nitrogen (NE1), and grain dry weight per unit plant N (NE2) for selection. The objective of this study was to evaluate alternative criteria for determining relative NUEs in sorghum genotypes to reduce the need for costly and time-consuming plant N analysis. From an assessment of three sets of trials, they concluded that TDW could be used to substitute for PN, and GDW to substitute for GN. Both of these dry weight measures would select for NE1 and NE2, but would best be used in a large-scale initial screen of diverse germplasm.

Buah et al. (1998) compared NUE and N responsiveness in eight sorghum accessions and five hybrids at three N levels. The genotypes had been previously selected on the basis of five criteria; harvest index, grain yield potential, N responsiveness, NE1, and NE2. Comparisons for grain yield and total dry weight response to increased N level, NE1 and NE2 were made between the various genotype groups. The hybrids differed from the lines for overall grain yield. The N-responsive and high-NUE groups were also higher yielding at all N levels. Hybrids and lines differed significantly in their responses to N-level with a consistent two-fold increase in hybrids. Responsive and moderately responsive genotypes differed significantly from non-responsive genotypes for linear yield increase to N. There was no difference in the high- and low-NUE groups in their linear response to N. Thus the prior classification of genotypes to NUE and N responsiveness was predictive of their performance in this study, even though the hybrids versus lines comparison was consistently greater than any other group comparison.

Blum (1988) cited a serious limitation to the use of yield as a selection index, particularly in a stress environment. In a complex trait such as yield, environmental variance and variance due to genotype × environment interaction make a significant contribution to the non-genetic component of phenotypic variance. Blum (1988) stated that, "It must be apparent that yield as such is not under direct control. It is the multitude of physiological and biochemical processes at any level of organization that are under genetic control, when the total is integrated towards an effect on yield". The inflation of non-additive variance can be attributed to the complexity of interactions between numerous physiological processes that contribute to yield. Greene (1992) evaluated the use of five physiological traits: biomass production, leaf area index, relative growth rate, rate of photosynthesis, and rate of transpiration, to evaluate non-domesticated and unadapted sorghum germplasm for drought tolerance. These traits were chosen to overcome the limitations inherent in grain yield, which is situated at the end of a hierarchy of physiological processes. The best candidate traits were relative growth rate, and photosynthetic rate under drought conditions. The level of domestication and adaptation influenced neither of these traits.

The choice of an evaluation method also depends upon the level of agronomic adaptation of the exotic accession. Gerloff and Smith (1988) concluded that the evaluation of performance per se is satisfactory for qualitatively inherited traits so long as the selection environment is conducive to the growth of the accession, and the trait is expressed. The evaluation of testcross progeny may be necessary to predict the genetic potential of an exotic accession if there is a significant amount of genotype × environment interaction and if non-additive genetic variance is a principal constituent of the total genetic variance. Unadapted traits such as photoperiod sensitivity often preclude evaluation of performance per se in such highly unadapted germplasm as short-day primitive landraces or wild subraces. Potentially useful alleles may be confounded with traits related to growth habit or plant phenology.

In pearl millet, Bramel-Cox et al. (1986) found that testcross progeny derived from exotic × adapted inbred crosses clearly indicated that exotic accessions could contribute useful alleles for improving grain yield and growth rate despite the lack of adaptation and domestication of the exotic parent. Abel and Pollack (1991) concluded that grain yield per se ranked unadapted maize populations differently from a testcross evaluation. Cox and Frey (1984) suggested that useful alleles in the wild races of sorghum were masked by regulatory genes present in the wild races. They concluded that progeny evaluation may be necessary to unmask desirable alleles in the BC_1 or BC_2 populations.

Although more time and labor are required to conduct testcross evaluations, they have some distinct advantages. An influx of agronomically adapted alleles will dilute the effect of the undesirable alleles that will prevent the negative bias on yield of the effects of domestication and adaptation. The suppressive effect of 'wild' regulatory genes can also be mitigated. In crops where hybrids are routinely produced, e.g., sorghum, testcross evaluation can yield early information about the combining ability of exotic lines under evaluation, in addition to providing information about inbred testers and their interactions with exotic lines. Abel and Pollack (1991) concluded that these were considerations that need to be taken into account when backcrossing or introgressing exotic material into an adapted genepool.

Greene (1992) compared the use of per se performance with two levels of backcrossing, BC_0 and BC_1, to determine which methods were most appropriate to evaluate wild races and unadapted sorghum germplasm for traits related to drought tolerance and grain yield. This comparison also demonstrated that per se evaluation had little predictive value for the two levels of backcrossing. The use of two levels of testcross overcame the effects of the undesirable alleles, and provided useful information on the utility of a wide range of germplasm. The study demonstrated the impact of the levels of heterozygosity and heterogeneity on the evaluation of a particular plant response in a genotype. Heterozygosity has an important role in heterosis and its well-documented effects on productivity. It also has impact on the response to environmental constraints, and on the plasticity of response. Bramel-Cox (1996) concluded that population buffering due to both heterozygosity and heterogeneity would modify the response of a genotype to environmental fluctuations. This impact of population type on productivity and plant responses will impact the evaluation of genotypes for NUE. It will be a particular consideration when grain yield or biomass partitioning are used as the selection criteria. Thus, care should be taken to compare hybrids with hybrids, variable populations with variable populations, and inbred lines with inbred lines. Caution should be taken when comparing different population types unless the trait under evaluation is known to be inherited in a simple additive manner.

Conclusion

The key to progress in breeding nutrient use efficient cultivars will be in the enhanced utilization of genetic resources. Intra-specific strategies will concentrate on the use of landraces or farmer varieties and wild relatives in the primary genepool. The use of these sources will be direct, but may involve backcrossing to minimize the impact of undesirable alleles. The secondary and tertiary genepools could serve as very important sources of unique alleles, but there will be a limit to the crossability and their use by direct introgression. One option is the use of gene mapping and isolation. Progress in gene isolation will depend upon the availability of both sensitive and resistant responses within the species.

References

Abel B C and L M Pollack 1991 Rank comparisons of unadapted maize populations by testers and per se evaluation. Crop Sci. 31, 650–654.

Blum A 1988 Plant Breeding for Stress Environments. CRC Press, Boca Raton, Florida, USA.

Bramel-Cox P J, Andrews D J and Frey K J 1986 Exotic germplasm for improving grain yield and growth rate in pearl millet. Crop Sci. 26, 687–690.

Bramel-Cox P J 1996 Breeding for reliability of performance across unpredictable environments. *In* New Perspectives on Genotype-by-Environment Interaction: New Perspectives. CRC Press, Boca Raton, Florida, USA.

Buah S S J, Maranville J W, Traore A and P J Bramel-Cox 1998 Response of nitrogen use efficient sorghums to nitrogen fertilizer. J. Plant Nutr. 21, 303–2318.

Cox T S and Frey K J 1984 Genetic variance for yield and related traits in sorghum introgression populations. Theoret. Appl. Gen. 68, 145–153.

Evans L T 1993 Crop Evolution, Adaptation and Yield. pp 266–268. Cambridge University Press, Cambridge, UK.

Hetrick B A D, Wilson G W T, and Cox T S 1992 Mycorrhizal dependence of modern wheat varieities, landraces and ancestors. Can. J. Bot. 70, 2032–2040.

Hetrick A D, Wilson G W T, and Cox T S 1993 Mycorrhizal dependence of modern wheat cultivars and ancestors: a synthesis. Can. J. Bot. 71, 12–518.

Gerloff J E and O S Smith 1988 Choice of method for identifying germplasm with superior alleles. I. Theoretical results. Theoret. Appl. Gen. 76, 209–216.

Greene S 1992 Evaluation of the potential usefulness of exotic sorghum (*Sorghum bicolor* L. Moench). germplasm. Kansas State University. PhD Dissertation. Manhattan, Kansas, USA.

Reddy Podduturi S 1990 Nitrogen use efficiency of sorghum and pearl millet at extremely low soil levels. MS Thesis. Kansas State University. Manhattan, Kansas, USA.

Youngquist J B, Bramel-Cox P and Maranville J W 1992 Evaluation of alternative screening criteria for selecting nitrogen-use efficient genotypes in sorghum. Crop Sci. 32, 1310–1313.

J.J. Adu-Gyamfi (Ed.), Food security in nutrient-stressed environments: exploiting plants' genetic capabilities: 201–213, 2002.

Root exudates as mediators of mineral acquisition in low-nutrient environments

Felix D. Dakora[1,3] & Donald A. Phillips[2]
[1]*Research Development, Cape Technikon, Room 2.8 Admin Bldg., P.O. Box 652, Cape Town 8000, South Africa.* [2]*Department of Agronomy and Range Science, University of California, Davis, CA 95616, USA.* [3]*Corresponding author**

Received 10 May 2000. Accepted in revised form 29 January 2001

Key words: low nutrients, *nod* genes, organic acids, phenolics, phytosiderophores, root exudates, symbiotic microbes

Abstract

Plant developmental processes are controlled by internal signals that depend on the adequate supply of mineral nutrients by soil to roots. Thus, the availability of nutrient elements can be a major constraint to plant growth in many environments of the world, especially the tropics where soils are extremely low in nutrients. Plants take up most mineral nutrients through the rhizosphere where micro-organisms interact with plant products in root exudates. Plant root exudates consist of a complex mixture of organic acid anions, phytosiderophores, sugars, vitamins, amino acids, purines, nucleosides, inorganic ions (e.g. HCO_3^-, OH^-, H^+), gaseous molecules (CO_2, H_2), enzymes and root border cells which have major direct or indirect effects on the acquisition of mineral nutrients required for plant growth. Phenolics and aldonic acids exuded directly by roots of N_2-fixing legumes serve as major signals to Rhizobiaceae bacteria which form root nodules where N_2 is reduced to ammonia. Some of the same compounds affect development of mycorrhizal fungi that are crucial for phosphate uptake. Plants growing in low-nutrient environments also employ root exudates in ways other than as symbiotic signals to soil microbes involved in nutrient procurement. Extracellular enzymes release P from organic compounds, and several types of molecules increase iron availability through chelation. Organic acids from root exudates can solubilize unavailable soil Ca, Fe and Al phosphates. Plants growing on nitrate generally maintain electronic neutrality by releasing an excess of anions, including hydroxyl ions. Legumes, which can grow well without nitrate through the benefits of N_2 reduction in the root nodules, must release a net excess of protons. These protons can markedly lower rhizosphere pH and decrease the availability of some mineral nutrients as well as the effective functioning of some soil bacteria, such as the rhizobial bacteria themselves. Thus, environments which are naturally very acidic can pose a challenge to nutrient acquisition by plant roots, and threaten the survival of many beneficial microbes including the roots themselves. A few plants such as Rooibos tea (*Aspalathus linearis* L.) actively modify their rhizosphere pH by extruding OH^- and HCO_3^- to facilitate growth in low pH soils (pH 3 – 5). Our current understanding of how plants use root exudates to modify rhizosphere pH and the potential benefits associated with such processes are assessed in this review.

Introduction

Plant developmental processes are controlled by internal molecular signals that depend on the adequate supply of mineral nutrients by soil to roots. Changes in the levels of certain nutrients are reported to alter tissue concentrations of cytokinins and abscisic acid (Moorby and Besford, 1983), two major signals controlling plant growth. Consequently, seedling development has a high nutrient demand not only for growth of individual organs, synthesis of new cytoplasm as well as sub-cellular organelles and cell walls,

* FAX No: 27-21-650 4041. E-mail: dakora@botzoo.uct.ac.za

This paper previously appeared in *Plant and Soil* **245**, 2002, pp. 35–47.

but also for both cell division and expansion (Moorby and Besford, 1983). Nitrogen, a major component of DNA and proteins in all cells, is together with Mg also the key elemental constituent of chlorophyll, the light-harvesting pigment of photosynthetic plants (Lehninger, 1970). In addition, specific concentrations of N and other nutrient elements (P, S, Ca, Mg, Fe and Cu) elicit the production of isoflavonoids in plants, and these molecules function as signals to mutualistic soil microbes and/or phytoalexins against infecting pathogens (Dakora and Phillips, 1996). Calcium is involved in the closure of guard cells and also mediates the response of plants to ethylene, a gaseous plant hormone that controls important physiological processes including seedling development, flowering, fruit ripening and senescence (Raz and Fluhr, 1992). Besides their involvement in signalling, some elements also play crucial physiological roles in plant life. Low concentrations of nutrients such as K^+, Na^+ and Mg^{++} also readily stimulate the activity of major enzymes of the glycolytic pathway, namely phosphofructokinase and pyruvate kinase, which together regulate glycolysis in plant cells (Plaxton, 1996). Osmoregulation, the opening and closure of stomatal guard cells, and daily changes in leaf orientation are controlled by K^+ mobility in plant cells (Hsiao and Lauchli, 1986). Individual micronutrients are similarly important components of major enzymes, which regulate all biological processes in plants.

It is clear from these considerations that low nutrient availability can constrain plant growth in many environments of the world, especially the tropics where soils are extremely deficient in nutrients. Yet, the tendency in modern agriculture has been to select crop species for high soil fertility. This has resulted in the wide use of crop varieties that require high doses of applied fertilizer in order to meet optimal plant growth and grain yield. Plant species growing in naturally fertile soils also tend to respond to nutrient supply in a manner similar to agricultural cultivars. Because of the high concentrations of plant-available nutrients in fertilized or naturally fertile soils, root uptake rates are also high. This is in sharp contrast to nutrient-poor sites where root uptake rates are usually low due to low or poor nutrient availability (Marschner, 1995).

Plant uptake of nutrients from soil is more marked in the 'rhizosphere' surrounding the root than outside this zone (Darrah, 1993). Root exudation of various chemical molecules into the rhizosphere is largely dependent on the nutritional status of the plant, with some species exuding organic acid anions in response to P and Fe deficiency or phytosiderophores due to Fe and Zn defiency (Haynes, 1990; Jones and Darrah, 1994). Consequently, the released compounds can cause some nutrient elements to be relatively more available for uptake by plants. The rate of exudation itself is increased by the presence of microbes in the rhizosphere (Gardner et al., 1983) and promoted by the uptake and assimilation of certain nutrient elements. As a result, the composition of root exudates can be complex, and often ranges from mucilage, root border cells, extracellular enzymes, simple and complex sugars, phenolics, amino acids, vitamins, organic acids, nitrogenous macromolecules such as purines and nucleosides to inorganic or gaseous molecules such as HCO_3^-, OH^-, H^+, CO_2 and H_2 (Marschner, 1995; Rovira, 1969; Uren and Reisennuer, 1988; also see Table 1). Many of these organic substrates excreted into the rhizosphere, particularly amino acids, organic acids, proteins, carbohydrates and vitamins, promote microbial biosynthesis of ethylene (Arshad and Frankenberger, 1990), a powerful plant signal controlling development. That apart, these components all play different roles that ultimately affect nutrient acquisition by plants.

Enhancement of nutrient supply by root exudate effects on symbiotic microbes

The components of plant root exudates are many and complex (Table 1), and serve not only as a source of carbon substrate for microbial growth, but also contain chemical molecules that promote chemotaxis of soil microbes to the rhizosphere. Although the root exudates of N_2-fixing legumes are known generally for their capacity to attract rhizobia to root hairs, it is in fact their individual chemical components such as flavonoids (Caetano-Anolles et al., 1988), aromatic acids (Parke et al., 1985), amino acids and dicarboxylic acids (Barbour et al., 1991) that function as specific chemoattractants for micro-organisms (Table 2). Once in the rhizosphere, many bacteria may multiply rapidly in response to growth stimulation by quercetin or other flavonoid molecules released by plants (Hartwig et al., 1991), and in turn promote further exudation of new or existing flavonoids into the rhizosphere (Dakora et al., 1993a,b; Recourt et al., 1991). Besides their direct effects on rhizobia, root exudates can also attract pathogenic microbes (Morris and Ward, 1992) and promote the growth of plants, mutualistic fungi (Siqueira et al., 1991) and rhizobac-

Table 1. Organic compounds and enzymes identified in root exudates of different plant species[a]

Amino acids	Organic acids	Sugars	Vitamins	Purines/nucleosides	Enzymes	Inorganic ions and gaseous molecules
α-alanine	citric	glucose	biotin	adenine	acid/alkaline-	HCO_3^-
β-alanine	oxalic	fructose	thiamin	guanine	phosphatase	OH^-
asparagine	malic	galactose	niacin	cytidine	invertase	H^+
aspartate	fumaric	maltose	pantothenate	uridine	amylase	CO_2
cystein	succinic	ribose	rhiboflavin		protease	H_2
cystine	acetic	xylose				
glutamate	butyric	rhamnose				
glycine	valeric	arabinose				
isoleucine	glycolic	raffinose				
leucine	piscidic	desoxyribose				
lysine	formic	oligosaccharides				
methionine	aconitic					
serine	lactic					
threonine	pyruvic					
proline	glutaric					
valine	malonic					
tryptophan	aldonic					
ornithine	erythronic					
histidine	tetronic					
arginine						
homoserine						
phenylalanine						
γ-Aminobutyric acid						
α-Aminoadipic acid						

[a]Compiled from West (1939), Fries and Forsman (1951), Rovira and Harris (1961), Vancura (1964), Vancura and Hovadik (1965), Boutler et al. (1966), Rovira (1969), Gardner et al. (1983), Lipton et al. (1987), Fox and Comerford (1990), Ae et al. (1990), Ohwaki and Hirata (1992), Hoffland et al. (1992) and Gagnon and Ibrahim (1998). The root exudates of plants studied include Bison and Novelty flax, barley, wheat, oat, cucumber, tomato, red pepper, turnip cabbage, pea, soybean, chickpea, peanut, lupin, alfalfa, slash pine, pigeon pea and rape.

teria antagonistic to these pathogenic micro-organisms (Cook et al., 1995). In nutrient poor soils, such rapidly intense colonization of the rhizosphere can itself lead to stiff competition between microbes and the plant for nutrient resources.

In addition to serving as chemotactic signals and growth promoters of rhizosphere bacteria (Table 2), root exudates also control the N nutrition of symbiotic legumes. These nodulating plants routinely use flavonoid molecules in root exudates to induce transcription of nodulation (*nod*) genes in rhizobia, leading to nodule formation and N_2 fixation. The compounds involved are typically flavonoid in nature, although roles for betaines (Phillips et al., 1992) and aldonic acids (Gagnon and Ibrahim, 1998) have also been reported (Table 3). The flavonoid *nod* gene inducers are either exuded directly by roots of N_2-fixing legumes (Phillips, 1992) or by root border cells present in exudates (Zhu et al., 1998). It has been shown that root exudates of legumes growing in an extremely acidic medium cause reduced induction of *nod* genes (Richardson et al., 1988), clearly indicating that the quality of root exudate determines the level of *nod* gene transcription. In fact, legumes which are inhibited in the biosynthesis and exudation of phenolic *nod*-gene inducers, exhibit reduced nodulation and N_2 fixation for their N nutrition (Phillips et al., 1994). However, because high soil N inhibits nodulation and N_2 fixation, and low concentrations of certain nutrients stimulate the biosynthesis of isoflavone *nod* gene inducers in symbiotic legumes (Dakora and Phillips, 1996), low-nutrient soils would seem to be the ideal for stimulating N_2 fixation and increased N nutrition in nodulated legumes. Whatever the case, root exudates contain the signals for transcription of *nod* genes in symbiotic rhizobia, and therefore control the extent of legume dependence on N_2 fixation for its N nutrition.

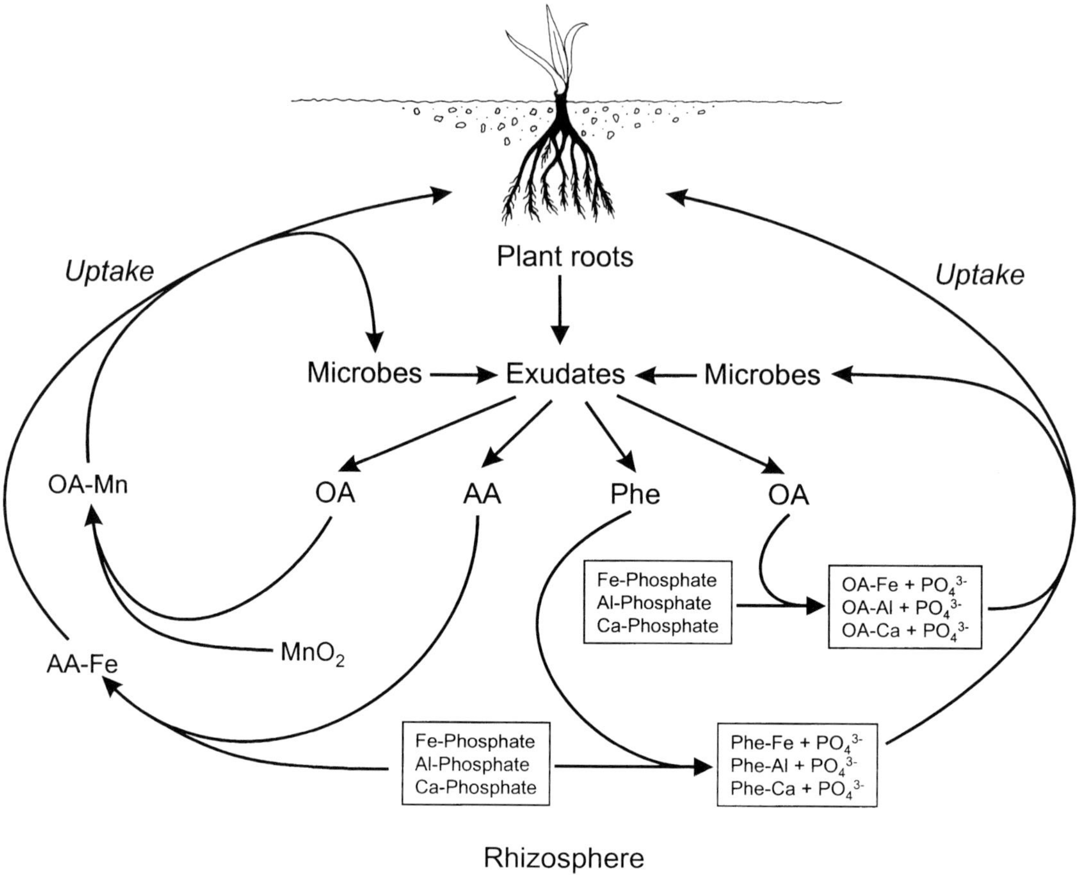

Figure 1. Effects of root exudate components on nutrient availability and uptake by plants and rhizosphere microbes. OA = organic acids; AA = amino acids including phytosiderophores, Phe = phenolic compounds.

An important consideration is the persistence of these signals in soil. Depending on their stability, the activity of individual molecules could be short-lived in the rhizosphere due to rapid utilization as substrates by microbes (Barz, 1970). Also, the efficacy of both aglycones and their conjugates could be altered either following hydrolysis or chemical modification in the rhizosphere (Rao et al., 1991). For example, bacterial hydrolysis of the inactive compound luteolin-7-O-glucoside can produce the active *nod* gene inducer luteolin (Hartwig and Phillips, 1991). However, the compounds released from decomposition of organic matter such as dead roots and nodules or fallen leaves could potentially supplement root exudates in maintaining a steady concentration of flavonoids and mineral nutrients in the rhizosphere.

Root exudates also contain chemical molecules that govern the development of plant-fungal symbiosis (Table 3), as they provide powerful signals that alert the mycorrhizal fungi to the presence of a host plant. The molecules involved are the same group of flavonoid compounds used for signalling rhizobia. They induce spore germination and/or hyphal growth in vesicular-arbuscular fungi (Becard et al., 1992; Gianinazzi-Pearson et al., 1989). In alfalfa root exudates, quercetin and 4′,7-dihydroxyflavone are the specific molecules used to stimulate spore germination and fungal growth in *Glomus* and *Gigaspora* species (Tsai and Phillips, 1991), two mycorrhizal symbionts of this legume. Because of the increased growth of fungal hyphae which extend outwards into the soil environment and the network of plant and fungal nutrient-specific transporters, mycorrhizae promote increased acquisition of water and nutrients, especially P and N, by the host plant (Harrison, 1999). However, its efficacy in supplying nutrients, especially P, to the plant is more marked under nutrient-poor conditions than in fertile soils. The fynbos biome of the Western Cape in South Africa is characterized by P-deficient soils, consequently the dominant field

Table 2. Functional role of root exudate components in the rhizosphere

Component	Rhizosphere function
Phenolics	nutrient source
	chemoattractant signals to microbes
	microbial growth promoters
	nod gene inducers in rhizobia
	nod gene inhibitors in rhizobia
	resistance inducers against phytoalexins
	chelators of poorly soluble mineral nutrients
	detoxifiers of Al
	phytoalexins against soil pathogens
Organic acids	nutrient source
	chemoattractant signals to microbes
	chelators of poorly soluble mineral nutrients
	acidifiers of soil
	detoxifiers of Al
	nod gene inducers
Amino acids and	nutrient source
phytosiderophores	chelators of poorly soluble mineral nutrients
	chemoattractant signals to microbes
Vitamins	promoters of plant and microbial growth
	nutrient source
Purines	nutrient source
Enzymes	catalysts for P release from organic molecules
	biocatalysts for organic matter transformation in soil
Root border cells	produce signals that control mitosis
	produce signals controlling gene expression
	stimulate microbial growth
	release chemoattractants
	synthesize defense molecules for the rhizosphere
	act as decoys that keep root cap infection-free
	release mucilage and proteins

Compiled from Hawes et al. (1998) and Dakora and Phillips (1996).

legumes are infected with mychorrizae to enhance P nutrition (Allsop, 1992).

Solubilization of nutrients by root exudate enzymes and cells

Plant species growing in low-nutrient environments also employ root exudates in ways other than as symbiotic signals to soil microbes involved in nutrient procurement. The activity of extracellular enzymes is one source of available P. Low P concentration in roots as a result of P deficiency induces *de novo* synthesis of extracellular and intracellular acid phosphatases (Duff et al., 1994), followed by an increase in the release of the extracellular phosphatases into root exudates (Dinkelaker and Marschner, 1992; Goldstein et al., 1988; Lee, 1988). These ectoenzymes of root exudates include root-borne acid phosphatase, fungal acid/alkaline phosphatase, and bacterial alkaline phosphatase (Table 1), and they function by hydrolyzing and mobilizing inorganic P from monoester soil organic phosphates (Table 2; Duff et al., 1994), which are estimated to account for about 30–80% of total P in agricultural soils (Gilbert et al., 1999). The role of these phosphatases in P nutrition is evidenced by the fact that P-stressed lupin plants secrete about 20-fold more acid phosphatases from roots compared to P-sufficient control plants (Tadano and Sakai, 1991).

A recent study (Gilbert et al., 1999) with white lupin has also shown that low P concentration in tissues elicits the synthesis and exudation into the rhizosphere of a novel acid phosphatase associated with cluster roots of minus-P plants. This suggests that, in addition to normal acid phosphatases, plants with the ability to form cluster roots release novel acid phosphatases for hydrolyzing P in the vicinity of the proteoid root hairs. Apyrases (enzymes that cleave γ- and β-phosphates on ATP) are another group of new enzymes involved in P nutrition of plants. Apyrases from pea are reported to improve P nutrition of this species through P release from extracellular ATP and other organic phosphate molecules (Thomas et al., 1999) commonly present in the rhizosphere.

In germinating seedlings, phytases hydrolyze phytic acid to release inorganic P for plant utilization (Duff et al., 1994). A recent study (Li et al., 1997) showed a strong correlation between P-deficiency and release of phytases by several plant species. Many other intracellular acid phosphatases exist in plants, with specific and non-specific substrates, but whether these hydrolases are released into root exudates has not yet been reported. The discovery that root border cells release extracellular enzymes, *nod*-gene-inducing compounds, signals controlling gene expression and substrate for mycorrhizal development (Hawes et al., 1998; Zhu et al., 1998) clearly suggest that they may also play a major role in nutrient acquisition in the rhizosphere. However, experiments involving the incubation of root border cells with poorly soluble nutrient compounds are needed to provide direct evidence of this function.

Table 3. Legume root exudate compounds that induce rhizobial *nod* genes and/or VA fungal development

Legume species	Inducer molecule	Reference
Rhizobium symbiosis		
Alfalfa	4,4′-dihydroxy-2′-methoxychalcone	Maxwell et al. (1989)
	4′-7-dihydroxyflavone	
	liquiritigenin	
Cowpea	daidzein	Dakora (2000)
	genistein	
	coumestrol	
Common bean	genistein	Dakora et al. (1993)
	genistein-3-Oglucoside	Hungria et al. (1991)
	eriodictyol	
	naringenin	
	daidzein	Dakora et al. (1993)
	coumestrol	
Kersting's bean	daidzein	Dakora (2000)
	genistein	
	coumestrol	
Soybean	isoliquiritigenin	Kape et al. (1992)
	genistein	Smit et al. (1992)
	genistein-7-O-glucoside	
	genistein-7-O-(6″-O-malonylglucoside)	
	daidzein	
	daidzein-7-O-(6″-O-malonylglucoside)	
Vetch	3,5,7,3′-tetrahydroxy-4′-methoxyflavanone	Recourt et al. (1991)
	7,3′-dihydroxy-4′-methoxyflavanone	
	2′,4′,4-trihydroxychalcone	
	4′,4-dihydroxy-2′-methoxychalcone	
	naringenin	
	liquiritigenin	
	7,4′-dihydroxy-3′-methoxyflavanone	
	5,7,4′-trihydroxy-3′-methoxyflavanone	
	5,7,3′-trihydroxy-4′-methoxyflavanone	
White clover	7,4′-dihydroxyflavone	Djordjevic et al. (1987)
	umbelliferone[a]	
	formononetin[a]	
Bambara groundnut	daidzein	Dakora and Muofhe (1996)
	genistein	
	coumestrol	
Sesbania	liquiritigenin	Messens et al. (1991)
Lupin	erythronic acid	Gagnon and Ibrahim (1998)
	tetronic acid	
VA fungal symbiosis		
Alfalfa	4′,7-dihydroxyflavone	Tsai and Phillips (1991)
	4′,7-dihydroxyflavanone	

[a] *nod* gene inhibitors.

Mobilization of nutrients by root exudate organic compounds

Phenolics are chemical components of root exudates (D'Arcy, 1982; D'Arcy-Lameta, 1986) that solubilize Fe, P and other nutrients from unavailable sources for uptake by plants (Table 2). When challenged with Fe deficiency, plants, especially dicots, release phenolic molecules that influence Fe and P mobility (Römheld, 1987). An example is the release by Fe-deficient alfalfa plants of the isoflavonoid phytoalexin [2-(3′,5′-dihydroxyphenyl)-5,6-dihydroxyben zofuran], which intensely dissolves ferric phosphate, thus making Fe and P available for plant utilization, and also serves as phytoalexin against infecting pathogens (Masaoka et al., 1993). Interestingly, Fe-sufficient plants produced root exudates that were limited in ferric phosphate-dissolving capacity. Fe-deficient tomato plants also exude caffeic acid for solubilizing Fe from insoluble Fe sources (Olsen et al., 1981). These phenolics make Fe and P available by forming relatively stable chelates with Fe and Al present in insoluble Fe- and Al-phosphates thereby increasing the solubility of Fe and P for plant uptake (Figure 1).

A more effective strategy used by Fe-deficient grasses and cereals is the release of phytosiderophores (Table 2), a group of hydroxy- and amino-substituted iminocarboxylic acids, with an ability to solubilize ferric compounds for uptake by roots (Römheld, 1987). Phytosiderophores also solubilize other micronutrients such as Mn, Cu and Zn (Treeby et al., 1989), indicating that phytosiderophore release induced by Fe-deficiency can increase micronutrient concentrations in the rhizosphere. The rate of phytosiderophore exudation can increase by about 20 times in cereal plants confronted with iron deficiency in contrast to an increase in uptake rate of only 5 times (Römheld and Marschner, 1990), suggesting that the release and uptake systems are under different genetic control. Also, there are cultivar, species and genotypic differences in response to iron and micronutrient deficiencies (Römheld and Marschner, 1990), in terms of rates of exudation, chemical nature of phytosiderophores produced, and the size of microbial populations at the apical root zone (Von Wirén et al., 1993). According to Römheld (1991), Fe or Zn deficiency in soil induces *de novo* synthesis of phytosiderophore in the cytoplasm, followed by enhanced release across the plasma membrane into the rhizosphere, where the phytosiderophore forms a complex with the insoluble ferric compound and the entire complex transported into the cytoplasm via a specific translocator in the membrane. H^+ extrusion into the rhizosphere promotes ferric reductase enzyme activity in the plasma membrane, leading to increased reduction of ferric to ferrous ion, which is preferentially absorbed by roots (Römheld, 1987). Apparently, microbial siderophores are also capable of solubilizing ferric compounds in the rhizosphere (Jurkevitch et al., 1986), although they are the least preferred in uptake. Iron is toxic when in excess inside cells. However, unlike in microbes, its regulation in plants still remains unknown. Microbial uptake of Fe is regulated by a chromosomal repressor, which shuts down expression of all components of the siderophore biosynthetic pathway, thus restricting further uptake of Fe into the cell (Neilands, 1987). Given that mainly grasses and cereal crops exude phytosiderophores, different plant species must have different ways of regulating internal Fe concentration than that observed in microbes.

Such Krebs Cycle intermediates as citric, malic, succinic, fumaric, and aconitic acids accumulate markedly in root exudates of different plant species (Table 1) suffering from nutrient starvation (Gardner et al., 1983; Hoffland et al., 1992; Jones, 1998; Lipton et al., 1987; Ohwaki and Hirata, 1992). But acetic acid, glycolic, malonic, oxalic, formic, and piscidic acid have also been identified in root exudates of a number of plants (Ae et al., 1990; Fox and Comerford, 1990; Smith, 1969, 1976; Vancura, 1965). These acids play a crucial role in nutrient acquisition (P, Fe and Mn) by plants growing in low nutrient soils (Table 2; Figure 1), and their release in response to nutrient starvation differs between plant species (Table 4). In one study (Lipton et al., 1987), symptons of P-starvation was exemplified by 82% increase in citrate exudation. However, not all plant species respond the same way to P deficiency (Table 4). While P-deficient rape typically releases malic acid near its root tips or at sites in contact with insoluble rock phosphate, P-starved hedge mustard does not secrete organic acids (Hoffland et al., 1992), possibly due to problems associated with synthesis and transport across membranes. Legume species also respond differently to P deficiency, in terms of exudation rates and the spectrum of organic acids released. For example, chickpea produced 11 and 24 times more root exudates than pigeon pea and soybean, respectively, while peanut produced only 8 and 17 times more than the two species (Ohwaki and Hirata, 1992). Although the concentrations of fumaric, malic and citric acids were also lower in exudates of soybean and pigeon pea

Table 4. Concentration of citric and malic acids in root exudates of P-deficient or P-sufficient plant species. Wheat, tomatoe and chickpea initially received 0.25 mm P, and root exudates collected following a period of P deprivation. For these three species, 0.0 acid concentration ranges from actual zero to negative values relative to the control. Values followed by different letters for *P* treatments of each plant species are significant different at P < 0.05. The other values had standard deviations ranging from 0 to 25%

Plant Species	P supply (mol m^{-3})	Organic acid (μmol h^{-1} g^{-1} fwt)		Reference
		Citrate	Malate	
White lupin	1.0	0.02a	0.03a	Hohnson et al. (1996)
	0.0	0.57b	0.51b	
White lupin	0.25	0.02a	0.08a	Neumann et al. (1999)
	0.0	0.09b	0.61a	
Wheat	0.25	0.0	0.25	Neumann and Römheld (1999)
	0.0	0.0	0.0	
Tomato	0.25	0.13	0.14	Neumann and Römheld (1999)
	0.0	0.0	0.0	
Chickpea	0.25	0.001	0.001	Neumann and Römheld (1999)
	0.0	0.0	0.002	

(Ohwaki and Hirata. 1992), the latter is very efficient at P acquisition in acidic soils and apparently uses piscidic acid, a strong chelator of iron, to mobilize P (Ae et al., 1990). The organic acid anions in root exudates can also chelate Fe and Mn in iron and manganese oxides (i.e. Fe_2O_3 and MnO_2), thus making them available for plant uptake (Figure 1; Marschner, 1995). But whether the chelated metal is taken up is still open to debate. Similarly, these acid anions form complexes with Ca, Al and Fe present in soil as insoluble phosphates of calcium, iron and aluminium, and liberate P for uptake by roots (Figure 1; Marschner, 1995). Additionally, these acids can desorb P from sesquioxide surfaces by anion exchange (Bolan et al., 1994; Jones, 1998; Jones and Darrah, 1994; Parfitt, 1979), and also maintain sulphate mobility in rhizosphere soil through competitive displacement from adsorption sites (Evans and Anderson, 1990).

In some plants, the synthesis and release of organic acids represents a mechanism for detoxifying excess Al internally and externally via chelation of the active Al into a non-phytotoxic form. A number of studies show that in response to Al stress: (1) roots of Al-resistant snapbean release higher levels of citric acid into the rhizosphere than the Al-sensitive cultivar (Miyasaka et al., 1991), (2) Al-resistant wheat exuded 5–10-fold more malic acid than Al-sensitive genotypes (Delhaize et al., 1993), (3) citric acid exudation was enhanced in Al-resistant maize (Pellet et al., 1995) and Al-resistant buckwheat (Ma et al., 1998; Zheng et al., 1998). In these Al-resistant plants, there was a correlation between organic acid release and resistance to Al, and substituting other elements for Al did not elicit organic acid exudation (Delhaize et al., 1993). In buckwheat, oxalic acid forms an Al-oxalate complex, which renders the Al non-phytotoxic to the plant (Ma et al., 1998), suggesting that this might be the mechanism employed by plants to detoxify active Al species.

Root exudate effects on rhizosphere pH and nutrient availability

Another nutritional effect that organic acids have in root exudates is acidification of the rhizosphere (Table 2). Root exudation of high concentrations of organic acid anions as a result of P deficiency (Hoffland et al., 1989) does lower rhizosphere pH, making P (Haynes, 1990; Jones and Darrah, 1994) and micronutrients such as Mn, Fe and Zn to be more available in calcareous soils (Dinkelaker et al., 1989). However, the relationship between organic acid exudation and rhizosphere acidification is not that simple as the extrusion of H^+ would depend on the amounts of anions absorbed by roots relative to cations (Haynes, 1990; Jones and Darrah, 1994). Whatever the case, acidification below pH 5.5 can cause even major macronutrients to become limiting. Because micronutrients such as Mn, Fe and Al occur in high concentrations below pH 5.5 (Brady, 1990), any further acidification by organic acids below this level can result in phytotoxic effects on plant roots and beneficial microbes. Intriguingly, the white lupin can mobilize P from both acid and alkaline soils by using citric acid in its proteoid root exudates to acidify even the alkaline soil, and thus solubilize P as well as Fe, Mn, Cu and Zn for uptake by roots (Gardner et al., 1983).

Root excretion of inorganic ions (e.g. HCO_3^-, OH^-, H^+) is also important in the mineral nutrition of plants. Plant uptake of anions in excess of cations often causes the roots to secrete HCO_3^- in order to maintain electrical neutrality, a process that leads to increased rhizosphere pH. Conversely, the uptake of cations in excess of anions can cause roots to exude H^+ and lower the rhizosphere pH. The change in pH with HCO_3^- extrusion tends to increase nutrient supply in acidic soils, as happens with H^+ exudation in calcareous soils.

H^+ extrusion is largely governed by cation/anion balance and is thus very much influenced by the N source. Plants growing on nitrate generally maintain electronic neutrality by releasing excess anions, including OH^-, which cause an increase in rhizosphere pH and an enhanced nutrient (P, Mo, etc.) availability in acidic soils (Marschner, 1995). An exception is the study by Gahoonia et al. (1992) which showed no effect on P mobilization in a luvisol soil following nitrate-induced increase in rhizosphere pH. However, when ammonium is supplied, there is a large excess of cations and a resulting enhanced H^+ extrusion and rhizosphere acidification, thereby making P, Mn, Fe, Cu and Zn more toxic in the acidic range, or readily available in the alkaline range (Aguilar and Van Diest, 1981; Gahoonia et al., 1992; Gardner et al., 1983; Gillespie and Pope, 1990; Runge and Rode, 1991). Sometimes, however, acidification from ammonium nutrition does not result in increased P mobilization, especially in acidic oxisols (Gahoonia et al., 1992). Furthermore, H^+ extrusion occurs during N_2 fixation by symbiotic legumes (Raven et al., 1990), and this can lead to rhizosphere acidification and increased availability of limiting nutrient elements like P, Mo and Fe (Aguilar and Van Diest, 1981; Gahoonia et al., 1992; Gillespie and Pope, 1990; Runge and Rode, 1991) which are much needed in diazotrophy. Additionally, there are many reports of enhanced H^+ extrusion under Fe defiency and P deficiency, both leading to acidification of localized areas around the root tips (Bienfait, 1985, 1988; Gardner et al., 1983; Hoffland et al., 1989; Römheld and Marschner, 1986) and a consequent improvement in the availability of these nutrients. Higher cation/anion uptake ratios as occurs during Zn deficiency (Cakmak and Marschner, 1990) can also acidifiy the rhizosphere and influence nutrient uptake by plants.

Environments which are naturally very acidic can pose a challenge to nutrient acquisition by plant roots, and threaten the survival of many beneficial microbes and that of the roots themselves (Runge and Rode, 1991). Under those conditions, rhizobia develop tolerance of low pH through expression of acid tolerance genes (Tiwari et al., 1996). As found with a transgenic plant (Degenhardt et al., 1998), a few species such as the Rooibos tea plant (*Aspalathus linearis*), actively modify their rhizosphere pH by extruding OH^- and HCO_3^- (Muofhe and Dakora, 2000) to facilitate growth in low pH soils (pH 3–5). By raising the pH of the rhizosphere, major nutrients such as K, Ca, Mg, P, S and Mo that are low in Cederberg soils (Muofhe, 1997), become readily available for uptake by plant roots. With a favourable pH, soil N levels can also rise from increased activity of nitrifying bacteria. However, an elevation in rhizosphere pH caused by *Aspalathus linearis* plants growing in very acidic soils can also be a strategy for reducing the toxicity of Al, Mn and Fe (Runge and Rode, 1991). Whereas net alkalinization of the rhizosphere by nitrate assimilation in plants is understood, the mechanisms underlying rhizosphere alkalinization by root exudates of *Aspalathus linearis* are only now being clarified.

Increased soil acidification is generally accompanied by Al toxicity as both the concentration and activity of this element are increased with a lowering of pH in the rhizosphere (Gahoonia, 1993) often resulting in reduced Ca and Mg uptake and root growth (Kochian, 1995; Ryan et al., 1992). One of the ways for reducing Al toxicity is to increase rhizosphere pH. The findings of a recent study (Degenhardt et al., 1998) involving the use of transgenic plants revealed a two-fold increase in H^+ influx, which resulted in increased rhizosphere pH. So, in addition to internal or external Al detoxification by organic acids (Larsen et al., 1998; Zheng et al., 1998), an increase in rhizosphere pH can also reduce Al toxicity. Plant genotypes however differ in their handling of excess Al. Calba and Jaillard (1997) showed that a maize genotype that was most tolerant of Al toxicity caused a lesser rhizosphere acidification because NO_3^- uptake was less affected than for the Al-sensitive genotype.

Gaseous root exudates affecting nutrient supply

Root excretion of gaseous molecules (e.g. CO_2, H_2) is also important in the mineral nutrition of plants. Actively-growing young roots release a lot of CO_2 into the soil environment from carbohydrate respiration (Zwarun, 1972) and stimulation by lumichrome, a plant and bacterial exudate molecule (Phillips et al., 1999). If its escape is prevented by high soil water, CO_2 can accumulate up to 17.5% in the root zone (Jackson, 1979). Such increased levels of CO_2 (and hence of carbonic acid which dissociates to form H^+ at neutral to alkaline pH) can enhance the dissolution of soil $CaCO_3$ to produce Ca^{2+} for plant uptake. Similarly, HCO_3^- excreted directly by plant roots into soil can aid the dissolution of calcite to yield soluble supplies of Ca^{2+} for plant nutrition (Rendig and Taylor, 1989). CO_2 release by roots can also have other indirect effects on N and P nutrition of plants that form symbiosis with *Rhizobium* bacteria and VA fungi. An exogenous supply of CO_2 is required for better growth of rhizobia and VA fungi in the rhizosphere (Becard et al., 1992; Lowe and Evans, 1962); the increase in microsymbiont population should have the potential to provide adequate symbiotic N and P for the host plant.

Hydrogen gas (H_2) is a major byproduct of N_2 fixation in legumes, and its production consumes about 5% of net photosynthesis. In rhizobia with uptake hydrogenase (HUP^+) system, H_2 produced by nitrogenase is oxidised to yield more energy. However, with the HUP^- strains, the H_2 is evolved as a gas into the rhizosphere environment. Although this was previously thought to be wasteful of C, a recent study (Kettlewell et al., 2000) has shown that pre-treating soil with H_2 can stimulate growth of agricultural crops by 10–30%. In the H_2-treated soils, 60% of the reducing power was found to flow to O_2 and 40% to CO_2, resulting in net chemoautrotrophic CO_2 fixation. Although the specific elicitor of plant growth promotion remains unknown, nodule release of H_2 does affect the growth and nutrition of symbiotic legumes.

Beyond their role in mineral acquisition, root exudates also appear to have potential in phytoremediation. A number of studies (Banuelos et al., 1993; De Souzaet al., 1999; Mench and Martin, 1991; Nanda Kumar et al., 1995; Terry et al., 1992; Zayed et al., 1998) have shown the ability of certain plant species to accumulate and release heavy metals as volatile root exudates, thus reducing their concentrations in the soil environment.

Conclusions

Through diverse mechanisms, root exudates play a fundamental role in the mineral nutrition of plants. They either contain signals that act as regulators of microbial growth and function, or they possess molecules which directly control the rhizosphere processes that enhance nutrient uptake and assimilation (Figure 1). No doubt, manipulating some or all of these rhizosphere processes is likely to hold the key to improved plant nutrition, and hopefully increased crop yields in the next millenium.

Acknowledgements

This study was conducted with grants awarded by the National Research Foundation, Pretoria, and the URC, University of Cape Town (to FDD), and with financial assistance from the US Department of Agriculture National Research Initiative Competitive Grants Program and from BARD, the US-Israel Binational Agricultural Research & Development Fund (to DAP).

References

Ae N, Arihara J, Okada K, Yoshihara T and Johansen C 1990 Phosphorus uptake by pigeon pea and its role in cropping systems of the Indian subcontinent. Science 248, 477–480.

Aguilar A S and Van Diest A 1981 Rock-phosphate mobilization induced by the alkaline uptake pattern of legumes utilizing symbiotically fixed nitrogen. Plant Soil 61, 27–42.

Allsop N 1992 The Occurrence and Ecophysiology of Vesicular-Arbuscular Mycorrhizal Plants in the Cape Floristic Region. PhD Thesis, Botany Department, University of Cape Town.

Arshad M and Frankenberger W T Jr. 1990 Ethylene accumulation in soil in response to organic amendments. Soil Sci. Soc. Am. J. 54, 1026–1031.

Banuelos G S, Cardon G, Mackey B, Ben-Asher J, Wu L, Beuselinck P, Akohouse S and Zambrzuski S 1993 Boron and selenium removal in boron laden soils by four sprinkler irrigated plant species. J. Environ. Qual. 22, 786–792.

Barbour W M, Hattermann D R and Stacey G 1991 Chemotaxis of *Bradyrhizobium japonicum* to soybean exudates. Appl. Environ. Microbiol. 57, 2635–2639.

Barz W 1970 Isolation of rhizosphere bacterium capable of degrading flavonoids. Phytochemistry 9, 1745–1749.

Becard G, Douds D D and Pfeffer P E 1992 Extensive *in vitro* hyphal growth of vesicular-arbuscular mychorrhizal fungi in the presence of CO_2 and flavonols. Appl. Environ. Microbiol. 58, 821–825.

Bienfait H F 1985 Regulated redox processes at the plasmalemma of plant root cells and their function in iron uptake. J. Bioenerg. Biomembr. 14, 191–205.

Bienfait H F 1988 Mechanisms in Fe efficiency reactions of higher plants. J. Plant Nutr. 11, 605–629.

Bolan N S, Naidu R, Mahimairaja S and Baskaran S 1994 Influence of low-molecular-weight organic acids on the solubilization of phosphates. Biol. Fert. Soils 18, 311–319.

Boutler D, Jeremy J J and Wilding M 1966 Amino acids liberated into the culture medium by pea seedling roots. Plant Soil 24, 121–127.

Brady N C 1990 The Nature and Properties of Soils, Tenth Edition. Macmillan Publishing Company, New York. 621 p.

Caetano-Anolles G, Crist-Estes D K and Bauer D W 1988 Chemotaxis of *Rhizobium meliloti* to the plant flavone luteolin requires functional nodulation genes. J. Bacteriol. 170, 3164–3169.

Cakmak I and Marschner H 1990 Decrease in nitrate uptake and increase in proton release in zinc deficient cotton, sunflower and buckwheat plants. Plant Soil 129, 261–268.

Calba H and Jaillard B 1997 Effect of aluminium on ion uptake and H^+ release by maize. New Phytol. 137, 607–616.

Cook R J, Thomashow L S, Weller D M, Fujimoto D, Mazzola M, Bangera G and Kim D 1995 Molecular mechanisms of defence by rhizobacteria against root disease. Proc. Natl. Acad. Sci. USA 92, 4197–4201.

Dakora F D 2000 Commonality of root nodulation signals and nitrogen assimilation in tropical grain legumes belonging to the tribe Phaseoleae. Aust. J. Plant Physiol. 27, 885–892.

Dakora F D, Joseph C M and Phillips D A 1993 Common bean root exudates contain elevated levels of daidzein and coumestrol in response to *Rhizobium* inoculation. Mol. Plant-Microbe Interact. 6, 665–668.

Dakora F D and Phillips D A 1996 Diverse functions of isoflavonoids in legumes transcend anti-microbial definitions of phytoalexins. Physiol. Mol. Plant Pathol. 49, 1–20.

Dakora F D and Muofhe M L 1996 Molecular signals involved in nodulation of the African Bambara groundnut (*Vigna subterranea* L. Verdc.). *In* Proceedings of the International Bambara groundnut Symposium, University of Nottingham, UK, 23–25 July 1996. pp. 171–179.

Darrah P R 1993 The rhizosphere and plant nutrition: A quantitative approach. Plant Soil 155/156, 1–20.

D'Arcy A L 1982 Study of soja and lentil exudates I. Kinetics of exudation of phenolic compounds, amino acids and sugars in the first days of plant growth. Plant Soil 68, 399–403.

D'Arcy-Lameta A 1986 Study of soybean and lentil root exudates. II. Identification of some polyphenolic compounds, relation with plantlet physiology. Plant Soil 92, 113–123.

Degenhardt J, Larsen P B, Howell S H and Kochian L V 1998 Aluminium resistance in the *Arabidopsis* mutant *alr*-104 is caused by an aluminium-induced increase in pH. Plant Physiol. 117, 19–27.

Delhaize E, Ryan P R and Randall P J 1993 Aluminium tolerance in wheat (*Triticum aestivum* L.) II. Aluminium stimulated excretion of malic acid from root apices. Plant Physiol. 103, 695–702.

De Souza M P, Chu D, Zhao M, Zayed A M, Ruzin S E, Schichnes D and Terry N 1999 Rhizosphere bacteria enhance selenium accumulation and volatilization by Indian mustard. Plant Physiol. 119, 565–573.

Dinkelaker B, Römheld V and Marschner H 1989 Citric acid excretion and precipitation of calcium citrate in the rhizosphere of white lupin (*Lupinus albus* L). Plant Cell Physiol. 12, 285–292.

Dinkelaker B and Marschner H 1992 *In vivo* demonstration of acid phosphatase activity in the rhizosphere of soil-grown plants. Plant Soil 144, 199–205.

Djordjevic M A, Redmond J W, Batley M and Rolfe B G 1987 Clovers secrete specific phenolic compounds which either stimulate or repress *nod* gene expression in *Rhizobium trifolii*. EMBO J. 6, 1173–1179.

Duff S M G, Sarath G and Plaxton W C 1994 The role of acid phosphatases in plant phosphorus metabolism. Physiol. Plant. 90, 791–800.

Evans A Jnr. and Anderson T J 1990 Aliphatic acids: influence on sulphate mobility in a forested cecil soil. Soil Sci. Soc. Am. J. 54, 1136–1139.

Fox T R and Comerford N B 1990 Low-molecular weight organic acids in selected forest soils of the southeastern USA. Soil Sci. Soc. Am. J. 54, 1139–1144.

Fries N and Forsman B 1951 Quantitative determination of certain nucleic acid derivatives in pea root exudate. Physiol. Plant. 4, 210–234.

Gagnon H and Ibrahim R K 1998 Aldonic acids: A novel family of *nod* gene inducers of *Mesorhizobium loti*, *Rhizobium lupini* and *Sinorhizobium meliloti*. Mol. Plant- Microbe Interact. 11, 988–998.

Gahoonia T S 1993 Influence of root-induced pH on the solubility of soil aluminium in the rhizosphere. Plant Soil 149, 289–291.

Gahoonia T S, Claassen N and Jungk A 1992 Mobilization of phosphate in different soils by ryegrass supplied with ammonium or nitrate. Plant Soil 140, 241–248.

Gardner W K, Barber D A and Parbery D G 1983 The acquisition of phosphorus by *Lupinus albus* L. III. The probabale mechanism by which phosphorus movement in the soil/root interface is enhanced. Plant Soil 70, 107–124.

Gianinazzi-Pearson V, Branzanti B and Gianinazzi S 1989 *In vitro* enhancement of spore germination and early hyphal growth of a vesicular-arbuscular mycorrhizal fungus by host root exudates and plant flavonoids. Symbiosis 7, 243–255.

Gilbert G A, Knight J D, Allan D L and Vance C P 1999 Acid phosphatase activity in phosphorus-deficient white lupin roots. Plant Cell Environ. 22, 801–810.

Gillespie A R and Pope P E 1990 Rhizosphere acidification increases phosphorus recovery of Black locust: I. Induced acidification and soil response. Soil Sci. Soc. Am. J. 54, 533–537.

Goldstein A H, Baertlein D A and McDaniel R G 1988 Phosphate starvation inducible metabolism in *Lycopersicon esculentum*. I. Excretion of acid phosphatase by tomato plants and suspension-cultured cells. Plant Physiol. 87, 711–715.

Harrison M J 1999 Molecular and cellular aspects of the arbuscular mycorrhizal symbiosis. Annu. Rev. Plant Physiol. Plant Mol. Biol. 50, 361–389.

Hartwig U A, Joseph C M and Phillips D A 1991 Flavonoids released naturally from alfalfa seeds enhance growth of *Rhizobium meliloti*. Plant Physiol. 95, 797–803.

Hartwig U A and Phillips D A 1991 Release and modification of *nod*-gene-inducing flavonoids from alfalfa seeds. Plant Physiol. 95, 804–807.

Hawes M C, Brigham L A, Wen F, Woo H H and Zhu Y 1998 Function of root border cells in plant health: Pioneers in the rhizosphere. Annu. Rev. Phytopathol. 36, 311–327.

Haynes R J 1990 Active ion uptake and maintenance of cation-anion balance: A critical examination of their role in regulating rhizosphere pH. Plant Soil 126, 247–264.

Hoffland E, Findenegg G R and Nelemans J A 1989 Solubilization of rock phosphate by rape. II Local root exudation of organic acids as a response to P-starvation. Plant Soil 113, 161–165.

Hoffland E, Van Den Boogaard R, Nelemans J and Findenegg G 1992 Biosynthesis and root exudation of citric and malic acids in phosphate-starved rape plants. New Phytol. 122, 675–680.

Hungria M, Joseph C M and Phillips D A 1991 *Rhizobium nod*-gene inducers exuded naturally from roots of common bean (*Phaseolus vulgaris* L.). Plant Physiol. 97, 759–764.

Hsiao T C and Lauchli A 1986 Role of potassium in plant-water relations. *In* Advances in Plant Nutrition Volume 2. Eds. B T inker and A Lauchli. pp. 281–312. Praeger Publishers, New York, USA.

Jackson M B 1979 Rapid injury to peas by soil waterlogging. J. Sci. Food Agric. 30, 143–152.

Jones D L 1998 Organic acids in the rhizosphere – a critical review. Plant Soil 205, 25–44.

Jones D L and Darrah P R 1994 Role of root derived organic acids in the mobilization of nutrients from the rhizosphere. Plant Soil 166, 247–257.

Jurkevitch E, Hadar Y and Chen Y 1986 The remedy of lime-induced chlorosis in peanuts by *Pseudomonas* sp. siderophores. J. Plant Nutr. 9, 535–545.

Kettlewell B, Dong Z, Wu L and Layzell D 2000 Why do legume nodules evolve H_2 gas? Abstract C10, 17th North American Conference on Symbiotic Nitrogen Fixation, 23–28 July 2000, Quebec City, Canada.

Kochian L V 1995 Cellular mechanisms of aluminium toxicity and resistance in plants. Annu. Rev. Plant Physiol. Plant Mol. Biol. 46, 237–260.

Larsen P B, Degenhadrt J, Tai C-Y, Stenzler L M, Howell S H and Kochian L V 1998 Aluminium-resistant Arabidopsis mutants that exhibit altered patterns of aluminium accumulation and organic acid release from roots. Plant Physiol. 117, 9–18.

Lee R B 1988 Phosphate influx and extracellular phosphatase activity in barley roots and rose cells. New Phytol. 109, 141–148.

Lehninger A L 1970 Biochemistry. Worth Publishers, 70 Fifth Avenue, New York.

Li M, Osaki M, Rao I M and Tadano T 1997 Secretion of phytase from the roots of sevaral plant species under phosphorus-deficient conditions. Plant Soil 195, 161–169.

Lipton D S, Blanchar R W and Blevins D G 1987 Citrate, malate, and succinate concentration in exudates from P-sufficient and P-stressed *Medicago sativa* L. seedlings. Plant Physiol. 85, 315–317.

Lowe R H and Evans H J 1962 Carbon dioxide requirement for growth of legume nodule bacteria. Soil Sci 94, 351–356.

Ma J F, Hiradate S and Matsumoto 1998 High aluminium resistance in buckwheat. II. Oxalic acid detoxifies aluminium internally. Plant Physiol. 117, 753–759.

Masaoka Y, Kojima M, Sugihara S, Yoshihara T, Koshino M and Ichihara A 1993 Dissolution of ferric phosphate by alfalfa (*Medicago sativa* L.) root exudates. Plant Soil 155/156, 75–78.

Marschner H 1995 Mineral Nutrition of Higher Plants. 2nd Edn. Academic Press, London. 889 pp.

Maxwell C A, Hartwig U A, Joseph C M and Phillips D A 1989 A chalcone and two related flavonoids released from alfalfa roots induce *nod* genes in *Rhizobium meliloti*. Plant Physiol. 91, 842–847.

Mench M and Martin E 1991 Mobilization of cadmium and other metals from two soils by root exudates of *Zea mays* L., *Nicotiana tabacum* L. and *Nicotiana rustica* L. Plant Soil 132, 187–196.

Messens E, Geelen D, Montagu M V and Holsters M 1991 7,4′-dihydroxyflavanone is the major *Azorhizobium nod* gene-inducing factor in *Sesbania rostrata* seedling exudate. Mol. Plant-Microbe Interact. 4, 262–267.

Miyasaka S C, Buta J G, Howell R K and Foy C D 1991 Mechanisms of aluminium tolerance in snapbeans. Root exudation of citric acid. Plant Physiol. 96, 737–743.

Moorby J and Besford R T 1983 Mineral nutrition and growth. *In* Encyclopedia of Plant Physiology New Series Vol. 15 B: Iorganic Plant Nutrition. Eds. A Laüchli and RL Bieleski. pp. 481–515. Springer-Verlag, Berlin.

Morris P F and Ward E W B 1992 Chemoattraction of zoospores of the soybean pathogen, *Phytophthora sojae*, by isoflavones. Physiol. Mol. Plant Pathol. 40, 17–22.

Muofhe M L 1997 N_2 Fixation and Rhizosphere Ecology of *Aspalathus linearis* subsp. *linearis* (Rooibos tea). MSc Thesis, Botany Department, University of Cape Town.

Muofhe M L and Dakora F D 2000 Modification of rhizosphere pH by the symbiotic legume *Aspalathus linearis* growing in a sandy acidic soil. Aust. J. Plant Physiol. 27, 1169–1173.

Nanda Kumar P B A, Dushenkov V, Motto H and Raskin I 1995 Phytoextraction: The use of plants to remove heavy metals from soils. Environ. Sci. Technol. 29, 1232–1238.

Neilands J B 1986 Biochemistry and genetics of microbial iron assimilation. *In* Biotechnology for Solving Agricultural Problems. Eds. PC Augustine, HD Danforth and MR Bakst. pp. 195–206. Martinus Nijhoff Publishers, Dordrecht, The Netherlands.

Neumann G and Römheld V 1999 Root excretion of carboxylic acids and protons in phosphorus-deficient plants. Plant Soil 211, 121–130.

Neumann G, Massonneau A, Martinoia E and Römheld V 1999 Physiological adaptations to phosphorus deficiency during proteoid root development in while lupin. Planta 208, 373–382.

Ohwaki Y and Hirata H 1992 Differences in carboxylic acid exudation among P-starved leguminous crops in relation to carboxylic acid contents in plant tissues and phospholipid level in roots. Soil Sci. Plant Nutr. 38, 235–243.

Olsen R A, Bennett J H, Blume D and Brown J C 1981 Chemical aspects of the Fe stress response mechanism in tomatoes. J. Plant Nutr. 3, 905–921.

Parfitt R L 1979 The availability of P from phosphate-geothite bridging complexes. Description and uptake by ryegrass. Plant Soil 53, 55–65.

Parke D, Rivelli M and Ornston 1985 Chemotaxis to aromatic and hydroaromatic acids: Comparison of *Bradyrhizobium japonicum* and *Rhizobium trifolii*. J. Bacteriol. 163, 417–422.

Pellet D M, Grunes D L and Kochian L V 1995 Organic acid exudation as an aluminium tolerance mechanism in maize (*Zea mays* L.) Planta 196, 788–795.

Phillips D A 1992 Flavonoids: Plant signals to soil microbes. *In* Phenolic Metabolism in Plants. Eds. HA Stafford and RK Ibrahim. 26, 201–231. Plenum Press, New York. Rec. Adv. Phytochem.

Phillips D A, Joseph C M and Maxwell C A 1992 Trigonelline and stachydrine released from alfalfa seeds activate NodD2 protein in *Rhizobium meliloti*. Plant Physiol. 99, 1526–1531.

Phillips D A, Dakora F D, Sande E, Joseph C M and Zon J 1994 Synthesis, release and transmission of alfalfa signals to rhizobial symbionts. Plant Soil 161, 69–80.

Phillips D A, Joseph C M, Yang G-P, Martinez-Romero E, Sanborn J R and Volpin H 1999 Identification of lumichrome as a *Sinorhizobium* enhancer of alfalfa root respiration and shoot growth. Proc. Natl. Acad. Sci. (USA) 22, 12275–12280.

Plaxton W C 1996 The organization and regulation of plant glycolysis. Annu. Rev. Plant Physiol. Plant Mol. Biol. 47, 185–214.

Rao J R, Sharma N D, Hamilton J T G, Boyd D R and Cooper J E 1991 Biotransformation of the pentahydroxy flavone quercetin by *Rhizobium loti* and *Bradyrhizobium* strains (*Lotus*). Appl. Environ. Microbiol. 57, 1563–1565.

Raven J A, Franco A A, De Jesus E L and Jacob-Neto J 1990 H^+ extrusion and organic-acid synthesis in N_2-fixing symbioses involving vascular plants. New Phytol. 114, 369–389.

Raz V and Fluhr R 1992 Calcium requirement for ethylene-dependent responses. Plant Cell 4, 1123–1130.

Recourt K, Schripsema J, Kijne J W, Van Brussel A A N and Lugtenberg B J J 1991 Inoculation of *Vicia sativa* subsp. *nigra* roots with *Rhizobium leguminosarum* biovar *viciae* results in release of *nod* gene activating flavanones and chalcones. Plant Mol. Biol. 16, 841–852.

Rendig V V and Taylor H M 1989 Principles of Soil-Plant Interrelationships. McGraw-Hill Publishing Company, New York, USA.

Richardson A E, Djordjevic M A, Rolfe B G and Simpson R J 1988 Effects of pH, Ca and Al on the exudation from clover seedlings of compounds that induce the expression of nodulation genes in *Rhizobium trifolii*. Plant Soil 109, 37–47.

Römheld V 1987 Different strategies for iron acquisition in higher plants. Physiol. Plant. 70, 231–234.

Römheld V 1991 The role of phytosiderophores in acquisition of iron and other micronutrients in graminaceous species: An ecological approach. Plant Soil 130, 127–134.

Römheld V and Marschner H 1986 Mobilization of iron in rhizosphere of different plant species. *In* Advances in Plant Nutrition, Volume 2. Eds. PBH Tinker and A Laüchli. pp. 155–204.

Praeger Publishers, New York, USA. Römheld V and Marschner H 1990 Genotypical differences among graminaceous species in release of phytosiderophores and uptake of iron phytosiderophores. Plant Soil 123, 147–153.

Rovira A D 1969 Plant root exudates. Bot. Rev. 35, 35–57.

Rovira A D and Harris J R 1961 Plant root excretions in relation to the rhizosphere effect. V. The exudation of B-group vitamins. Plant Soil 14, 199–214.

Runge M and Rode M W 1991 Effects of soil acidity on plant associations. *In* Soil Acidity. Eds. B Ulrich and ME Sumner. pp. 183–202. Springer-Verlag, Berlin.

Ryan P R, Shaff J E and Kochian L V 1992 Aluminium toxicity in roots. Correlation among ionic currents, ion fluxes and root elongation in aluminium-sensitive and aluminium- tolerant wheat cultivars. Plant Physiol. 99, 1193–1200.

Siqueira J O, Safir G R and Nair M G 1991 Stimulation of vesicular-arbuscular mycorrhiza formation and growth of white clover by flavonoid compounds. New Phytol. 118, 87–93.

Smit G, Puvanesarajah V, Carlson R W, Barbour M and Stacey G 1992 *Bradyrhizobium japonicum nod* D1 can be specifically induced by soybean flavonoids that do not induce the *nodYABCSUIJ* operon. J. Biol. Chem. 267, 310–318.

Tadano T and Sakai H 1991 Secretion of acid phosphatase by the roots of several crop species under phosphorus-deficient conditions. Soil Sci. Plant Nutr. 37, 129–140.

Terry N, Carlson C, Raab T K and Zayed A M 1992 Rates of selenium volatilization among crop species. J. Environ. Qual. 21, 341–344.

Thomas C, Sun Y, Naus K, Lloyd A and Roux S 1999 Apyrase functions in plant phosphate nutrition and mobilizes phosphate from extracellular ATP. Plant Physiol. 119, 543–551.

Tiwari R P, Reeve W G, Dilworth M J and Glenn A R 1996 An essential role for *act*A in acid tolerance of *Rhizobium meliloti*. Microbiology 142, 601–610.

Treeby M, Marschner H and Römheld V 1989 Mobilization of iron and other micronutrient cations from calcareous soil by plant-borne, microbial and synthetic metal chelators. Plant Soil 114, 217–226.

Tsai S M and Phillips D A 1991 Flavonoids releasaed naturally from alfalfa promote development of symbiotic *Glomus* spores *in vitro*. Appl. Environ. Microbiol. 57, 1485–1488.

Uren N C and Reisenauer H M 1988 The role of root exudates in nutrient acquisition. Adv Plant Nutr. 3, 79–114.

Vancura V and Hovadik A 1965 Root exudates of plants. II. Composition of root exudates of some vegetables. Plant Soil 22, 21–32.

Vancura V 1964 Root exudates of plants. I. Analysis of root exudates of barley and wheat in their initial phases of growth. Plant Soil 21, 231–248.

Von Wirén N, Römheld V, Morel J L and Guckert A 1993 Influence of micro-organisms on iron acquisition in maize. Soil Biol. Biochem. 25, 371–376.

West P M 1939 Excretion of thiamin and biotin by the roots of higher plants. Nature 144, 1050–1051.

Zayed A, Lytle C M and Terry N 1998 Accumulation and volatilization of different chemical species of selenium by plants. Planta 206, 284–292.

Zheng S J, Ma J F and Matsumoto H 1998 High aluminium resistance in buckwheat. I. Al-induced specific secretion of oxalic acid from root tips. Plant Physiol. 117, 745–751.

Zhu Y, Pierson L S III and Hawes M C 1998 Release of nodulation gene inducing chemicals from root border cells. Plant Physiol. 115, 1691–1698.

Zwarun A A 1972 Microbial competition for glucose in excised root experiments. Soil Sci. Soc. Am. Proc. 36, 968–969.

J.J. Adu-Gyamfi (Ed.), *Food security in nutrient-stressed environments: exploiting plants' genetic capabilities:* 215–226, 2002.

Genetic control of root exudation

Z. Rengel
Soil Science and Plant Nutrition, Faculty of Agriculture, The University of Western Australia, Nedlands WA 6907, Australia

Key words: aluminium toxicity, citrate synthase, *nod*-gene inducers, nutrient deficiency, phosphatase, phytase, phytosiderophores, root exudation

Abstract

The literature on genetics of root exudation and on genotypic differences in qualitative and quantitative composition of root exudates in crop and native plant species was critically assessed. Differences in exudation have been reported for genotypes that differ in tolerance to nutrient deficiencies, ion toxicities, and pathogen attack. The exudation profile of a limited number of genotypes (frequently only two genotypes with the contrasting response to the environmental stress) have been reported to date. Little is known about the variability in larger samples of the germplasm or about actual genetics behind differential qualitative and quantitative composition of root exudates. Changing the exudation profile of a given genotype may be achieved by manipulating the biosynthetic capacity and by increasing the capacity of the plasma membrane to transport the specific compound out into the rhizosphere. Overexpression of the bacterial citrate synthase gene in the cytoplasm of tobacco plants resulted in exudation of large quantities of citrate into the rhizosphere and partial alleviation of the aluminium (Al) toxicity stress. A similar strategy of transforming plants with citrate synthase gene is being tried as a way of improving plant capacity to extract phosphorus (P) from soils with notoriously low P availability.

More research into the genetic basis of qualitative and quantitative differences in root exudation is warranted. Understanding the genetic control of root exudation, followed by manipulation of qualitative and quantitative composition of root exudates, will result in better adaptation of plants to environmental conditions and a greater yield of crops.

Introduction

Plants exude a variety of organic compounds (carboxylate anions, phenolics, carbohydrates, amino acids, enzymes, other proteins, etc.) and inorganic ions (protons, phosphate, other nutrients, etc.) into the rhizosphere to change the chemistry and biology of the rhizosphere and enhance adaptation to a particular environment (Crowley and Rengel, 1999). Common conditions under which plants increase exudation of organic compounds into the rhizosphere encompass deficiency of various nutrients [e.g., phosphorus (P), iron (Fe), zinc (Zn), manganese (Mn); Azaizeh et al., 1995; Marschner and Römheld, 1994; Rengel, 1997, 1999; Rengel et al., 1998; Subbarao et al., 1997], ion toxicity (e.g., Al Basu et al., 1994, 1999; de la Fuente et al., 1997; Delhaize et al., 1993; Ma et al., 1997; Ma and Miyasaka, 1998; Ryan et al., 1995a, b; Ryan et al., 1997; Zheng et al., 1998a, b), pathogen attack (e.g., Mandal and Sinha, 1991; Mehta et al., 1992; Parashar et al., 1992; Stevenson et al., 1995), and nodulation of legume hosts (Bolanos-Vasquez and Werner, 1997; Gagnon and Ibrahim, 1998; Janczarek et al., 1997; Maxwell et al., 1989; Wojtaszek et al., 1993; Zuanazzi et al., 1998).

Understanding of the role that root exudates play in increasing plant adaptation to a given set of environmental conditions is sketchy at present. However, it is generally accepted that the type of exudates plants release into the rhizosphere has played a significant role in the distribution of plant species in various ecosystems. Calcicole plants exude mainly di- and tricarboxylic acids (with the former being good extractors of P and the latter good extractors of Fe and Mn from calcareous soils), while calcifuge plants exude mostly monocarboxylic acids (poor in mobilising

This paper previously appeared in *Plant and Soil* **245**, 2002, pp. 59–70.

P or Fe from calcareous soils) (Ström, 1997; Ström et al., 1994; Tyler and Ström, 1995). Similar distinctions also exist between crop plants (cf. Zhang et al., 1997).

Effective exudation of organic substances into the rhizosphere relies on at least three processes: 1, signalling sequence; 2, effective biosynthetic machinery producing a relatively large amount of potential exudate compounds, and 3, a membrane transporter that allows transfer of organic compounds into the rhizosphere.

Regulation of the complete exudation sequence and the underlying genetics is only in the initial stages of being unravelled at present. The two most popular strategies in increasing root exudation (overexpressing genes coding for critical enzymes in the biosynthesis of carboxylate anions, and engineering plants to increase exudation of phosphatase and phytase) (Delhaize, 1995) are promising approaches currently tested in several laboratories. In addition, genotypic differences among existing crop germplasm, including wild relatives should also be utilized (Cakmak et al., 1996a) and, particularly for cereals, rye as the cereal most tolerant to a wide range of nutrient deficiency and ion toxicity stresses (see Cakmak et al., 1997). Using genotypes from the local crop germplasm in the breeding programme would be advantageous because development of cultivars with tolerance to rhizosphere-related stresses could be expedited without severely disrupting the broad adaptation already selected and proved to work effectively, including regulation of relevant biochemical reactions involved in root exudation at an appropriately enhanced level.

This review will concentrate on the genetics underlying the capacity of plants to exude various organic compounds into the rhizosphere. More comprehensive treatises on other aspects dealing with root exudates can be found elsewhere (e.g., Crowley and Rengel, 1999; Dinkelaker et al., 1995; 1997; Jones, 1998; Rengel, 1999; Uren and Reisenauer, 1988).

Role of exudates in resistance to ion toxicity

Tolerant genotypes of various plant species exude organic acid anions when challenged with Al toxicity. An Al-tolerant *Phaseolus vulgaris* genotype exuded 70-fold more citrate under Al toxicity than under control conditions and also 10 times more citrate than an Al-sensitive genotype grown with Al. In contrast, Al stress did not cause an increase in citrate release in the Al-sensitive genotype (Miyasaka et al., 1991). Under Al toxicity, an Al-tolerant maize genotype exudes citrate and relatively smaller amounts of malate in a dose-dependent manner (Jorge and Arruda, 1997). Under similar conditions, wheat genotypes tolerant to Al toxicity exude mostly malate (Basu et al., 1994; Delaize et al., 1993; Pellet et al., 1996; Ryan et al., 1995a, b), but sometimes succinate and oxalate as well (Christiansen-Weniger et al., 1992). Taro (Ma and Miyasaka, 1998) and buckwheat seedlings exude oxalic acid under Al toxicity (Zheng et al., 1998a, b), while radish, canola, and oats exude citric and malic acids (Zheng et al., 1998a), and *Arabidopsis* exudes either malate or citrate (Larsen et al., 1998).

Exudation of malate by wheat (Basu et al., 1994; Delhaize et al., 1993; Ryan et al., 1995a, b) and citrate by maize roots (Pellet et al., 1995) is confined to the root tips. Calculations showed that the concentration of exuded organic acid anions at the root tip might be sufficiently high to complex and therefore detoxify Al ions around the root tips, especially if mucilage that surrounds the root tip, and increases the thickness of the unstirred layer at the root tip surface, is considered (Pellet et al., 1995; Ryan et al., 1995b). Therefore, exudation of a relatively small amount of carboxylate (organic acid) anions, concentrated in a small soil volume at the root tip, is an energetically effective means of overcoming inhibition of growth due to Al toxicity. Root tips would need to be protected for a relatively short period of time during which root tip cells would mature and become less sensitive to Al (Basu et al., 1994), even though exudation of carboxylate anions may continue for the duration of the Al stress (cf. Zheng et al., 1998a).

The three examples where the genetics behind the enhanced exudation process is, at least, partially known are malate exudation in wheat, citrate exudation in transgenic tobacco and papaya, and malate, citrate, and/or pyruvate exudation by *Arabidopsis* mutant plants.

Malate exudation in wheat

Isogenic wheat lines segregating at the *Alt1* locus differ in tolerance to Al: when challenged with toxic Al concentrations, the Al-tolerant isogenic line exudes large amounts of malate, while the Al-sensitive line does not (Delhaize et al., 1993; Ryan et al., 1995a). However, the two isogenic lines do not differ in concentrations of malate in their root tissue (Ryan et al., 1995a), indicating that the difference between them resides in the transport component of the exudation sequence. So, it appears likely that the *Alt1* locus codes for a protein that represents a malate channel or a non-specific anion channel embedded in the plasma membrane, or a regulator of such channels. Given the fast reaction to Al exposure (Delhaize et al., 1993; Ryan et al., 1997), it is likely that the *Alt1* locus is expressed constitutively. Electrophysiological studies have indicated that the Al-regulated plasma membrane anion channel is insensitive to La^{3+} and might be present in root tip cells, but not in more mature root cells (Ryan et al., 1997). These characteristics are in accordance with Al-induced malate exudation by intact root tips (Basu et al., 1994; Delhaize et al., 1993; Pellet et al., 1996; Ryan et al., 1995a). The potential difference in the presence and operation of the Al-regulated anion channel between root tip cells of Al-tolerant and Al-sensitive isogenic lines has yet to be tested. Similarly, the channel conductance was only measured for Cl^- (Ryan et al., 1997); it has yet to be shown that the channel can facilitate transfer of carboxylate anions (malate in particular). Finally, a gene coding for a putative malate/anion channel in the wheat root plasma membrane has yet to be cloned, with studies of the potential difference in the genetic code responsible for either the presence or a lack of regulation by Al yielding a potential breakthrough in understanding of the membrane efflux and root exudation into the rhizosphere.

At present, the exact function of the gene products of the *Alt1* locus is not certain (possibly a regulator of the plasma membrane malate-conductive channel allowing exudation of malate into the rhizosphere, or maybe the channel protein itself altered to become sensitive to Al). Despite the fact that the exact function of the gene is not known, it appears logical to suggest that transfer of that gene into Al-sensitive wheat genotypes may result in increased tolerance to Al, especially given the additive effects of Al tolerance genes (cf. Rengel, 1992). Such a transfer within the wheat germplasm can be done via traditional breeding techniques. In addition to such traditional techniques, molecular biology approaches can be used in transferring any genes associated with tolerance to Al from any plant or bacterial species into higher plant species, and especially crops. As reliable transformation systems are becoming common for all major crops, improvements in Al tolerance to crops by horizontal transfer of foreign genes can be expected. In case the crop genotype that is amenable to transformation is not the one that has desirable commercial characteristics, the transformed genotype will serve as a parent in the standard backcrossing programme aimed at transferring the Al-tolerance trait into commercially important breeding material.

Exudation of citrate by transgenic tobacco and papaya

The citrate synthase gene has been transferred from the bacterium *Pseudomonas aeruginosa* into tobacco and papaya and recorded an increased accumulation of citrate in the cytosol (de la Fuente et al., 1997). This accumulation of citrate was accompanied by increased citrate efflux into the rhizosphere and a somewhat increased tolerance of transformed plants to Al.

As in wheat, where exudation appears to be regulated by the membrane transport step (Ryan et al., 1995a), in transgenic tobacco plants transformed with the *P. aeruginosa* citrate synthase gene, saturation of the membrane transport systems regulated exudation of citrate into the rhizosphere because biosynthesis of citrate in transgenic lines increased up to 10-fold, but citrate exudation increased only up to 4-fold, compared to non-transformed parental lines (de la Fuente et al., 1997). The signalling sequence that existed in the parental genotype was adequate for transferring the toxicity signal via the biosynthetic machinery to the membrane efflux transporter to close the sequence. However, the report by de la Fuente et al. (1997), with the amount of citrate synthesised in the cytosol of transgenic tobacco lines representing up to 86% of the fresh weight of roots, casts some doubt on the validity of the results reported. Moreover, increased citrate exudation by transgenic plants compared to the parental line offered increased protection from Al applied at concentrations of 75 to 200 μM, but not to Al applied at 50 μM (de la Fuente et al., 1997). This makes mechanistic explanations hard to come by.

Despite all the problems, the report by de la Fuente et al. (1997) showed in principle that transformation of plants with foreign genes to increase tolerance to Al may be possible. That report may therefore represent an important step toward more widespread transformation of crop plants with genes known to be associated with Al tolerance in other plant species. Transferring a single gene of bacterial citrate synthase (de la Fuente et al., 1997) resulted in a number of events. Increased biosynthesis of citrate in the cytosol had to be accompanied by adjustments in the rate of biosynthesis of precursors of citrate and products downstream from citrate in the biosynthetic pathways, and by adjustments in the mitochondria-cytosol traffic of precursors and products. Citrate synthesis could occur in the mitochondria and the cytosol, but most of it is in mitochondria under physiological conditions. Functional regulation (and *de novo* biosynthesis?) of membrane transporters involved in transfer of citrate into the rhizosphere was also important. Given that no prior knowledge existed about the possible capacity of tobacco and papaya to increase exudation of citrate or any other carboxylate anion under stress conditions (e.g., ion toxicity or nutrient deficiency), de la Fuente et al. (1997) took a crude shot-gun approach that appears to have worked amazingly well. However, transgenic tobacco plants created by de la Fuente et al. (1997) exuded more citrate than non-transformed parents regardless of the presence or absence of Al from the medium. Further work will need to concentrate on creating transgenic plants that will increase exudation of citrate only after exposure to Al toxicity, thus minimizing expenditure of carbon in adaptation to the environment.

Exudation of organic acid anions by Arabidopsis *mutants*

Arabidopsis mutants tolerant to Al appeared at the low frequency of 6 in 100,000. Five of these mutations mapped to a single locus on chromosome 1; these mutants exuded more malate, citrate, and/or pyruvate than the wild type (Larsen et al., 1998). Such an altered pattern of carboxylate anion exudation does not appear to be induced by Al [in contrast to wheat (Basu et al., 1994; Delhaize et al., 1993; Pellet et al., 1996; Ryan et al., 1995a, b), and maize (Pellet et al., 1995)]; rather, it is constitutively expressed. The case of *Arabidopsis* mutants, with malate, citrate and/or pyruvate (organic acid anions that are accumulated in the cytosol upon transfer of citrate from mitochondria) being exuded from roots, points to a model in which increased exudation of carboxylate (organic acid) anions may simply be a consequence of their accumulation in the cytosol.

The genetics of carboxylate anion exudation related to Al-toxicity stress appears simple, i.e., only slightly different from the normal genetics coding for compounds and processes existing in plants not subjected to stress. Therefore, any future attempts to transform plants with a small number of genes (even a single one) should have a relatively high chance of success. There are several research groups in the world working on transforming crop plants (e.g., wheat) with citrate synthase to increase biosynthesis and exudation of citrate. This is expected to increase tolerance to Al toxicity and P deficiency (citrate is effective not only in chelating Al into non-toxic forms, but also in solubilizing water-insoluble P complexes). Successful reports are expected to appear soon.

Role of exudates in response to nutrient deficiency

Breeding wheat and possibly other cereals for greater tolerance to micronutrient deficiency is possible by transferring rye genes controlling these tolerance traits. Tolerance to Fe, Zn and Mn deficiency traits is reliant, at least to some extent, on root exudation (Cakmak et al., 1994; Marschner and Römheld, 1994; Rengel, 1997, 1999; Rengel et al., 1998; Timonin, 1946). Individual genes conferring tolerance to nutrient deficiencies have not been deciphered yet, but transferring rye chromosomes 1R and 7R into wheat has increased its tolerance to Zn deficiency (Cakmak et al., 1997), and transferring chromosomes 2R and 7R increased tolerance to Mn and Fe deficiency in wheat (Schlegel et al., 1997). In contrast, there is no report linking tolerance to copper (Cu) deficiency with root exudation. However, transferring the rye chromosome 5R (Schlegel et al., 1997) or transferring just a small piece of chromatin from the long arm of the chromosome 5R to wheat increases its tolerance to Cu deficiency (Graham et al., 1987; Schlegel et al., 1993). It is known that genes for mugineic acid synthase and 3-hydroxymugineic acid synthase, the enzymes involved in biosynthesis of common phytosiderophores, are located on the rye chromosome 5R (Mori et al., 1990), but they map outside the region conferring Cu deficiency tolerance to the wheat-rye addition line (Schlegel et al., 1993). These findings are in accordance with the fact that phytosiderophores do not appear to be involved in uptake and transport of Cu (see Schlegel et al., 1993), but a possibility cannot be excluded that the function of the gene products coded by the genes conferring tolerance to Cu deficiency may in some way be connected with root exudation.

Iron deficiency

Generally, there are two types of mechanisms plants employ under Fe deficiency (for reviews see Crowley and Rengel, 1999; Jolley et al., 1996; Marschner and Römheld, 1994; Rengel, 1999). Only strategy II (employed by grasses), which involves increased exudation of phytosiderophores into the rhizosphere, will be considered here.

Graminaceous species acquire Fe by releasing phytosiderophores and taking up ferrated-phytosiderophore complexes through a specific uptake system that is strongly activated under Fe deficiency (Römheld, 1991; von Wiren et al., 1994, 1995). The rate of phytosiderophore release differs between crops and between genotypes of a particular crop. The rate of release of phytosiderophores is positively related to the tolerance of a species to Fe deficiency. The relative effectiveness in phytosiderophore release, and thus tolerance to Fe deficiency, decreases in the order: barley > maize > sorghum (Römheld and Marschner, 1990) or oat > maize (see Mori, 1994). Under Fe-deficiency conditions, the Fe-deficiency-tolerant genotypes of wheat (Hansen et al., 1995, 1996) and oats (Hansen and Jolley, 1995; Jolley and Brown, 1989) exuded more phytosiderophores than genotypes sensitive to Fe deficiency, making a phytosiderophore-exudation test suitable for screening genotypes in the breeding programme.

The release of phytosiderophores and the subsequent uptake of the Fe-phytosiderophore complex are under different genetic control (Römheld and Marschner, 1990). Indeed, it was found that the Fe-deficiency-sensitive yellow-stripe mutant of maize (*ys1*) maintains a rate of phytosiderophore release similar to that of cultivars tolerant to Fe deficency (e.g., Alice), with sensitivity of *ys1* to Fe deficiency being the result of a defect in the uptake system for the Fe-phytosiderophore complex. Up to a 20-fold lower uptake rate in *ys1* than Alice maize (von Wiren et al., 1994) is due to the mutation that affects a high-affinity uptake component, leading to a decrease in activity and/or a number of Fe-phytosiderophore transporters (decreased I_{max}, unchanged K_m) (von Wiren et al., 1995).

A number of genes involved in biosynthesis of phytosiderophores and their precursor nicotianamine have been cloned (Mori, 1997), the most recent being a gene family coding for nicotianamine synthase (Higuchi et al., 1999). Nicotianamine synthase catalyses the polymerisation of three S-adenosylmethionine molecules, with the release of adenine and the azetidine ring formation. Since only strategy II plants (graminaceous monocots) exude phytosiderophores under Fe and Zn deficiency (for references see Rengel, 1999), it is interesting to note that nicotianamine synthase is regulated by Fe supply in these plants (Higuchi et al., 1999), but not in other species (Higuchi et al., 1996), in which nicotianamine may only be a putative chelator involved in long-distance transport of divalent cations in plants (see von Viren et al., 1999; Welch, 1995).

Zinc deficiency

While exudation of phytosiderophores has primarily been associated with the response of graminaceous monocots to Fe deficiency (Cakmak et al., 1994; Jolley et al., 1996; Marschner and Römheld, 1994; Rengel, 1999; Römheld and Marschner, 1990) similar effect can be initiated as a response of wheat to Zn deficiency (Cakmak et al., 1994, 1996b; Rengel et al., 1998; Walter et al., 1994; Zhang et al., 1989). More phytosiderophores being exuded under Fe than under Zn deficiency (see Rengel et al., 1998; Walter et al., 1994) may indicate that the triggering process is more direct in the case of Fe deficiency, while a more complicated and/or slower sequence of events is necessary to trigger exudation of phytosiderophores under Zn deficiency. Moreover, wheat genotype considered tolerant to both Zn deficiency (Cakmak et al., 1994; 1996b; Hopkins et al., 1998; Walter et al., 1994; Rengel et al., 1998) and Fe deficiency (Hansen et al., 1995, 1996) exude considerably larger amounts of phytosiderophores into the rooting medium than genotypes sensitive to Zn and Fe deficiencies, in order to increase mobilization of Zn and Fe from sparingly soluble sources.

There is usually a good relationship between root exudation of phytosiderophores and differential tolerance to Z deficiency when durum (sensitive) and bread wheats (tolerant) are compared (Cakmak et al., 1994, 1996b; Rengel, 1999; Rengel et al., 1998). However, the positive correlation between phytosiderophore exudation and tolerance of bread wheat genotypes to Zn deficiency may not always hold, as shown for seven (Erenoglu et al., 1996) and five bread wheat genotypes (Cakmak et al., 1998).

Phosphorus deficiency

Under P deficiency, plants exude a wide range of organic and inorganic compounds to increase mobilization of P from sparingly soluble sources (for references see Rengel, 1999). Similar amounts of exudates were collected from roots with or without arbuscular mycorrhizae (AM) (Azaizeh et al., 1995), indicating that an extra carbon cost of supporting AM colonization did not adversely affect the capacity of plants to partition a proportion of assimilates to root exudation.

Species and genotypes within species that differ in tolerance to P deficiency also differ in the solubilizing activity of their root exudates (e.g., for acid phosphatase see Caradus, 1995; for unspecified Fe-P solubilizing activity of root exudates see Subbarao et al., 1997). A comparison of durum and flax genotypes of unknown levels of tolerance to P deficiency has shown large differences in the amounts of organic acids released into the rhizosphere (Cieslinski et al., 1997).

The activity of both phytase and acid phosphatase increased in root exudates from a number of species and genotypes within species grown under low-P supply (e.g., Ascencio, 1997; Asmat, 1997; Li et al., 1997). However, kinetic properties of enzymes from different species are different (Ascencio, 1997), indicating the importance of optimizing enzyme properties, rather than just increasing enzyme synthesis and exudation in any future attempt to manipulate phosphatase activity to increase adaptation of genotypes to low-P environments.

Acid phosphatase is exuded from roots under P deficiency, with the relative rate of increase in exudation greatly varying among various plant species (Tadano et al., 1993). The P-deficiency-tolerant genotypes of bean (Helal, 1990) and barley (Asmar et al., 1995) had a greater activity of extracellular phosphatases in the rhizosphere soil than genotypes sensitive to P deficiency. In rice, genotypes differed in acid phosphatase exudation under P deficiency (Ni et al., 1996).

There is a wide variability in both the absolute amounts and the type of organic acid (carboxylate) anions exuded by roots of Australian native plants [various species from cluster-root-forming (Lamont, 1982; Skene, 1998) genera from the family Proteaceae: *Hakea, Banksia, Dryandra*, etc.] adapted to low-fertility soils where deficiency of available P and micronutrients is common (Dinkelaker et al., 1995, 1997; Grierson, 1992; Roelofs, Rengel, Lambers and Kuo, unpublished results). L-malate, malonate, lactate, acetate, citrate, succinate, and trans-aconitate accounted for the largest proportions of carboxylate anions exuded by entire root systems of most of the species tested. In contrast, for cluster roots of all the species tested, L-malate, malonate, lactate, citrate, and trans-aconitate represented large proportions of total carboxylate anion exudation (Roelofs, Rengel, Lambers and Kuo, unpublished results). Qualitative and quantitative differences in carboxylate anion exudation among Proteaceae species may correspond with the distribution of these species in the south-west Australia, and may contribute to the species richness in the kwongan vegetation (cf. Cowling and Lamont, 1998). The hypothesis is that species use P from various soil pools and thus, instead of competing among themselves, complement each other and coexist in the same ecosystem.

The average rate of exudation of carboxylate (organic acid) anions by cluster roots of Australian native species varied between 1.0 (*Hakea petiolaris*) and 2.7 nmol g^{-1} fresh weight s^{-1} (*H. undulata*) (Roelofs, Rengel, Lambers and Kuo, unpublished results). These rates of root exudation are at least up to one order of magnitude greater than rates published for crop plants (e.g., Hoffland et al., 1992; Subbarao et al., 1997), with a possible exception of white lupin (*Lupinus albus*) where exudation rates were about a half those shown by native Australian species (Keerthisinghe et al., 1998; Neumann et al., 1999). The genetic basis of regulation of carboxylate anion exudation, once deciphered, will provide important knowledge to use in breeding and transforming crop species for enhanced root exudation.

White lupin is an excellent model plant for studying organic acid anion exudation into the rhizosphere under P deficiency. Depending on the developmental stage of cluster roots, they may exude only malate (Braum, 1995), almost exclusively citrate (Gardner et al., 1983), or various proportions of malate and citrate (Johnson et al., 1996b; Neumann et al., 1999). Mature root clusters exude primarily citrate (Neumann et al., 1999; Sas, Tang and Rengel, unpublished results). Increased accumulation of citrate is due not only to increased activity of citrate synthase (Johnson et al., 1994) but also to decreased decomposition of citrate because of decreased activity of aconitase (Neumann et al., 1999). Regulation of exudation appears to reside in the membrane transport step rather than in accumulation of citrate, because old cluster roots stop exuding even though they have a similar internal concentration to somewhat younger cluster roots that exude large quantities of citrate (Neumann et al., 1999). Similar regulation of malate exudation from wheat roots was observed under Al toxicity (cf. Ryan et al., 1995a), indicating a possible common regulatory mechanism for exudation of organic acid anions regardless of the stress factor causing increased exudation. Citrate accumulated in cluster roots of white lupin (Keerthisinghe et al., 1998; Neumann et al., 1999) as well as malate accumulated in wheat roots subjected to Al toxicity (Ryan et al., 1995a, 1997) are likely to be transported across the plasma membrane into the rhizosphere via an anion channel. Exudation of other organic acid anions is probably also mediated by an anion channel (cf. Jones, 1998).

In white lupin cluster roots subjected to P deficiency, the carbon required for relatively large amounts of citrate exuded comes primarily from non-photosynthetic CO_2 fixation (Hoffland et al., 1992; Johnson et al., 1996b) brought about by increased activity of phosphoenolpyruvate carboxylase (PEPC) (Johnson et al., 1996a; Neumann et al., 1999). Malate dehydrogenase and citrate synthase may also be increased (Allan and Vance, 1999), even though Neumann et al. (1999) did not find an increase in citrate synthase activity under P deficiency. The PEPC gene has been cloned from white lupin cluster roots subjected to P deficiency; *in situ* hybridization showed increased expression of that gene in the cortex of cluster roots (Allan and Vance, 1999), thus confirming its involvement in organic acid anion accumulation and exudation. Intensive work is in progress in several laboratories to decipher the sequence of events leading to increased exudation of organic acid anions and to cloning of the genes involved.

In addition to cloning genes involved in production of organic acid anions in white lupin, genes involved in the biosynthesis of acid phosphatase are being cloned at an increasing pace (e.g., Allan and Vance, 1999; Ozawa et al., 1995; Wasaki et al., 1997). Cloned genes will allow transformation of various plant species with the intention to increase exudation of acid phosphatase and thus increase plant capacity to utilize organic forms of P.

Role of root exudates in nodulation of legumes

An important part in the signalling process involved in the rhizobia-legume symbiosis is exudation of flavonoids by host roots to induce *nod* genes in rhizobia and mediate the nodulation process (e.g., Bolanos-Vasquez and Werner, 1997; Gagnon and Ibrahim, 1998; Janczarek et al., 1997; Maxwell et al., 1989; McKhann et al., 1997; Rosenblueth et al., 1998; Wojtaszek et al., 1993; Zuanazzi et al., 1998). Soybean roots exude glutamate and aspartate and dicarboxylic organic acids that serve as natural chemoattractants for *Bradyrhizobium japonicum* (Barbour et al., 1991). It is interesting that specific Nod factors of *B. japonicum* increase flavonoid exudation by soybean roots (Schmidt et al., 1994), indicating the mutual character of the signalling pathway.

Knowledge of the genetics behind exudation of specific *nod*-gene inducers in host root exudates is sketchy at present. Advancements may come from studying legume host genotypes that differ in components of the signalling pathway (e.g., in rice, Soejima et al., 1992; bean *Phaseolus vulgaris*, Hungria and Phillips, 1993; lucerne *Medicago sativa*, Hernandez et al., 1995), but considerably more genotypic testing needs to be done, including the development of isogenic lines (see Hungria and Phillips, 1993), to provide an excellent genetic material for studies aimed at discerning genetics of *nod*-gene inducer exudation.

Role of exudates in resistance to plant pathogens

In a number of experimental systems plant genotypes tolerant to fungal and other pathogens exude specific compounds that are inhibitory to spore germination or pathogen growth, while root exudates from susceptible genotypes either do not contain these inhibitory compounds, or contain other compounds that have a stimulating effect on pathogen growth. Constitutively synthesized toxic compounds present in roots and root exudates of tolerant tomato genotypes contribute towards host resistance to *Fusarium oxysporum* f.sp. *lycopersici* by inhibiting growth of the pathogen at the infection site and subsequently inside the host after penetration. In contrast, root exudates from sensitive genotypes were stimulatory rather than inhibitory to the pathogen (Mandal and Sinha, 1991).

Wilt-resistant chickpea genotypes exuded an unidentified apolar, ethyl-acetate-extractable compound that had antifungal properties and inhibited germination of *F. oxysporum* f.sp. *ciceri* spores, thus conferring resistance to the pathogen to these genotypes. Wilt-susceptible genotypes did not exude any compounds with antifungal properties against *F. oxysporum* wilt (Stevenson et al., 1995), and there was even stimulation of fungal growth upon exposure to root exudates from susceptible chickpea genotypes (Shinde and Deshmukh, 1989).

The root exudates from a chickpea genotype relatively more tolerant to *Rhizoctonia solani* contained a higher concentration and more types of phenolic compounds than the root exudates from a more-susceptible genotype (Mehta et al., 1992; Parashar et al., 1992). Growth of *R. solani* hyphae was repressed with root exudates from some relatively tolerant chickpea genotypes (Mehta et al., 1992), but not others (Parashar et al., 1992), indicating the complexity of the relationships between exudation of phenolic compounds and tolerance to *R. solani*.

There is good potential for genetically engineering plants to exude into the rhizosphere inhibitors acting against various microorganisms, including pathogens, because these inhibitors are frequently coded for by single genes that in some cases have already been successfully transferred into plants (Jach et al., 1995). In addition, engineering plants to exude specific nutrients required by biocontrol microorganisms may increase plant tolerance to pathogens. For example, *Bacillus cereus* UW85 is auxotroph for threonine (Milner et al., 1995), and plants overexpressing specific genes and accumulating free threonine have been produced (Shaul and Galili, 1992).

Breeding for increased root exudation

The role of genotypes tolerant to nutrient deficiency in modern agriculture has been discussed by Lynch (1998). Some of these genotypes (e.g., tolerant to P or micronutrient deficiencies) can be used in conventional breeding as a source of valuable genes conferring the increased capacity for exudation of organic compounds into the rhizosphere. However, the relatively slow progress in deciphering the genetics, physiology, and biochemistry behind the mechanisms of tolerance to nutrient deficiency has hampered development of genotypes of superior tolerance to nutrient deficiency through conscious breeding efforts geared specifically toward that purpose.

In general, tolerance to Zn and Fe deficiencies that may, at least partly, rely on root exudation as a mechanism of tolerance, is simply inherited, with a relatively small number of genes involved, and tolerance to deficiency being dominant (Cianzio, 1999; Graham et al., 1992). Examples of increased exudation after transformation of plants with a single gene testify to the potential for producing new genotypes with improved adaptation to environmental stresses due to enhanced exudation of specific compounds. The availability of promoters whose regulation is cell- or tissue-specific in roots (e.g., de Pater and Schilperoort, 1992) are important in that respect. In addition, genetic engineering of border cells that are released from roots into the rhizosphere (O'Connell et al., 1996) offers potential to influence the rhizosphere processes in addition to any exudation of specific organic and inorganic compounds from roots.

Transgenic plants have been developed that exude into the rhizosphere specific compounds that only selected microorganisms can catabolise (Guyon et al., 1993; O'Connell et al., 1996), the strategy based on the 'opine concept' (Rossbach et al., 1994). If transgenic plants can be produced that exude specific compounds that will selectively promote growth of targeted microorganisms (for an example of differential tolerance to Mn deficiency see Timonin, 1946), such plants would have enhanced adaptation to particular environments. Another exciting prospect is that transgenic plans that could exude compounds acting as inducers of gene expression in microorganisms will be produced (see O'Connell et al., 1996).

The author is not aware of any field trials reported to date with plants that have been transformed or bred to express an altered exudation pattern. This remains an exciting future research area.

Screening for increased root exudation

Increasing root exudation capacity may adversely affect crop yield because of increased partitioning of assimilates toward roots. Indeed, it appears that selection of genotypes in breeding programmes over the last decades resulted in crops' decreased capacity to exude organic compounds into the rhizosphere (e.g., in rice, Waschutza et al., 1992). So, advocating selection toward increased exudation may appear a step backwards. However, instead of generally increased rates of exudation, a better (and more economic) solution would obviously be to increase exudation of only those targeted compounds that perform a specific function in the rhizosphere, a function that will result in improved tolerance to nutrient deficiencies, ion toxicities, and plant pathogens, or in a more effective nodulation process. For such a selection pressure to be included in the breeding programme, breeders would need to have well-defined screening procedures that would rely heavily on understanding the physiological processes involved. Such knowledge at present appears to be inadequate.

Simple, fast, and inexpensive screening techniques would permit the assessment of a large number of genotypes in segregating populations, without relying on expensive field testing (Graham and Rengel, 1993). Screening techniques based on characterizing root exudation are not in routine use at present, even though HPLC determination of phytosiderophores exuded under Fe deficiency has potential (Hansen and Jolley, 1995; Hansen et al., 1995, 1996; Jolley and Brown, 1989). The better the exudation sequence is understood, the more successful a screening technique will be developed.

Screening various plant species or large numbers of genotypes of some plant species may identify a source of genes important in controlling root exudation. A screening system can be devised that would estimate genotypic differences among various cultivars and/or breeding lines in response to limiting environments by measuring the relative rate of root exudation of specific compounds.

Numerous technical problems in collecting and analysing root exudates still hamper serious attempts at using screening techniques reliant on root exudate characterization in plant breeding (Crowley and Rengel, 1999), despite the recently published report on total analysis of organic compounds in the root exudate (Fan et al., 1997). Moreover, quantitative collection of root exudates from soil-grown plants has a little chance of ever being achieved. As a consequence, plants are often grown in nutrient solution and root exudates are collected from the nutrient solution, or even worse, in distilled water. However, mechanical impedance increases root exudation significantly (Boeuf-Tremblay et al., 1995), suggesting that the rates of exudation calculated for soil-grown plants from the data on exudation by solution-grown plants have been underestimated.

A further problem in collecting root exudates may be the presence of microorganisms. The rate of root exudation depends on the specific microorganisms growing in the rhizosphere (Meharg and Killham, 1995), with the relationship being influenced by the genotype of the host plant (Liljeroth et al., 1994) and by soil compaction (Ikeda et al., 1997). When the effect of bacterial microflora was minimized by the addition of an antibiotic to the agar sheets collecting exudates, 3- to 6-fold greater amounts of carbohydrates, amino acids, and phenolics were recovered (Azaizeh et al., 1995).

In the near future, molecular methods will dominate selection for increased root exudation. Molecular markers will be identified for regions of the genome segregating with the trait of increased root exudation of a particular compound, leading to rapid and efficient selection. However, the molecular markers may be population-dependent (e.g., markers for tolerance to Fe deficiency in soybeans; Cianzio, 1999), making them unsuitable for use in breeding programmes.

Conclusions

The knowledge on the genetics underlying root exudation of organic compounds is expanding rapidly. Such knowledge will allow selection of superior genotypes and deliberate breeding of crop plants for increased tolerance to nutrient deficiency and ion toxicity, efficiency of nodulation, tolerance to various diseases, etc. Traditional breeding techniques will be combined with transformation of plants with genes coding for specific enzymes in the biosynthesis of organic compounds and/or those involved in regulation of efflux of organic compounds across the membrane. All these developments will allow better adaptation of crop genotypes to environment and more sustainable and profitable farming in the future.

References

Allan DL and Vance 1999 Altered metabolism and gene expression in cluster roots of phosphorus-deficient white lupin. *In* Abstracts of the 16th International Botanical Congress. Aug. 1–7, 1999. pp. 191. St. Louis, Missouri, USA.

Ascencio J 1997 Root secreted acid phosphatase kinetics as a physiological marker for phosphorus deficiency. J. Plant Nutr. 20, 9–26.

Asmar F 1997 Variation in activity of root extracellular phytase between genotypes of barley. Plant Soil 195, 61–64.

Asmar F, Gahoonia T S and Nielsen N E 1995 Barley genotypes differ in activity of soluble extracellular phosphatase and depletion of organic phosphorus in the rhizosphere soil. Plant Soil 172, 117–122.

Azaizeh H A, Marschner H, Römheld V and Wittenmayer L 1995 Effects of a vesicular-arbuscular mycorrhizal fungus and other soil microorganisms on growth, mineral nutrient acquisition and root exudation of soil-grown maize plants. Mycorrhiza 5, 5321–5327.

Barbour W M, Hattermann D R and Stacey G 1991 Chemotaxis of *Bradyrhizobium japonicum* to soybean exudates. Appl. Environ. Microbiol. 57, 2635–2639.

Basu U, Godbold D and Taylor G J 1994 Aluminum resistance in *Triticum aestivum* associated with enhanced exudation of malate. J. Plant Physiol. 144, 747–753.

Basu U, Good A G, Aung T, Slaski J J, Basu A, Briggs K G and Taylor G J 1999 A 23-kDa, root exudate polypeptide co-segregates with aluminum resistance in *Triticum aestivum*. Physiol. Plant. 106, 53–61.

Boeuf-Tremblay V, Plantureux S and Guckert A 1995 Influence of mechanical impedance on root exudation of maize seedlings at two development stages. Plant Soil 172, 279–287.

Bolanos-Vasquez M C and Werner D 1997 Effects of *Rhizobium tropici, R. etli* and *R. leguminosarum* bv. *phaseoli* on nod gene-inducing flavonoids in root exudates of *Phaseolus vulgaris*. Mol. Plant-Microbe Interact. 10, 330–346.

Braum S M 1995 Mobilization of phosphorus and other mineral nutrients by citrate in the rhizosphere of *Lupinus albus* L. Ph.D. thesis, University of Wisconsin, Madison, Wisconsin, USA.

Cakmak I, Derici R, Torun B, Tolay I, Braun H I and Schlegel R 1997 Role of rye chromosomes in improvement of zinc efficiency in wheat and triticale. Plant Soil 196, 249–253.

Cakmak I, Gülüt K Y, Marschner H and Graham R D 1994 Effect of zinc and iron deficiency on phytosiderophore release in wheat genotypes differing in zinc efficiency. J. Plant Nutr. 17, 1–17.

Cakmak I, Öztürk L, Karanlik S, Marschner H and Ekiz H 1996a Zinc-efficient wild grasses enhance release of phytosiderophores under zinc deficiency. J. Plant Nutr. 19, 551–563.

Cakmak I, Sari N, Marschner H, Ekiz H, Kalayci M, Yilmaz A and Braun H J 1996b Phytosiderophore release in bread and durum wheat genotypes differing in zinc efficiency. Plant Soil 180, 183–189.

Cakmak I, Torun B, Erenoglu B, Öztürk L, Marschner H, Kalayci M, Ekiz H and Yilmaz A 1998 Morphological and physiological differences in the response of cereals to zinc deficiency. Euphytica 100, 340–357.

Caradus J R 1995 Genetic control of phosphorus uptake and phosphorus status in plants. *In* Genetic Manipulation of Crop Plants to Enhance Integrated Nutrient Management in Cropping Systems. 1. Phosphorus. Eds C Johansen, K K Lee, K K Sharma, G V Subbarao and E A Kueneman. pp. 55–74. International Crops Research Institute for the Semi-Arid Tropics, Patancheru, Andhra Pradesh, India.

Cianzio S R 1999 Breeding crops for improved nutrient efficiency: soybean and wheat as case studies. *In* Mineral Nutrition of Crops: Fundamental Mechanisms and Implications. Ed. Z Rengel. pp. 267–287. The Haworth Press, New York, USA.

Cieslinski G, van Rees K C J, Szmigielska A M and Huang P M 1997 Low molecular weight organic acids released from roots of durum wheat and flax into sterile nutrient solutions. J. Plant Nutr. 20, 753–764.

Cowling R M and Lamont B B 1998 On the nature of Gondwana species flocks: Diversity of Proteaceae in Mediterranean Southwestern Australia and South Africa. Aust. J. Bot. 46, 335–355.

Christiansen-Weniger C, Groneman A F and van Veen J A 1992 Associative N_2 fixation and root exudation of organic acids from wheat cultivars of different aluminium tolerance. Plant Soil 139, 167–174.

Crowley D E and Rengel Z 1999 Biology and chemistry of rhizosphere influencing nutrient availability. *In* Mineral Nutrition of Crops: Fundamental Mechanisms and Implications. Ed. Z Rengel. pp. 1–40. The Haworth Press, New York, USA

de la Fuente J M, Ramirez-Rodriguez V, Cabrera-Ponce J L and Henera-Estrella L 1997 Aluminum tolerance in transgenic plants by alteration of citrate synthesis. Science 276, 1566–1568.

Delhaize E 1995 Genetic control and manipulation of root exudates. *In* Genetic Manipulation of Crop Plants to Enhance Integrated Nutrient Management in Cropping Systems. 1. Phosphorus. Eds C Johansen, K K Lee, K K Sharma, G V Subbarao and B A Kueneman. pp. 145–152. International Crops Research Institute for the Semi-Arid Tropics, Patancheru, Andhra Pradesh, India.

Delhaize E, Ryan P R and Randall P J 1993 Aluminum tolerance in wheat (*Triticum aestivum* L.). II. Aluminum-stimulated excretion of malic acid from root apices. Plant Physiol. 103, 695–702.

de Pater B S and Schilperoort R A 1992 Structure and expression of a root-specific rice gene. Plant Mol. Biol. 18, 161–164.

Dinkelaker B, Hengeler C and Marschner H 1995 Distribution and function of proteoid roots and other root clusters. Bot. Acta 108, 181–200.

Dinkelaker B, Hengeler C, Neumann G, Eltrop L and Marschner H 1997 Root exudates and mobilization of nutrients. *In* Trees – contributions to Modern Tree Physiology. Eds H Rennenberg, W Eschrich and H Ziegler. pp. 441–452. Backhuys Publishers, Leiden, The Netherlands.

Erenoglu B, Cakmak I, Marschner H, Römheld V, Eker S, Daghan H, Kalayci M and Ekiz H 1996 Phytosiderophore release does not relate well with Zn efficiency in different bread wheat genotypes. J. Plant Nutr. 19, 1569–1580.

Fan T W-M, Lane A N, Pedler J, Crowley D and Higashi R M 1997 Comprehensive analysis of organic ligands in whole root exudates using nuclear magnetic resonance and gas chromatograph – mass spectrometry. Anal. Biochem. 251, 57–68.

Gagnon H and Ibrahim R K 1998 Aldonic acids: a novel family of nod gene inducers of *Mesorhizobium loti, Rhizobium lupini*, and *Sinorhizobium meliloti*. Mol. Plant-Microbe Interact. 11, 988–998.

Gardner W K, Barber D A and Parberry K G 1983 The acquisition of phosphorus by *Lupinus albus* L. III. The probable mechanism by which phosphorus movement in the soil/root interface is enhanced. Plant Soil 70, 107–124.

Graham R D and Rengel Z 1993 Genotypic variation in zinc uptake and utilization. *In* Zinc in Soils and Plants. Ed. A D Robson. pp. 107–118. Kluwer Academic Publishers, Dordrecht, The Netherlands.

Graham R D, Ascher J S, Ellis P A E and Shepherd K W 1987 Transfer to wheat of the copper efficiency factor carried on rye chromosome arm 5RL. Plant Soil 99, 107–114.

Graham R D, Ascher J S and Hynes S C 1992 Selecting zinc-efficient cereal genotypes for soils and low zinc status. Plant Soil 146, 241–250.

Grierson P F 1992 Organic acids in the rhizosphere of *Banksia integrifolia* L. Plant Soil 144, 259–265.

Guyon P, Petit A, Tempe J and Dessaux Y 1993 Transformed plants producing opines specifically promote growth of opine-degrading agrobacteria. Mol. Plant-Microbe Interact. 6, 92–98.

Hansen N C and Jolley V D 1995 Phytosiderophore release as a criterion for genotypic evaluation of iron efficiency in oat. J. Plant Nutr. 18, 455–465.

Hansen N C, Jolley V D, Berg W A, Hodges M E and Krenzer E G 1996 Phytosiderophore release related to susceptibility of wheat to iron deficiency. Crop Sci. 36, 1473–1476.

Hansen N C, Jolley V D and Brown J C 1995 Clipping foliage differentially affects phytosiderophore release by two wheat cultivars. Agron. J. 87, 1060–1063.

Helal H M 1990 Varietal differences in root phosphatase activity as related to utilisation of organic phosphates. Plant Soil 123, 161–163.

Hernandez G, Ramirez M, Suarez R and Fuentes S I 1995 Root exuded *nod*-gene inducing signals limit the nodulation capacity of different alfalfa varieties with *Rhizobium meliloti*. Plant Cell Reports 14, 626–629.

Higuchi K, Nishizawa N K, Römheld V, Marschner H and Mori S 1996 Absence of nicotianamine synthase activity in the tomato mutant ‘chloronerva’. J. Plant Nutr. 19, 1235–1239.

Higuchi K, Suzuki K, Nakanishi H, Yamaguchi H, Nishizawa N-K and Mori S 1999 Cloning of nicotianamine synthase genes, novel genes involved in the biosynthesis of phytosiderophores. Plant Physiol. 119, 471–479.

Hoffland E, van den Boogaurd R, Nelemans J and Findenegg G 1992 Biosynthesis and root exudation of citric and malic acids in phosphate-starved rape plants. New Phytol. 122, 675–680.

Hopkins B G, Whitney D A, Lamond R E and Jolley V D 1998 Phytosiderophore release by sorghum, wheat, and corn under zinc deficiency. J. Plant Nutr. 21, 2623–2637.

Hungria M and Phillips D A 1993 Effects of a seed color mutation on rhizobial *nod*-gene-inducing flavonoids and nodulation in common bean. Mol. Plant-Microbe Interact. 6, 418–422.

Ikeda K, Toyota K and Kimura M 1997 Effects of soil compaction on the microbial populations of melon and maize rhizoplane. Plant Soil 189, 91–96.

Jach G, Gornhardt B, Mundy J, Logemann J, Pinsdorf E, Leah R, Schell J and Maas C 1995 Enhanced quantitative resistance against fungal disease by combinatorial expression of different barley antifungal proteins in transgenic tobacco. Plant J. 8, 97–109.

Janczarek M, Urbanik-Sypniewska T and Skorupska A 1997 Effect of authentic flavonoids and the exudate of clover roots on growth rate and inducing ability of *nod* genes of *Rhizobium leguminosarum* bv. *trifolii*. Microbiol. Res. 152, 93–98.

Jolley V D and Brown J C 1989 Iron inefficient and efficient oat cultivars. II. Characterization of phytosiderophore released in response to iron deficiency stress. J. Plant Nutr. 12, 923–937.

Jolley V D, Cook K A, Hansen N C and Stevens W B 1996 Plant physiological responses for genotypic evaluation of iron efficiency in strategy I and strategy II plants. A review. J Plant Nutr. 19, 1241–1255.

Johnson J F, Allan D L and Vance C P 1994 Phosphorus stress-induced proteoid roots show altered metabolism in *Lupinus albus*. Plant Physiol. 104, 657–665.

Johnson J F, Vance C P and Allan D L 1996a Phosphorus deficiency in *Lupinus albus*. Altered lateral root development and enhanced expression of phosphoenolpyruvate carboxylase. Plant Physiol. 112, 31–41.

Johnson J F, Allan D L, Vance C P and Weiblen G 1996b Root carbon dioxide fixation by phosphorus-deficient *Lupinus albus*. Contribution to organic acid exudation by proteoid roots. Plant Physiol. 112, 19–30.

Jones D L 1998 Organic acids in the rhizosphere – a critical review. Plant Soil 205, 25–44.

Jorge R A and Arruda P 1997 Aluminum-induced organic acids exudation by roots of an aluminum-tolerant tropical maize. Phytochemistry 45, 675–681.

Keerthisinghe G, Hocking P J, Ryan P R and Delhaize E 1998 Effect of phosphorus supply on the formation and function of proteoid roots of white lupin (*Lupinus albus* L.). Plant Cell Environ. 21, 467–478.

Lamont B 1982 Mechanisms for enhancing nutrient uptake in plants, with particular reference to mediterranean South Africa and Western Australia. Bot. Rev. 48, 597–689.

Larsen P B, Degenhardt J, Tai C-Y, Stenzler L M, Howell S H and Kochian L V 1998 Aluminum-resistant *Arabidopsis* mutants that exhibit altered patterns of aluminum accumulation and organic acid release from roots. Plant Physiol. 117, 9–18.

Li M, Osaki M, Rao I M and Tadano T 1997 Secretion of phytase from the roots of several plant species under phosphorus-deficient conditions. Plant Soil 195, 161–169.

Liljeroth E, Kuikman P and van Veen J A 1994 Carbon translocation to the rhizosphere of maize and wheat and influence on the turnover of native soil organic matter at different soil nitrogen levels. Plant Soil 161, 233–240.

Lynch J 1998 The role of nutrient-efficient crops in modern agriculture. *In* Nutrient Use in Crop Production, Ed. Z Rengel. pp. 241–264. The Haworth Press, New York, USA.

Ma J F, Zheng S J and Matsumoto H 1997 Specific secretion of citric acid induced by Al stress in *Cassia tora* L. Plant Cell Physiol. 38, 1019–1025.

Ma Z and Miyasaka S C 1998 Oxalate exudation by taro in response to Al. Plant Physiol. 118, 861–865.

Mandal N C and Sinha A K 1991 Differential effects of root exudates and extracts from tomato cultivars on *Fusarium oxysporum* f.sp. *lycopersici* in relation to their disease reaction. J. Mycopathol. Res. 29, 9–16.

Marschner H and Römheld V 1994 Strategies of plants for acquisition of iron. Plant Soil 165, 261–274.

Maxwell C A, Hartwig U A, Joseph C M and Phillips D A 1989 A chalcone and two related flavonoids released from alfalfa roots induce *nod* genes of *Rhizobium meliloti*. Plant Physiol. 91, 842–847.

McKhann H I, Paiva N L, Dixon R A and Hirsch A M 1997 Chalcone synthase transcripts are detected in alfalfa root hairs following inoculation with wild-type *Rhizobium meliloti*. Mol. Plant-Microbe Interact. 10, 50–58.

Meharg A A and Killham K 1995 Loss of exudates from the roots of perennial ryegrass inoculated with a range of micro-organisms. Plant Soil 170, 345–349.

Mehta S, Sharma S and Sindhan G S 1992 Analysis of root exudates of cowpea and their influence on the growth of *Rhizoctonia solani*. Indian J. Mycol. Plant Pathol. 22, 227–231.

Milner J L, Raffel S J, Letbbridge B J and Handelsman J 1995 Culture conditions that influence accumulation of zwittermicin A by *Bacillus cereus* UW85. Appl. Microbiol. Biotechnol. 43, 685–691.

Miyasaka S C, Buta J G, Howell R K and Foy C D 1991 Mechanism of aluminum tolerance in snapbeans. Root exudation of citric acid. Plant Physiol. 96, 737–743.

Mori S 1994 Mechanisms of iron acquisition by graminaceous (strategy II) plants. *In* Biochemistry of Metal Micronutrients in the Rhizosphere. Eds. J A Manthey, D E Crowley and D G Luster. pp. 225–249. Lewis Publishers, Boca Raton, Florida, USA.

Mori S 1997 Reevaluation of the genes induced by iron deficiency in barley roots. Soil Sci. Plant Nutr. 43, 975–980.

Mori S, Nishizawa N K and Fujigaki J 1990 Identification of rye chromosome 5R as a carrier of the genes for mugineic acid synthetase and 3-hydroxymugineic acid synthetase using wheat-rye addition lines. Jpn. J. Genet. 65, 343–352.

Neumann G, Massonneau A, Martinoia E and Römheld V 1999 Physiological adaptations to phosphorus deficiency during proteoid root development in white lupin. Planta 208, 373–382.

Ni J J, Wu P, Lou A C, Zhang Y S and Tao Q N 1996 Low phosphorus effects on the metabolism of rice seedlings. Commun. Soil Sci. Plant Anal. 27, 3073–3084.

O'Connell K P, Goodman R M and Handelsman J 1996 Engineering the rhizosphere: expressing a bias. Trends Biotechnol. 14, 83–88.

Ozawa K, Osaki M, Matsui H, Honma M and Tadano T 1995 Purification and properties of acid phosphatase secreted from lupin roots under phosphorus-deficiency conditions. Soil Sci. Plant Nutr. 41, 461–469.

Parashar R D, Hooda I and Sindhan G S 1992 Penetration and infection of different chickpea cultivars by *Rhizoctonia solani*. Plant Dis. Res. 7, 60–63.

Pellet D M, Grunes D L and Kochian L V 1995 Organic acid exudation as an aluminum-tolerance mechanism in maize (*Zea mays* L.). Planta 196, 788–795.

Pellet D M, Papernik L A and Kochian L V 1996 Multiple aluminum-resistance mechanisms in wheat. Roles of root apical phosphate and malate exudation. Plant Physiol. 112, 591–597.

Rengel Z 1992 Role of calcium in aluminium toxicity. New Phytol. 121, 499–513.

Rengel Z 1997 Root exudation and microflora populations in rhizosphere of crop genotypes differing in tolerance to micronutrient deficiency. Plant Soil 196, 255–260.

Rengel Z 1999 Physiological mechanisms underlying differential nutrient efficiency of crop genotypes. *In* Mineral Nutrition of Crops: Mechanisms and Implications. Ed. Z Rengel. pp. 227–265. The Haworth Press, New York, USA.

Rengel Z, Römheld V and Marschner H 1998 Uptake of zinc and iron by wheat genotypes differing in zinc efficiency. J. Plant Physiol. 152, 433–438.

Römheld V 1991 The role of phytosiderophores in acquisition of iron and other micronutrients in graminaceous species: an ecological approach. Plant Soil 130, 127–134.

Römheld V and Marschner H 1990 Genotypical differences among graminaceous species in release of phytosiderophores and uptake of iron phytosiderophores. Plant Soil 123, 147–153.

Rosenblueth M, Hynes M F and Martinez-Romero E 1998 *Rhizobium tropici* teu genes involved in specific uptake of *Phaseolus vulgaris* bean-exudate compounds. Mol. Gen. Genet. 258, 587–598.

Rossbach S, Kulpa D A, Rossbach U and de Bruijn F J 1994 Molecular and genetic characterization of the rhizopine catabolism (*mocABRC*) genes of *Rhizobium meliloti* L5-30. Mol. Gen. Genet. 245, 11–24.

Ryan P R, Delhaize E and Randall P J 1995a Characterization of Al-stimulated efflux of malate from the apices of Al-tolerant wheat roots. Planta 196, 103–110.

Ryan P R, Delhaize E and Randall P J 1995b Malate efflux from root apices and tolerance to aluminium are highly correlated in wheat. Aust. J. Plant Physiol. 22, 531–536.

Ryan P R, Skerrett M, Findlay G P, Delhaize E and Tyerman S D 1997 Aluminum activates an anion channel in the apical cells of wheat roots. Proc. Natl. Acad. Sci. USA 94, 6547–6552.

Schlegel R, Cakmak I, Torun B, Eker S and Köleli N 1997 The effect of rye genetic information on zinc, copper, manganese and iron concentation of wheat shoots in zinc-deficient soil. Cereal Res. Commun. 25, 177–184.

Schlegel R, Kynast R, Schwarzacher T, Römheld V and Walter A 1993 Mapping genes for copper efficiency in rye and the relationship between copper and iron efficiency. Plant Soil 154, 61–65.

Schmidt P E, Broughton W J and Werner D 1994 Nod factors of *Bradyrhizobium japonicum* and *Rhizobium* sp. NGR234 induce flavonoid accumulation in soybean root exudate. Mol. Plant-Microbe Interact. 7, 384–390.

Shaul O and Galili G 1992 Threonine overproduction in transgenic tobacco plants expressing a mutant desensitized aspartate kinase of *Escherichia coli*. Plant Physiol. 100, 1157–1163.

Shinde N V and Deshmukh R B 1989 Effect of chickpea root exudates on *Fusarium oxysporum* f.sp. *ciceri*. Indian J. Pulses Res. 2, 90–92.

Skene K R 1998 Cluster roots: some ecological considerations. J. Ecol. 86, 1060–1064.

Soejima H, Sugiyama T and Ishihara K 1992 Changes in cytokinin activities and mass spectrometric analysis of cytokinins in root exudates of rice plants (*Oryza sativa* L.). Comparison between cultivars Nipponbare and Akenohoshi. Plant Physiol. 100, 1724–1729.

Stevenson P C, Padgham D E and Haware M P 1995 Root exudates associated with the resistance of four chickpea cultivars (*Cicer arietinum*) to two races of *Fusarium oxysporum* f.sp. *ciceris*. Plant Pathol. 44, 686–694.

Ström L 1997 Root exudation of organic acids: importance to nutrient availability and the calcifuge and calcicole behaviour of plants. Oikos 80, 459–466.

Ström L, Olsson T and Tyler G 1994 Differences between calcifuge and acidifuge plants in root exudation of low-molecular organic acids. Plant Soil 167, 239–245.

Subbarao G V, Ae N and Otani T 1997 Genotypic variation in iron- and aluminum-phosphate solubilizing activity of pigeonpea root exudates under P deficient conditions. Soil Sci. Plant Nutr. 43, 295–305.

Tadano T, Ozawa K, Sakai H, Osaki M and Matsui H 1993 Secretion of acid phosphatase by the roots of crop plants under phosphorus-deficient conditions and some properties of the enzyme secreted by lupin roots. Plant Soil 155/156, 95–98.

Timonin M I 1946 Microflora of the rhizosphere in relation to the manganese-deficiency disease of oats. Soil Sci. Soc. Amer. Proc. 11, 284–292.

Tyler G and Ström L 1995 Differing organic acid exudation pattern explains calcifuge and acidifuge behaviour of plants. Ann. Bot. 75, 75–78.

Uren N C and Reisenauer H M 1988 The role of root exudates in nutrient acquisition. Adv. Plant Nutr. 3, 79–114.

von Wiren N, Klair S, Bansal S, Briat J-F, Khodr H, Shioiri T, Leigh R A and Hider R C 1999 Nicotianamine chelates both Fe^{III} and Fe^{II}. Implications for metal transport in plants. Plant Physiol. 119, 1107–1114.

von Wiren N, Marschner H and Römheld V 1995 Uptake kinetics of iron-phytosiderophores in two maize genotypes differing in iron efficiency. Physiol. Plant. 93, 611–616.

von Wiren N, Mori S, Marschner H and Römheld V 1994 Iron inefficiency in maize mutant *ys1* (*Zea mays* L cv. yellow-stripe) is caused by a defect in uptake of iron phytosiderophores. Plant Physiol. 106, 71–77.

Walter A, Römheld V, Marschner H and Mori S 1994 Is the release of phytosiderophores in zinc-deficient wheat plants a response to impaired iron utilization? Physiol. Plant. 92, 493–500.

Wasaki J, Michiko A, Ozawa K, Omura M, Osaki M, Ito H, Natsui H and Tadano T 1997 Properties of secretory acid phosphatase from lupin roots under phosphorus-deficient conditions. Soil Sci. Plant Nutr. 43, 981–986.

Waschutza S, Hofmann N, Niemann E G and Fendrik I 1992 Investigations on root exudates of Korean rice. Symbiosis 13, 181–189.

Welch R M 1995 Micronutrient nutrition of plants. CRC Crit. Rev. Plant Sci. 14, 49–82.

Wojtaszek P, Stobiecki M and Gulewicz K 1993 Role of nitrogen and plant growth regulators in the exudation and accumulation of isoflavonoids by roots of intact white lupin (*Lupinus albus* L.) plants. J. Plant Physiol. 142, 689–694.

Zhang F, Römheld V and Marschner H 1989 Effect of zinc deficiency in wheat on the release of zinc and iron mobilizing root exudates. Z. Pflanzenernähr. Bodenk. 152, 205–210.

Zhang F S, Ma J and Cao Y P 1997 Phosphorus deficiency enhances root exudation of low-molecular weight organic acids and utilization of sparingly soluble inorganic phosphates by radish (*Raphanus sativus* L.) and rape (*Brassica napus* L.) plants. Plant Soil 196, 261–264.

Zheng S J, Ma J F and Matsumoto H 1998a Continuous secretion of organic acids is related to aluminium resistance during relatively long-term exposure to aluminium stress. Physiol. Plant. 103, 209–214.

Zheng S J, Ma J F and Matsumoto H 1998b High aluminum resistance in buckwheat. I. Al-induced specific secretion of oxalic acid from root tips. Plant Physiol. 117, 745–751.

Zuanazzi J A S, Clergeot P H, Quirion J C, Husson H P, Kondorosi A and Ratet P 1998 Production of *Sinorhizobium meliloti* nod gene activator and repressor flavonoids from *Medicago sativa* roots. Mol. Plant-Microbe Interact. 11, 784–794.

J.J. Adu-Gyamfi (Ed.), Food security in nutrient-stressed environments: exploiting plants' genetic capabilities: 227–234, 2002.

Transporter genes to enhance nutrient uptake: Opportunities and challenges

James Dunlop[1] & Thai Phung
AgResearch, Grasslands Research Centre, Private Bag 11008, Palmerston North, New Zealand. [1]*Corresponding author**

Received 10 May 2000. Accepted in revised form 24 October 2000

Key words: breeding, fuzzy cognitive mapping, genetic modification, metabolic control analysis, phosphate

Abstract

Recently, there has been very rapid progress in understanding the transport of mineral nutrients across plant membranes. Genes for a number of primary ion pumps, cotransporters and ion channels have been cloned and the characteristics of their function are being investigated. While these advances have yet to produce cultivars that are better able to cope with nutrient deficient soils, they provide powerful tools to address some important gaps in our knowledge, particularly the regulation of transporter genes. While the current focus is on nutrient influx into roots, other processes are also significant in determining nutrient acquisition. This is illustrated by research on the relationship between the high affinity phosphate absorption mechanism and phosphate transport to the shoot, and with data on relating to the importance of phosphate efflux relative to influx. Molecular biology and plant physiology are providing information and technology that will be valuable in improving the suitability of crops for nutrient deficient environments. However, there is an urgent need for new ways to integrate this information so that the significance of individual processes to the performance of the whole system can be understood.

Introduction

The cloning of genes coding ion transporters i.e. proteins that provide ways by which nutrients can cross membranes, may seem to validate and provide a fitting climax to the pioneering work on the kinetics of nutrient ion absorption by Epstein (1966). However, Epstein foresaw another conclusion for his work – the development of new cultivars which were better able to cope with salinity or low nutrient availability. Very few cultivars which decrease fertiliser requirements or extend the range of low nutrient environments in which crops can be productive have been developed (Maranville, 1993). In most cases, the suite of characters that would confer nutrient efficiency is still unclear. However, during the past decade, much more detailed information about the molecular structures involved in influx into, and efflux from, cells and movement from cell to cell through plasmodesmata has been obtained (Chrispeels et al., 1999; Lucas, 1997). Genes that code for transporter proteins for all major inorganic nutrients and some of the minor nutrients have been cloned from plants (Chrispeels et al., 1999). The availability of these genes gives rise to the expectation that they will facilitate development of cultivars with improved nutritional properties.

Plant transporter genes

There are three broad types of transporters – primary ion pumps, ion channels and cotransporters (Chrispeels et al., 1999). Primary ion pumps use ATP directly as an energy source to pump ions across membranes, usually against a potential energy (diffusion) gradient. The most important examples of these are H^+-ATPases which pump H^+ ions to produce potential energy in the form of pH and electrical potential gradients for driving many other secondary transport

* FAX No: +64 63518042. Tel.: +64 3518108.
E-mail: dunlopj@agresearch.cri.nz

This paper previously appeared in *Plant and Soil* **245**, 2002, pp. 115–122.

processes (Sze et al., 1999). H^+-ATPases were among the first transporter genes cloned. Ion channels are protein structures that span membranes and enable the diffusion of ions down potential energy gradients (Hedrich and Schroeder, 1989). Flow through a single ion channel is usually rapid – 10^6–10^7 molecules per channel per second. Channels are not static pores but are dynamic molecules whose function can be regulated by a variety of factors including voltage, pH and various metabolites. Their activity can be investigated by studying the currents that are associated with the flows of ions. Co-transporters transport by utilising gradients in electrical potential or concentration generated by the H^+-ATPases (Tanner and Caspari, 1996). Anion transporters are commonly symporters that obtain the necessary energy by simultaneously co-transporting the second chemical ion, normally H^+, down its gradient. Antiporters, also called exchangers, obtain their energy by the simultaneous transport of a secondary chemical species in the opposite direction to the primary substrate. Cotransporters function at slower rates than ion channels, 10^2–10^4 per molecule per second. It was originally thought that cotransporters were quite distinct from channels, but recent work has shown that cotransporters can function like ion channels under certain conditions (Cammack and Schwartz, 1996; Fairman et al., 1995; Gassmann et al., 1996). Physiological and molecular studies of transporters commonly distinguish between them on the basis of their affinity (value for K_m) for their substrate. Epstein (1966) demonstrated that roots possess multiple mechanisms for the absorption of specific ions that are primarily distinguished by their different affinities for that ion. This foreshadowed the cloning of genes coding for transporters with different affinities (see Table 1 below). Recently, it has been demonstrated that the wheat K^+ transporter, *HKT1*, can function in two modes with different affinities (Gassmann et al., 1996; Rubio et al., 1995).

The identification of transporter genes from angiosperms has relied heavily on yeast mutants that are deficient in the ability to transport a specific nutrient (Chrispeels et al., 1999; Dreyer et al., 1999). When the expression of a gene in the mutant restores its ability to transport, it is taken as evidence that the gene codes for a transporter. *E. coli*, insect cells and oocytes from *Xenopus laevis*, have also been used as tools in testing genes for a transport function (Dreyer et al., 1999). In the case of oocytes, cDNA is injected and, following incubation, the oocyte is examined for membrane electrical currents that are consistent with transport of the ion being studied. Another approach has been to search for plant genes that share a high degree of identity with transporter genes from other species. Hydrophobicity plots for the amino acid sequences of a putative transporter protein that may show strongly hydrophobic zones are indicative of a protein that spans a membrane and therefore may have a transport function (Tanner and Caspari, 1996). The number of membrane spanning regions can vary from three up to 12. If expression of the gene is regulated in a manner parallel to regulation of the putative ion transport mechanism, confidence that the gene codes for a transporter is increased (Chrispeels et al., 1999).

The work involved in identifying, cloning and characterising transporter genes is very considerable but notwithstanding this, an impressive number have been cloned. The range of transporter genes that have been reported is indicated in Table 1, which gives a representative list but not a complete catalogue. There are frequent examples of multiple genes from a single species coding for very similar transporters (i.e. sharing a high degree of homology) e.g. *SHST1* and *SHST2*, genes from *Stylosanthes hamata* which code for high affinity sulphate transporters, share 95% expressed sequence identity (Smith et al., 1995a), and *APT1* and *APT2,* which code for high affinity phosphate transporters in *Arabidopsis thaliana*, are 99% identical (Smith et al., 1997). In some cases, the various versions of transporters may be expressed in different parts of the plant. The genes *LeNRT1;1* and *LeNRT1;2*, cloned from tomato, both code nitrate transporters expressed only in the root. Expression of *LeNRT1;2* is nitrate inducible and restricted to root hairs; *LeNRT1;1* is constitutive and restricted to roots, but not only root hairs (Lauter et al., 1996). However, the significance of multiple versions of genes is not always clear.

Based on their amino acid sequence and/or function, many transporter genes can be grouped into families that may span genera and even the plant and animal kingdoms, indicating their highly conserved nature. Examples of these noted in Table 1 include the P-types ATPases, primary ion pumps, of which the genes *ACA1*, *ACA2* and *BCA1*, which code for Ca^{2+} transporters in endomembranes, and *PAA1* and *AMA1* putative transporters of Cu from the cytosol (Fox and Guerinot, 1998). The Shaker K^+ channels, originally identified in *Drosophylla*, are ion channels represented in plants by *KAT1* and *AKT1*, which are involved in low affinity absorption of K^+ (Schachtman et al., 1992; Sentenac et al., 1992). There are two famil-

Table 1. Transporter genes cloned from plants. This is a representative sample of genes, not a complete list

Substrates	Examples	Function	Characteristics[a]
NH_4^+	*AMT1*	High affinity NH_4^+ transport Located in root hairs Constitutive Expression affected by NO_3^-	9-12 MSRs
NO_3^-, peptides, basic amino acids	*AtNRT1(CHL1)* *AtNRT3* *BnNRT1;2*	Low or dual affinity, NO_3^- transport Located primarily in roots	**NRT1** or nitrate transporter family1 12 MSRs[b]
NO_3^-, (and NO_2^- in one case)	*HvNRT2A* *AtNRT2;1 & 2* *GmNRT2* *NpNRT2;1*	High affinity, NO_3^- inducible, transport Mainly located in roots Down regulated by reduced N	**NRT2** or nitrate transporter family 2 12 MSRs
$H_2PO_4^-$	*LePT1, LePT2,* *APT1, APT2* *MtPT1, StPT1*	High affinity $H_2PO_4^-$ transport Derepressed in low P	12 MSRs
SO_4^{2+}	*SHST1* *SHST2* *HVST1*	High affinity SO_4^{2+} transport Increased expression in low S Root specific	12 MSRs
SO_4^{2+}	*SHST3* *AST56* *AST68*	Internal SO_4^{2+} transport Lower affinity	12 MSRs
K^+, Na^+	*HKT1* *KEA1*	High affinity transport Located in root cortex and shoot and leaf vascular tissue	10–12 MSRs
K^+, Na^+	*KAT1* *AKT1*	Low affinity inward transport i.e. K^+ channel Locations include guard cells, vascular tissue and flowers	**Shaker K^+ channels** family 6 MSRs Cyclic nucleotide binding region
K^+, Na^+ and, possibly, Ca^{2+}	*LCT1*	Low affinity transport of K^+ and Na^+ Present in both leaves and roots	**Unique to plants** 8–10 MSRs Sequences enriched in Pro, Ser, Thr and Glu (PEST)
K^+	*KCO1*	Outward K^+ movement Outward rectifying K^+ channel Present in seedlings, leaves and flowers	4MSRs
Mg^{2+}	*ALR1* *ALR2*	Mg^{2+} transport	3 MSRs

Table 1. Continued

Substrates	Examples	Function	Characteristics[a]
Ca^{2+}	*ACA1* *ACA2* *BCA1*	PM-type Ca^{2+}-ATPases Ca^{2+} transport in endo- membranes, e.g. tonoplast Stimulated by calmodulin	**P-type ATPase** family
Ca^{2+}	*LCA1* *PH27* *ACA3*	ER-type Ca^{2+}-ATPases, Ca^{2+}transport in endo- membranes and plasmalemma Not stimulated by calmodulin	**P-type ATPase** family
Ca^{2+}	*CAX1* *CAX2*	Ca^{2+}/H^{+} antiporters Extrusion of Ca^{2+} from cytoplasm into the vacuole *CAX1 is high affinity,* *CAX2 is low affinity.*	11 MSRs
Cu	*COPT1*	High affinity Cu transport Present in flowers, stems, leaves	3 MSRs
Cu	*PAA1* **AMA1**	Cu-ATPases Probable function is transport of Cu from the cytosol.	**P-type ATPase** family
Mn (?)	*OsNramp1* *OsNramp2* *OsNramp3*	Probable function is high affinity Mn transport. *OsNramp1* is located in roots and *OsNramp2* in shoots.	**Nramp** family 10 MSRs
Zn	*ZIP1* *ZIP2*	Transport of Zn with affinities (Km) of 10–100 nM.	**ZIP transporter** family
Fe Fe(III)-chelate reductase	*frohA frohB* *frohC*	Reduction of chelated Fe(III)	
Fe Fe(II) transport	*IRT1*	Transport of reduced Fe(III)	**ZIP transporter** family

[a]MSRs = Hydrophobic membrane spanning regions.
[b]Hydrophobic membrane spanning regions.

ies of nitrate transporters, *NRT1*, which has low or dual affinity transport, and *NRT2* which has high affinity transport. The *NRT2* family is conserved among higher plants, algae and fungi (Crawford and Glass, 1998).

The examples listed in Table 1 include transporters for all the major plant nutrients and four trace elements, Cu, Mn, Zn and Fe. Comprehensive accounts of the molecular biology of transporters are available in the reviews by Crawford and Glass (1998), Fox and Guerinot (1998), Chrispeels et al. (1999) and Schachtman and Liu (1999). The cloning of so many plant nutrient transporters represents a very important advance in our understanding of processes involved in the mineral nutrition of plants. However, the job is not yet complete (Somerville and Somerville, 1999). The function of a significant number of these genes is best described as deduced rather than demonstrated.

For many of these genes, the location of their expression in plants is not known and their exact role in the physiology of plants is not yet clear (Chrispeels et al., 1999; Fox and Guerinot, 1998). Furthermore, there are some substantial aspects of nutrient absorption and transport in plants that are not yet elucidated in molecular terms. One such area is the regulation of expression of transporter genes. This gap in our knowledge may be a particularly significant impediment to designing and engineering plants better suited for nutrient deficient conditions. This can be illustrated by pointing to the development of modern maize from its ancestor, teosinte. Research on the domestication of maize (Wang et al., 1997) indicates that changes in the regulation of gene expression rather than the introduction of new genes have produced the modern, productive, agronomic form of maize. There are few reports of the identification and function of promoters that control expression of transporter genes.

The ability of plant roots to absorb nutrients is seldom dominated by a single process such as the maximum rate or affinity of an ion transporter (Crawford and Glass, 1998). The efflux of previously absorbed nutrients, their transport to the shoot, growth of the shoot, the compartmentation and metabolism of nutrients, and the morphology of the root can all affect the net rate of nutrient absorption. The following examples from our laboratory are used to illustrate recent work on these processes.

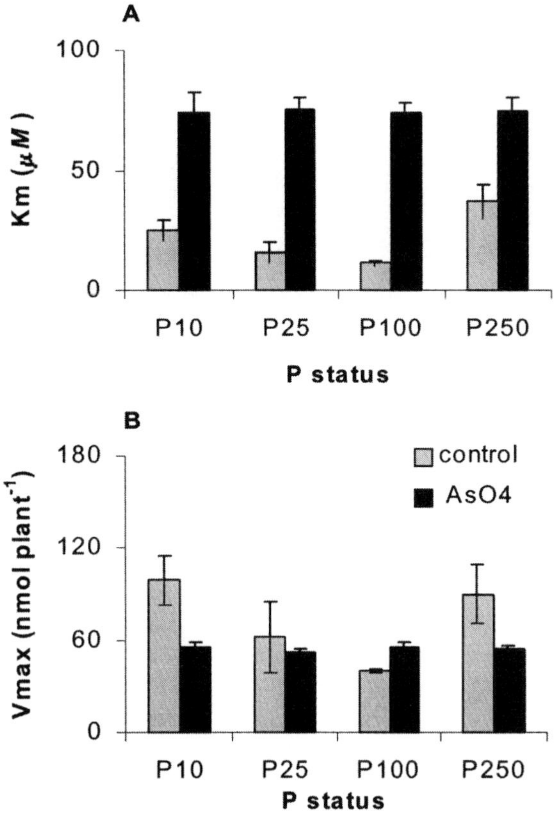

Figure 1. The influence of P status and 20 μM KH_2AsO_4 on values for the constants K_m and V_{max} for ^{32}P-labelled phosphate absorption by *Arabidopsis thaliana.* (A) K_m. (B) V_{max}. Data are means of four replicates ±± s.d. and are derived from Clark et al. (1999). Pretreatments range from very P deficient (P10) to P sufficient (P250).

Transport to the shoot controls absorption of phosphate by roots

Evidence that ions or metabolites translocated from the shoot by phloem can regulate ion absorption by the root has been reported by Dong et al. (1998) for phosphate and Laparttient et al. (1999) for sulphate. In a recent study of phosphate absorption by *Arabidopsis thaliana*, we demonstrated that inhibiting phosphate transport to the shoot inhibits phosphate absorption by the root with high affinity transport being the most severely affected (Clark et al., 1999). Phosphate absorption by the roots of the intact, 21-day old plants of differing phosphorus (P) status was measured by following the depletion of ^{32}P-phosphate from a 5 μM solution. Different P pre-treatments resulted in a range of rates of absorption with the highest rate for the most P deficient plants. The differences in rates of absorption were reflected in differences in both K_m (affinity) and V_{max} (maximum velocity) of the transport system (see Figure 1). The values for K_m were indicative of a high affinity transport system. When 20 μM arsenate was added to the absorption solutions of a parallel set of plants, there were two dramatic effects. Firstly, there were large decreases in the rates of $^{32}PO_4$ absorption, the extent of which depended on the P status of the plant – P deprived plants experienced a larger decrease such that the rates for all treatments were the same and the depletion curves overlay each other. This is indicated by Figure 1 in which K_m and V_{max} for absorption in the presence of arsenate were the same for all phosphorus levels. K_m in the presence of arsenate was significantly higher than for the controls. The second effect of arsenate was on the distribution of ^{32}P label between the roots and shoots. The amounts of ^{32}P in roots were not significantly affected by arsenate but in shoots it decreased by amounts ranging from 92% in the most phosphorus deficient plants to 67%

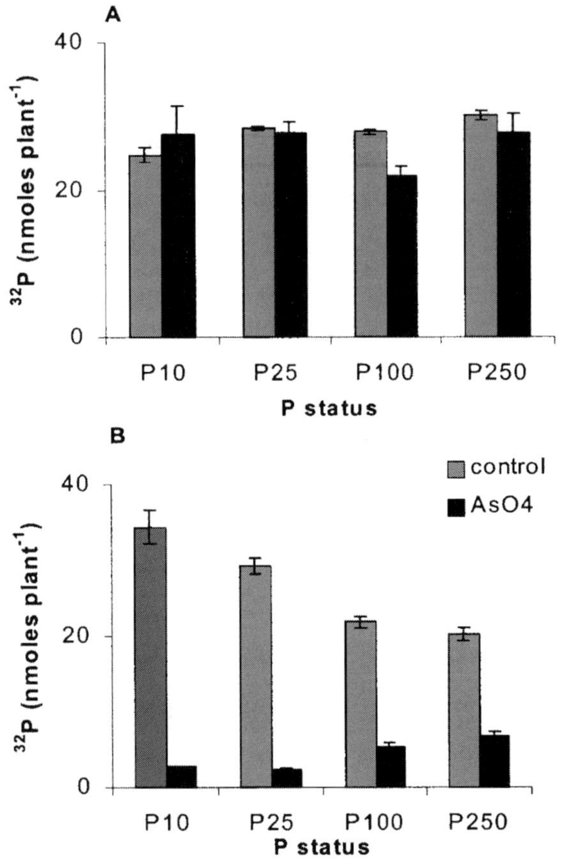

Figure 2. The influence of P status and 20 μM KH_2AsO_4 on contents of ^{32}P-labelled phosphate in roots and shoots of *Arabidopsis thaliana* plants following 12 h in solutions initially containing ^{32}P-labelled 10 μM KH_2PO_4. (A) Roots. (B) Shoots. Data are means of four replicates ±± s.d. and are derived from Clark et al. (1999). Pretreatments range from very P deficient (P10) to P sufficient (P250).

in phosphorus sufficient plants (Figure 2). This result was unexpected, as a previous study of Arabidopsis had shown only a slight inhibition of phosphate uptake by arsenate (Dunlop et al., 1997). This and other earlier studies, which concluded that arsenate is a weak competitive inhibitor of phosphate absorption, used conditions which minimised transport from the root to the shoot. However, Figure 2 demonstrates a severe inhibition of phosphate transport from roots to shoots by a low level of arsenate which coincides with a strong inhibition of absorption via the roots and elimination of the differences in absorption caused by differing P status. The associated increase in K_m indicated inhibition of the high affinity transport system induced by P deficiency. These results indicate that the ability of roots to absorb phosphate by the high affinity transport mechanism and the ongoing transport of absorbed phosphate to the shoots are closely linked. This is pertinent to strategies to increase the ability of roots to take up scarce phosphate by increasing expression of high affinity phosphate transporters as it suggests it may be necessary to simultaneously increase transport to the shoot.

The significance of efflux

The use of radioactive isotopes to measure nutrient absorption emphasises unidirectional influx and frequently ignores the effect of efflux, but comparison of phosphate absorption (net flux), measured chemically, with influx measured using ^{32}P, illustrates that efflux can be very significant. Phosphate efflux is a normal process of healthy plants with rates that are higher when plants have a high internal phosphorus status or roots are bathed in high concentrations of phosphate (Bieleski and Läuchli, 1992). In some situations, the rate of efflux approaches the rate of influx (McPharlin and Bieleski, 1989; Macklon et al., 1996). In *Arabidopsis*, influx resulted in an immediate rapid depletion of ^{32}P from the solution, whereas the chemical measurements showed no net influx of phosphate for at least 1 h and as long as 3 h for P sufficient plants (Clark et al., 2000).

In another demonstration of the significance of efflux, the characteristics of phosphate influx and efflux of two pastoral species, *Trifolium repens* L., a weak competitor for phosphate, and *Lolium perenne* L., a strong competitor, were measured simultaneously by double labelling with ^{32}P and ^{33}P (Dunlop and Phung, 1999). Rates of influx for *L. perenne* grown under phosphorus deficiency were 32–91% higher than the corresponding values for *T. repens* L., but in phosphorus with sufficient conditions, there were no significant differences between the two species. There were larger relative differences in rates of efflux between species with the values for phosphorus deficient *T. repens* being 303–322% higher than *L. perenne*. Efflux for P sufficient plants was 5–30 fold higher than for P deficient plants. After 2 h, 87% of the total phosphate influx in low P *T. repens* plants was offset by efflux compared with 26% in *L. perenne*. If there are corresponding differences in plants growing in soil, the higher rates of phosphate efflux in *T. repens* are likely to be an important factor in its poorer ability to compete with *L. perenne* for phosphate.

Bridging the gap between cell biology and plant breeding

The relationships between high affinity phosphate influx and phosphate transport to the shoot and the balance between influx and efflux are just two examples of many individual biophysical and metabolic processes that interconnect in extremely complex ways to provide the functions collectively known as plant nutrition. These complex networks of interacting processes make it very difficult to predict the outcome of manipulating a single component of the system. "Metabolic systems are networks with great density of connections and counter-intuitive properties. The counter-intuitive character of network behaviour is that properties are shared throughout the network by virtue of the connections... Parts do not function in isolation, but only in a context or framework of the whole of metabolism" (Trewavas, 1986). The individual processes associated with transporter genes are connected in a complex way with each other and other processes that make it impossible to predict the consequences of altering the expression of any of these genes with certainty. Consequently, genetic modification is frequently unable to produce plants that can reliably give the desired outcome. Plant physiologists have a similar difficulty in generating information from research that can be applied to plant breeding (Jackson et al., 1996). To deal with this problem, Wallace and Yan (1998) have developed a 'whole system' approach to crop physiology that emphasises measurement of the 'penultimate outputs' which are "the plant processes whose further integration accounts in full for the system output". However, while this approach may produce results, it would receive little benefit from the rapidly burgeoning information on plant molecular biology and physiology. There is need for a method to integrate the information these disciplines produce and evaluate the influence that various components exert on the properties being sought (Dunlop, 1999). Such a method will have to cope with the disparate experimental conditions and species used in producing plant physiological and molecular biological knowledge, and it also needs to capture the insight developed by the experienced practitioners of these sciences. Techniques such as fuzzy cognitive mapping (Taber, 1991) or metabolic control analysis (Cornish-Bowden, 1999) may provide an initial approach.

Conclusions

Molecular biology has hastened progress in understanding the absorption of essential nutrients by roots and has identified genes that code for many membrane transport proteins involved, including some for all of the major elements and some of the trace elements. However, the use of these genes to develop plants better suited to infertile soils is not a simple matter. Cloning of the promoters controlling expression of these genes is likely to prove very important, but it is less advanced. There is evidence that influx mechanisms do not constitute the over-riding determinant of nutrient acquisition. The function of the high affinity phosphate co-transporters in roots requires the onward transport of phosphate to the shoot. Furthermore, loss of previously absorbed phosphate through efflux is substantial and can be a significant factor in the competitive acquisition of phosphorus from soils. Other factors not discussed in this paper will also have an important role in determining nutrient acquisition, the most important of which is probably root morphology. Genetic techniques will play a crucial role in researching the questions that must be answered before we can directly employ transporter genes to engineer plants better suited for nutrient stressed environments.

References

Bieleski R L and Läuchli A 1992 Phosphate uptake, efflux and deficiency in the water fern, *Azolla*. Plant Cell Environ. 15, 665–673.

Cammack J N and Schwartz E A 1996 Channel behavior in a γ-aminobutyrate transporter. Proc. Natl. Acad. Sci. USA 93, 723–727.

Clark G T, Dunlop J and Phung H T 2000 Phosphate absorption by *Arabidopsis thaliana*: Interactions between phosphorus status and inhibition by arsenate. Aust. J. Plant Physiol. 27, (in press).

Chrispeels M J, Crawford N M and Schroeder J I 1999 Proteins for transport of water and mineral nutrients across the membranes of plant cells. Plant Cell 11, 661–675.

Clarkson D T and Grignon C 1991 The phosphate transport system and its regulation in roots. *In* Phosphorus Nutrition of Grain Legumes in the Semi-arid Tropics. Eds. C Johansen, KK Lee and KL Sarawat. pp. 49–62. ICRISAT, Pancheru, A.P., India.

Cornish-Bowden A 1999 Metabolic control analysis in biotechnology and medicine. Nature Biotech. 17, 641–643.

Crawford N M and Glass A D M 1998 Molecular and physiological aspects of nitrate uptake in plants. Trends Plant Sci. 3, 389–395.

Dong B, Rengel Z and Delhaize E 1998 Uptake and translocation of phosphate by *pho2* mutant and wild-type seedlings of *Arabidopsis thaliana*. Planta 205, 251–256.

Dreyer I, Horeau C, Lemaillet G, Zimmerman S, Bush D R, Rodriguez-Navarro A, Schachtman D P, Spalding E P, Sentenac H and Gaber R 1999 Identification and characterisation of plant transporters using heterologous expression systems. J. Exp. Bot. 50, 1073–1087.

Dunlop J 1999 The physiology of phosphorus nutrition in plants – can we assemble the components into a practical model? *In* Plant Nutrition: Molecular Biology and Genetics. Eds. G Gissel-Nielsen and A Jensen. pp. 371–377. Kluwer Academic Publishers, Dordrecht, The Netherlands.

Dunlop J and Phung H T 1999 Efflux and influx as factors determining the relative abilities of ryegrass and white clover to compete for phosphate. *In* Plant Nutrition: Molecular Biology and Genetics. Eds. G Gissel-Nielsen and A Jensen. pp. 105–110. Kluwer Academic Publishers, Dordrecht, The Netherlands.

Epstein E 1966 Dual pattern of ion absorption by plant cells and by plants. Nature 212, 1324–1327.

Fairman W A, Vandenberg R J, Arriza J L, Kavanaugh M P and Amara S G 1995 An excitatory amino-acid transporter with the properties of a ligand-gated chloride channel. Nature 375, 599–603.

Fox T C and Guerinot M L 1998 Molecular biology of cation transport in plants. Annu. Rev. Plant Physiol. Plant Mol. Biol. 49, 669–696.

Gassmann W, Rubio F and Schroeder J I 1996 Alkali cation selectivity of the wheat root high affinity potassium transporter HKT1. Plant J. 10, 869–882.

Hedrich R and Schroeder J I 1989 The physiology of ion channels and electrogenic pumps in higher plants. Ann. Rev. Plant Physiol. 40, 539–569.

Jackson P, Robertson R, Cooper M and Hammer G et al. 1996 The role of physiological understanding in plant breeding: from a breeding perspective. Field Crops Res. 49, 11–37.

Lappartient A G, Vidmar J J, Leustek T, Glass A D M and Touraine B 1999 Inter-organ signalling in plants: Regulation of ATP sulfurylase and sulfate transporter genes expression in roots mediated by phloem-translocated compound. Plant J. 18, 89–95.

Lauter R-R, Ninneman O, Bucher M, Riesmeier J W and Frommer W B 1996 Preferential expression of an ammonium transporter and two putative nitrate transporters in root hairs of tomato. Proc. Natl. Acad. Sci. 93, 8138–8144.

Lucas W J 1997 Application of microinjection techniques in plants. *In* Plant Nutrition for Sustainable Food Production and Environment. Developments in Plant and Soil Sciences, Volume 78. Eds. T Ando, K Fujita, T Mae, H Matsumoto, S Mori and J Sekiya. pp. 25–39. Kluwer Academic Publishers, Dordrecht, The Netherlands.

Macklon A E S, Lumsdon D G, Sim A and McHardy W J 1996 Phosphate fluxes, compartmentation and vacuolar speciation in root cortex cells of intact *Agrostis capillaris* seedlings: Effects of non-toxic levels of aluminium. J. Exp. Bot. 47, 793–803.

McPharlin I R and Bieleski R L 1989 P_i efflux and influx by P-adequate and P-deficient *Spirodela* and *Lemna*. Aust. J. Plant Physiol. 16, 391–399.

Maranville J W 1993 Adaptation of plants to soil stresses. Proceedings of a workshop at University of Nebraska, Lincoln, Nebraska. INTSORMIL Publication No. 94–2.

Rubio F, Gassmann W and Schroeder J I 1995 Sodium driven potassium uptake by the plant potassium transporter *HKT1* and mutations conferring salt tolerance. Science 270, 1660–1663.

Scachtman D and Liu W 1999 Molecular pieces to the puzzle of the interaction between potassium and sodium uptake in plants. Trends Plant Sci. 4, 281–287.

Schachtman D P, Schroeder J I, Lucas W J, Anderson J A and Gaber R F 1992 Expression of an inward-rectifying potassium channel by the *Arabidopsis KAT1* cDNA. Science 258, 1654–1658.

Sentenac H, Bonneaud N, Minet M, Lacroute F and Salmon J-M 1992 Cloning and expression in yeast of a plant potassium ion transport system. Science 256, 663–665.

Smith F W, Ealing P M, Hawkesford M J and Clarkson D T 1995a Plant members of a family of sulfate transporters reveal functional subtypes. Proc. Natl. Acad. Sci. USA 92, 9373–9377.

Smith F W, Ealing P M, Dong B and Delhaize E 1997 The cloning of two Arabidopsis genes belonging to a phosphate transporter family. Plant J. 11, 83–92.

Somerville C and Somerville S 1999 Plant functional genomics. Science 285, 380–383.

Sze H, Li X and Palmgren M J 1999 Energization of plant cell membranes by H^+-pumping ATPases: Regulation and biosynthesis. Plant Cell 11, 677–689.

Taber R 1991 Knowledge processing with fuzzy cognitive maps. Exp. Syst. Appl. 2, 83–87.

Tanner W and Caspari T 1996 Membrane transport carriers Annu. Rev. Plant Physiol. Plant Mol. Biol. 47, 595–626.

Trewavas A 1986 Understanding the control of plant development and the role of growth substances. Aust. J. Plant Physiol. 13, 447–457.

Wallace D H and Yan W 1998 Plant breeding and whole-system crop physiology. CAB International, Wallingford, Oxon, UK. 631 p.

Wang R-L, Stec A, Hey J, Lukens L and Doebley J 1997 The limits of selection during maize domestication. Nature 398, 236–239.

J.J. Adu-Gyamfi (Ed.), *Food security in nutrient-stressed environments: exploiting plants' genetic capabilities*: 235–244, 2002.

The phosphate uptake mechanism

Frank W. Smith
*CSIRO Tropical Agriculture, Long Pocket Laboratories, 120 Meiers Rd, Indooroopilly, Qld 4068 Australia**

Received 10 May 2000. Accepted in revised form 5 December 2000

Key words: cotransporters, ion transport, membrane proteins, plasmalemma, phosphate transporters, phosphate uptake

Abstract

The slow rate of diffusion of phosphate in soil results in a zone of depletion of phosphate ions in solution around the roots of plants in low phosphate soils. Transfer of phosphate to the site of uptake into the root symplasm limits phosphate uptake in such soils. This transfer involves movement across the depletion zone and through the root apoplasm. The apoplasm is made up of the cell walls of epidermal and cortical cells, together with the associated intercellular spaces. Although the pores in the open latticework of these cell walls permit movement of nutrients around cells, they increase the path length across which phosphate ions have to diffuse. The structural components and net negative charges of the cell walls also influence the effective concentrations of phosphate in the apoplasm. This concentration may be further modified by excreted organic compounds around cell walls and the presence of micro-organisms that use such compounds as carbon sources. A membrane on the inner surface of the cell wall, the plasmalemma, separates the apoplasm from the symplasm. Uptake of nutrients into the root symplasm occurs through transporter proteins embedded in this membrane. Understanding of the mechanisms by which phosphate is transported across the plasmalemma into the plant symplasm has advanced considerably over the past 4 years due to the application of molecular techniques. Genes encoding the transporters involved in this process have been isolated from a number of plant species. These transporters belong to a family of membrane proteins characterized by having 12 membrane-spanning domains arranged in a '6+6' configuration. $H_2PO_4^-$ ions, together with protons, are transported through this protein. This transport process is driven by the potential across the membrane maintained by the action of a H^+-ATPase, the 'proton pump', that extrudes protons to the outer surface of the membrane. The expression of genes encoding high-affinity root phosphate transporters is regulated by the phosphorus (P) status of the plant. The transduction pathway involved in this regulation is not known at present. It is a systemic response rather than a localized response, however, the overall phosphate status of the plant being the controlling factor. Under phosphate stress, the expression of genes encoding these phosphate transporters is up-regulated. This results in a greater number of transporter proteins in the plasmalemma and enhanced phosphate uptake rates, if phosphate is available at the membrane surface. Uptake occurs around the root tip, into epidermal cells with their associated root hairs and into cells in the outer layers of the root cortex. Further back along the root axis, phosphate can also be taken up by transfer from mycorrhizal fungi to root cortical cells.

Strategies for increasing nutrient uptake by overexpressing genes encoding high-affinity phosphate transporters are likely to be mainly applicable to situations where a reasonable phosphate concentration can be maintained at the outer surface of the plasmalemma. Maintaining such a concentration is a major problem in the phosphate deficient soils of the semi-arid tropics (SAT), so emphasis in these soils is on strategies to improve the movement of phosphate to the surface of the plasmalemma. There may be scope, however, for manipulating the expression of genes involved in the internal mobilisation of phosphate within the plant, thereby improving phosphate utilisation.

* FAX No: +617 3214 2288. E-mail: frank.smith@pi.csiro.au

This paper previously appeared in *Plant and Soil* **245**, 2002, pp. 105–114.

Introduction

Whilst soil moisture and nitrogen (N) are major limitations to agricultural production systems in the SAT, phosphorus (P) deficiency also limits crop growth on many soils. The cost and availability of phosphatic fertilizers to the majority of farmers in the region restrict their use. Attention has, therefore, turned to making more efficient use of the soil phosphate reserves by seeking crop genotypes and management systems that result in more effective uptake and utilization of soil-P. A number of promising strategies are being explored, many of which are presented in this Workshop. To be effectively developed, all of them require an understanding of the mechanisms of phosphate uptake and utilization by crop plants. Use of molecular tools by nutritional physiologists in recent years has considerably enhanced the understanding of these mechanisms and provided new opportunities for manipulating nutrient uptake and utilization. Key genes involved in the process have been identified and information on their role and regulation is accumulating. This paper provides a summary of the phosphate uptake process and highlights some of the important molecular mechanisms involved.

The external phosphate concentration

Plant roots acquire their phosphate from the external soil solution where it is in equilibrium with phosphate sorbed onto soil minerals and colloids. These sorption reactions maintain low concentrations of phosphate in soil solution whilst buffering the amount of phosphate in solution. The movement of phosphate ions to the sites through which it is taken up into root cells occurs by diffusion. This is a relatively slow process and, in P-deficient soils, results in the concentration of phosphate in solution being depleted around plant roots. Thus, many of the strategies for improving phosphate uptake are aimed at reducing this depletion zone and increasing the solution phosphate concentration immediately adjacent to the sites of phosphate uptake in the roots.

Extension of roots into undepleted regions of soil provides the root tip with external P concentrations similar to those in the bulk soil solution. Further back along the root axis extension of root hairs from epidermal cells in many plant species considerably increases the volume of soil explored for phosphate. Still further back, the soil volume explored by some species growing in low phosphate soils may be enhanced by the presence of hyphae of mycorrhizal fungi which can extend several centimetres from the root surface. A cone of soil in which the concentration of phosphate in solution is depleted thus develops back from the root tip. Within this zone the equilibrium of the phosphate sorption will have shifted towards release of sorbed phosphate ions into solution. Distance to the uptake sites within the root and any barriers to phosphate diffusion determine whether the plant can access these ions.

The root apoplasm

The walls of root epidermal and cortical cells and the associated intercellular spaces make up the apoplasm. In young roots, these walls are composed of interlaced fibres that form an open latticework (Peterson and Cholewa, 1998). Soil solution can therefore, move radially towards the central stelar region of the root through the pores in this latticework and the intercellular spaces. The suberised Casparian band around the tangential walls of endodermal cells prevents radial movement into the central stele of nutrients in the soil solution. The band also restricts nutrients within the stele from leaking out into the apoplasm. Older areas of some roots have another layer of suberised cells in the outer layers of cortical cells that form the exodermis. This layer further restricts apoplastic movement of external soil solution in these regions of the root. In slower growing roots, such as those on plants subjected to stress, the exodermis may be formed closer to the tip than in rapidly growing roots (Perumalla and Peterson, 1986). Movement of solutes through the apoplasm also appears to be restricted near the meristematic region close to the root tip where the microfibrils of the cell walls appear densely packed (Peterson and Cholewa, 1998).

The interlacing fibres of cell walls in the apoplasm serve to filter soil solution. They also increase the path length over which phosphate ions must diffuse to the underlying uptake sites on the plasmalemma. The presence of carboxyl groups associated with the pectic polysaccharides of the cell wall fibres results in an overall negative charge. Anions such as phosphate are repelled by this charge and restricted to the larger pores within the apoplasm. Mucilages, excreted into cell walls and surrounding many roots, carry negatively charged hydroxyl groups which can further modify the flow of anions. These, and other root excretions, provide substrates for rhizosphere micro-organisms that can influence nutrient concentrations close to the uptake sites. The net effect is that movement of phosphate may be impeded within the apoplast, further modifying the concentration of phosphate at the outer surface of the plasmalemma, particularly in cells in the inner cortex. Even in soils well supplied with phosphate this concentration is likely to be less than 2 micromolar. In the P-deficient soils of the SAT, the concentration will be much lower than this.

Uptake of phosphate into the symplasm

The plasmalemma of root epidermal and cortical cells provides the boundary between the apoplasm and the symplasm. Once inside the symplasm, nutrient ions in the cytoplasm can move radially through to the stele via plasmodesmata connections without encountering further membrane barriers (Clarkson, 1993). Transport of ions across the semipermeable plasmalemma is, therefore, a critical step that mediates and regulates the uptake of nutrients into the plant. The physiology and kinetics of transport of nutrients across the plasmalemma has been known for a long time. Epstein and colleagues (Epstein and Hagen, 1952; Epstein, 1953) conducted classical experiments over 40 years ago that showed that ion uptake by plant roots could be described by first order kinetics in a similar manner to many enzyme reactions. They also showed that, for the major nutrients studied, the process could be described by two phases – a high-affinity system operating at low external nutrient concentrations and a low-affinity system operating at higher external concentrations. An implication arising from these experiments was that uptake through the plasmalemma was mediated by proteins embedded in this membrane. However, isolation and identification of the specific proteins involved proved to be very difficult until nutritional physiologists began to apply molecular techniques to the study of the mechanisms of ion transport in plants. With the aid of this new technology over the past 8 years, many of the specific proteins involved in transport of a number of nutrient ions in plants have been characterized, the genes encoding these proteins identified, and the complex regulatory systems involved have begun to be untaggled. Genes encoding the phosphate transporter proteins responsible for influx of phosphate into the cells of roots and some other tissues have been isolated, and the roles of some of these have been defined.

Uptake of phosphate into the root symplasm involves transport from concentrations less than 2 micromolar in the surrounding apoplasm across the membrane to the cytoplasm where phosphate concentrations are maintained in the millimolar range. This, together with the net negative charge on the inside of the plasmalemma, necessitates that strong electrochemical gradients need to be overcome for successful transfer of phosphate anions into root cells. Transport of phosphate across the plasmalemma, therefore, requires a high-affinity, energy driven transport mechanism. The genes encoding such transporters have been isolated from a number of plant species during the past 4 years and the sequence and topology of the encoded transporter proteins inferred from the DNA sequences.

Identification of plant phosphate transporters

An Expressed Sequence Tag from an *Arabidopsis* clone containing similarities to the sequences of genes encoding phosphate transporters that had been isolated from yeast and fungi led to the isolation of the first reported genes encoding plant phosphate transporters (Muchhal et al., 1996: Smith et al., 1997a). These genes were isolated from *Arabidopsis*. They now form part of the rapidly growing *Pht1* family of plant phosphate transporters which includes members isolated from tomato (Daram et al., 1998; Liu et al., 1998a), potato (Leggewie et al., 1997), *Catharanthus* (Kai et al., 1997), *Medicago* (Liu et al., 1998b), barley (Smith et al., 1999) and additional genes from *Arabidopsis* (Mitsukawa et al., 1997a). Eight different members of this family of phosphate transporters have been isolated from the barley genome to date (Smith et al., 1999). A member of a second family of phosphate transporters, *Pht2*, that has similarities to the quite different family of phosphate transporters represented by some mammalian Na^+/phosphate cotransporters has recently been isolated from *Arabidopsis* (Daram et al., 1999). This transporter, which functions as a $H^+/H_2PO_4^-$ cotransporter in plants, is primarily expressed in *Arabidopsis* shoot tissues. It appears to be involved in the internal cycling of phosphorus within the plant.

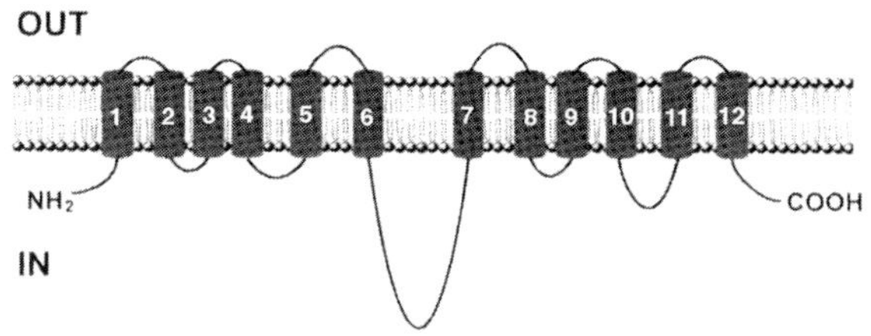

Figure 1. Topology of a typical transporter protein responsible for the movement of phosphate from the external soil solution across the plasmalemma and into the cytoplasm of root cells. These are characterised by 12 membrane spanning domains arranged in a '6+6' configuration.

The topology of plant phosphate transporters

The deduced sequences of all members of the *Pht1* family of plant phosphate transporter proteins identified to date are very similar. Members of this family of transporter proteins are 520–550 amino acids in length and approximately 58 KDa in size. All contain a number of highly hydrophobic regions that traverse the membrane. These hydrophobic regions are known as membrane-spanning domains (MSDs), each of which is usually composed of 17–25 amino acids arranged in a helix that crosses the membrane. This family of plant phosphate transporters is predicted to have 12 MSDs arranged in a well defined 6+6 configuration (Figure 1). The MSDs are separated by extracellular and intracellular loops of more highly charged hydrophilic amino acids. It is predicted that the N-terminal, C-terminal and the long central hydrophilic loop are oriented towards the inner surface of the membrane. A number of other membrane transporters in the major facilitator superfamily of proteins (Pao et al., 1998; Saier and Reizer, 1991) share this configuration. These include transporters involved in transfer of organic acids, amino acids, sugars and inorganic ions across biological membranes. The topology of this family of phosphate transporters is shared by some fungal, yeast, plant and mammalian phosphate transporters (Smith et al., 1997a) so appears to have been well conserved during evolution.

At present, there is little detailed information available on the structure/function relationships of plant phosphate transporters. Research is required on those structural characteristics that define their specificity for a particular ion and their affinity for the transport of that ion. These properties are likely to be influenced by amino acid residues in the hydrophylic loops between the MSDs and structural and charge characteristics on those inner surfaces of the MSDs that form the central pore of the transporters. Now that clones of phosphate transporters are available it should be possible to study these structure/function relationships through use of site directed mutagenesis to alter potentially important domains.

The functional analysis of plant phosphate transporters

The yeast *Saccharomyces cerevisiae* has provided a useful model for developing heterologous expression systems to aid the functional analysis of plant sulfate transporters. This system, that relies upon complementation of a mutation in the yeast by expression of a plant cDNA, has enabled the sulfate transporter function to be verified, the kinetics of transport to be determined and the effects of such factors as pH and inhibitors to be assessed (Smith et al., 1995; Smith et al., 1997b). The PHO84 yeast mutant, defective in a gene encoding one of the yeast phosphate transporters, has been available for a number of years (Bunya et al., 1991). Results of complementation studies using this mutant to characterise plant phosphate transporters have generally been disappointing. Michaelis Constant (K_m) values for plant phosphate transporters derived by heterologous expression in this mutant have generally been much higher than expected (Leggewie et al., 1998; Liu et al., 1998a) making it difficult to obtain reliable kinetic data. Thus, less is known about the affinity sub-groupings of plant phosphate transporters than there is about the families of sulphate transporters. Recently, a yeast double mutant, PAM2, has become available. This mutant has disruptions in the genes encoding both the PHO84 phosphate transporter and a Na^+-coupled PHO89 phosphate transporter (Martinez and Persson, 1998). It offers prospects for improved functional analysis of plant phosphate transporters. An apparent K_m for phosphate of 31 micromolar was measured for the tomato LePT1 phosphate transporter (Daram et al., 1998) using this mutant. Whilst this may still be higher than expected for a high-affinity phosphate transporter, it is an order of magnitude lower than some of the measurements on plant phosphate transporters that have been made with the single PHO84 mutant. Expression of a cDNA encoding a plant phosphate transporter in cultured tobacco cells has yielded the most reliable functional analysis of a plant phosphate transporter. Using this expression system, Mitsukawa et al. (1997b) proved that the PHT1 phosphate transporter from *Arabidop-*

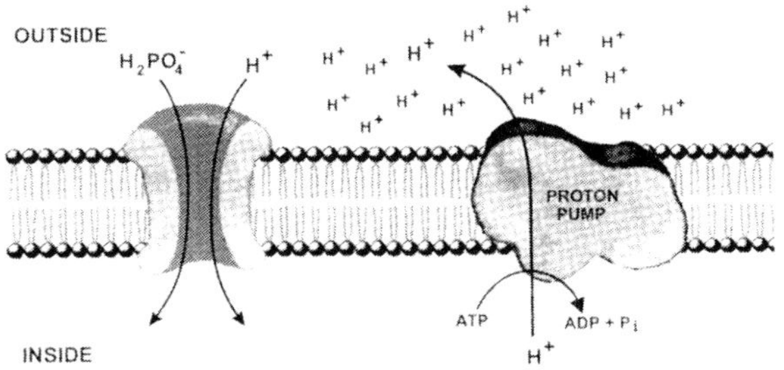

Figure 2. A model depicting the mechanism of phosphate transport across the plasmalemma of cells in the root epidermis or outer layers of the root cortex.

sis was a high-affinity phosphate transporter with an apparent K_m for phosphate of 3.1 micromolar. The sequences and expression patterns of some of the phosphate transporters that have been isolated from *Arabidopsis* (Smith et al., 1997a) and barley (Smith et al., 1999) are almost identical and suggest these transporters have similar functions and roles. It seems that at least in some diploid plant genomes, redundancy of genes encoding phosphate transporters that are critical for the uptake of phosphate, and hence plant survival, may have developed.

Complementation studies with sulfate transporters have shown that sulfate uptake rates are sensitive to the external pH. In a similar manner, the phosphate uptake rates by the potato StPT1 and StPT2 phosphate transporters were reduced when the pH was increased over the range 4.5–7.5 (Leggewie et al., 1997) and uptake by the tomato LePT1 phosphate transporter was dependent upon the proton gradient across the plasma membrane (Daram et al., 1998). Physiological measurements indicate that during phosphate uptake the cytoplasmic pH decreases (Mimura et al., 1992; Ulrich and Novacky, 1992) and that agents that collapse the membrane potential inhibit phosphate uptake. These data provide evidence that these transporters function as proton cotransporters and their transport function relies upon a H^+ pump to maintain a proton gradient across the membrane.

A molecular model for phosphate uptake through the plasmalemma

The predicted topology of this family of phosphate transporters with 12 MSDs, together with the proton coupled cotransport function suggest a molecular model for the operation of these transporters (Figure 2). This model indicates that the MSD helices combine to form a doughnut-shaped structure in the membrane having an inner pore through which $H_2PO_4^-$ ions and protons move from the outside of the membrane to the inside. The process is energized by the electrochemical gradient maintained across the membrane by the action of a H^+-ATPase that 'pumps' protons to the outer surface of the membrane. The outer hydrophilic loops and possibly the inner surfaces of the pore formed by the MSDs may contain critical domains that define the specificity of the transporter for $H_2PO_4^-$ ions and the 'binding capacity', and hence the affinity and K_m of the process for phosphate.

Regulation of plant phosphate transporters

The rate of transport of an ion across a plant membrane is determined by the number of transporters in the membrane and the activity of those transporters. The number of transporters is ultimately determined by the rates of transcription of the genes encoding the transporters, the rates of decay of the mRNAs, and the rate of turnover of the transporter proteins themselves. Ion homeostasis requires that the concentration of inorganic nutrients within plant tissues be maintained within certain limits (Mimura, 1995, 1999). Close regulation of both the influx and efflux processes by which ions move across plant membranes is required to maintain this homeostasis. To date, little is known about the molecular mechanisms of phosphate efflux. However, studies using molecular probes derived from genes encoding plant phosphate transporters have established that transcriptional control is an important regulatory mechanism for at least some of the genes involved in phosphate influx.

Studies on the regulation of plant sulfate transporters have shown that the sulfur status of the plant is a major factor controlling the expression of genes encoding the high-affinity SHST1, SHST2 and HVST1 sulphate transporters involved in uptake of sulfate by roots from the external solution (Hawkesford and Smith, 1997; Smith et al., 1995, 1997b). Similar transcriptional regulation by negative feedback of those members of the *Pht1* family of plant phosphate transporters that are expressed in roots has been reported. Expression studies with the genes encoding the APT1 (Figure 3) and APT2 phosphate transporters of *Arabidopsis* have shown that they are only expressed in roots and that their expression is considerably enhanced by phosphate deprivation (Smith et al., 1997a). Upon resupplying phosphate to phosphate-deprived

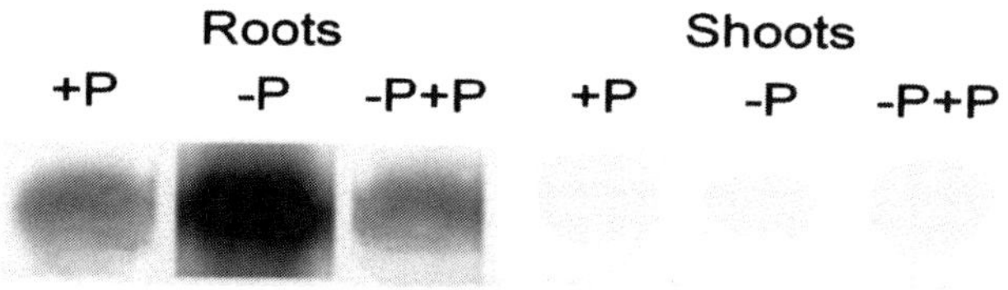

Figure 3. Northern blot indicating feedback regulation of expression of the gene encoding the *Arabidopsis* high-affinity phosphate transporter APT1 by the phosphate status of the plant. Similar levels of total RNA were loaded in each lane and confirmed by monitoring the rRNA band (not shown). The *APT*1 gene is only expressed in plant roots. There is a four- to five-fold increase in steady-state levels of mRNA corresponding to *APT*1 when plants previously adequately supplied with phosphous (+P) are deprived of phosphate for 72 h (-P). Following restoration of the external phosphate supply for 24 h (-P+P), steady-state mRNA levels corresponding to *APT*1 decline to low levels.

plants the expression of the *APT1* and *APT2* genes was down-regulated and the steady-state mRNA levels corresponding to these two genes decreased rapidly. Many of the expression studies on genes encoding root phosphate transporters isolated from tomato (Daram et al., 1998; Liu et al., 1998a), potato (Leggewie et al., 1997), *Medicago* (Liu et al., 1998b) and barley (Smith et al., 1999) have indicated that these are similarly regulated. Data related to these experiments have been summarised in a recent review by Mimura (1999).

Some very nice *in situ* hybridisation studies of Daram et al. (1998) on the tomato *LePT*1 high-affinity phosphate transporter have indicated that, although mRNA corresponding to this transporter could be detected in all vegetative organs of tomato seedlings, this gene was primarily expressed in the peripheral cell layers of the root including the root cap, root hairs, epidermis and outer layers of the cortex. It was further shown that *LePT*1 was also expressed in some cells in the stelar region when plants were deprived of phosphate. With the aid of a specific antibody to LePT1 it has also been shown that LePT1 protein levels increased when plants were deprived of phosphate (Muchhal and Raghothama, 1999), and that the enriched levels of LePT1 protein were associated with plasma membrane preparations from phosphate-starved roots. These data indicate that the enhanced levels of mRNA arising from phosphate deprivation are translated into LePT1 protein and this protein is targeted to the plasma membrane. Thus, the total number of phosphate transporters in the plasma membrane of root cells in close proximity to the soil solution increases when the external phosphate supply is sub-optimal. Feedback regulation of the expression of these genes is consistent with physiological measurements showing increased capacity for phosphate uptake by roots of phosphate-deprived plants (Clarkson and Scattergood, 1982; Clarkson and Luttge, 1991; Cogliatti and Clarkson, 1983; Lefebvre and Glass, 1982).

Calculations based upon changes in the rates of sulfate uptake and steady-state levels of mRNA following changes in the sulfate status of plants suggest the barley high-affinity sulphate transporter HVST1 turns over rapidly with a half-life of approximately 2.5 h. A shorter half-life of less than 2 h has been suggested for the mRNA (Smith et al., 1997b). Western blot analyses indicate that phosphate transporters also turn over rapidly (Muchhal and Raghothama, 1999) thereby enabling fast modulation of phosphate transport rates by direct transcriptional control.

Another potential mean of regulation of phosphate transport is via post-translational modification of plant phosphate transporters. Most of the *Pht1* family of phosphate transporters have highly conserved sites that could be phosphorylated or glycosylated (Leggewie et al., 1997; Muchhal et al., 1996; Smith et al., 1997a). Studies targeting these sites are being conducted at a number of research centres. Another avenue for regulating the activity of these plant phosphate transporters may be through interaction of the transporters with other proteins. Evidence has accumulated that interactions between a number of proteins are involved in regulating phosphate uptake by yeast (Bun-ya et al., 1992, 1996; Yompakdee et al., 1996a, b). Plant homologues of the yeast genes encoding these interacting proteins have not yet been identified. The suggestion has been made that poor complementation of the yeast PHO84 phosphate transporter mutant by plant phosphate transporters may arise from inadequate interaction between plant phosphate transporters and these yeast proteins (Smith, 1999). It is likely that fine-tuning of the phosphate transport process may be exercised by post-transcriptional modifications of the transporter proteins and by allosteric regulation of the activity of transporters.

Split root studies indicate that phosphate uptake rates by a root deprived of phosphate respond to the overall phosphate status of the plant (Drew and Saker, 1984) rather than to the localised external phosphate concentration adjacent to that section of the root. This is being confirmed by molecular studies that indicate that transcriptional regulation of genes encoding phosphate transporters primarily respond to the phosphate status of the whole plant (Liu et al., 1998a). Feed back regulation of phosphate uptake into plant roots is, therefore, a systemic response rather than a localised response. Another important observation is the failure of root-specific sulfate and phosphate transporters to be up-regulated during some dual nutrient deficiencies (Smith et al., 1999). It appears that when growth is also limited by a second nutrient deficiency, the capacity of the roots to take up phosphate may remain down-regulated. This may have a bearing on the success of attempts to increase phosphate acquisition from low fertility-soils in the SAT. Additional soil phosphate made available to the plant by some of the processes suggested in this Workshop may not be taken up into the plant unless some other nutrient deficiencies are also alleviated.

The roles of different phosphate transporters in plants

Studies on the localization of expression of specific transporters, together with information on their characteristics, provide a useful guide to their probable roles. As indicated earlier in situ hybridisation studies with the high-affinity LePT1 phosphate transporter from tomato indicate that this transporter is mainly expressed in root hairs, epidermal cells, outer layers of the cortex and the root cap. These plant tissues are in contact with the soil solution and are known to be active in ion uptake. Furthermore, the low concentration of phosphate in most soil solutions requires the involvement of high-affinity phosphate transporters for uptake of these ions. It is, therefore, very likely that high-affinity phosphate transporters of the LePT1 and PHT1 (Mitsukawa et al., 1997b) types from the *Pht1* family mediate the initial uptake of these ions into roots from the soil solution. Once inside the symplast, phosphate ions can move radially via plasmodesmata connections through the cortical cells and the suberised endodermis to the central stele of the root without having to traverse further plasma membranes (Clarkson, 1993).

The *in situ* hybridisation micrographs showing expression of the gene encoding the LePT1 phosphate transporter in tomato roots (Daram et al., 1998) shows that, when induced by phosphate deprivation, *LePT*1 was also expressed in the young stelar tissues. The gene encoding the AST68 sulfate transporter is also expressed in a number of cell types in the stelar region of the root (Takahashi et al., 1997). These findings raise the question of the role of these transporters within the stele. They could be involved in unloading the millimolar concentrations of phosphate or sulfate within cells into the xylem. This would only require a low-affinity transport system. However, such a role appears unlikely for these transporters since unloading into the xylem is an efflux process and presumably only involves the xylem parenchyma cells adjacent to the xylem vessels. It is possible that these transporters expressed in many of the cells in the stelar region of the root recover nutrients leaked from the vascular tissues. Such a scavenging process would involve influx and could be mediated by either high-affinity or low-affinity transporters operating at the millimolar concentrations found in vascular tissues. Such a mechanism may be particularly important in nutrient deficient plants and thus be up-regulated under conditions of low external nutrient supply.

Although this review is directed at the mechanism of phosphate uptake into the plant from the external soil solution, it should not be forgotten that phosphate is transported across plant membranes at many other locations within the plant. These transport processes are important for plant metabolism, growth and development. Some of them are likely to be key determinates of the efficiency with which the phosphorus that is acquired by the plant is utilized. They are also likely to play a role in defining the quality of plant products produced for human and livestock nutrition. It might be expected that many of these processes could be mediated by low-affinity transporters or by ion channels. A number of phosphate transporters that are expressed in shoot tissues have been isolated. At present, however, there is not enough information available on the cell types in which they are expressed to permit likely roles to be ascribed to these. Some of these phosphate transport processes include xylem loading, the loading of leaf cells, phloem loading, transfer of nutrients to meristems and storage tissues and the efflux of phosphate from plant cells, including efflux from the root into the external solution. The transporters or channels involved in the movement of phosphate into and out of vacuoles through the tonoplast membrane have yet to be identified. Other transporters are likely to play a role in the transfer of nutrients to organelles such as chloroplasts. Several phosphate transporters with quite a different topology to the H^+/phosphate cotransporters shown in Figure 1 have been cloned from plastids (Fischer etal., 1994, 1997; Kammerer et al., 1998).

Phosphate uptake into plants through vesicular-arbuscular mycorrhizal fungi

Vesicular-arbuscular mycorrhizal (VAM) fungi form symbiotic associations with the root systems of many plants and play a very important role in the acquisition of phosphate by the plant (Harrison, 1999). These fungi colonise cortical cells from which they extend a network of hyphae several centimetres out into the surrounding soil, thereby expanding the effective volume of soil that the plant can exploit. The hyphae gather nutrients from the soil solution and transfer them back to the cortical cells of the host plant. The fungi develop specialised structures known as arbuscules within infected cortical cells. Materials are exchanged between the symbionts through these arbuscules. The acquisition of phosphate through VAM associations involves transport of phosphate from the soil solution across the membrane of the fungal hyphae, movement of that phosphate along the hyphae to the arbuscule, unloading the phosphate from the fungal arbuscules at the arbuscule-cortical cell interface and uptake of that phosphate by the plant cortical cells. A gene encoding a high-affinity phosphate transporter has been isolated from the VAM fungus *Glomus versiforme* (Harrison and Van Buuren, 1995). This gene complements mutations in the gene encoding the yeast PHO84 transporter and has similar characteristics to PHO84 and to members of the *Pht1* family of plant phosphate transporters that are expressed in roots. It is expressed in the fungal hyphae and is most likely to be a high-affinity phosphate transporter involved in the initial uptake of phosphate from the soil solution into the hyphae. At least one other ion channel or transporter of fungal origin should be involved in unloading phosphate from the fungus at the arbuscule/cortical cell interface but this has not yet been identified. Similarly, phosphate transporters of plant origin that are responsible for uptake from this interface into the cortical cells have not been specifically identified. The phosphate concentration at the arbuscule/cortical cell interface is not known at present, but if it were high enough a low-affinity phosphate transporter could suffice to transfer phosphate into the cortical cells. Such a transporter could also be among the root phosphate transporters already isolated, but regulated in such a way as to enable the plant to exploit the mycorrhizal symbiosis to best advantage.

Two genes encoding phosphate transporters have been isolated from the roots of *Medicago truncatula* (Liu et al., 1998b). The expression of these genes is subject to feedback regulation by the internal phosphate status of the plant as discussed earlier. However, following mycorrhizal infection of *M. truncatula*, the expression of these phosphate transporters and other phosphate starvation inducible genes such as Mt4 is down-regulated (Burleigh and Harrison, 1999). Data suggest that this down-regulation is a systemic response to colonisation by the VAM fungus, and may occur prior to significant transport of phosphate into the plant (Burleigh and Harrison, 1997, 1999; Harrison and Van Buuren, 1995). These studies imply that a translocatable shoot factor induced by VAM colonization may be involved in the down-regulation of Mt4-like genes and possibly the plant phosphate transporters. If this does prove to be a general phenomenon, it raises important questions about the contribution of the high-affinity plant phosphate transporters to phosphate uptake by roots of mycorrhizal plants. The implications should be considered when designing strategies to increase the availability of phosphate in soil close to the root surface since mycorrhizal fungi may extend beyond this zone.

Conclusion

Efforts are being made to increase the uptake of a number of major nutrients, including phosphate, by overexpressing genes encoding nutrient transporter proteins. Preliminary data indicate that this may be successful for enhancing sulfate uptake and sulfur accumulation in some plants (Ealing and Smith, unpublished). Similar experiments with phosphate also suggest enhanced phosphate levels in plants overexpressing high-affinity phosphate transporters (Ealing and Smith, unpublished). However, further experimentation is required on the balance between influx and efflux in these transgenic plants. There are concerns that this strategy may lead to accumulation of toxic levels of phosphate in plants where phosphate is readily available from the growth medium. If successful, the strategy is likely to be most suitable to situations where phosphate ions are available at the root surface of young plants. These conditions often exist when phosphate fertilizer is applied. Growth of plants with a high capacity to take up phosphate due to overexpression of genes encoding phosphate transporters may offer such plants an opportunity to rapidly acquire phosphate, thereby improving the efficiency of fertilizer use. In a related strategy, manipulation of genes involved in the internal transfer of phosphate within plants has potential for increasing the efficiency of phosphate utilization by plants and, in some situations, the quality of plant products.

Whilst there is little doubt that increased phosphatic fertilizer inputs are required to improve water use efficiency and production of crops growing on the phosphate deficient soils of the SAT, economic and logistic factors will preclude this in many situations. At this time, therefore, priorities for research aimed at improving the phosphate nutrition of cropping systems in this region should be directed towards improving the availability of soil-P. Important strategies include enhancing mycorrhizal associations, enhancing root development, increasing organic anion excretion by roots and increasing the release of soil organic phosphorus by excretion of stable phosphatases and phytases. Many of the presentations at this Workshop address these strategies and provide evidence that a number of the important crop species of the region have the genetic potential to use such mechanisms. These species need to be built into farming systems that maximize crop production and improve the efficiency of phosphate uptake and utilization. Where the genetic potential does not exist, molecular technologies offer a means to introduce appropriate genes and regulatory systems into key components of the cropping systems.

Acknowledgements

Research at Commonwealth Scientific and Industrial Research Organization (CSIRO) Tropical Agriculture is supported in part by the Australian Grains Research and Development Corporation.

References

Bun-ya M, Nishimura M, Harashima S and Oshima Y 1991 The PHO84 gene of *Saccharomyces cerevisiae* encodes an inorganic phosphate transporter. Mol. Cell. Biol. 11, 3229–3238.

Bun-ya M, Harashima S and Oshima Y 1992 Putative GTP-binding protein, *Grt1*, associated with the function of the PHO84 inorganic phosphate transporter in *Saccharomyces cerevisiae*. Mol. Cell. Biol. 12, 2958–2966.

Bun-ya M, Shikata K, Nakade S, Yompakdee C, Harashima S and Oshima Y 1996 Two new genes, *PHO86* and *PHO87*, involved in inorganic phosphate uptake in *Saccharomyces cerevisiae*. Curr. Genet. 29, 344–351.

Burleigh S H and Harrison M J 1997 A novel gene whose expression in *Medicago truncatula* roots is suppressed in response to colonization by vesicular-arbuscular mycorrhizal (VAM) fungi and to phosphate nutrition. Plant Mol. Biol. 34, 199–208.

Burleigh S H and Harrison M J 1999 The down-regulation of *Mt4*-like genes by phosphate fertilization occurs systemically and involves phosphate translocation to the shoots. Plant Physiol. 119, 241–248.

Clarkson D T and Scattergood C B 1982 Growth and phosphate transport in barley and tomato plants during development of and recovery from phosphate stress. J. Exp. Bot. 33, 865–875.

Clarkson D T and Lüttge U 1991 Mineral nutrition: Inducible and repressible nutrient transport systems. Prog. Bot. 52, 61–83.

Clarkson D T 1993 Roots and the delivery of solutes to the xylem. Phil. Trans. R. Soc. Lond. B 341, 5–17.

Cogliatti D H and Clarkson D T 1983 Physiological changes in phosphate uptake by potato plants during development of and recovery from phosphate deficiency. Physiol. Plant. 58, 287–294.

Daram P, Brunner S, Persson B L, Amrhein N and Bucher M 1998 Functional analysis and cell-specific expression of a phosphate transporter from tomato. Planta 206, 225–233.

Daram P, Brunner S, Rausch C, Steiner C, Amrhein N and Bucher M 1999 *Pht2;1* encodes a low affinity phosphate transporter from *Arabidopsis*. Plant Cell 11, 2153–2166.

Drew M C and Saker L R 1984 Uptake and long-distance transport of phosphate, potassium and chloride in relation to internal ion concentrations in barley: Evidence of allosteric regulation. Planta 160, 500–507.

Epstein E 1953 Mechanism of ion absorption by roots. Nature 171, 83–84.

Epstein E and Hagen C E 1952 A kinetic study of the absorption of alkali cations by barley roots. Plant Physiol. 27, 457–474.

Fischer K, Arbinger B, Kammerer B, Busch C, Brink S, Wallmeier H, Sauer N, Eckerskorn C and Flugge U I 1994 Cloning and *in vivo* expression of functional triose phosphate/phosphate translocators from C_3- and C_4-plants: Evidence for the putative participation of specific amino acid residues in the recognition of phospho*enol*pyruvate. Plant J. 5, 215–226.

Fischer K, Kammerer B, Gutensohn M, Arbinger B, Weber A, Hausler R E and Flugge U I 1997 A new class of plastidic phosphate translocators: A putative link between primary and secondary metabolism by the phospho*enol*pyruvate/phosphate antiporter. Plant Cell 9, 453–462.

Harrison M J 1999 Molecular and cellular aspects of the arbuscular mycorrhizal symbiosis. Annu. Rev. Plant Physiol. Plant Mol. Biol. 50: 361–389.

Harrison M J and Van Buuren M L 1995 A phosphate transporter from the mycorrhizal fungus *Glomus versiforme*. Nature 378, 626–629.

Hawkesford M J and Smith F W 1997 Molecular biology of higher plant sulphate transporters. *In* Sulphur Metabolism in Higher Plants. Eds. WJ Cram, LJ De Kok, I Stulen, C Brunold and H Rennenberg. pp. 13–25. Backhuys Publishers, Leiden, The Netherlands.

Kai M, Masuda Y, Kikuchi Y, Osaka M and Tadano T 1997 Isolation and characterization of a cDNA from *Catharanthus roseus* which is highly homologous with phosphate transporter. Soil Sci. Plant Nutr. 83, 227–235.

Kammerer B, Fischer K, Hilpert B, Schubert S, Gutensohn M, Weber A and Flugge U I 1998 Molecular characterization of a carbon transporter in plastids from heterotrophic tissues: The glucose 6-phosphate/phosphate antiporter. Plant Cell 10, 105–107.

Lefebvre D D and Glass A D M 1982 Regulation of phosphate influx in barley roots: Effects of phosphate deprivation and reduction in influx with provision of orthophosphate. Physiol. Plant. 54, 199–206.

Leggewie G, Willmitzer L and Reismeier J W 1997 Two cDNAs from potato are able to complement a phosphate uptake-deficient yeast mutant: Identification of phosphate transporters from higher plants. Plant Cell 9, 381–392.

Liu C, Muchhal U S, Uthappa M, Kononowicz A K and Raghothama K G 1998a Tomato phosphate transporter genes are differentially regulated in plant tissues by phosphorus. Plant Physiol. 116, 91–99.

Liu H, Trieu A T, Blaycock L A and Harrison M J.1998b Cloning and characterization of two phosphate transporters from *Medicago truncatula* roots: Regulation in response to phosphate and to colonization by arbuscular mycorhizal (AM) fungi. Mole Plant-Microbe Int. 11, 14–22.

Martinez P and Persson B L 1998. Identification, cloning and characterization of a derepressible Na^+-coupled phosphate transporter in *Saccharomyces cerevisiae*. Mol. Gen. Genet. 258, 628–638.

Mimura T, Yin Z-H, Wirth E and Dietz K-J 1992 Phosphate transport and apoplastic phosphate homeostasis in barley leaves. Plant Cell Physiol. 33, 563–568.

Mimura T 1995 Homeostasis and transport of inorganic phosphate transport in plants. Plant Cell Physiol. 36, 1–7.

Mimura T 1999 Regulation of phosphate transport and homeostasis in plant cells. Int. Rev. Cytology 191, 149–200.

Mitsukawa N, Okumura S and Shibata D 1997a High-affinity phosphate transporter genes of *Arabidopsis thaliana*. *In* Plant Nutrition – For Sustainable Food Production and Environment. Proceedings of the XIII International Plant Nutrition Colloquium, 13–19 September 1997, Tokyo, Japan. Development in Plant and Soil Sciences 78. Eds. T Ando, K Fulita, T Mae, H Matsumoto, S Mori and J Sekiya. pp. 187–190. Kluwer Academic Publishers, Dordrecht, The Netherlands.

Mitsukawa N, Okumura S, Shirano Y, Sato S, Kato T, Harashima S and Shibata D 1997b Overexpression of an *Arabidopsis thaliana* high-affinity phosphate transporter gene in tobacco cultured cells enhances cell growth under phosphate-limited conditions. Proc. Natl. Acad Sci. USA 94, 7098–7102.

Muchhal U S, Pardo J M and Raghothama K G 1996 Phosphate transporters from the higher plant *Arabidopsis thaliana*. Proc. Nat. Acad. Sci. USA 93, 10519–10523.

Muchhal U S and Raghothama K G 1999 Transcriptional regulation of plant phosphate transporters. Proc. Natl. Acad. Sci USA 96, 5868–5872.

Pao S S, Paulsen I T and Saier M H 1998 Major facilitator superfamily. Microb. Mol. Biol. Rev. 62, 1–34.

Perumalla C J and Peterson C A 1986 Deposition of Casparian bands and suberin lamellae in the exodermis and endodermis of young corn and onion roots. Can. J. Bot 64, 1873–1878.

Peterson C A and Cholewa E 1998 Structural modifications of the apoplast and their potential impact on ion uptake. Z. Pflanzenernahr. Bodenk. 161, 521–531.

Saier M H and Reizer J 1991 Families and superfamilies of transport proteins common to prokaryotes and eukaryotes. Current Opinion in Structural Biology 1, 362–368.

Smith F W, Ealing P M, Hawkesford M J and Clarkson D T 1995 Plant members of a family of sulfate transporters reveal functional subtypes. Proc. Natl. Acad. Sci USA 92, 9373–9377.

Smith F W, Ealing P M, Dong B and Delhaize E. 1997a The cloning of two *Arabidopsis* genes belonging to a phosphate transporter family. Plant J. 11, 83–92.

Smith F W, Hawkesford M J, Ealing P M, Clarkson D T, Vanden Berg P J, Belcher A R and Warrilow A G S 1997b Regulation of expression of a cDNA from barley roots encoding a high affinity sulphate transporter. Plant J. 12, 875–884.

Smith F W 1999 Molecular biology of nutrient transporters in plant membranes. *In* Mineral Nutrition of Crops: Fundamental Mechanisms and Implications. Ed. Z Rengel. pp. 67–89. Haworth Press, Binghamton, USA.

Smith F W, Cybinski D and Rae A L 1999 Regulation of expression of genes encoding phosphate transporters in barley roots. *In* Plant Nutrition – Molecular Biology and Genetics. Eds. G Gissel-Nielsen and A Jensen. pp. 145–150. Kluwer Academic Publishers, Dordrecht, The Netherlands.

Takahashi H, Yamazaki M, Sasakura N, Watanabe A, Leustek T, De Almeida Engler J, Engler G, Van Montagu M and Saito K 1997 Regulation of sulfur assimilation in higher plants: A sulfate transporter induced in sulfate starved roots plays a central role in *Arabidopsis thaliana*. Proc. Natl. Acad. Sci. 94, 11102–11107.

Ulrich C I and Novacky A J 1992 Extra and intracellular pH and membrane potential changes induced by K^+, Cl^- $H_2PO_4{}^-$ and $NO_3{}^-$ uptake and fusicoccon in root hairs of *Limnobium stoloniferum*. Plant Physiol. 94, 1561–1567.

Yompakdee C, Bun-ya M, Shikata K, Ogawa N, Harashima S and Oshima Y 1996a A putative new membrane protein, Pho86p, in the inorganic phosphate uptake system of *Saccharomyces cerevisiae*. Gene 171, 41–47.

Yompakdee C, Ogawa N, Harashima S and Oshima Y 1996b A putative membrane protein, Pho88p, involved in inorganic phosphate transport in *Saccharomyces cerevisiae*. Mol. Gen. Genet. 251, 580–590.

J.J. Adu-Gyamfi (Ed.), *Food security in nutrient-stressed environments: exploiting plants' genetic capabilities*: 245–260, 2002.

Breeding for better symbiosis

Z. Rengel

*Soil Science and Plant Nutrition, Faculty of Agriculture, The University of Western Australia, 35 Stirling Highway, Crawley WA 6009, Australia**

Key words: associative N_2 fixation, dinitrogen fixation, genotype, mycorrhiza, nodulation, rhizobia, root exudation, screening, selection, symbiosis

Abstract

The present review gives a critical assessment of the literature dealing with symbiosis between rhizobia and legumes and between AM fungi and most plants. Associative N_2 fixation (even though strictly speaking not a symbiotic relationship) does have some characteristics of symbiosis due to mutualistic dependence and usefulness of the relationship, and is therefore covered in this review. Nodulation in the rhizobia–legume symbiosis may be limited by an insufficient amount of the *nod*-gene inducers released from seed and/or roots. However, there is genotypic variation in the germplasm of legume species in all components of the signalling pathway, suggesting a prospect for improving nodulation by selecting and/or transforming legume genotypes for increased exudation of flavonoids and other signalling compounds. Deciphering chromosomal location as well as cloning *nod*, *nif* and other genes important in nodulation and N_2 fixation will allow manipulation of the presence and expression of these genes to enhance the symbiotic relationship. Increased efficacy of symbiotic N_2 fixation can be achieved by selecting not only the best host genotypes but by selecting the best combination of host genotype and nodule bacteria. As flavonoids exuded by legume seedlings may not only be *nod*-gene inducers, but also stimulants for hyphal growth of the AM fungi, selecting and/or transforming plants to increase exudation of these flavonoids may result in a double benefit for mycorrhizal legumes. Mutants unable to sustain mycorrhizal colonisation are instrumental in understanding the colonisation process, which may ultimately pay off in breeding for the more effective symbiosis. In conclusion, targeted efforts to breed genotypes for improved N_2 fixation and mycorrhizal symbiosis will bring benefits in increased yields of crops under a wide range of environmental conditions and will contribute toward sustainability of agricultural ecosystems in which soil-plant-microbe interactions will be better exploited.

Introduction

Two most important types of symbioses are N_2 fixation and acquisition of P and other nutrients by mycorrhizae. They will be discussed with particular reference to signalling pathways involved and genotypic differences in the capacity of symbionts to sustain the symbiosis, especially under unfavourable environmental conditions.

Dinitrogen fixation provides more N to the agricultural ecosystems worldwide than the total amount of fertiliser N applied. However, a significant potential exists to further improve symbiotic and associative N_2 fixation by breeding genotypes with a greater capacity to sustain these interactions with bacteria. Since taxonomy of root and stem nodule bacteria has been expanding recently (six genera have been distinguished so far: *Rhizobium, Bradyrhizobium, Sinorhizobium, Mesorhizobium, Azorhizobium* and *Allorhizobium*), the generic term 'rhizobium' will be used in this article to refer to members of these six genera.

Better N_2-fixing symbiosis may be brought about by manipulating both rhizobia and plant hosts and by eventually creating an artificial rhizosphere. An important aim is also to improve the symbiotic relationship in suboptimal environmental situations related to soil-borne or environmental stress. However, it is

* FAX No.: +61-8-9380-1050; Tel.: +61-8-9380-2557; E-mail: zrengel@agric.uwa.edu.au

This paper previously appeared in *Plant and Soil* **245**, 2002, pp. 147–162.

still unclear whether attempts to increase the symbiotic capacity where symbiosis is already fully functional ('overexpression', supernodulation, etc.) will be beneficial in terms of increasing crop yields.

Symbiotic or even mutualistic relationships involving rhizobia depend on chemical signals between the two organisms. These signals define the rhizobia–host specificity in the relationship. Selecting for the optimal combination of the rhizobium and the host usually results in more effective symbiosis and better growth of the host plant.

Many crop plants may form symbiotic associations with mycorrhizae and thus gain access to soil pools of P that are unavailable to non-mycorrhizal plants (e.g., Smith et al., 1994). This is especially important in soils where P-sorption capacity is high, resulting in poor agronomic efficiency of P fertilisers applied. On P-sorbing soils, only 10–20% of applied P fertilisers may be utilised by crops in the year of application, while a major portion of applied fertilisers becomes fixed in soils and unavailable to non-mycorrhizal crops (cf. Zhang and Rengel, 1999). Mycorrhizal hyphae are thinner than roots, and they create a greater P concentration gradient between the bulk soil and the hyphal surface, thus allowing more substantial movement of P toward hyphae and greater uptake across the membrane.

This review will concentrate on genetics and breeding related to N_2 fixation via symbiotic rhizobia and associative diazotroph bacteria as well as to AM symbiotic relationship. Ectomycorrhiza will not be covered here, and readers are referred to papers by Godbold et al. (1998) and Tagu and Martin (1996) for more information on that topic.

Associative N_2 fixation

Associative diazotroph bacteria are more numerous in the rhizosphere than in the bulk soil, indicating dependence on the organic compounds exuded by plant roots (Natsvaladze et al., 1992). Therefore, associative N_2 fixation, even though strictly speaking not a symbiotic relationship, does have some characteristics of such a relationship due to mutualistic dependence of, and usefulness to, the partners in the relationship. Indeed, recent evidence suggests that N_2 fixation in some grasses (e.g., sugarcane) may be symbiotic (James, 2000).

Under certain conditions, biological N_2 fixation via associative bacteria may contribute up to 30% of N needs of wheat and up to 150 kg N/ha per year in sugar cane fields in Brazil. It has been suggested that Brazilian cereal and sugar cane genotypes have inadvertently been selected for efficient utilisation of N fixed by associative diazotrophs (Dobereiner, 1997), thus significantly decreasing or completely eliminating the need for N fertilisation of these crops. Single or dual inoculation of wheat with *Azotobacter chroococcum*, *Azospirillum brasilense* and *Streptomyces mutabilis* in sterile soil allowed these microorganisms to establish well in the rhizosphere and to increase N content both in soil and in the wheat shoots (El-Shanshoury, 1995).

Associative N_2 fixation can provide between 10 and 80 kg N/ha per cropping season in rice depending on the cultural practices and the rice variety grown (Rao et al., 1998). Rice genetic factors contributing to efficient associative N_2 fixation in the rhizosphere have been identified and can be used in the breeding programme. Four putative QTL loci all showed a recessive gene action. Individual markers explained between 8 and 16% of the total phenotypic variation (Wu et al., 1995).

Differences among cereal species as well as among genotypes within cereal species in their capacity to support associative N_2 fixation have been clearly established (Jagnow, 1990). In rice, appropriate combinations of auxin action, *Acetobacter diazotrophicus* and rice genotype are needed for the relationship to be effective: some plants grew for up to 12 months on N-free medium getting N from the associative fixation (Rolfe et al., 1995). Further research on designing suitable screening methods to select and breed crop cultivars with desirable responses to beneficial diazotroph bacteria is warranted.

Rice genotypes have differential ability to support associative N_2 fixation, with the percentage of N derived from air (Ndfa) ranging from 1.5 to 21% in various genotypes (Shrestha and Ladha, 1996). Maize genotypes also showed differences in the capacity to support associative N_2 fixation, with some genotypes having similar yield when inoculated with diazotroph *Azospirillum* bacteria and when fertilised with 100 kg N/ha (Garcia de Salamone et al., 1997). Wheat genotypes also differed in the degree of support of associative *Azospirillum* bacteria in the rhizosphere and the rhizoplane (Ramanathan et al., 1997; Schloter and Hartmann, 1998).

The effectiveness of associative N_2 fixation may be diminished under environmental stress. Under saline conditions, differential tolerance of Azospirillum strains to salt stress resulted in the significant interaction between plant genotypes (finger millet, *Eleusine coracana*) and *Azospirillum brasilense* strains (Rai, 1991a).

Genes important in initial steps of Azospirillum attachment to roots (chemotaxis and adhesion) are being deciphered (Moens et al., 1997). However, chemotaxis in Azospirillum may be a general response rather than a specific host-dependent one (Fedi et al., 1992). Nevertheless, increased knowledge about genes mediating associative N_2 fixation will be instrumental in breeding crop plants for more effective utilisation of N from associative fixation. New insights into the process of associative N_2 fixation can also be gained from studies employing genetically engineered bacteria and cyanobacteria. Nitrogenase-derepressed mutant of *Anabaena variabilis* inoculated onto wheat roots resulted in high nitrogenase activity in the growing zones of young roots and increased wheat yield (Spiller et al., 1993).

There is a possibility that the rhizosphere competence of associative diazotrophs may be improved by increasing the capacity to degrade polygalacturonic acid (pectin) by overexpressing the polygalacturonase (pectinase) gene as it was shown that the rhizosphere competence, at least in a rhizobacterium *Burkholderia pickettii* MSP3Rif, depends on the activity of that particular enzyme (Ikeda et al., 1998). Another exciting strategy for enhancing growth of associative N_2-fixing microorganisms is to engineer plants to exude into the rhizosphere compounds specifically catabolised by the targeted associative N_2 fixers (see O'Connell et al., 1996). This strategy is based on the rhizopine concept (Rossbach et al., 1994; Murphy et al., 1995; Vlassak and Vanderleyden, 1997).

In addition to the associative diazotrophs in the rhizosphere of non-legume plants, these non-legumes may benefit from endophytic N_2-fixing bacteria. For example, N_2-fixation-competent endophytic strains of *R. leguminosarum* bv. *trifolii* significantly enhanced shoot and root growth of rice in growth chamber experiments (albeit no evidence of N_2 fixation was found), and increased grain yield as well as agronomic fertilizer N-use efficiency of hybrid rice in a field inoculation experiment conducted in the Nile Delta, Egypt (Yannai et al., 1997). These endophytic strains multiplied and moved within root tissue, indicating that they possess necessary genes to colonise and get established in rice roots and promote rice growth (Prayitno et al., 1999). However, possible N_2 fixation by these strains in rice roots is yet to be confirmed. Other possible causes of improved growth caused by endophytic bacteria may be through production of plant growth regulators and/or chelation of nutrients that have poor availability to plants. Further work on clarifying mechanisms underlying positive contribution of endophytic bacteria to plant growth is eagerly awaited.

The nodulation process in the rhizobia-legume symbiosis

Inheritance of the process of nodulation may be relatively simple. Recent discovery of non-fixing mutants with altered patterns of nodulation has revealed that genes involved in the control of the nodulation process are recessive and may not be too numerous (Tsyganov et al., 1998). Duc (1995) found that faba bean mutants unable to nodulate, as well as those with ineffective nodulation and supernodulation, had mutations of single recessive genes. However, grafting experiments showed that supernodulating mutant was influenced by the shoot genotype, while the non-nodulating and ineffective nodulation mutants were influenced by the root genotype, indicating that the regulation of gene expression may be more complicated.

A common translocatable factor facilitating hypernodulation in various legume species appears to exist in their shoots (e.g., in soybean, mung bean, *Vigna radiata*, and hyacinth bean, *Lablab purpureus*, Harper et al., 1997). Isolation and characterisation of such a factor may prove instrumental in increasing efficiency of the nodulation and thus ultimately of the N_2 fixation process.

Soybean genotypes with loci that restrict nodulation by native ineffective strains allow nodulation only by inoculant superior N_2-fixing strains of specific serogroups (Keyser and Cregan, 1987). However, more recent work shows that the host–rhizobia relationship is more complicated (Lohrke et al., 1995), indicating that better understanding of molecular mechanisms involved in selectivity of the nodulation process is needed. Work with other plant species (e.g., clover and *R. leguminosarum* bv. *trifolii*, Rolfe et al., 1995, and references therein) also indicates that positive and negative control mechanisms in the nodulation process are quite complex.

Lipo-chitooligosaccharide nodulation factors (Nod factors) produced by rhizobia determine the host range. These factors play a pivotal role in the molecular signal exchange, infection and induction of symbiotic developmental responses in legumes leading to the formation of a nodule in which rhizobia carry out N_2 fixation (Reddy et al., 1998). Subnanomolar concentrations of membrane lipo-chitooligosaccharides from *R. leguminosarum* bv. *trifolii* induce root hair deformation that leads to nodulation on white clover roots (Orgambide et al., 1996). There is fairly strong rhizobia–legume specificity because the same compounds could not induce any changes in lucerne, and more than two orders of magnitude greater concentration was required to induce a measurable response in vetch.

Lucerne genotypes capable of exuding *nod*-gene inducers into the rhizosphere were nodulated by *R. meliloti* to a larger extent and had a better growth than genotypes that lacked the capacity to exude *nod*-gene inducers. An addition of root exudates from highly-nodulating genotypes significantly improved nodulation of otherwise poorly nodulating genotypes (Hernandez et al., 1995). Root exudates turned on the proline dehydrogenase gene in *R. meliloti* during root invasion and nodule formation (Jimenez-Zurdo et al., 1997).

Initial nodulation may be limited by an insufficient amount of the *nod*-gene inducers released from seed, as shown for bean *Phaseolus vulgaris* (Hungria and Phillips, 1993) and lucerne (Hernandez et al., 1995). Isogenic bean genotypes differing in the quantitative and qualitative composition of flavonoid released from seed also differed in nodulation, with a positive relationship between the amount of flavonoids and nodulation (Hungria and Phillips, 1993). Testing a range of lucerne genotypes also detected the variability in the flavonoid exudation (Phillips et al., 1995). These results suggest a prospect of improving nodulation by appropriately selecting legume genotypes, or altering them through transformation and breeding, for increased exudation of flavonoids.

In considering improvement of the symbiotic relationships by changing the host plant to increase exudation from roots, the extra carbon cost of such exudation will need to be counterbalanced with potential benefits that such a change in root exudation may give in a particular set of environmental conditions. Indeed, it appears that selection of genotypes in breeding programmes resulted in a decreased capacity to exude organic compounds into the rhizosphere (e.g., in rice, Waschutza et al., 1992). To minimise extra carbon costs, only increased exudation of targeted compounds that enhance symbiosis should be pursued in selection and breeding.

Rhizobial mutants (e.g., Tn5 mutants with decreased production of acidic exopolysaccharides, Tn5 mutants of *R. fredii* USDA257 with increased nodulation capacity, etc.) are important in elucidating processes and characterising products required for efficient symbiosis (e.g., Eisenschenk et al., 1994; Heron et al., 1989). This knowledge will also be important in manipulating rhizobial inoculum to improve symbiosis. However, an assessment of the genotypic variation among a large number of *Sinorhizobium meliloti* strains isolated from nodules developed on various lucerne genotypes showed that the host genotype had a major influence on the shaping of the genetic structure of nodulating bacterial strains (Paffetti et al., 1998).

Transgenic crops will provide excellent material for studying various aspects of the symbiotic processes and will thus prove valuable in breeding for more effective symbioses. For example, transgenic clover *Trifolium repens* (Rolfe et al., 1997) has been used to study genetic programmes activated during nodule initiation and development.

Isogenic lines, with more than 95% similarity in the genetic background and a difference only in genes of interest, represent an excellent plant material for studying relevant physiological processes. Shelp and colleagues (1998) have produced a range of isogenic pea lines differing in production of effective nodules with *R. leguminosarum* bv. *viceae* and tolerance of nodulation to nitrate. These lines should be used in future studies of nodulation and N_2 fixation.

Signalling pathways in the rhizobia-legume symbiosis

There are three distinct parts in the signalling process involved in the rhizobia–legume symbiosis. First, flavonoids in plant root exudates induce *nod* genes in rhizobia and mediate the nodulation process. Secondly, the product of the regulatory rhizobial *nodD* gene induced transcription of rhizobial structural *nod*, *nol* and *noc* genes which code for extracellular lipo-chitooligosaccharides, protein Nod O, and other extracellular Nod factors. Thirdly, endogenous plant hormones, as well as factors produced by rhizobia, are important mediators of nodule initiation. It should, however, be emphasised that the current knowledge of the signalling process involved in the rhizobia–legume symbiosis is somewhat fragmented and hampered by the complexity of the plant–soil–microbe interactions that exist in agricultural ecosystems (Van Rhijn and Vanderleyen, 1995; Vlassak and Vanderleyen, 1997).

Nod factors are lipo-chitooligosaccharides varying in the oligosaccharide chain length, the nature of the fatty acids and substitutions on the oligosaccharide. Allelic and non-allelic genes coding for Nod factors have co-evolved together, while the host plant divergence ensures variability in Nod factors between different rhizobial strains and the specificity of the rhizobia–host relationship. Nod factor genes are located on the transmissible plasmids and/or 'symbiosis islands', which are symbiotic regions associated with movable elements, thus allowing for easy horizontal gene transfer (Mergaert et al., 1997). Rhizobia have a range of *nodD* genes producing NodD factors to cater for a variety of hosts (e.g., *R. meliloti* has three *nodD* genes; Honma et al., 1990). Functional analysis of *nodD* alleles is a useful way of ascertaining the differences between various NodD products (Van Rhijn et al., 1994).

A large body of literature has been accumulated on the flavonoid signalling pathway in the *nod* gene induction (e.g., Bolanos-Vasquez and Werner, 1997; Janczarek et al., 1997; Maxwell et al., 1989; Zuanazzi et al., 1998). It appears that expression of *nod* genes in various species of symbiotic N_2-fixing rhizobia may be increased by a number of flavonoid components (e.g., aldonic acids, Gagnon and Ibrahim, 1998; 4′,7-dihydroxyflavone and 4,4′-dihydroxy-2′-methoxychalcone, McKhann et al., 1997), even though different flavonoids may also suppress the *nod* gene expression (Zuanazzi et al., 1998). These flavonoid compounds appear to be synthesised predominantly in roots (Armero et al., 1994). It is interesting that specific Nod factors of *B. japonicum* increase flavonoid exudation by soybean roots (Schmidt et al., 1994), indicating the mutualistic character of the signalling pathway.

Transconjugants of rhizobial isolates having nodABC–lacZ fusion are frequently used for studying activation of the *nod* genes by the flavonoid/root exudate compounds by monitoring β-galactosidase activity (e.g., Bolanos-Vasquez and Werner, 1997; Dakora et al., 1993; Janczarek et al., 1997; Maxwell et al., 1989; Pandya et al., 1998). The potential host range of the rhizobial isolates (for example in the case of cowpea miscellany *Rhizobium* spp.; Pandya et al., 1998) can be determined in such a way.

Flavonoid exudation by the host seed/seedlings can also be regulated by the nutritional status of the growing medium: increasing N supply decreases exudation of phenolics (e.g., isoflavonoid and aglycone) (Wojtaszek et al., 1993). Since these isoflavones may serve as signalling molecules in the rhizobial–legume symbiosis, it is obvious that N supply may act as an additional control mechanism in such exudation.

Specificity of the *Rhizobium*-legume symbiosis is also governed by specific genes on the bacterial chromosome that code for proteins involved in recognition and uptake of specific signal molecules present in root exudates. *R. tropici* contains *teu* genes (*t*ropici *e*xudate *u*ptake) activated by root exudates of *Phaseolus vulgaris* and *Macroptilium artropurpureum* as hosts for *R. tropici*, but not by root exudates of other non-host legumes (Rosenblueth et al., 1998). Localisation of *nod*, *nif* and other genes important in development of the symbiotic relationship (e.g., on the large plasmid in the slow-growing cowpea *Rhizobium* spp. S2; Pandya and Desai, 1998) will create new avenues in using molecular biology tools to manipulate the presence and expression of these genes to enhance the symbiotic relationship.

Not only symbiotic colonisation (infection stress), but other types of stresses may induce root exudation of similar compounds (e.g., genistenin and a range of isoflavonoids) (Gagnon and Ibrahim, 1997), indicating some non-specificity in the rhizobia–host signalling sequence and suggesting that transformation of non-host plants to produce compounds that act as inducers in the effective symbiotic relationship may be quite feasible. It should be emphasised that isoflavonoids also act as phytoalexins (Dakora et al., 1993), indicating some similarity in response to pathogenic infection and infection with rhizobia. However, the level of phytoalexin production caused by *B. japonicum* in soybean roots was considerably lower than that caused by pathogenic infection by *Phytophthora megasperma* f.sp. *glycinea* (Schmidt et al., 1992).

There is genotypic variation in the germplasm of legume species in all components of the signalling pathway. Such a statement is based on the literature in which testing of only a limited number of genotypes has been reported (frequently only two, e.g., Soejima et al., 1992) and would therefore need to be ascertained on a large number of genotypes. Improved understanding of the signalling process will allow improving the symbiosis by more targeted breeding efforts. There is generally a gene(s)-for-gene interaction between rhizobia and the host. The genotypic specificity of the nodulation process depends on such an interaction. For example, McCall genotype of soybean cannot be nodulated when inoculated with *R. fredii*

USDA257 that nodulates other soybean genotypes tested (Trese, 1995). Even though host-genotype-specific flavonoid signals have not been found, it was clear that the strain- and host-genotype-specific interactions are characterised by unique patterns of signal release and response (Pueppke et al., 1998). Genotypic variation in flavonoids, betaines and other nod-gene inducers indicates a potential for genotype selection for increased exudation of these beneficial, but sometimes limiting, compounds (see Phillips et al., 1995).

Different flavonoids are exuded by germinating lucerne seeds (3′,4′,5,7-substituted flavonoids) compared with roots (5-deoxy molecules) (Hartwig et al., 1991). These compounds apparently have a different function, with 5-deoxy molecules being more effective in stimulating *R. meliloti* growth, and the other types being more effective as *nod*-gene inducers. These results pose another level of difficulty in breeding for better symbiosis because different metabolic pathways obviously predominate during different stages in the plant growth cycle.

Competitiveness of rhizobial strains

Under most soil conditions, odds are stacked against the inoculant strain. These strains have a limited gene pool for adapting to local conditions as opposed to indigenous strains which are well adjusted to prevailing conditions at the site and, moreover, can adapt to just about any change in conditions (see Vlassak and Vanderleyden, 1997). It is therefore of utmost importance that inoculant rhizobia have strong nodulation effectiveness for the host being grown as well as that inoculant strains are well adapted to the relevant soil conditions.

Better N_2 fixation can be achieved by selecting superior rhizobia. However, selection of these rhizobia would need to take into consideration not only their N_2-fixing capacity, but also competitive ability against native rhizobia which are frequently ineffective in N_2 fixation. Superior N_2-fixing strains have to outcompete native rhizobia and occupy a significant proportion of the nodules. For this to be achieved, rhizobia have to be selected under natural conditions in competition with the native rhizobia.

Better rhizobial symbiosis can be achieved by more effective rhizobia that will establish well in the soil and the rhizosphere (saprophytic competence), will cause greater nodulation, occupy a greater proportion of nodules, and have greater activity of nitrogenase and associated enzymes. More effective rhizobial strains can be selected from native populations (Howieson, 1995) or constructed in various ways (for references see Vlassak and Vanderleyden, 1997). However, achieving a large occupancy of nodules by these effective rhizobia represents one of the most significant practical problems. Mass inoculation does not always enhance the nodule occupancy (Kuykendall, 1989), and the amount of inoculum needed to overwhelm the ineffective native rhizobia is frequently excessive and uneconomical (Vlassak and Vanderleyden, 1997). However, the number of inoculant bacteria in comparison to the native rhizobia is not the most important parameter of competitiveness because other more qualitative criteria play a part (e.g., rhizobia mobility may influence their capacity to nodulate lateral roots in addition to the crown where the frequency of nodulation is the greatest due to inoculum most frequently being applied to seed, McLoughlin et al., 1990).

Competitiveness of introduced rhizobia strains can be increased if they are engineered to produce compounds that are inhibitory to *nod* gene expression in native rhizobia. For example, a repressor of the *nod* gene expression, NolA, was identified in USDA110; *nolA* gene transferred to *B. japonicum* serocluster 123 resulted in depression of Nod factor production (Dockendorff et al., 1994; Vlassak and Vanderleyden, 1997).

In contrast to *nod* genes and surface determinants, the role of an increasing number of genes and gene products that determine strain competitiveness has been elucidated (e.g., genes controlling catabolism of specific legume secondary metabolites like homoserine and trigonelline (Boivin et al., 1991), genes regulating nodule formation efficiency in *R. meliloti* (Sanjuan and Olivares, 1991), etc. — for further examples see Vlassak and Vanderleyden, 1997). These genes and gene products increase competitiveness as they allow nutritional utilisation of specific legume-produced substrates. However, even a larger number of genes and gene products remain to be characterised.

Competitiveness may also be increased by transferring the rhizopine *moc* and *mos* genes to good N_2 fixers (Vlassak and Vanderleyden, 1997). In the presence of rhizopine-producing bacteria, the rhizobia strains that are capable of catabolising rhizopines occupy a higher percentage of nodules (Murphy et al., 1995).

The specific compatibility between *nodX* gene carried on the symbiotic plasmid and *sym-2* gene of pea cv. Afghanistan (this cultivar could be nodulated only by strains containing *nodX* gene) has been recognised as offering the capacity to increase competitiveness of desired rhizobia–host combination and exclude all other combinations. Therefore, the *sym-2* gene was crossed into other pea cultivars and *nodX* was transferred into good N_2 fixers (Fobert et al., 1991), establishing superbly competitive host-rhizobia combination.

Rhizobia strains have been genetically modified to increase their competitiveness (e.g., Toro et al., 1998). Genes conferring capacity to produce trifolitoxin, a compound with bactericidal and/or bacteriostatic effects on other bacteria (see Vlassak and Vanderleyden, 1997), have been transferred to the effective nodulator *R. leguminosarum* bv. *trifolii* strain, which showed increased competitiveness for nodulation of clover when compared with the trifolitoxin-sensitive strain (Leung et al., 1994).

Expression of some genes in *Bradyrhizobium japonicum* was induced by exposure to soil extract as opposed to the basal medium (Bhagwat and Keister, 1992). The strain that was a better nodulator had more genes expressed; the two genes cloned by a subtractive RNA hybridisation from the better nodulating strain increased the competitive nodulation capacity in the otherwise poorly nodulating strain.

Genotypic differences in symbiotic N_2 fixation

A continuous and coordinated selection of the most effective combinations of host and microbial symbionts is a prerequisite for profitable and sustainable agricultural systems. Different legume genotypes differ in efficacy of symbiotic N_2 fixation they can support (e.g., white clover, *Trifolium repens*, Ledgard, 1989; faba bean, Caba et al., 1993; soybean, Ofosu Budu et al., 1993; Song et al., 1993; De Chueire and Hungria, 1997; mung bean, *Vigna radiata*, Espiritu et al., 1993; common bean, *Phaseolus vulgaris*, Hardarson et al., 1993; pea, Evans et al., 1995; Fesenko et al., 1995; groundnut, Bell et al., 1994, Ibrahim et al., 1995; lucerne, Hernandez et al., 1995; chickpea, Sattar et al., 1995; tree *Gliricidia sepium*, Sanginga et al., 1992) as well as in the response of N_2 fixation to fertiliser N addition (e.g., faba bean, Caba et al., 1993; soybean, Song et al., 1993; Zhang et al., 1997; pea, Evans et al., 1995). Genotypic differences in activities of glutamine synthetase (glutamate-ammonia ligase) and glutamate synthase (GOGAT) in nodules were related to genotypic differences in symbiotic N_2 fixation efficiency, at least in faba bean (Caba et al., 1993). Therefore, screening for increased activities of these enzymes might be used in a breeding programme to increase symbiotic efficiency of faba bean genotypes.

A number of other symbiotic characteristics can be considered in breeding for better symbiosis. For example, chickpea genotypes differed in the total number and weight of nodules per plant as well as the weight of individual nodules (Dangaria et al., 1994). Other criteria related to the functioning of nodules may also be considered. In pea cultivars, N accumulation in shoots is a more suitable criterion than the shoot mass for selection of either highly effective or highly competitive *R. leguminosarum* bv. *viceae* strains in the pot experiments (Fesenko et al., 1995).

Over the course of evolution, rhizobial strains developed that could nodulate only one or, at best, a limited number of legume species. Various rhizobia strains also differ in N_2 fixation capacity; it is therefore possible to select for more efficient strains (Hartmann et al., 1998; Hungria et al., 1998). As the effectiveness of the nodulation and the N_2-fixation process varies widely in different rhizobia–host combinations, it would be worthwhile to identify the highly effective rhizobial strains, followed by identifying responsible genes that regulate high effectiveness of the symbiosis. These genes could then be transferred into commercial strains of rhizobia to increase efficiency of N_2 fixation or their host range. In addition, there is generally a strong host genotype × rhizobia strain interaction that determines the efficiency of the nodulation and N_2 fixation processes (Luna and Planchon, 1995). However, in bean genotypes of Mezo-American and Andean origin tested with 10 genetically diverse bean rhizobia, no host × rhizobia interaction was detected (Buttery et al., 1997).

It would be particularly important to develop genotypes that maintain high efficiency of N_2 fixation in the presence of small or moderate levels of nitrate. Otherwise, nitrate produced from mineralisation of the organic matter may not be taken up by legumes (Blumenthal and Russelle, 1996), but instead would leach into the lower horizons or ground water (see Anderson et al., 1998) and/or may hinder the symbiotic process. This is especially important as recently released, high-yielding cultivars may contain high N concentrations in the harvestable product, thus removing large amounts of N from the field. Genotypic variability exists among host genotypes in the degree of tolerance of the N_2-fixation process to nitrate (Raffin and Roumet, 1994). Additional genotypic variability and tolerance to nitrate can be introduced by mutagenesis (e.g., in soybean, Neo et al., 1996). Also, there is a genetic variability within nodule bacteria germplasm in tolerance of the N_2 fixation process to nitrate (e.g., Leung et al., 1994; Nour et al., 1994), which needs to be taken into account when creating a breeding programme for better symbioses. Such genotypic variability clearly indicates a potential for greatly increasing efficacy of symbiotic N_2 fixation by selecting the best combination of nitrate-tolerant plant genotype and rhizobia.

Screening genotypes for nodulation and N_2 fixation in hydroponics under controlled environmental conditions needs to be calibrated against field performance of standard cultivars because nodulation of, for example, bean genotypes may be different between controlled conditions and the field, even when growth in the two environments appears similar (Hernandez et al., 1993). Early screening of soybean genotypes for nodulation and N_2 fixation (20–60 days after planting) in both glasshouse and field may generate useful parameters for breeding programmes aimed at improving efficiency of the legume-rhizobia symbiosis (Pazdernik et al., 1997).

Wild genotypes of crop plants may serve as a source of genes for improved and more effective N_2 fixation (Andriolo et al., 1994). Similarly, host mutants (e.g., white sweetclover, *Melilotus alba*, Utrup et al., 1993), unable to achieve complete and functional symbiosis with rhizobia, should be studied more to increase our understanding of the symbiotic process.

Adaptation of legume-rhizobia symbiosis to unfavourable environments

Nodulation of a range of pasture and crop legumes (e.g., white clover, subclover, lucerne, pea, cowpea, bean, etc.) is reduced in acid soils, mainly because of sensitivity of early nodulation events, such as attachment, root hair curling and initiation of infection thread formation (for references see Vlassak and Vanderleyden, 1997). Problems with *nod* gene expression, more notable in acid-sensitive than in acid-tolerant rhizobial strains, may account for these deleterious acidity-related effects (McKay and Djordjevic, 1993). It is also important to mention that legume hosts grown under conditions of environmental stresses differ in the quantity and quality of *nod*-gene inducers found in their root exudates (e.g., Raghuwanshi et al., 1994).

Low P (Mullen et al., 1988) and low Ca (Munns, 1970), the soil properties associated with acidic soils, deleteriously affect the nodulation process independently of pH. Rhizobial strains with relatively higher capacity to nodulate their hosts under low P (Leung and Bottomley, 1987) or in acidic soils (Howieson, 1995) have been isolated. These strains are essential for good symbiosis under such adverse conditions that are prevalent in many tropical soils farmed by subsistence farmers who are unable to invest substantial funds required to alleviate soil acidity.

Considerable efforts over the years have been directed toward selecting the optimal combination of rhizobial inoculum and the legume genotype for acidic soils where high H^+, Al or Mn may limit the effectiveness of some rhizobia–host combinations (e.g., in lucerne, Munns, 1970; white clover, Wood et al., 1984; chickpea, Rai, 1991b; *Lens culinaris*, Rai, 1992). In general, slow-growing bradyrhizobia are more acid-tolerant than fast-growing rhizobia (Cooper et al., 1985). The mechanisms behind such a difference have yet to be elucidated. Nevertheless, screening of rhizobia for tolerance to acidic conditions in the laboratory did result in strains that prove more effective in acid soils under field conditions (e.g., using acid-tolerant *R. meliloti* increased the area under pasture in Australia by at least 350 000 ha; Howieson et al., 1988). However, there are also examples of poor correlation between tolerance to acidity in the laboratory and in the field (e.g., Gemell and Roughley, 1993).

When grown in the acidic environments, some acid-tolerant Medicago species (*M. murex, M. polymorpha*) can nodulate better than others (*M. truncatula, M. littoralis*) (Howieson et al., 1992) because these acid-tolerant species maintained the composition of root exudates suitable for inducing *nod* genes in *R. meliloti* and thus fostering nodulation. *Biserulla pelecinus*, a pasture legume adapted to acidic soils, has a strain of nodule bacteria that is different from other known symbiotic N_2-fixing bacteria, which apparently represents an adaptation to acidic soils (Howieson et al., 1995).

Eight characteristics required in an acid-tolerant pasture legume symbiosis for use in ley-farming have been described by Howieson (1995). Among others, the capacity of rhizobia (preferably of the genus *Bradyrhizobium*) to regulate internal pH and thus maintain saprophytic function at low pH were emphasised together with stability of the nodABC gene products as well as surface polysaccharides and proteins at acidic pH. The capacity of host roots to produce pH-stable organic acids for linkage with rhizobial cell surface structures is also important.

Large differences exist among rhizobial strains in tolerance to Al or Mn toxicity (for references see Vlassak and Vanderleyden, 1997), with generally a poor correlation between tolerance to acidity and tolerance to Al or Mn. However, strains showing tolerance to Al in laboratory conditions also performed better in Al-toxic soils (e.g., Graham et al., 1982).

Al-tolerant wheat genotypes exuded a greater amount of organic acid anions into the rhizosphere than Al-sensitive genotypes exposed to Al toxicity (for the review see Rengel, 1996). As organic acid anions are a good substrate for *Azospirillum brasilense*, these associative diazotroph fixers contributed considerably more fixed nitrogen to the Al-tolerant than to the Al-sensitive wheat genotype (Christiansen-Weniger et al., 1992).

Tn5-induced mutants varying in acid tolerance have been instrumental in isolating several acid-tolerant chromosome-located loci. Some of the acid-tolerant loci can be transferred into acid-sensitive rhizobia to increase their tolerance to soil acidity (Tiwari et al., 1992).

Tolerance of the symbiotic relationship was tested under saline stress (e.g., in faba bean, Cordovilla et al., 1995; chickpea, Zurayk et al., 1998; soybean, Abd-Alla et al., 1998), low temperature (e.g., in faba bean, Fyson and Sprent, 1982; sainfoin, Prevost and Bromfield, 1991; soybean, Zhang et al., 1997), drought (e.g., in subclover, Leung and Bottomley, 1994; bean, Castellanos et al., 1996; soybean, Serraj and Sinclair, 1998) and waterlogging (e.g., in *Trifolium semipilosum*, Lupwayi et al., 1997). Genotypic differences in tolerance of the legume-rhizobia symbiosis to these various stresses have been documented.

Symbiotic N_2 fixation in non-legume hosts

While there are several avenues for increasing the effectiveness of the naturally occurring symbiotic relationships (e.g *Rhizobium*–legume), it appears quite difficult at present to create conditions conducive to other symbioses (e.g., *Rhizobium*–rice) (Ladha et al., 1997; Reddy et al., 1997). Among some notable problems, rice root exudates did not induce expression of the *nod* genes in various rhizobial species, and neither the wild-type *Rhizobium* nor purified lipo-chitooligosaccharide Nod factors induced root hair deformation or true nodulation on rice roots (Reddy et al., 1997). However, continued improvement in our understanding of the physiological and molecular basis of the *Rhizobium*–legume symbiosis may allow transfer and expression of required genes into non-host species, thus improving chances that the viable symbiosis between rhizobia and (otherwise) non-host plants can be achieved. Indeed, early results on transforming rice with the *nodulin* gene (see below) indicate that a rice line may soon be developed that can fix nitrogen from the atmosphere. Results with rice root endophytes fixing atmospheric N_2 are also encouraging (e.g., Hurek et al., 1999).

When the early nodulin gene *ENOD12* from *Medicago truncatula* was transferred into rice, the gene was expressed in root cortical parenchyma upon induction by rhizobial lipo-chitooligosaccharide nodulation factors (Reddy et al., 1998). These results have indicated that rice possesses at least a part of the signal transduction pathway required for legume root nodulation. Further work has shown that species from genera *Oryza* as well as from a range of other grasses (including maize and sugarcane) possess homologues of *ENOD* early nodulation genes (Reddy et al., 1999). Moreover, rice *ENOD40* promoter activity is essentially the same as that of soybean *ENOD40*, indicating a similar role of these genes in the differentiation and/or function of nodular vascular bundles (Kouchi et al., 1999).

Symbiosis between plants and arbuscular mycorrhizal fungi

Mycorrhizal symbiosis is sensitive to the soil P status. If soils are fertilized with P, mycorrhizal symbiosis becomes ineffective (Smith et al., 1994), but also unnecessary as plants can access sufficient P. However, there are at least two situations when an improvement in mycorrhizal symbiosis may be worth considering:

(1) P-sorbing soils where only a small proportion (10-20%) of fertilizer P is available to plants, while most of fertilizer P is sorbed and complexed in soils (cf. Zhang and Rengel, 1999); and

(2) Low-input farming, mostly in developing countries, where little or no P fertilizer is used because of prohibitive costs.

Efficiency of AM symbiosis can be evaluated by considering the benefit:cost ratio, with the benefit represented by the difference in shoot dry matter between mycorrhizal and non-mycorrhizal plants under stress conditions and cost represented by the same difference under optimal conditions (Al-Karaki, 1998). Various host genotypes can be evaluated in such a way to optimise AM symbiosis under a range of environmental conditions.

There is genetic variability in the AM colonisation capacity of various genotypes of host species (e.g., bell pepper and tomato, Nemec and Datnoff, 1993; barley, Boyetchko and Tewari, 1995; grapevine, Karagiannidis et al., 1995). Wide variability also exists in populations of mycorrhizal fungi in their hyphal growth and thus competitive ability (De la Bastide et al., 1995). In addition, the genotype of the fungus is important because some degree of host specificity exists in AM fungi (Boyetschko and Tewari, 1995).

Wheat genotypes differ in the capacity to sustain AM colonisation, with the yield responses varying from zero to positive or negative values (Xavier and Germida, 1998). In durum wheat genotypes the benefit arising from the AM symbiosis is not proportional to the extent of the root colonisation because various genotypes have a various degree of dependence on mycorrhiza (Al-Karaki and Al-Raddad, 1997). Similar results were obtained for barley genotypes varying in P efficiency (Baon et al., 1993).

Using differential RNA display, Martin-Laurent et al. (1997) cloned one of the plant genes involved in early events leading to the successful colonisation of pea roots by *Glomus moseae*. Expression of that gene was independent of the rhizobial bacteria. Knowledge on genetics of the colonisation process will be instrumental in designing screening procedures and molecular markers to breed genotypes for more efficient AM symbiosis. Mutants unable to sustain mycorrhizal colonisation (e.g., in pea, Balaji et al., 1995; tomato, Barker et al., 1998) are going to be important in increasing such knowledge.

Mutants resistant to AM colonisation were introduced in pea and *Medicago truncatula* (for references see Gianinazzi-Pearson, 1996). The Myc^- mutation is recessive, genetically stable and controlled by the same single gene as Nod^- (Duc et al., 1989). The product of the wild-type alleles of Myc^- mutated loci may be involved in the biosynthesis of a plant susceptibility factor that negatively regulates the defence response (Gianinazzi-Pearson, 1996). So, without this susceptibility factor, plant root cells of Myc^- mutants have thick, reinforced cell walls loaded with defence-related molecules, thus preventing AM colonisation of such cells. In contrast to Myc^- mutation, Myc^{2-} loci may be involved in metabolic specialisation of the AM-containing cells (Gianinazzi-Pearson et al., 1995).

Clearly, it would be too dangerous to try to enhance the symbiosis by down-regulating plant defence response as susceptibility to the pathogen attack may increase. Even though Myc^- and Nod^- mutants described so far do not appear to suffer from increased susceptibility to the pathogen attack, indicating considerable specificity in the infection pattern and localisation of changes in plant defence responses of AM fungi and rhizobia as opposed to pathogens. In contrast, the fact that Myc^- mutants are also Nod^- mutants (at least in pea) indicate that there are common mechanisms regulating the plant–microbe interactions in the two symbioses (Gianinazzi-Pearson, 1996), thus raising the possibility that breeding efforts to improve one symbiosis may fortuitously result in improvements to the other.

A number of genes involved in AM or rhizobial symbioses have been cloned (see Gianinazzi-Pearson, 1996; Murphy et al., 1997). However, it would be fair to say that genes known so far represent just a tip of the iceberg of what is involved in the effective symbioses. With the expanding number of genes cloned and with insights into transcription regulation and product function, we should see that knowledge translated into breeding for better symbioses.

AM signalling

Hydrophobic components in the root exudates of P-deficient onion facilitate appressorium growth and mycorrhizal colonisation of onion plants (Tawaraya et al., 1998). Flavonoid compounds in root exudates of carrot seedlings stimulate hyphal growth of the AM fungus *Gigaspora margarita* (Poulin et al., 1993). In contrast, root exudates from pea inhibited hyphal growth of the same fungus (Balaji et al., 1995).

Polyamines may be signalling factors exuded by host roots and allowing colonisation by the AM fungi, with a positive correlation between the polyamine chain length and the fungal development (El-Ghachtouli et al., 1995). Non-host plants appear rather to lack factors necessary for stimulation of the AM colonisation than to have inhibitory factors (cf. Becard and Piche, 1990). Therefore, introduction of these stimulatory factors into non-host plants via genetic transformation to study possibilities of extending the AM host range appears to be an appealing possibility. Flavonoids that elicit rhizobial *nod* genes also stimulate mycorrhizal colonisation (Phillips and Tsai, 1992), indicating commonality of at least some mechanisms involved in plant-microbe interactions in the two symbioses. In addition, nodulating genotype of lucerne developed AM mycorrhizae with genera *Glomus* and *Gigaspora*, but the non-nodulating and the poorly nodulating genotype did not, as the aborted appressoria could not penetrate the cortical cells of these genotypes (Bradbury et al., 1993).

As flavonoids exuded by legume seedlings represent *nod*-gene inducers (see above) and may also be stimulants for hyphal growth of the AM fungi (Kape et al., 1992; Phillips et al., 1995; Poulin et al., 1993), there is a possibility that needs to be explored about a potential stimulating effect of several common flavonoids for a range of AM and rhizobial hosts because selecting and/or transforming plants to increase exudation of these flavonoids may result in a double benefit for mycorrhizal legumes. Preliminary research has shown that flavonones and flavones that induce *nod* gene expression in rhizobia also stimulate spore germination and hyphal growth of the AM fungus *Gigaspora margarita*, probably through facilitating the contact between the AM symbionts (Gianinazzi-Pearson et al., 1989). In contrast, a mutation causing a change in the morphology of root hairs in pea plants resulted in inability to form nodules with Rhizobium or to become colonised by AM hyphae (Borisov et al., 1994).

Tripartite symbioses

Improvements in the symbiotic relationship between rhizobia and legumes may come from introducing the third partner: *Azospirillum brasilense* increased exudation of the flavonoid *nod* gene inducers into the bean rhizosphere, thus increasing nodulation by rhizobial inoculum (Burdman et al., 1996). Similarly, there was a positive interaction among cassava, diazotrophic *Azospirillum* and the AM *Glomus clarum* fungus, with root exudates from cassava stimulating mycorrhizal growth (Balota et al., 1995). N_2 fixation was enhanced by AM colonisation in comparison to non-mycorrhizal lucerne plants (Toro et al., 1998).

In the tripartite symbiosis (AM + nodule bacteria + groundnut as a host plant species), there was a significant genotype effect (Ibrahim et al., 1995), indicating a genetic variability in host capacity to sustain effective symbiosis with AM and nodule bacteria. Therefore, including AM inoculum together with the rhizobial inoculum into the groundnut breeding programme should be considered.

Helper bacteria

Helper bacteria may be involved in facilitating rhizobia nodulation of non-host plants (e.g., in canola plants infected with *B. japonicum* the presence of bacteria *Erwinia carotovora* that produce cell wall-degrading enzymes cellulase and pectinase allowed nodule formation; Katsunori et al., 1995). Experiments like that appear to have dubious practical significance (as *E. carotovora* causes plant death if present in higher numbers), but the direction of research appears promising if combinations of the non-host plant, rhizobia and helper bacteria can be found that will be suitable.

Mycorrhization helper bacteria show specificity in their action, i.e. they may support only a single host-symbiont interaction (Duponnois et al., 1993). P-solubilising bacteria (e.g., *Bacillus subtilis* or *Enterobacter* sp.) may act as mycorrhization helper bacteria when inoculated together with *Glomus intraradices* on lucerne. These P-solubilising bacteria released P from rock phosphate or indigenous plant-unavailable P sources in soils, while external AM mycelium took up that released P (Toro et al., 1997, 1998).

Conclusions

Advances in the basic knowledge about processes and gene products involved in plant–microbe interactions in the N_2 fixation and AM symbioses will be instrumental in improving these relationships. Molecular biology together with screening genotypes and classical breeding techniques will increase productivity of symbioses and eventually result in increased maximum economic yields of crop plants. Utilisation of N_2 fixation and AM symbioses described here will be extremely important in increasing the world food supplies required to feed the rapidly increasing population. Such an importance is emphasised by the fact that the area of good agricultural land is shrinking, thus forcing agriculture onto more marginal land where nutrient availability may be jeopardised by soil chemical and physical conditions. Moreover, the expansion in food production will be required, especially in developing nations because they will experience the greatest increase in population, yet they lack resources to supply sufficient fertilizers to allow adequate crop nutrition. Thus, increasing capacity of plants to scavenge nutrients from soil as well as fix N_2 from air will be of immense importance.

References

Abd-Alla M H, Vuong T D and Harper J E 1998 Genotypic differences in dinitrogen fixation response to NaCl stress in intact and grafted soybean. Crop Sci. 38, 72–77.

Al-Karaki G N 1998 Benefit, cost and water-use efficiency of arbuscular mycorrhizal durum wheat grown under drought stress. Mycorrhiza 8, 41–45.

Al-Karaki G N and Al-Raddad A 1997 Drought stress and VA mycorrhizal fungi effects on growth and nutrient uptake of two wheat genotypes differing in drought resistance. Crop Res. (Hisar) 13, 245–257.

Anderson G C, Fillery I R P, Dunin F X, Dolling P and Asseng S 1998 Nitrogen and water flows under pasture-wheat and lupin-wheat rotations in deep sands in Western Australia. 2. Drainage and nitrate leaching. Austr. J. Agric. Res. 49, 345–361.

Andriolo J, Pereira P A A and Henson R A 1994 Variability among wild *Phaseolus vulgaris* lines for traits related to biological N_2 fixation. Pesquisa Agropecuaria Brasileira 29, 831–837.

Armero J, Lopez-Valbuena R, Jorrin J and Tena M 1994 Contribution of different plant organs to the accumulation and secretion of phenolic compounds in glutathione-treated chickpea (*Cicer arietinum*) seedlings. Acta Hortic. 381, 506–509.

Balaji B, Poulin M J, Vierheilig H and Piche Y 1995 Responses of an arbuscular mycorrhizal fungus, *Gigaspora margarita*, to exudates and volatiles from the Ri T-DNA-transformed roots of nonmycorrhizal and mycorrhizal mutants of *Pisum sativum* L. Sparkle. Exp. Mycol. 19, 275–283.

Balota E L, Lopes E S, Hungria M and Dobereiner J 1995 Interactions and physiological effects of diazotrophic bacteria and arbuscular mycorrhizas on cassava. Pesquisa Agropecuaria Brasileira 30, 1335–1345.

Baon J B, Smith S E and Alston A M 1993 Mycorrhizal responses of barley cultivars differing in P efficiency. Plant Soil 157, 97–105.

Barker S J, Stummer B, Gao L, Dispain I, O'Connor P J and Smith S E 1998 A mutant in *Lycopersicon esculentum* Mill. with highly reduced VA mycorrhizal colonization: isolation and preliminary characterisation. Plant J. 15, 791–797.

Becard G and Piche Y 1990 Physiological factors determining vesicular-arbuscular mycorrhizal formation in host and nonhost Ri T-DNA transformed roots. Can. J. Bot. 68, 1260–1264.

Bell M J, Wright G C, Suryantini and Peoples M B 1994 The N_2-fixing capacity of peanut cultivars with differing assimilate partitioning characteristics. Aust. J. Agric. Res. 45, 1455–1468.

Bhagwat A A and Keister D L 1992 Identification and cloning of *Bradyrhizobium japonicum* genes expressed strain selectively in soil and rhizosphere. Appl. Environ. Microbiol. 58, 1490–1495.

Blumenthal J M and Russelle M P 1996 Subsoil nitrate uptake and symbiotic dinitrogen fixation by alfalfa. Agron. J. 88, 909–915.

Boivin C, Barran L R, Malpica C A and Rosenberg C 1991 Genetic analysis of a region of the *Rhizobium meliloti* pSym plasmid specifying catabolism of trigonelline, a secondary metabolite present in legumes. J. Bacteriol. 173, 2809–2817.

Bolanos-Vasquez M C and Werner D 1997 Effects of *Rhizobium tropici*, *R. etli*, and *R. leguminosarum* bv. *phaseoli* on *nod* gene-inducing flavonoids in root exudates of *Phaseolus vulgaris*. Mol. Plant-Microbe Interact. 10, 339–346.

Borisov A Y, Rozov S M, Tsyganov V E, Kulikova O A, Kolycheva A N, Yakobi L M, Ovtsyna A O and Tikhonovich I A 1994 Identification of symbiosis genes in pea (*Pisum sativum* L.) by means of induced mutagenesis. Genetika (Moskva) 30, 1484–1494.

Boyetchko S M and Tewari J P 1995 Susceptibility of barley cultivars to vesicular-arbuscular mycorrhizal fungi. Can. J. Plant Sci. 75, 269–275.

Bradbury S M, Peterson R L and Bowley S R 1993 Further evidence for a correlation between nodulation genotypes in alfalfa (*Medicago sativa* L.) and mycorrhiza formation. New Phytol. 124, 665–673.

Burdman S, Volpin H, Kigel J, Kapulnik Y and Okon Y 1996 Promotion of *nod* gene inducers and nodulation in common bean (*Phaseolus vulgaris*) roots inoculated with *Azospirillum brasilense* Cd. Appl. Environ. Microbiol. 62, 3030–3033.

Buttery B R, Park S J, Van Berkum P 1997 Effects of common bean (*Phaseolus vulgaris* L.) cultivar and rhizobium strain on plant growth, seed yield and nitrogen content. Can. J. Plant Sci. 77, 347–351.

Caba J M, Lluch C and Ligero F 1993 Genotypic differences in nitrogen assimilation in *Vicia faba*: effect of nitrate. Plant Soil 151, 167–174.

Castellanos J Z, Pena-Cabriales J J and Acosta-Gallegos J A 1996 ^{15}N-determined dinitrogen fixation capacity of common bean (*Phaseolus vulgaris*) cultivars under water stress. J. Agric. Sci. 126, 327–333.

Christiansen-Weniger C, Groneman A F and Van Veen J A 1992 Associative N_2 fixation and root exudation of organic acids from wheat cultivars of different aluminium tolerance. Plant Soil 139, 167–174.

Cooper J E, Wood M and Bjourson A J 1985 Nodulation of *Lotus peduncululatus* in acid rooting solution by fast- and slow-growing rhizobia. Soil Biol. Biochem. 17, 487–492.

Cordovilla M P, Ligero F, Lluch C and Cordovilla M P 1995 Influence of host genotypes on growth, symbiotic performance and nitrogen assimilation in faba bean (*Vicia faba* L.) under salt stress. Plant Soil 172, 289–297.

Dakora F D, Joseph C M and Phillips D A 1993 Common bean root exudates contain elevated levels of daidzein and coumestrol in response to *Rhizobium* inoculation. Mol. Plant-Microbe Interact. 6, 665–668.

Dangaria C J, Parameshwarappa R, Salimath P M and Annigeri B S 1994 Genetic divergence for nodulating characters in chickpea. Legume Research 17, 32–36.

De Chueire L M and Hungria M 1997 N_2-fixation ability of Brazilian soybean cultivars with *Sinorhizobium fredii* and *Sinorhizobium xinjiangensis*. Plant Soil 196, 1–5.

De la Bastide P Y, Kropp B R and Piche Y 1995 Population structure and mycelial phenotypic variability of the ectomycorrhizal basidiomycete *Laccaria bicolor* (Maire) Orton. Mycorrhiza 5, 371–379.

Dobereiner J 1997 Biological nitrogen fixation in the tropics: social and economic contributions. Soil Biol. Biochem. 29, 771–774.

Dockendorff T C, Sanjuan J, Grob P and Stacey G 1994 NolA represses *nod* gene expression in *Bradyrhizobium japonicum*. Mol. Plant-Microbe Interact. 7, 596–602.

Duc G 1995 Mutagenesis of faba bean (*Vicia faba* L.) and the identification of five different genes controlling no nodulation, ineffective nodulation or supernodulation. Euphytica 83, 147–152.

Duc G, Trouvelot A, Gianinazzi-Pearson V and Gianinazzi S 1989 First report of non-mycorrhizal plant mutants (Myc^-) obtained in pea (*Pisum sativum* L.) and faba bean (*Vicia faba* L.). Plant Sci. 60, 215–222.

Duponnois R, Garbaye J, Bouchard D and Churin J L 1993 The fungus-specificity of mycorrhization helper bacteria (MHBs) used as an alternative to soil fumigation for ectomycorrhizal inoculation of bare-root Douglas-fir planting stocks with *Laccaria laccata*. Plant Soil 157, 257–262.

Eisenschenk L, Diebold R, Perez-Lesher J, Peterson A C, Peters N K and Noel K D 1994 Inhibition of *Rhizobium etli* polysaccharide mutants by *Phaseolus vulgaris* root compounds. Appl. Environ. Microbiol. 60, 3315–3322.

El-Ghachtouli N, Paynot M, Morandi D, Martin-Tanguy J and Gianinazzi S 1995 The effect of polyamines on endomycorrhizal infection of wild-type *Pisum sativum*, cv. Frisson (nod+myc+) and two mutants (nod-myc+ and nod-myc-). Mycorrhiza 5, 189–192.

El-Shanshoury A R 1995 Interactions of *Azotobacter chroococcum,Azospirillum brasilense* and *Streptomyces mutabilis*, in relation to their effect on wheat development. J. Agron. Crop Sci. 175, 119–127.

Espiritu B M, Lales E H and Palacpac N Q 1993 Interaction effects of Bradyrhizobium strain and cultivar in mungbean (*Vigna radiata* (L.) Wilczek). Philippine J. Biotechnol. 4, 61–68.

Evans J, Chalk P M and O'Connor G E 1995 Potential for increasing N_2 fixation of field pea through soil management and genotype. Biol. Agric. Hortic. 12, 97–112.

Fedi S, Montaini P and Favilli F 1992 Chemotactic response of *Azospirillum* toward root exudates of C_3 and C_4 plants. Symbiosis 13, 101–105.

Fesenko A N, Provorov N A, Orlova I F, Orlov V P and Simarov B V 1995 Selection of *Rhizobium leguminosarum* bv. *viceae* strains for inoculation of *Pisum sativum* L. cultivars: analysis of symbiotic efficiency and nodulation competitiveness. Plant Soil 172, 189–198.

Fobert P R, Roy N, Nash J H E and Iyer V N 1991 Procedure for obtaining efficient root nodulation of a pea cultivar by a desired *Rhizobium* strain and preempting nodulation by other strains. Appl. Environ. Microbiol. 57, 1590–1594.

Fyson A and Sprent J I 1982 The development of primary root nodules on *Vicia faba* L. grown at two temperatures. Ann. Bot. 50, 681–692.

Gagnon H and Ibrahim R K 1997 Effects of various elicitors on the accumulation and secretion of isoflavonoids in white lupin. Phytochemistry 44, 1463–1467.

Gagnon H and Ibrahim R K 1998 Aldonic acids: a novel family of nod gene inducers of *Mesorhizobium loti*, *Rhizobium lupini*, and *Sinorhizobium meliloti*. Mol. Plant-Microbe Interact. 11, 988–998.

Garcia de Salamone I E, Dobereiner J, Urquiaga S and Boddey R M 1997 Biological nitrogen fixation in *Azospirillum* strain–maize genotype associations as evaluated by the ^{15}N isotope dilution technique. Biol. Fertil. Soils 23, 249–256.

Gemell L G and Roughley R J 1993 Field evaluation in acid soils of strains of *Rhizobium leguminosarum* bv. *trifolii* selected for their tolerance or sensitivity to acid soil factors in agar medium. Soil Biol. Biochem. 25, 1447–1452.

Gianinazzi-Pearson V 1996 Plant cell responses to arbuscular mycorrhizal fungi: getting to the roots of the symbiosis. Plant Cell 8, 1871–1883.

Gianinazzi-Pearson V, Branzanti B and Gianinazzi S 1989 In vitro enhancement of spore germination and early hyphal growth of a vesicular-arbuscular mycorrhizal fungus by host root exudates and plant flavonoids. Symbiosis 7, 243–255.

Gianinazzi-Pearson V, Gollote A, Lherminier J, Tisserant B, Franken P, Dumas-Gaudot E, Lemoine M C, Van Tuinen D and Gianinazzi S 1995 Cellular and molecular approaches in the characterization of symbiotic events in functional arbuscular mycorrhizal associations. Can. J. Bot. 73, S526–S532.

Godbold D L, Jentschke G, Wintr S and Marschner P 1998 Ectomycorrhizas and amelioration of metal stress in forest trees. Chemosphere 36, 757–762.

Graham P H, Viteri S E, Mackie F, Vargas A T and Palacios A 1982 Variation in acid soil tolerance among strains of *Rhiozobium phaseoli*. Field Crops Res. 5, 121–128.

Hardarson G, Bliss F A, Cigales-Rivero M R, Henson R A, Kipe-Nolt J A, Longeri L, Manrique A, Pena-Cabriales J J, Pereira P A A, Sanabria C A and Tsai S M 1993 Genotypic variation in biological nitrogen fixation by common bean. Plant Soil 152, 59–70.

Harper J E, Corrigan K A, Barbera A C and Abd-Alla M H 1997 Hypernodulation of soybean, mung bean, and hyacinth bean is controlled by a common shoot signal. Crop Sci. 37, 1242–1246.

Hartmann A, Giraud J J and Catroux G 1998 Genotypic diversity of *Sinorhizobium* (formerly *Rhizobium*) *meliloti* strains isolated directly from a soil and from nodules of alfalfa (*Medicago sativa*) grown in the same soil. FEMS Microbiol. Ecol. 25, 107–116.

Hartwig U A, Joseph C M and Phillips D A 1991 Flavonoids released naturally from alfalfa seeds enhance growth rate of *Rhizobium meliloti*. Plant Physiol. 95, 797–803.

Hernandez G, Vasquez H, Toscano V, Sanchez M, Penate-T, Franchi-Alfaro A, Mendez N and Drevon J J 1993 Nodulation and growth of common bean (*Phaseolus vulgaris* L.) cultivars in hydroponic culture and in the field. Trop. Agric. 70, 230–234.

Hernandez G, Ramirez M, Suarez R and Fuentes S I 1995 Root exuded *nod*-gene inducing signals limit the nodulation capacity of different alfalfa varieties with *Rhizobium meliloti*. Plant Cell Rep. 14, 626–629.

Heron D S, Ersek T, Krishnan H B and Pueppke S G 1989 Nodulation mutants of *Rhizobium fredii* USDA257. Mol. Plant-Microbe Interact. 2, 4–10.

Honma M A, Asomaning M and Ausubel F 1990 *Rhizobium meliloti nodD* genes mediate host-specific activation of *nodABC*. J. Bacteriol. 172, 901–911.

Howieson J G 1995 Characteristics of an ideotype acid tolerant pasture legume symbiosis in Mediterranean agriculture. Plant Soil 171, 71–76.

Howieson J G, Ewing M A and D'Anuono M F 1988 Selection for acid tolerance in *Rhizobium meliloti*. Plant Soil 105, 179–188.

Howieson J G, Robson A D and Abbott L K 1992 Acid-tolerant species of Medicago produce root exudates at low pH which induce the expression of nodulation genes in *Rhizobium meliloti*. Aust. J. Plant Physiol. 19, 287–296.

Howieson J G, Loi A and Carr S J 1995 *Biserrula pelecinus* L. — a legume pasture species with potential for acid, duplex soils which is nodulated by unique root-nodule bacteria. Aust. J. Agric. Res. 46, 997–1009.

Hungria M and Phillips D A 1993 Effects of a seed color mutation on rhizobial *nod*-gene-inducing flavonoids and nodulation in common bean. Mol. Plant-Microbe Interact. 6, 418–422.

Hungria M, Boddey L H, Santos M A and Vargas M A T 1998 Nitrogen fixation capacity and nodule occupancy by *Bradyrhizobium japonicum* and *B. elkanii* strains. Biol. Fertil. Soils 27, 393–399.

Hurek T, Tan Z, Egener T, Engelhard M, Gyaneshwar P, Ladha J K and Reinhold-Hurek B 1999 Novel nitrogen-fixing bacteria associated with the root interior of rice. *In* Proceedings of the Meeting of the Third Working Group on Biological Nitrogen Fixation in Rice, August 9–12, 1999. IRRI, Manila, Philippines.

Ibrahim K K, Arunachalam V, Rao P S K and Tilak K V B R 1995 Seasonal response of groundnut genotypes to arbuscular mycorrhiza – *Bradyrhizobium* inoculation. Microbiol. Res. 150, 218–224.

Ikeda K, Toyota K and Kimura M 1998 Role of extracellular pectinases in the rhizoplane competence of a rhizobacterium *Burkholderia pickettii* MSP3Rif. Soil Biol. Biochem. 30, 323–329.

Jagnow G 1990 Differences between cereal crop cultivars in root-associated nitrogen fixation, possible causes of variable yield response to seed inoculation. Plant Soil 123, 255–259.

James E K 2000 Nitrogen fixation in endophytic and associative symbiosis. Field Crops Res. 65, 197–209.

Janczarek M, Urbanik-Sypniewska T and Skorupska A 1997 Effect of authentic flavonoids and the exudate of clover roots on growth rate and inducing ability of *nod* genes of *Rhizobium leguminosarum* bv. *trifolii*. Microbiol. Res. 152, 93–98.

Jimenez-Zurdo J I, Garcia-Rodriguez F M and Toro N 1997 The *Rhizobium meliloti putA* gene: its role in the establishment of the symbiotic interaction with alfalfa. Mol. Microbiol. 23, 85–93.

Kape R, Wex K, Parniske M, Gorge E, Wetzel A and Werner D 1992 Legume root metabolites and VA-mycorrhiza development. J. Plant Physiol. 141, 54–60.

Karagiannidis N, Nikolaou N and Mattheou A 1995 Influence of three VA-mycorrhiza species on the growth and nutrient uptake of three grapevine rootstocks and one table grape cultivar. Vitis 34, 85–89.

Katsunori I, Makie K, Perigio F B Jr and Shoichiro A 1995 Induction of nodule-like structures on oilseed rape by inoculation with rhizobia in the presence of helper bacteria. Soil Sci. Plant Nutr. 41, 313–320.

Keyser H H and Cregan P B 1987 Nodulation and competition for nodulation of selected soybean genotypes among *Bradyrhizobium japonicum* serogroup 123 isolates. Appl. Environ. Microbiol. 53, 2631–2635.

Kouchi H, Takane K, So R B, Ladha J K and Reddy P M 1999 Rice ENOD40: isolation and expression analysis in rice and transgenic soybean root nodules. Plant J. 18, 121–129.

Kuykendall L D 1989 Influence of *Glycine max* nodulation on the persistence in soil of a genetically marked *Bradyrhizobium japonicum* strain. Plant Soil 116, 275–277.

Ladha J K, De Bruijn F J and Malik K A (eds.) 1997 Opportunities for Biological Nitrogen Fixation in Rice and Other Non-Legumes. Kluwer Academic Publishers, Dodrecht.

Ledgard S F 1989 Nitrogen fixation and transfer to associated grasses by white clover cultivars under dairy cow grazing. *In* Proceedings of the XVI International Grassland Congress, 4–11 October 1989, Nice, France. pp. 169–170.

Leung K and Bottomley P J 1987 Influence of phosphate on the growth and nodulation characteristics of *Rhizobium trifolii*. Appl. Environ. Microbiol. 53, 2098–2105.

Leung K and Bottomley P J 1994 Growth and nodulation characteristics of subclover (*Trifolium subterraneum* L.) and *Rhizobium leguminosarum* bv. *trifolii* at different soil water potentials. Soil Biol. Biochem. 26, 805–812.

Leung K, Strain S R, De Bruijn F J and Bottomley P J 1994 Genotypic and phenotypic comparisons of chromosomal types within an indigenous soil population of *Rhizobium leguminosarum* bv. *trifolii*. Appl. Environ. Microbiol. 60, 416–426.

Lohrke S M, Orf J H, Martinez-Romero E and Sadowsky M J 1995 Host-controlled restriction of nodulation by *Bradyrhizobium japonicum* strains in serogroup 110. Appl. Environ. Microbiol. 61, 2378–2383.

Luna R and Planchon C 1995 Genotype × *Bradyrhizobium japonicum* strain interactions in dinitrogen fixation and agronomic traits of soybean (*Glycine max* L. Merr.). Euphytica 86, 127–134.

Lupwayi N Z, Haque I and Holl F B 1997 Strain-specific response of *Trifolium semipilosum* to inoculation with *Rhizobium* and the significance of waterlogging in Vertisols. J. Agric. Sci. 129, 439–446.

Martin-Laurent F, Van Tuinen D, Dumas-Gaudot E, Gianinazzi-Pearson V, Gianinazzi S and Franken P 1997 Differential display analysis of RNA accumulation in arbuscular mycorrhiza of pea and isolation of a novel symbiosis-regulated plant gene. Mol. Gen. Genet. 256, 37–44.

Maxwell C A, Hartwig U A, Joseph C M and Phillips D A 1989 A chalcone and two related flavonoids released from alfalfa roots induce *nod* genes of *Rhizobium meliloti*. Plant Physiol. 91, 842–847.

McKay I A and Djordjevic M A 1993 Production and excretion of Nod metabolites by *Rhizobium leguminosarum* bv. *trifolii* are disrupted by the same environmental factors that reduce nodulation in the field. Appl. Environ. Microbiol. 59, 3385–3392.

McKhann H I, Paiva N L, Dixon R A and Hirsch A M 1997 Chalcone synthase transcripts are detected in alfalfa root hairs following inoculation with wild-type *Rhizobium meliloti*. Mol. Plant-Microbe Interact. 10, 50–58.

McLoughlin T J, Alt S G and Merlo P A 1990 Persistence of introduced *Bradyrhizobium japonicum* strains in forming nodules in subsequent years after inoculation in Wisconsin soils. Can. J. Microbiol. 36, 794–800.

Mergaert P, Van Montagu M and Holsters M 1997 Molecular mechanisms of Nod factor diversity. Mol. Microbiol. 25, 811–817.

Moens S, Bastelaere E V, Broek A V, Lambrecht M, Keijers V, Revers L F, Passaglia L M P, Schrank I S and Vanderleyden J 1997 *Azospirillum* genes involved in chemotaxis and adhesion to plant roots. *In* Biological Fixation of Nitrogen for Ecology and Sustainable Agriculture. A Legocki, H Bothe and A Puhler (Eds.). pp. 123–127. NATO Advanced Research Workshop, Poznan, Poland.

Mullen M D, Israel D W and Wollum A G I I 1988 Effects of *Bradyrhizobium japonicum* and soybean (*Glycine max* (L.) Merr.) phosphorus nutrition on nodulation and dinitrogen fixation. Appl. Environ. Microbiol. 54, 2387–2392.

Munns N D 1970 Nodulation of *Medicago sativa* in solution culture. V. Calcium and pH requirements during infection. Plant Soil 32, 90–102.

Murphy P J, Wexler W, Grzemski W, Rao J P and Gordon D 1995 Rhizopines — their role in symbiosis and competition. Soil Biol. Biochem. 27, 525–529.

Murphy P J, Langridge P and Smith S 1997 Cloning plant genes differentially expressed during colonisation of *Hordeum vulgare* (L.) roots by the vesicular arbuscular-mycorrhizal fungus *Glomus intraradices* Schenck & Smith. New Phytol. 135, 291–301.

Natsvaladze M Y, Nitse L K and Sakharova S N 1992 Free-living and associative diazotrophs in the rhizocoenosis of tea. Izvestiya Timiryazevskoi Sel'skokhozyaistvennoi Akademii 2, 95–102.

Nemec S and Datnoff L 1993 Pepper and tomato cultivar responses to inoculation with *Glomus intraradices*. Adv. Hortic. Sci. 7, 161–164.

Neo H H, Hunt S and Layzell D B 1996 Can genotypes of soybean (*Glycine max*) selected for nitrate tolerance provide good 'models' for studying the mechanism of nitrate inhibition of nitrogenase activity? Physiol. Plant. 98, 653–660.

Nour S M, Cleyet-Marel J C, Beck D, Effosse A and Fernandes M P 1994 Genotypic and phenotypic diversity of Rhizobium isolated from chickpea (*Cicer arietinum* L.). Can. J. Microbiol. 40, 345–354.

O'Connell K P, Goodman R M and Handelsman J 1996 Engineering the rhizosphere: expressing a bias. Trends Biotechnol. 14, 83–88.

Ofosu Budu K G, Fujita K, Gamo T and Akao S 1993 Dinitrogen fixation and nitrogen release from roots of soybean cultivar Bragg and its mutants nts1116 and nts1007. Soil Sci. Plant Nutr. 39, 497–506.

Orgambide G G, Philip-Hollingsworth S, Mateos P F, Hollingsworth R I and Dazzo F B 1996 Subnanomolar concentrations of membrane chitolipooligosaccharides from *Rhizobium leguminosarum* biovar *trifolii* are fully capable of eliciting symbiosis-related responses on white clover. Plant Soil 186, 93–98.

Paffetti D, Daguin F, Fancelli S, Gnocchi S, Lippi F, Scotti C and Bazzicalupo M 1998 Influence of plant genotype on the selection of nodulating *Sinorhizobium meliloti* strains by *Medicago sativa*. Antonie van Leeuwenhoek 73, 3–8.

Pandya S and Desai A 1998 Localization of nod, nif, and acidic exopolysaccharide determinants on a large plasmid in slow-growing cowpea *Rhizobium* sp. S2. Curr. Microbiol. 36, 36–40.

Pandya S, Uchil P, Subramanian M and Desai A 1998 Determination of host specificity of cowpea miscellany Rhizobium spp. by nodABC-lacZ fusion. Curr. Microbiol. 36, 361–364.

Pazdernik D L, Graham P H and Orf J H 1997 Variation in the pattern of nitrogen accumulation and distribution in soybean. Crop Sci. 37, 1482–1486.

Phillips D A and Tsai S M 1992 Flavonoids as plant signals to rhizosphere microbes. Mycorrhiza 1, 55–58.

Phillips D A, Wery J, Joseph C M, Jones A D and Teuber L R 1995 Release of flavonoids and betaines from seeds of seven *Medicago* species. Crop Sci. 35, 805–808.

Poulin M J, Bel-Rhlid R, Piche Y and Chenevert R 1993 Flavonoids released by carrot (*Daucus carota*) seedlings stimulate hyphal development of vesicular-arbuscular mycorrhizal fungi in the presence of optimal CO_2 enrichment. J. Chem. Ecol. 19, 2317–2327.

Prayitno J, Stefaniak J, McIver J, Weinman J J, Dazzo F B, Ladha J K, Barraquio W, Yanni Y G and Rolfe B G 1999 Interactions of rice seedlings with bacteria isolated from rice roots. Aust. J. Plant Physiol. 26, 521–535.

Prevost D and Bromfield E S P 1991 Effect of low root temperature on symbiotic nitrogen fixation and competitive nodulation of *Onobrychis viciifolia* (sainfoin) by strains of arctic and temperate rhizobia. Biol. Fertil. Soil 12, 161–164.

Pueppke S G, Bolanos-Vasquez M C, Werner D, Bec-Ferte M P, Prome J C and Krishnan H B 1998 Release of flavonoids by the soybean cultivars McCall and Peking and their perception as signals by the nitrogen-fixing symbiont *Sinorhizobium fredii*. Plant Physiol. 117, 599–608.

Raffin A and Roumet P 1994 Shoot-root control of nitrate tolerance of N_2 fixation in spontaneously tolerant soybean lines: reciprocal grafting experiments. Agronomie 14, 473–480.

Raghuwanshi A, Dudeja S S and Khurana A L 1994 Effect of temperature on flavonoid production in pigeonpea (*Cajanus cajan* (L.) Millsp) in relation to nodulation. Biol. Fertil. Soils 17, 314–316.

Rai R 1991a Strain-specific salt tolerance and chemotaxis of *Azospirillum brasilense* and their associative N-fixation with finger millet in saline calcareous soil. Plant Soil 137, 55–59.

Rai R 1991b Effects of soil acidity factors on interaction of chickpea (*Cicer arietinum* L.) genotypes and Rhizobium strains: symbiotic N-fixation, grain quality and grain yield in acid soils. *In* Plant-Soil Interactions at Low pH. R J Wright, V C Baligar and R P Murrmann (Eds.). pp. 619–631. Kluwer Academic Publishers, Dordrecht.

Rai R 1992 Effect of acidity factors on aspects of symbiotic N_2 fixation of *Lens culinaris* in acid soils. J. Gen. Appl. Microbiol. 38, 391–406.

Ramanathan N, Ramamurthy R and Balasubramanian V 1997 Nitrogenase activity and *Azospirillum* population in the rhizosphere of rice cultivars. Indian J. Microbiol. 37, 211–212.

Rao V R, Ramakrishnan B, Adhya T K, Kanungo P K and Nayak D N 1998 Review: current status and future prospects of associative nitrogen fixation in rice. World J. Microbiol. Biotechnol. 14, 621–633.

Reddy P M, Aggarwal R K, Ramos M C, Ladha J K, Brar D S and Kouchi H 1999 Widespread occurrence of the homologues of the early nodulin (ENOD) genes in *Oryza* species and related grasses. Biochem. Biophys. Res. Commun. 258, 148–154.

Reddy P M, Ladha J K, So R B, Hernandez R J, Ramos M C, Angeles O R, Dazzo F B and De Bruijn F J 1997 Rhizobial communication with rice roots: induction of phenotypic changes, mode of invasion and extent of colonization. Plant Soil 194, 81–98.

Reddy P M, Ladha J K, Ramos M C, Maillet F, Hernandez R J, Torrizo L B, Oliva N P, Datta S K and Datta K 1998 Rhizobial lipochitooligosaccharide nodulation factors activate expression of the legume early nodulin gene *ENOD12* in rice. Plant J. 14, 693–702.

Rolfe B G, Djordjevic M A, Weinman J J, McIver J, Gärtner E, Chen C, Creaser E H, Britt K, Lawson C G R, De Boer M H, McKay I A, Shoobridge M V, De Majnik J, Pittock C, Broderick K and Delbridge T 1995 Molecular genetic analysis of subterranean clover-microbe interactions. Soil Biol. Biochem. 27, 485–490.

Rolfe B G, Djordjevic M A, Weinman J J, Mathesius U, Pittock C, Gärtner E, Ride K M, Dong Z M, McCully M and McIver J 1997 Root morphogenesis in legumes and cereals and the effect of bacterial inoculation on root development. Plant Soil 194, 131–144.

Rosenblueth M, Hynes M F and Martinez-Romero E 1998 *Rhizobium tropici teu* genes involved in specific uptake of *Phaseolus vulgaris* bean-exudate compounds. Mol. Gen. Genet. 258, 587–598.

Rossbach S, Kulpa D A, Rossbach U and De Bruijn F J 1994 Molecular and genetic characterization of the rhizopine catabolism (*mocABRC*) genes of *Rhizobium meliloti* L5-30. Mol. Gen. Genet. 245, 11–24.

Sanginga N, Danso S K A and Bowen G D 1992 Variation in growth, sources of nitrogen and N-use efficiency by provenances of *Gliricidia sepium*. Soil Biol. Biochem. 24, 1021–1026.

Sanjuan J and Olivares J 1991 Nifa-NtrA regulatory system activates transcription of *nfe*, a gene locus involved in nodulation competitiveness of *Rhizobium meliloti*. Arch. Microbiol. 155, 543–548.

Sattar M A, Quader M A and Danso S K A 1995 Nodulation, N_2 fixation and yield of chickpea as influenced by host cultivar and *Bradyrhizobium* strain differences. Soil Biol. Biochem. 27, 725–727.

Schloter M and Hartmann A 1998 Endophytic and surface colonization of wheat roots (*Triticum aestivum*) by different *Azospirillum brasilense* strains studied with strain-specific monoclonal antibodies. Symbiosis 25, 159–179.

Schmidt P E, Parniske M and Werner D 1992 Production of the phytoalexin glyceollin I by soybean roots in response to symbiotic and pathogenic infection. Bot. Acta 105, 18–25.

Schmidt P E, Broughton W J and Werner D 1994 Nod factors of *Bradyrhizobium japonicum* and *Rhizobium sp.* NGR234 induce flavonoid accumulation in soybean root exudate. Mol. Plant-Microbe Interact. 7, 384–390.

Serraj R and Sinclair T R 1998 Soybean cultivar variability for nodule formation and growth under drought. Plant Soil 202, 159–166.

Shelp B J, Kaiser B N and Deschesne A M 1998 Registration of five near-isogenic genetic stocks of 'Juneau' pea with altered nodulation and nitrate reductase deficiency: A317I, nod3I, A317nod3I, E135I, and R25I. Crop Sci. 38, 554.

Shrestha R K and Ladha J K 1996 Genotypic variation in promotion of rice dinitrogen fixation as determined by nitrogen-15 dilution. Soil Sci. Soc. Am. J. 60, 1815–1821.

Smith S E, Gianinazzi-Pearson V, Koide R and Cairney J W G 1994 Nutrient transport in mycorrhizas: structure, physiology and consequences for efficiency of the symbiosis. Plant Soil 159, 103–113.

Soejima H, Sugiyama T and Ishihara K 1992 Changes in cytokinin activities and mass spectrometric analysis of cytokinins in root exudates of rice plants (*Oryza sativa* L.). Comparison between cultivars Nipponbare and Akenohoshi. Plant Physiol. 100, 1724–1729.

Song L, Carroll B J, Gresshoff P M and Herridge D F 1993 Field assessment of supernodulating genotypes of soybean for yield, N_2 fixation and benefit to subsequent crops. Soil Biol. Biochem. 27, 563–569.

Spiller H, Stallings W J, Woods T and Gunasekaran M 1993 Requirement for direct association of ammonia-excreting *Anabaena variabilis* mutant (SA-1) with roots for maximal growth and yield of wheat. Appl. Microbiol. Biotechnol. 40, 557–566.

Tagu D and Martin F 1996 Molecular analysis of cell wall proteins expressed during the early steps of ectomycorrhiza development. New Phytol. 133, 73–85.

Tawaraya K, Hashimoto K and Wagatsuma T 1998 Effect of root exudate fractions from P-deficient and P-sufficient onion plants on root colonisation by the arbuscular mycorrhizal fungus *Gigaspora margarita*. Mycorrhiza 8, 67–70.

Tiwari R P, Reeve W G, Glenn A R 1992 Mutations conferring acid sensitivity in the acid-tolerant strains *Rhizobium meliloti* WSM419 and *Rhizobium leguminosarum* biovar *viciae* WSM710. FEMS Microbiol. Lett. 100, 107–112.

Toro M, Azcon R and Barea J M 1997 Improvement of arbuscular mycorrhiza development by inoculation of soil with phosphate-solubilizing rhizobacteria to improve rock phosphate bioavailability and nutrient cycling. Appl. Environ. Microbiol. 63, 4408–4412.

Toro M, Azcon R and Barea J M 1998 The use of isotopic dilution techniques to evaluate the interactive effects of *Rhizobium* genotype, mycorrhizal fungi, phosphate-solubilizing rhizobacteria and rock phosphate on nitrogen and phosphorus acquisition by *Medicago sativa*. New Phytol. 138, 265–273.

Trese A T 1995 A single dominant gene in McCall soybean prevents effective nodulation with *Rhizobium fredii* USDA257. Euphytica 81, 279–282.

Tsyganov V E, Morzhina E V, Stefanov S Y, Borisov A Y, Lebsky V K and Tikhonovich I A 1998 The pea (*Pisum sativum* L.) genes *sym33* and *sym40* control infection thread formation and root nodule function. Mol. Gen. Genet. 295, 491–503.

Utrup L J, Cary A J and Norris J H 1993 Five nodulation mutants of white sweetclover (*Melilotus alba* Desr.) exhibit distinct phenotypes blocked at root hair curling, infection thread development, and nodule organogenesis. Plant Physiol. 103, 925–932.

Van Rhijn P and Vanderleyen J 1995 The *Rhizobium*-plant symbiosis. Microbiol. Rev. 59, 124-142.

Van Rhijn P, Desair J, Vlassak K and Vanderleyden J 1994 Functional analysis of *nodD* genes of *Rhizobium tropici* CIAT899. Mol. Plant-Microbe Interact. 7, 666–677.

Vlassak K M and Vanderleyden J 1997 Factors influencing nodule occupancy by inoculant rhizobia. Crit. Rev. Plant Sci. 16, 163–229.

Waschutza S, Hofmann N, Niemann E G and Fendrik I 1992 Investigations on root exudates of Korean rice. Symbiosis 13, 181–189.

Wood M, Cooper J E and Holding A J 1984 Soil acidity factors and nodulation of *Trifolium repens*. Plant Soil 78, 367–79.

Wojtaszek P, Stobiecki M and Gulewicz K 1993 Role of nitrogen and plant growth regulators in the exudation and accumulation of isoflavonoids by roots of intact white lupin (*Lupinus albus* L.) plants. J. Plant Physiol. 142, 689–694.

Wu P, Zhang G, Ladha J K, McCouch S R and Huang N 1995 Molecular-marker-facilitated investigation on the ability to stimulate N_2 fixation in the rhizosphere by irrigated rice plants. Theor. Appl. Genet. 91, 1177–1183.

Xavier L J C and Germida J J 1998 Response of spring wheat cultivars to *Glomus clarum* NT4 in a P-deficient soil containing arbuscular mycorrhizal fungi. Can. J. Soil Sci. 78, 481–484.

Yanni Y G, Rizk R Y, Corich V, Squartini A, Ninke K, Philip-Hollingsworth S, Orgambide G, De Bruijn F, Stoltzfus J, Buckley D, Schmidt T M, Mateos P F, Ladha J K and Dazzo F B 1997 Natural endophytic association between *Rhizobium leguminosarum* bv. *trifolii* and rice roots and assessment of its potential to promote rice growth. Plant Soil 194, 99–114.

Zhang X-K and Rengel Z 1999 Gradients of pH and ammonium and phosphorus concentration between the banded fertilizer and wheat roots. Aust. J. Agric. Res. 50, 365-373.

Zhang F, Dijak M, Smith D L, Lin J, Walsh K, Voldeng H, Macdowell F and Layzell D B 1997 Nitrogen fixation and nitrate metabolism for growth of six diverse soybean (*Glycine max* (L.) Merr.) genotypes under low temperature stress. Environ. Exp. Bot. 38, 49–60.

Zuanazzi J A S, Clergeot P H, Quirion J C, Husson H P, Kondorosi A and Ratet P 1998 Production of *Sinorhizobium meliloti* nod gene activator and repressor flavonoids from *Medicago sativa* roots. Mol. Plant-Microbe Interact. 11, 784–794.

Zurayk R, Adlan M, Baalbaki R and Saxena M C 1998 Interactive effects of salinity and biological nitrogen fixation on chickpea (*Cicer arietinum* L.) growth. J. Agron. Crop Sci. 180, 249–258.

J.J. Adu-Gyamfi (Ed.), *Food security in nutrient-stressed environments: exploiting plants' genetic capabilities*: 261–265, 2002.

Genetic approaches for manipulating crop plants to enhance root exudation and access low nutrients in the rhizosphere

F. D. Dakora[1] & T. R. Sinclair[2]
[1]*Research Development, Cape Technikon, Room 2.8 Admin Bldg., P.O. Box 652, Cape Town 8000, South Africa.*
[2]*United States Department of Agriculture-Agricultural Research Service (USDA-ARS), and Agronomy, Physiology, and Genetics Laboratory, University of Florida, PO Box 11095 Gainesville, FL 32611-0965, USA*

Introduction

Opportunities exist for promoting plant nutrient uptake through genetic manipulation of rhizosphere processes. An assessment of the genotypic differences in plants such as in the excretion of nod-gene inducers by symbiotic legumes, the release of protons and phytosiderophores by crop varieties suffering from iron (Fe) and zinc (Zn) deficiencies, the extrusion of organic acid anions under aluminium (Al) and phosphorus (P) stress, and the release of root border cells and/or enzymes (phosphatases and phytases) concerned with P nutrition, is essential for identifying crop cultivars adapted to low-nutrient conditions. Approaches for creating nutrient-efficient crops suitable for low-nutrient environments include: (1) the screening of adapted landraces for a better understanding of the genetic basis of root exudation, (2) the selection of plants with enhanced symbiotic performance when infected with stem/root nodule bacteria (rhizobia) and/or vesicular arbuscular (VA) mycorrhizae, (3) increased experimentation on ion-channels, transporter genes and the promoters controlling their expression, and (4) the use of molecular markers to select useful plant traits. Additionally, we need to understand the operational relationship of these various genes to other processes involved with mineral nutrition, and to integrate the latter with such whole-plant processes as photosynthesis in order to obtain higher crop yields in low-nutrient environments. But the real challenge with these low-nutrient environments lies in the production of food crops with clinically acceptable levels of nutrients and vitamins that, when consumed by humans, do not cause malnutrition. Breeding for micro-nutrient enriched crops could be one way to produce foods that meet the dietary requirements of human nutrition. This summary presents the tools currently available for manipulating rhizosphere processes and plant genes for increased nutrient uptake and assimilation in low-nutrient environments.

To promote nutrient uptake and increase yields of crop plants in low-nutrient environments requires a multi-pronged approach, involving selection and breeding for useful nutrition-related biological traits, genetic manipulation of cellular processes involved with nutrient uptake and assimilation, a deeper understanding of how plants growing in very low-nutrient soils actually optimize growth and productivity in ecosystems that would otherwise not support growth and yield of conventional crop varieties, and a costing of carbon and other inputs that enhance nutrient uptake and its use efficiency.

Assessing germplasm collections for traits useful in plant nutrition

The key to progress in breeding nutrient use efficient cultivars will be in the enhanced utilization of genetic resources. Intra-specific strategies will concentrate on the use of landraces or farmer varieties and wild relatives in the primary genepool. The use of these sources could be direct, but could involve backcrossing to minimize the impact of undesirable alleles. The secondary and tertiary genepools might serve as very important sources of unique alleles, but there will be a limit to the crossability and direct introgression from these accessions.

One option for the future will be the use of gene mapping and isolation. Once isolated and sequenced, the gene can be directly introduced into the crop cultivars with transformation. This same approach could be used in unrelated species where mechanisms are carefully evaluated using comparative approaches (Bramel-Cox, 2002).

Plant selection for low-nutrient environments

Large differences exist among plant cultivars, species, and genotypes in their tolerance of low nutrients, and this is due largely to genetic variation in root characteristics and nutrient acquisition. An understanding of these biological differences could help develop crop plants that enhance growth from efficient uptake and utilization of nutrients in low-nutrient environments (Zhu et al., 2002). The use of classical plant breeding techniques has contributed significantly to the development of high-yielding disease-, drought-, and pest-resistant crop varieties, and can similarly be applied to the selection of crop plants that grow well in low-nutrient situations. Selection criteria based on root architecture, cluster root formation, massiveness of root hair development, and VA mycorrhizal infection which all enhance the absorptive surface are important in addition to the use of suitable bioassays (Zhu et al., 2002).

Some plants have evolved effective mechanisms for attaining good growth under low-nutrient conditions. These include release of root exudates which modify the rhizosphere for increased nutrient uptake and for the establishment of symbiotic interactions that enhance nutrient acquisition (Dakora and Phillips, 2002; Rengel, 2002a). Plants growing in low-nutrient environments employ root exudates as symbiotic signals to such soil microbes as *Rhizobium* bacteria and VA fungi (Phillips, 1992), that are involved in nutrient procurement. Selecting cultivars that are enhanced in the biosynthesis and release of these signal molecules is likely to produce crop plants with the ability to increase nutrient acquisition from symbiotic interactions in nutrient-poor soils (Dakora and Phillips, 2002; Rengel, 2002b). Root exudation of organic acids, phenolics, amino acids, and extracellular enzymes by plants adapted to low-nutrient environments serves to improve nutrient uptake from direct modification of the rhizosphere. Cereal plants use phytosiderophores in root exudates to promote Fe acquisition, while organic acids and phenolics chelate and solubilize sparingly soluble Fe, calcium (Ca), and Al phosphates. The selection of cultivars and species that can produce quality exudates would be crucial for promoting plant growth for higher yields in low-nutrient environments.

But to increase plant growth also requires higher photosynthetic rates in addition to efficient nutrient uptake and utilization. It is known that plants grown under elevated CO_2 generally show greater biomass accumulation than ambient controls; and this occurs under conditions of increased photosynthetic activity and lowered levels of nutrients (e.g., N) in tissues (Drake et al., 1997). Thus, with the rising CO_2 concentration in the atmosphere, it is to be expected that plant species will respond naturally, and perhaps gradually, by increasing their photosynthetic rates even in low-nutrient environments. In fact, plant species that photosynthesize efficiently in low-N regimes have been identified. Some sorghum cultivars grow well under highly limiting soil-N conditions by maintaining high rates of photosynthesis via a sustained increase in the activity of PEP carboxylase enzyme (Maranville and Madhavan, 2002). Such plants are characterized by a low concentration of chlorophyll in their tissues, which in itself is an adaptive mechanism, and therefore offers scope for selecting species under low-nutrient conditions. While achieving high productivity in crop plants is therefore feasible in low-nutrient environments, food quality becomes a problem in terms of dietary balance of nutrients in the edible grain. So although plants growing in low-nutrient conditions can increase their biomass from an efficient PEP carboxylase enzyme system that leads to increased grain yield, the challenge is how to maintain clinically acceptable levels of nutrients and vitamins in the grain of such plants (Welch and Graham, 2002). No doubt, conventional breeding techniques can improve the protein and micronutrient contents of food crops especially cereals, but the task becomes the more difficult if the plants are growing in low-nutrient environments.

Manipulation of plant genes for enhanced nutrient acquisition

Nutrient uptake and assimilation involves transport across root hair membranes, loading into xylem, and transport across the membranes that occur between sources, sinks, and such storage units as vacuoles. Absorbed nutrients are generally partitioned between a metabolically active cytoplasmic pool and a vacuolar storage pool, with the former being maintained at relatively constant concentration at the expense of the latter. When nutrient supply is limited, as in low-nutrient environments, plants grow more roots, increase uptake rates from the soil, retranslocate from older leaves, and deplete vacuolar pools for use in new metabolic processes. So, given the multitude of cellular membranes that must be traversed by nutrient ions during uptake and translocation, the movement of any single mineral nutrient from root hairs to storage depots and metabolic sinks must involve many different membrane-embedded transport proteins. In the case of P, an increase in its uptake rates following P-starvation of the roots has been observed in many plant species, with the enhanced uptake being attributed to increased synthesis of P-transporter proteins in response to the P starvation.

Modern biology has afforded us ample opportunities for gene manipulation in plants, and these tools can be applied by plant breeders and geneticists to promote nutrient acquisition in crops (Johansen et al., 1995). Molecular markers such as isozymes, restriction fragment length polymorphism (RFLP), and random amplified polymorphic DNA (RAPD) are routinely used to map genes concerned with plant functions. They can similarly be used to map genes controlling root development and nutrient uptake, and in so doing identify and characterize each quantitative trait loci. Genetic maps revealing quantitative trait loci for water use efficiency and tolerance of P-stress in tomato have been developed (Johansen et al., 1995) using RFLP techniques. These can be applied to the screening of new cultivars and new germplasm collections for nutrient use efficiency.

In addition to manipulating native genes in plants, there are many successful gene transfer programmes that offer opportunities to produce and screen clones for such specific traits as tolerance of low nutrients. Through the use of these techniques, some important genes controlling the synthesis of transporter proteins have been cloned from plants for a number of essential mineral nutrients (Chrispeels et al., 1999). The recent cloning and characterization of potassium (K), selenium (Se), P, magnesium (Mg), Ca, and N transporters support the existence of both high-affinity and low-affinity ion transport components in plants (Dunlop and Phung, 2002). The expression of some of those genes specifically in roots (the organs involved in nutrient acquisition by plants) suggests their involvement in mineral nutrient uptake. However, some of these transporter genes are not root-specific, and can be located in shoot, leaves, and flowers (Dunlop and Phung, 2002). Such transporter genes are probably concerned with routine nutrient translocation between organs of a plant, and not direct uptake from soil. While P-deficient rape typically releases malic acid near its root tips or at the site of contact with insoluble rock phosphate, P-starved hedge mustard, in contrast, does not secrete organic acids; and this has been attributed to problems associated with synthesis and transport across membranes (Hoffland et al., 1992). Clearly, the manipulation of these transporter genes has potential for 'creating' crop plants capable of producing high yields from improved mineral nutrition, especially in low-nutrient environments. However, to achieve this, requires an equally deeper understanding of the ion channels that control the influx and efflux of mineral nutrients. In general, influx increases and efflux decreases under conditions of nutrient deficiency, while with nutrient sufficiency efflux increases to avoid the nutrient accumulating to toxic levels. While the tools of modern biology have started to provide an insight into our current understanding of plant nutrition, a multidisciplinary approach is needed if results are to be translated into crop yields of field plants.

Insights from nutrient acquisition by novel plant systems

A new study (Ae and Shen, 2002) has shown that the root cell walls of groundnut are capable of solubilizing insoluble Fe-P and Al-P for uptake by the species. In contrast, the root cell walls of soybean and sorghum are ineffective in mobilizing P from sparingly insoluble forms. Although the trait might be genotype-specific, it clearly indicates the diversity of strategies employed by plants to improve mineral nutrition. Root border cells are pioneers in the rhizosphere of plants, and their quality and quantity control many rhizosphere processes including the release of extracellular enzymes, proteins, antibiotics, mucilage, and molecular signals concerned with symbiotic establishment (Hawes et al., 1998). It is therefore likely that future research may find that root border cells play a major role in nutrient acquisition, especially in nutrient-poor environments.

Another recent study (Dakora et al., 1999) has also provided very puzzling data on the N nutrition of a desert legume. The African marama bean (*Tylosema esculentum* L.) is a non-nodulating legume that grows in very nutrient-poor sandy soils including those of the Kalahari desert. However it can accumulate up to 3% N in leaves, tubers and seeds; N-concentrations comparable to those found in effective N_2-fixing nodules. As with the groundnut (Ae and Shen, 2002), the marama bean must have a special mechanism for enhancing its N nutrition in these nutrient-poor soils. Additionally, a few plants, such as Rooibos tea (*Aspalathus linearis*), which grows in naturally very acidic sands of the Western Cape in South Africa, can actively modify their rhizosphere pH to enhance nutrient uptake in the low pH soils (pH 3–5). They do so by extruding OH^- and HCO_3^- to increase the rhizosphere pH, and thus make nutrients more available (Dakora and Phillips, 2002).

These three examples, involving the nutrient-solubilizing ability of groundnut root cell walls, the high N-acquiring capacity of marama bean plants in nutrient-poor soils, and the active modification of rhizosphere pH for enhanced nutrient uptake by *Aspalathus linearis* plants, together offer an opportunity for increased understanding of mineral nutrition in low-nutrient environments. So, in addition to selecting crop varieties for root exudation and nutrient use efficiency, future studies should consider wild landraces that are naturally adapted to low-nutrient environments. That way, the chance of creating or identifying crop plants with the ability to increase growth and provide higher yields under low-nutrient conditions would be greater.

Future research

This synthesis has identified four key areas for future research: (1) screening large collections of genotypes, cultivars, and their wild relatives for tolerance of low-nutrient conditions, (2) development of simple bioassays for quantifying root exudation of organic acid anions, phenolics and phytosiderophores, followed by an assessment of their respective roles in solubilizing unavailable nutrients (e.g., P, Fe, etc.) in low-nutrient environments, (3) increased experimentation on symbiotic systems involving rhizobia and VA fungi that enhance N and P supply to crop plants in nutrient-poor soils, and (4) an aggressive application of molecular tools to increase our knowledge of ion channels, transporter genes, and their promoters of gene expression in order to enhance root exudation and promote nutrient uptake through gene transfer and genetic manipulation in plants.

References

Ae N and Shen R-F 2002 Root cell-wall properties are proposed to contribute to phosphorus (P) mobilization by groundnut and pigeonpea. Plant Soil 245, 95–103.

Bramel P J 2002 Assessing germplasm collections for traits useful in plant nutrition. *In* Food Security in Nutrient-Stressed Environments: Exploiting Plants' Genetic Capabilities, Proceedings of an International Workshop, 27–30 September 1999, Patancheru, India. Developments in Plant and Soil Sciences, Volume 95. Ed. J J Adu-Gyamfi. pp 193–200. Kluwer Academic Publishers, Dordrecht, The Netherlands.

Chrispeels M, Crawford N M and Schroeder J 1999 Proteins for transport of water and mineral nutrients across the membranes of plant cells. Plant Cell 11, 661–675.

Dakora F D and Phillips D A 2002 Root exudates as mediators of mineral acquisition in low nutrient environments. Plant Soil 245, 35–47.

Dakora F D, Lawlor D W and Sibuga K P 1999 Assessment of symbiotic nitrogen nutrition in marama bean (*Tylosema esculentum* L.), a tuber-producing underutilized grain legume. Symbiosis (in press).

Drake B G, Gonzalez-Meler M A and Long S P 1997 More efficient plants: a consequence of rising atmospheric CO_2? Annu. Rev. Plant Physiol. Mol. Biol. 48, 609–639.

Dunlop J and Phung T 2002 Transporter genes to enhance nutrient uptake: opportunities and challenges. *In* Food Security in Nutrient-Stressed Environments: Exploiting Plants' Genetic Capabilities, Proceedings of an International Workshop, 27–30 September 1999, Patancheru, India. Developments in Plant and Soil Sciences, Volume 95. Ed. J J Adu-Gyamfi. pp 227–234. Kluwer Academic Publishers, Dordrecht, The Netherlands.

Fox T C and Guerinot M L 1998 Molecular biology of cation transport in plants. Ann. Rev. Plant Physiol. Plant Mol. 49, 669–696.

Hawes M C, Brigham L A, Wen F, Woo H H and Zhu Y 1998 Functions of root border cells in plant health: pioneers in the rhizosphere. Ann. Rev. Phytopathol. 36, 311–327.

Hoffland E, van den Boogaard R, Nelemans J and Findenegg G 1992 Biosynthesis and root exudation of citric and malic acids in phosphate-starved rape plants. New Phytol. 122, 675–680.

Johansen C, Subbarao G V, Lee K K and Sharma K K 1995 Genetic manipulation of crop plants to enhance integrated nutrient management in cropping systems – the case of phosphorus. *In* Genetic Manipulation of Crop Plants to Enhance Integrated Nutrient Management in Cropping Systems – 1. Phosphorus. Eds. C Johansen, K K Lee, K K Sharma, G V Subbarao and E A Kueneman. pp 9–29. International Crops Research Institute for the Semi-Arid Tropics, Patancheru, Andhra Pradesh, India.

Maranville J W and Madhavan S 2002 Physiological adaptations for nitrogen use efficiency in sorghum. Plant Soil 245, 25–34.

Phillips D A 1992 Flavonoids: Plant signals to soil microbes. *In* Phenolic Metabolism in Plants. Eds. H A Stafford and R K Ibrahim. Plenum Press, New York. Rec. Adv. Phytochem. 26, 201–231.

Rengel Z 2002a Breeding for better symbiosis. Plant Soil 245, 147–162.

Rengel Z 2002b Genetic control of root exudation. Plant Soil 245, 59–70.

Welch R M and Graham R D 2002 Breeding crops for enhanced micronutrient content. Plant Soil 245, 205–214.

Zhu Y-G, Huang C, Howes N K and Smith S E 2002 Breeding for low-nutrient environments. *In* Food Security in Nutrient-Stressed Environments: Exploiting Plants' Genetic Capabilities, Proceedings of an International Workshop, 27–30 September 1999, Patancheru, India. Developments in Plant and Soil Sciences, Volume 95. Ed. J J Adu-Gyamfi. pp 47–57. Kluwer Academic Publishers, Dordrecht, The Netherlands.

J.J. Adu-Gyamfi (Ed.), *Food security in nutrient-stressed environments: exploiting plants' genetic capabilities*: 267–276, 2002.

Breeding crops for enhanced micronutrient content

Ross M. Welch[1,3] & Robin D. Graham[2]

[1]*USDA-ARS, U.S. Plant, Soil and Nutrition Laboratory, Tower Road, Ithaca, New York, 14853 USA.* [2]*Department of Plant Science, Waite Campus, University of Adelaide, Glen Osmond, 5064, South Australia.* [3]*Corresponding author**

Received 10 May 2000. Accepted in revised form 27 September 2000

Key words: iron, nutritional quality, plant breeding, staple foods, vitamin A, zinc

Abstract

Micronutrient malnutrition (e.g. Fe, Zn and vitamin A deficiencies) now afflicts over 40% of the world's population and is increasing especially in many developing nations. Green revolution cropping systems may have inadvertently contributed to the growth in micronutrient deficiencies in resource-poor populations. Current interventions to eliminate these deficiencies that rely on supplementation and food fortification programs do not reach all those affected and have not proven to be sustainable. Sustainable solutions can only be developed through agricultural system approaches. One agricultural approach is to enrich major staple food crops (e.g. rice, wheat, maize, beans and cassava) in micronutrients through plant breeding strategies. Available research has demonstrated that micronutrient enrichment traits are available within the genomes of these major staple crops that could allow for substantial increases in Fe, Zn and provitamin A carotenoids without negatively impacting yield. Furthermore, micronutrient-dense seeds can increase crop yields when sowed to micronutrient-poor soils. The enrichment traits appear to be stable across various soil types and climatic environments. Further research is required to determine if increasing levels of micronutrients in staple foods can significantly improve the nutritional status of people suffering from micronutrient deficiencies.

Introduction

While deficiencies of dietary energy (i.e. calories) and protein currently affects more than 800 million people in food insecure regions, incredibly, micronutrient malnutrition (i.e. 'hidden hunger') now afflicts over two billion people, especially resource-poor women, infants and children in the developing world and the numbers are increasing (McGuire, 1993; Yip and Scanlon, 1994). Today, deficiencies of iron, vitamin A and iodine are of most concern to the nutrition community and healthcare officials although other nutrient deficiencies, including zinc, selenium, calcium, magnesium and other vitamins (e.g. riboflavin, vitamin C, vitamin E and folic acid), may be prevalent in some global regions. The consequences of malnutrition create immense economic and societal costs to nations. Micronutrient malnutrition greatly increases mortality and morbidity rates, diminishes congitive abilities of children and lowers their educational attainment, reduces labor productivity, stagnates national development efforts, contributes to continued high population growth rates and reduces the livelihood and quality of life for all those affected (Combs et al., 1996; Combs and Welch, 1998; Welch et al., 1997; Welch and Graham, 1999).

'Hidden hunger' is a capacious, public health issue among nearly all developing nations affecting primarily underprivileged people (Buyckx, 1993; Ramalingaswami, 1995). Disturbingly, the massive increases in the number of people suffering from micronutrient malnutrition over the last four decades (e.g. Fe deficiency anemia has grown from about 30% of the world's population in the 1960s to over 40% during the late 1990s) coincide with the expansion of 'green revolution' production systems. This immense public health problem requires a new paradigm for

* FAX No: 607-255-1132. Tel.: 607-255-5434.
E-mail: RMW1@CORNELL.EDU

This paper previously appeared in *Plant and Soil* **245**, 2002, pp. 205–214.

agriculture; a paradigm that no longer only focuses agricultural research on maximizing food production at minimum expense as the primary goal, but directs some research efforts towards producing enough food crop diversity and nutritional quality of agricultural products to meet human dietary demands. Food systems that feed the world must be changed in ways that will insure that balanced nutrient supplies are available continuously to all people in adequate, affordable amounts (Combs et al., 1996, 1997; Welch et al., 1997).

Past programs to combat micronutrient malnutrition have relied primarily on interventions directed at food fortification or nutrient supplementation programs (Yip, 1997). Unfortunately, these approaches have not proven to be sustainable for various reasons and do not reach all the people at highest risk of developing micronutrient malnutrition. Remarkably, the nutrition community has never embraced agriculture as an important 'tool' to use in fighting 'hidden hunger'. Nor has the agricultural industry viewed agricultural systems as playing an important and explicit role in human nutrition and health. Their efforts have been primarily focused on productivity, efficiency and profit margin issues to drive research and national agricultural programs without regard to either the nutritional quality of the products produced or the adequacy of such systems to meet all the nutritional needs of people (Ross, 1996). Furthermore, institutional and governmental structures and programs have not attempted to link closely agricultural production to human nutrition and health needs, although current knowledge and logic suggest that this should be a high priority for all nations (McGuire, 1993).

Within the agricultural community, plant breeding efforts greatly contributed to advances in staple plant food productivity (mostly cereal crops) during the 'green revolution'. Such breeding efforts, along with improved agriculture technologies, succeeded in providing enough calories and protein to prevent the threatening massive starvation and famines predicted in the early 1960s in many world regions. Importantly, plant breeding can again be used as a powerful weapon to use in fighting 'hidden hunger'. Breeding for micronutrient-enriched staple plant foods is a possibility that should be pursued (Bouis, 1996; Graham et al., 1998, 1999; Graham and Welch, 1996). Success in such a breeding effort would target those groups of people most at risk of developing micronutrient malnutrition because these sectors of societies are dependent on these foods for their sustenance. Furthermore, a plant breeding approach would be sustainable; once micronutrient-dense lines of staple plant foods are developed, there is little additional cost to continue their lineage in ongoing breeding programs for the foreseeable future (Bouis, 1996).

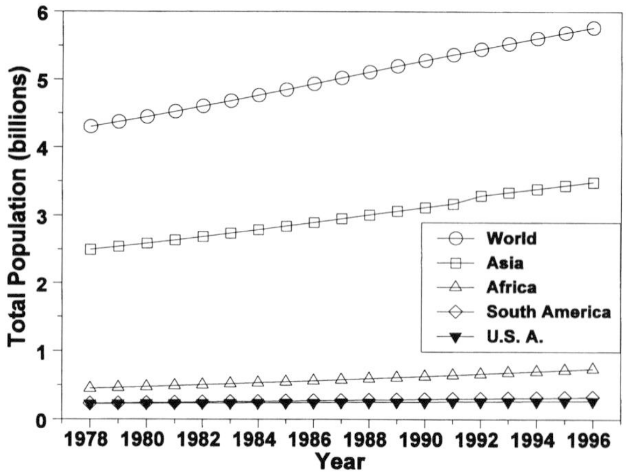

Figure 1. Trends in total population growth in the world, Asia, Africa, South America and the United States between 1978 and 1996 (data from FAO, FAOSTAT Database, 1998).

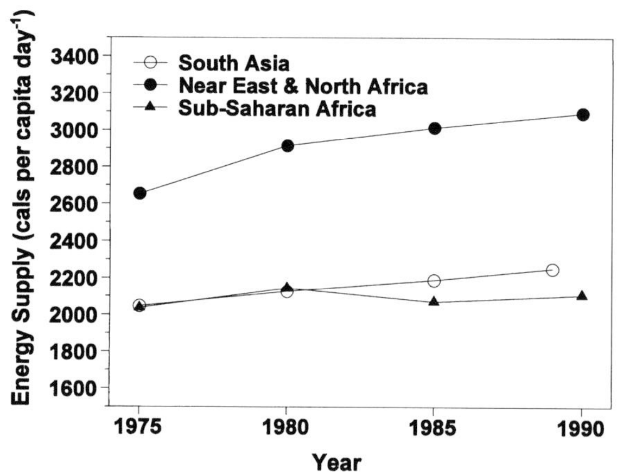

Figure 2. Trends in dietary Energy Supplies from 1975 to 1990 in some world regions (data from UNACCSN, 1992).

Reasons for concern

The extraordinary success of agricultural research during the last five decades provided many developing countries with the tools necessary to increase dramatically cereal production and feed a rapidly growing population (see Figure 1). For example, in the Near East/North Africa, in Sub-Saharan Africa and in South Asia food supply (in terms of dietary energy supply per capita per day) has kept up with population growth

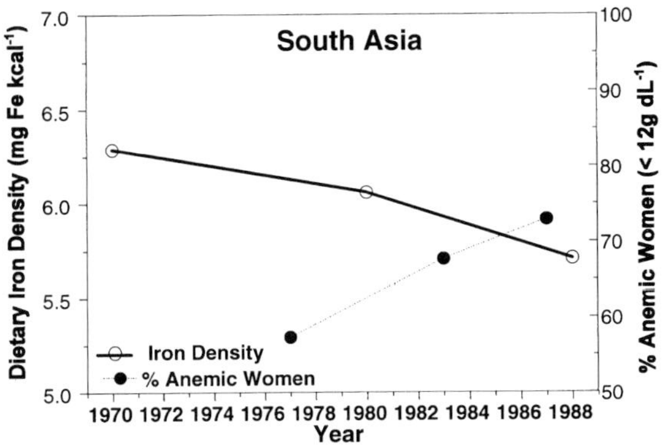

Figure 3. South Asian trends in dietary Fe density and in Fe deficiency anemia in women from 1970 to 1988 (data from FAO, 1990; UNACCSN, 1992).

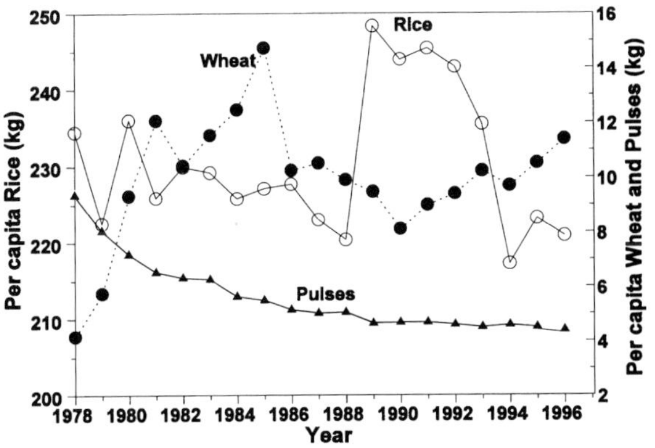

Figure 4. Per capita trends in total rice, wheat and pulses production in Bangladesh from 1978 to 1996 (data from FAO, FAOSTAT database, 1998).

(see Figure 2). However, only just producing enough calories (i.e. energy) has resulted in unforseen nutritional problems for nearly 50% of the world's people, especially among the poor (UNACCSN, 1992).

'Green revolution' cropping systems may have contributed to some unforeseen negative consequences on human nutrition and health. For example, in South Asia, the introduction of high input, modern wheat/rice cropping systems are associated with time trends in the growth of Fe deficiency anemia among poor pre-menopausal women, and negatively correlated to dietary Fe density (mg Fe $kcal^{-1}$ of available food; Figure 3). Data from China, Sub-Saharan Africa, South America, Middle America/Caribbean and Southeast Asia also show the same trends (UNACCSN, 1992).

What are the causal factors responsible for the incredible growth in 'hidden hunger' worldwide? Has agriculture unintendedly contributed to the growth in micronutrient malnutrition? Answers to these questions are not clear. However, changes in cropping systems may have contributed significantly to the growth in 'hidden hunger'. Clearly, the use of modern cereal cropping systems in many developing nations has been paralleled by a decreased per capita production of traditional edible legume crops which contain much higher levels of most micronutrients (Welch et al., 1997). This has resulted in lower availability and higher prices of many micronutrient-rich pulses at least in some world regions (Anonymous, 1993, 1994; Combs et al., 1996; Uvin, 1994; WHO, 1996; Yip, 1997). For instance, on the one hand, from 1978 and 1996, per capita rice and wheat production in Bangladesh kept pace with or exceeded population growth, while on the other, total pulse production dramatically decreased during this time frame (Figure 4). Interestingly, data complied by the Food and Agricultural Organization, United Nations (FAO), show negative trends in Fe density in South Asian diets (FAO, 1990) and an increase in Fe deficiency anemia among women (UNACCSN, 1992; WHO, 1996) during the development of green revolution cropping systems. Possibly, those changes in agricultural systems that occurred in South Asia during the green revolution resulted in dramatic increases in rice and wheat production, but at the expense of pulse production, iron density in the diet and women's Fe status. Possibly, losses in the availability of pulses for human consumption may be an important contributing factor to the rise in 'hidden hunger' in South Asia because edible legume seeds are a much richer source of micronutrients than cereals, especially after the cereal grains have been milled and/or polished before consumption (Welch et al., 1997).

Current data collected on the extent of micronutrient malnutrition in humans on earth only cover Fe, I and/or vitamin A deficiencies (Mason and Garcia, 1993; Ramalingaswami, 1995; UNACCSN, 1992; Uvin, 1994). Most certainly, several other micronutrients, including Zn, Cu, Se and other vitamins (e.g. riboflavin, vitamin C and vitamin B_{12}), are limiting in the diets of large numbers of people in many regions of the world (although little is known about the actual extent of these deficiencies because of technical and clinical problems in determining their numbers and distribution globally) (WHO, 1996).

The world nutrition community has made combating 'hidden hunger' a high priority (Anonymous, 1992, 1995, 1996). Interestingly, certain micronutrient deficiencies (especially Fe deficiency) are an issue even in developed countries such as the United States and Canada (Frazao, 1996; Welch et al., 1997).

Consequences of 'hidden hunger' to human health and felicity

Health consequences

Micronutrient malnutrition diminishes the motivation and curiosity of infants and young children, thereby reducing their exploratory activities including their development (Grantham-McGregor and Ani, 1999; Mongeau and Larivee, 2000). Consequently, this ailment impairs mental and cognitive abilities by reducing interactions of children with their environment, their peers and their family. These children, deprived of their full genetic potential for mental and physical development, become adults with lower intellectual and physical abilities. They are less able to provide for themselves and their families. Often they are less productive with higher rates of chronic disease and associated disabilities. National economic losses to some countries resulting from 'hidden hunger' can be as high as 5% of their gross national product in lost lives, disabilities, healthcare costs and labor productivity. In Bangladesh and in India, a total of $18 billion was calculated to be lost from their economies each year as a result of micronutrient malnutrition (Islam and Tori, 1998).

Rapid growth during fetal development, infancy, childhood and adolescents demand greater intakes of micronutrients. Growth failure or deficiency diseases will develop without increased intakes. Micronutrient deficiency symptoms are most prevalent during these stages of life. Also, interactions between micronutrient deficiencies and infectious diseases are very important. These interactions can confound healthcare efforts to control various diseases (such as the synergistic interaction between vitamin A deficiency and measles; increased severity of measles leading to vitamin A-deficiency-blindness and death). Malaria and hookworm infections are also associated with Fe deficiency anemia. Not withstanding the difficulties in associating micronutrient deficiencies directly to many diseases, clearly, prevention of these deficiencies can contribute to infectious disease control (Ramalingaswami, 1995).

A high risk of tissue hypoxia and heart failure, which can lead to death in young children and pregnant women, is associated with Fe deficiency (Viteri, 1998). Maternal anemia, aggravated by blood loss during child birth, is reported to be responsible for most of the maternal mortality during birth in the world (20% of all maternal deaths are attributed to Fe deficiency anemia (Maberly et al., 1994) and 30% of all children who enter hospitals with severe anemia die (Anonymous, 1994). Babies born to mothers that are iron-deficient are commonly stunted and unhealthy. Children suffering from Fe deficiency have poor attention spans, impaired fine motor skills and less capacity for memory (Walter et al., 1997). Iron deficiency in pregnant women may cause irreversible damage to fetal brain development leading to irreversible damage to intellectual development in their babies (Gordon, 1997). Iron deficiency in pregnant women is correlated to infant prematurity and low birth weight; this can result in long-term frailties such as immune system dysfunctions and growth failure (McGuire, 1993). Iron deficiency reduces both physical performance and work productivity. For example, plantation workers in Indonesia, cotton mill workers in China and workers in Sri Lanka when provided Fe supplements increased their productive work output by 20–50% (Anonymous, 1994; Li et al., 1994; Maberly et al., 1994). Some economic studies suggest that the calculated benefit/cost ratio from giving iron supplements to anemic workers is potentially as high as 260:1. In Fe fortification studies, the value of benefits, in terms of labor output, from fortification was reported to be 7, 42 and 70 times the cost of the fortification programs in field trials in Indonesia, Kenya and Mexico, respectively (Sanghvi, 1996).

Iron, vitamin A and I deficiencies are associated with increased mortality rates among resource-poor women, infants and children (McGuire, 1993). Additionally, Vitamin A deficiency is the most important cause of childhood blindness in developing nations. Vitamin A deficiency impairs the integrity of epithelial tissue barriers (e.g. skin) to infection. The immune system is also affected leading to increased severity of certain infections and increased risk of childhood death.

High rates (about 0.5–1%) of neonatal deaths and still births are associated with maternal I deficiency. Additionally, severe I deficiency in pregnant women results in irreversible mental retardation and neurological disorders (cretinism) in their babies. Other I deficiency disorders (IDD) include deafness, muteness and mild to moderate mental retardation. All of these outcomes are all irreversible. They limit children's ability to learn, their educational attainment, their occupational choices and ultimately, their future livelihoods and welfare. In nations with prevalent IDDs, there is an immense cost to societies in terms of lost human potential and its economic ramifications (WHO, 1996).

Acute Zn deficiency in humans is relatively rare resulting in hypogonadism and dwarfism in men, growth retardation in infants and children, orificial and acral dermatitis, diarrhea, alopecia, impaired reproductive performance and difficulty in parturition (Gibson, 1994; Prasad, 1993, 1996; WHO, 1996). Importantly, marginal Zn deficiency may be wide spread, but unreported globally because there is no established clinical method for determining marginal Zn deficiency in humans (Endre et al., 1990; Larsen, 1997; Reinhold, 1988; Shrimpton, 1993). Mild Zn deficiency reduces growth rates in infants and children, impairs resistance to infection, reduces taste acuity, increases the severity and duration of diarrhea and delays wound healing (WHO, 1996). Additionally, evidence is accumulating that mild Zn deficiency also impairs brain function and behavior in children (Penland, 1997). Furthermore, several reports suggest that Zn is an antioxidant involved in antioxidant defense systems, but the exact mechanism is not known. Kim et al. (1998) showed that marginal zinc deficiency lowers the lymphatic absorption of vitamin E (α-tocopherol) in rats. Thus, intestinal absorption of vitamin E is reduced by low-Zn status. Therefore, Zn status may have a profound effect on the bioavailability of lipid soluble vitamins like vitamin E.

Much attention has recently focused on the interaction between I and Se because of Se's role in thyroid hormone metabolism (as a constituent of iodothyronine 5′-deiodinase). Combined deficiencies of Se and I may have adverse effects on infant growth, development and survival. Selenium deficiency could also be a problem in areas where I deficiency disorders are not clearly related to I intakes and where the diet is also low in available Se (Hofbauer et al., 1997; WHO, 1996).

Societal and development consequences

Linear conceptualizing can lead to the conclusion that improving the nutritional health of people will result in increased population growth rates. History, however, suggests that this is not true. In countries that have dramatically reduced the incidence of malnutrition have concurrently reduced their birth rates dramatically over the last century (such as Japan and other industrialized countries). Evidence is accumulating showing that improved nutritional health promotes reductions in the birth rate (Behrman, 1993; Sanghvi, 1996). This should be especially true for micronutrients that are known to play critical roles in fetal brain development and in congnitive ability during infancy and childhood (e.g. Fe, I, Zn, B). Unchecked micronutrient malnutrition in a society will lead to more unhealthy people, reduced labor productivity, lower educational attainments in children, reduced school enrollments and attendance, earlier marriages, increased morbidity and mortality, lower standards of living, higher health care costs and civil discontent. This may lead to governmental instability and political unrest. Adequate nutrition for pregnant women, infants and children will lead to better school attendance, higher educational attainment, and ultimately, higher family incomes, later marriages and improved worker productivity – all outcomes that enhance civilization and lower birth rates. Eliminating 'hidden hunger' should be a high priority for all nations.

Breeding for micronutrient-dense crops

Physiological bases of micronutrient accumulation

The physiological basis for micronutrient efficiency in crop plants and the processes controlling the accumulation of micronutrients in seeds is not understood with any certainty. Because of space limitations and the complexity and volume of literature available, these subjects will not be covered in this short review. For an in-depth discussion of these topics, the reader is referred to the following references (Graham and Welch, 1996; Marschner, 1995; Welch, 1995, 1999; Yang and Römheld, 1999). A brief outline of these processes that determine micronutrient concentrations in edible plant tissues follows.

There are several barriers to overcome in genetically modifying plants to accumulate more micronutrient metals (e.g. Fe and Zn) in their edible parts. These barriers to micronutrient uptake and distribution in plants are the result of tightly controlled homeostatic mechanisms that regulate micronutrient uptake and distribution in plants assuring adequate, but nontoxic levels of these nutrients to accumulate in plant tissues. The first and most important barrier to micronutrient uptake reside at the root-soil interface. To increase micronutrient metal uptake by roots, the available levels of the micronutrient in the rhizosphere must be increased to allow for more absorption by root cells. This could be enhanced by stimulating certain root-cell processes that modify micronutrient solubility and movement to root surfaces (Welch, 1995), such as by stimulating root-cell H^+, metal chelating compounds and reductants release rates, and increasing root absorptive surface area such as number and extent of fine roots and root hairs. Second, the root-cell plasma membrane absorption mechanisms (e.g. transporters and ion channels) must be sufficient and specific enough to allow for the accumulation of micronutrient metals once they enter the apoplasm of root cells from the rhizosphere. Third, once taken up by root cells, the micronutrients must be efficiently translocated to edible plant organs. For seeds and grains, phloem sap loading, movement and unloading rates are important characteristics to consider in increasing micronutrient metal accumulation in seeds and grains.

The international breeding effort

The current thrust of research is to determine the genetic potential for increasing the concentrations of bioavailable Fe, Zn and provitamin A carotenoids (as well as Se and I) in edible portions of several staple food crops including rice, wheat, maize, beans and cassava (Bouis, 1996; Graham et al., 1998, 1999; Graham and Welch, 1996). The following discussion reviews the findings to date that have been obtained in this on-going effort.

Breeding criteria for improving micronutrient density

Certain conditions must be met before new lines of micronutrient-rich staple food crops are distributed globally to national agricultural research programs. Meeting these conditions will assure that targeted people at risk of developing micronutrient malnutrition will benefit from such action. These conditions are listed below.

1. Crop productivity (i.e. yield) must be maintained or increased to guarantee widespread farmer acceptance.
2. The micronutrient enrichment levels achieved must have significant impact on human health.
3. The micronutrient enrichment traits must be relatively stable across various edaphic environments and climatic conditions.
4. Ultimately, the bioavailability[1] of micronutrients in enriched lines must be tested in humans to assure that they are of benefit to people preparing and eating them in traditional ways within normal household environments.
5. Consumer acceptance must be tested (taste and cooking quality must be acceptable to household members) to assure maximum impact on nutritional health.

Meeting these conditions will require a new way of thinking and performing research by most agriculturalists including plant breeders, a holistic food systems view of agricultural production. It will necessitate that researchers co-operate with various specialists in disciplines not normally associated with agricultural research, including nutritionists, public health officials, sociologists and economists, to assure that their efforts will have meaningful impact on human nutrition and health (Combs et al., 1996).

The question of yields

The effects of breeding for micronutrient-dense staple seeds and grains on crop yields have been addressed in a number of recent reviews (Bouis, 1996; Graham et al., 1998, 1999; Graham and Welch, 1996). Briefly, increasing the micronutrient stores in seeds results in more seedling vigor and viability enhancing the performance of seedlings when the seeds are planted in micronutrient-poor soils. This improved seed vigor allows for the production of more and longer roots under deficient conditions allowing seedlings to scavenge more soil volume for micronutrients and water early in growth, an advantage that can lead to improved yields compared to seeds with low micronutrient stores when grown under micronutrient stress conditions. Many of the countries where micronutrient deficiencies in humans are a problem are also countries that have large areas of micronutrient-poor/deficient soils (White and Zasoski, 1999). Thus, improving seed vigor with respect to micronutrient stores should be very beneficial to agricultural production in these countries. Additionally, disease resistance and stress tolerance are improved in micronutrient-dense seeds which would also aid agricultural production in target countries (Welch, 1986, 1999). Thus, selecting for these traits in staple food crops is a 'win-win' opportunity. It has potential to enhance crop yields without additional

[1] Bioavailability is defined as that amount of a nutrient in a food-constituent of a meal that is absorbable and utilizable by a person eating the meal.

farmer inputs and to improve their nutritional quality at the same time.

The genetic potential

Over the past several years, scientists have collecting data on the potential for breeding for significant levels of bioavailable Fe, Zn and provitamin A carotenoids in rice, wheat, maize, beans and cassava. The discussion below summarizes the findings to date.

Bean

Researchers at the *Centro Internacional de Agricultura Tropical* (International Center for Tropical Agriculture; CIAT) have been studying the degree of genetic variability that exists in Fe and Zn concentrations in seeds of common beans (*Phaseolus vulgaris* L.) (Graham et al., 1999). Beebe (pers. comm.) evaluated a core collection of over 1000 accessions of common beans and found a range in Fe concentrations from 34 to 89 $\mu g\ g^{-1}$ Fe (average = 55 $\mu g\ g^{-1}$ Fe). Zinc concentrations in these same accessions ranged from 21 to 54 $\mu g\ g^{-1}$ Zn (average = 35 $\mu g\ g^{-1}$ Zn). Some bean accessions from Peru were recently found to contain especially high levels of Fe averaging over 100 $\mu g\ g^{-1}$ Fe. The range in seed-Zn concentrations in the core collection was narrower than seed-Fe concentrations ranging from 21 to 54 $\mu g\ g^{-1}$ Zn. Wild types tended to have lower Zn concentrations than common cultivated types. Some seeds from genotypes originating in Guatemala were highest in Zn levels. The data collected suggest that there is sufficient genetic variability to increase significantly Fe (by about 80%) and Zn (by about 50%) concentrations in common beans. Results also indicate that the traits responsible for genetic improvements in Fe and Zn concentrations are stable across various environments. For both Fe and Zn seed concentrations, there were significant location and location × genotype effects demonstrating (as expected) that environments can influence the concentrations of Fe and Zn in bean seeds. However, high-Fe and high-Zn genotypes will accumulate more of these nutrients when compared to low-Fe and low-Zn genotypes grown at the same location during the same growing season.

Interestingly, CIAT researchers also found a very highly significant positive correlation of 0.52 between the concentrations of Fe and Zn across different genotypes. Thus, genetic factors for increasing Fe are co-segregating with genetic factors increasing Zn. Therefore, selecting for higher Fe level in bean seeds will also select for increased Zn levels in the seeds.

Table 1. The mean and range in concentrations (dry weight basis) of Fe and Zn in six sets of brown rice germplasm (939 genotypes) grown under similar conditions at IRRI, Los Bãnos, Philippines (Table modified form Graham et al., 1999)

Genetic sets	Fe ($\mu g\ g^{-1}$)		Zn ($\mu g\ g^{-1}$)	
	Mean ±SE[a]	Range	Mean ±SE	Range
Traditional & improved lines	13±2.6	9.1–22.6	24.0±4.7	13.5–41.6
IRRI breeding lines	10.7±1.6	7.5–16.8	25.0±7.6	15.9–58.4
Traditional & improved lines	12.9±3.1	7.8–24.4	24.4±4.7	16.5–37.7
Tropical japonicas	12.9±1.5	8.7–16.5	26.3±3.8	17.1–40.1
Popular lines & donors	13.0±2.5	7.7–19.2	25.7±4.6	15.3–37.3
Traditional & improved lines	13.8±2.3	10.8–18.0	24.2±4.1	19.9–33.3

[a] ±SE = ± standard error of the mean.

Rice

Since 1992, researchers at IRRI have been evaluating the genetic variability of Fe concentration in rice grain. In 1995, the research was expanded to include Zn (Graham et al., 1999). Table 1 shows the results obtained for some of this research. The range in Fe and Zn concentrations within the six sets of genotypes (n=939) tested in this study were 7.5 $\mu g\ g^{-1}$–24.4 $\mu g\ g^{-1}$ for Fe, and 13.5 $\mu g\ g^{-1}$–58.4 $\mu g\ g^{-1}$ for Zn. Thus, within those genotypes tested, there was about a four-fold difference in Fe and Zn concentrations suggesting some genetic potential to increase the concentrations of these micronutrients in rice grain.

Among the highest grain-Fe concentrations (i.e. ranging from about 18 $\mu g\ g^{-1}$ to 22 $\mu g\ g^{-1}$) found were in a number of aromatic rice varieties including Jalmagna, Zuchem and Xua Bue Nuo. Additionally, these same aromatic lines also contained the highest grain-Zn concentrations (ranging from about 24 $\mu g\ g^{-1}$ to 35 $\mu g\ g^{-1}$). Further research using F2-derived populations demonstrated that the aromatic trait was not pleiotropic for grain-Fe or grain-Zn concentrations and, therefore, this trait may be used to screen for high Fe and Zn levels in rice grain but the linkage is broken at a low frequency.

Several studies were carried out at IRRI to examine the effect of soil and climatic factors on grain-Fe and grain-Zn concentrations among genotypes. Factors studied included wet season-dry season, normal-saline soils, acid- neutral soils and N-supply. The data from these various studies demonstrated that high-Fe and high-Zn grain traits are expressed in all rice environments tested although there is some evidence of significant genotype × environment interactions that can ultimately affect Fe and Zn concentrations (Graham et al., 1999) in extreme environments.

These IRRI results indicate that there is significant genetic diversity in the rice genome to increase substantially Fe and Zn concentrations in rice grain. However, the effects of rice grain processing on Fe and Zn levels in the edible product (i.e. polished and parboiled rice grain), as well as the bioavailability of the Fe and Zn in the grain to humans still await final results from continuing evaluations of these factors.

Wheat

A wide range of wheat germplasm is being studied at CIMMYT with respect to the concentration of Fe and Zn in the whole grain and environmental interactions on their concentrations. In one study, the ranges in Fe and Zn concentrations (dry weight basis) in wheat grain from plants grown in El Batan, Mexico in 1994 were 28.8–56.5 $\mu g\ g^{-1}$ (mean = 37.2 $\mu g\ g^{-1}$; S.D. = 4.10 $\mu g\ g^{-1}$; n=132) for Fe and 25.2–53.3 $\mu g\ g^{-1}$ for Zn (mean = 35.0 $\mu g\ g^{-1}$; S.D. = 4.99 $\mu g\ g^{-1}$; n=132) (Graham et al., 1999). Clearly, enough genetic variation exits within the wheat germplasm to substantially increase Fe and Zn concentrations in wheat grain. Among all wheat germplasm studied, species *Triticum dicoccum* Schrank had the highest concentrations of Fe and Zn, which warrants further study.

There was a high correlation between grain-Fe and grain-Zn concentrations in the wheat lines studied. While there was significant genotype × environmental interactions obtained for Fe and Zn grain concentrations, there was still a strong genetic component to Fe and Zn accumulation in the grain. This finding indicates that it should be possible to improve Fe and Zn levels in wheat grain simultaneously through plant breeding. Additional research has also shown that there is no negative linkage between grain yield and Fe and Zn density in the grain.

Maize

Current data being collected by CIMMYT suggests that the range in Fe and Zn concentrations in maize kernels is not as great as that found for other cereal crops, although more data are needed to confirm this finding (M. Banziger, pers. comm. CIMMYT). A Southern African germplasm collection containing 20 lines was evaluated at Harare, Zimbabwe in 1996–97. The range in Fe and Zn concentrations was 16.4–22.9 $\mu g\ g^{-1}$ for Fe (mean of 19.6 $\mu g\ g^{-1}$), and 14.7–24.0 $\mu g\ g^{-1}$ for Zn (mean of 19.8 $\mu g\ g^{-1}$). In earlier studies, CIMMYT reported a much wider range in Fe and Zn concentrations during evaluations of 1486 germplasm from 1995 to 1998 that included a wide range of landraces, varieties and breeding germplasm from CIMMYT and 57 germplasm from Southern Africa. In the earlier trials, genotypes high in grain-Fe and grain-Zn concentrations exceeded trial means by as much as 50%. For example, in an evaluation of 1045 breeding lines from Zimbabwe during the 1995/96 growing season, the range in Fe concentrations was from 17.7 to 61.8 $\mu g\ g^{-1}$. Zn concentrations in these same lines varied from 12.9 to 28.5 $\mu g\ g^{-1}$. The discrepancy between the earlier and later data sets may be the result of Fe and Zn contamination of the maize kernels analyzed, but further research is needed to verify this possibility.

Cassava

The variation in β-carotene concentration in cassava roots from a CIAT core collection (630 genotypes) from the global cassava germplasm collection (about 5500 genotypes) was reported by Iglesias et al. (1997). Additionally, the relationship between root color and heritability as well as the stability of root-β-carotene to different root-processing techniques was studied. They reported a range in β-carotene concentrations in fresh roots from 0.1 to 2.4 mg 100^{-1} g.

The inheritance of β-carotene root-concentration appears to be determined by two genes (one controlling transport of shoot precursors to roots and one responsible for the biochemical processes affecting the accumulation of β-carotene in the root). Furthermore, visual screening by using intensity of orange color seemed feasible. However, they also stated that there was a need to rely on quantitative screening techniques to increase the efficiency of the screening program. It is possible that other provitamin A carotenoids could also be responsible for the deep yellow colour observed in accessions that have intermediate β-carotene concentrations.

Iglesias et al. (1997) concluded that there is enough genetic variability within the available cassava germplasm that would make it possible to produce cassava-roots that contain enough β-carotene to meet the daily requirements of adults (i.e. 6 mg d^{-1} β-carotene) if the β-carotene in cassava roots is bioavailable. The genotypes containing the highest levels of β-carotene were collected from the Amazonian region of Brazil and Colombia where yellow-root lines are preferred by the indigenous farmers. Processing techniques were shown to have a large effect on the final β-carotene content in the food prepared from cassava roots with some genotypes being more stable to various forms of processing then others. This factor must also be included in any breeding program to increase β-carotene in cassava roots.

Summary

There is very compelling global human health and nutritional evidence to persuade plant breeders to include micronutrient density traits as primary objectives in their work targeted to the developing world (see also extended discussions of this issue in Combs et al., 1996; Welch et al., 1997; Welch and Graham, 1999). Furthermore, doing so should also improve crop productivity when micronutrient-enriched seeds and grains are planted to micronutrient-poor soils (Graham and Welch, 1996). Current evidence suggests that there is enough genetic diversity within the genomes of staple plant foods to accomplish this task. Succeeding in doing this would dramatically contribute to improving the health, livelihood and felicity of numerous resource-poor, micronutrient-deficient people in many developing countries, and would contribute greatly to sustaining national development efforts in these countries, and at the same time, should help to lower the birth rates in these countries. Without such a breeding effort, finding sustainable solutions to micronutrient malnutrition will not be forthcoming in the foreseeable future.

Acknowledgements

We thank Drs S. Beebe and C. Iglesias (CIAT), I. Monasterio and M. Banziger (CIMMYT), and D. Senadhira (deceased) and G. Gregorio (IRRI) for their valuable ideas, discussions and contributions we used in compiling this review.

References

Anonymous 1992 International Conference on Nutrition: World Declaration and Plan of Action for Nutrition. Food and Agricultural Organization/World Health Organization, United Nations, Rome. 42 p.

Anonymous 1993 Focus on Micronutrients. SCN News 9, 1–66.

Anonymous 1994 The Challenge of Dietary Deficiencies of Vitamins and Minerals. *In* Enriching Lives: Overcoming Vitamin and Mineral Malnutrition in Developing Countries. pp. 6–13. World Bank, Washington D.C.

Anonymous 1995 Progress report on the implementation of the ICN world plan of action for nutrition. FAO, United Nations Conference, Twenty-eighth Session, Rome, 20 October–2 November, 1995. Document C 95/INF/18 September, 1995. Food and Agricultural Organization of the United Nations. Rome 14 p.

Anonymous. Food security and nutrition. World Food Summit Technical Background Documents WFS 96/TECH/9, 1–45. 1996. Rome, Food and Agricultural Organization, United Nations. World Food Summit Technical Background Documents Vol. 1.

Behrman J R 1993 The Economic Rationale for Investing in Nutrition in Developing Countries. Monograph, US Agency for International Development / VITAL, Washington, D.C. 1–32.

Bouis H 1996 Enrichment of food staples through plant breeding: A new strategy for fighting micronutrient malnutrition. Nutrit. Rev. 54, 131–137.

Buyckx M 1993 The international community's commitment to combating micronutrient deficiencies. Food Nutr. Agr. 7, 2–7.

Combs G F Jr, Duxbury J M and Welch R M 1997 Food systems for improved health: linking agricultural production and human nutrition. Euro. J. Biochem. 51, S32–S33.

Combs G F, Jr and Welch R M 1998 Creating Healthful Food Systems: Linking Agriculture to Human Needs, Cornell International Institute for Food, Agriculture and Development, Ithaca, NY. 34 p.

Combs G F Jr, Welch R M, Duxbury J M, Uphoff N T, and Nesheim M C 1996 Food-Based Approaches to Preventing Micronutrient Malnutrition: An International Research Agenda, Cornell International Institute for Food, Agriculture and Development. Cornell University, Ithaca, NY. 68 pp.

Endre L, Beck F W J and Prasad A S 1990 The role of zinc in human health. J. Trace Elements Experl. Med. 3, 337–375.

FAO 1990 Food Production Yearbook, Food and Agricultural Organization, United Nations, Rome.

Frazao E 1996 The American diet: a costly health problem. FoodReview, January/April 2–6.

Gibson R S 1994 Zinc nutrition in developing countries. Nutr. Res. Rev. 7, 151–173.

Gordon N 1997 Nutrition and cognitive function. Brain Dev. 19, 165–170.

Graham R D, Senadhira D, Beebe S, Iglesias C and Monasterio I 1999 Breeding for micronutrient density in edible portions of staple food crops. Conventional approaches. Field Crops Res. 60, 57–80.

Graham R D, Senadhira D, Beebe S E and Iglesias C 1998 A strategy for breeding staple-food crops with high micronutrient density. Soil Sci. Plant Nutr. 43, 1153–1157.

Graham R D and Welch R M 1996 Breeding for staple-food crops with high micronutrient density International Food Policy Research Institute, Washington, D.C. 72 pp.

Grantham-McGregor D M and Ani C C 1999 The role of micronutrients in psychomotor and cognitive development Brit. Med. Bull. 55, 511–527.

Hofbauer L C, Spitzweg C, Magerstädt R A and Heufelder A E 1997 Selenium-induced thyroid dysfunction. Postgrad. Med. J. 73, 103–104.

Iglesias C, Mayer J, Chavez L and Calle F 1997 Genetic potential and stability of carotene content of cassava roots. Euphytica 94, 367–373.

Islam S and Tori T H 1998 The invisible adversary. Star Magazine, Jan. 9, 4–10.

Kim E-S, Noh S K and Koo S I 1998 Marginal zinc deficiency lowers the lymphatic absorption of α-tocopherol in rats. J. Nutr. 128, 265–270.

Larsen T 1997 Erythrocyte membrane enzymes as indicators of zinc status, *In* Trace Elements in Man and Animals – 9. Proceedings of the Ninth International Symposium on Trace Elements in Man and Animals. Eds. PWF Fischer et al., pp. 105–106. National Research Council of Canada, Ottawa.

Li R, Chen X, Yan H, Deurenberg P, Garby L and Hautvast J G A J 1994 Functional consequences of iron supplementation in iron-deficient female cotton mill workers in Beijing, China. Am. J. Clin. Nutr. 59, 908–913.

Maberly G F, Trowbridge F L, Yip R, Sullivan K M and West, C E 1994 Programs against micronutrient malnutrition: Ending hidden hunger. Ann. Rev. Public Health 15, 277–301.

Marschner H 1995 Mineral Nutrition of Higher Plants, Academic Press, London. 889 pp.

Mason J B and Garcia M 1993 Micronutrient deficiency – the global situation. SCN News 9, 11–16.

McGuire J 1993 Addressing micronutrient malnutrition. SCN News 9, 1–10.

Mongeau E and Larivee S 2000 Nutrition and intelligence 2000 Int. J. Psychol. 35, 10–23.

Penland J G 1997 Trace elements, brain function and behavior: Effects of zinc and boron. *In* Trace Elements in Man and Animals – 9. Proceedings of the Ninth International Symposium on Trace Elements in Man and Animals. Eds. PWF Fischer et al. pp. 213–216. National Research Council of Canada, Ottawa.

Prasad A S 1993 Essential and toxic trace elements in human health and disease: An update. Marginal deficiency of zinc and immunological effects. Prog. Clin. Biol. Res. 380, 1–22.

Prasad A S 1996 Zinc deficiency in women, infants and children: J. Am. Coll. Nutr. 15, 113–120.

Ramalingaswami V 1995 New global perspectives on overcoming malnutrition. Am. J. Clin. Nutr. 61, 259–263.

Reinhold J G 1988 Problems in mineral nutrition: a global perspective. *In* Trace Minerals in Foods. Ed. KT Smith. pp. 1–55. Marcel Dekker, Inc., New York and Basel.

Ross E B 1996 Malthusianism and agricultural development: False premises, false promises. Biotech. Develop. Monitor 26, 24.

Sanghvi T G 1996 Economic Rationale for Investing in Micronutrient Programs. A Policy Brief Based on New Analyses, Office of Nutrition, Bureau for Research and Development, United States Agency for International Development, Washington, D.C. 12 p.

Shrimpton R 1993 Zinc deficiency – is it widespread but under-recognized? SCN News 9, 24–27.

UNACCSN 1992 Second Report on the World Nutrition Situation. Vol. I. Global and Regional Results. United Nations Administrative Committee on Coordination, Subcommittee on Nutrition, Geneva. 80 pp.

Uvin P 1994 The state of world hunger. Nutr. Rev. 52, 151–161.

Walter T, Peirano P and Roncagliolo M 1997 Effect of iron deficiency anemia on cognitive skills and neuromaturation in infancy and childhood. *In* Trace Elements in Man and Animals – 9. Proceedings of the Ninth International Symposium on Trace Elements in Man and Animals. Eds. P WF Fischer et al. pp. 217–219. National Research Council of Canada, Ottawa.

Viteri F E V 1998 Prevention of iron deficiency. *In* Prevention of Micronutrient Deficiencies. Eds. CP Howson et al. pp. 45–102. National Academy Press, Washington, D. C.

Welch R M 1986 Effects of nutrient deficiencies on seed production and quality. Adv. Plant Nutr. 2, 205–247.

Welch R M 1995 Micronutrient nutrition of plants. Crit. Rev. Plant Sci. 14, 49–82.

Welch R M 1999 Importance of seed mineral nutrient reserves in crop growth and development. *In* Mineral Nutrition of Crops. Fundamental Mechanisms and Implications. Ed. Z Rengel. pp. 205–226. Food Products Press, New York.

Welch R M, Combs G F Jr and Duxbury J M 1997 Toward a 'Greener' revolution. Issues Sci. Tech. 14, 50–58.

Welch R M and Graham R D 1999 A new paradigm for world agriculture: Meeting human needs. Productive, sustainable, nutritious. Food Crops Res. 60, 1–10.

White J G and Zasoski R J 1999 Mapping soil micronutrients. Field Crops Res. 60, 11–26.

WHO 1996 Trace elements in human nutrition and health, World Health Organization, Geneva. 343 pp.

Yang X and Römheld V 1999 Physiological and genetic aspects of micronutrient uptake by higher plants. *In* Plant Nutrition – Molecular Biology and Genetics. Proceedings of the Sixth International Symposium on Genetics and Molecular Biology of Plant Nutrition. Eds. G Gissel-Nielsen and A. Jensen. pp. 151–186. Kluwer Academic Publishers, Dordrecht, The Netherlands.

Yip R 1997 The challenge of improving iron nutrition: Limitations and potentials of major intervention approaches. Euro. J. Clin. Nutr. 51, S16–S24.

Yip R and Scanlon K 1994 The burden of malnutrition: A population perspective. J. Nutr. 124 Suppl., 2043S–2046S.

J.J. Adu-Gyamfi (Ed.), *Food security in nutrient-stressed environments: exploiting plants' genetic capabilities*: 277–288, 2002.

Appropriate farm management practices for alleviating N and P deficiencies in low-nutrient soils of the tropics

Paul C. Smithson[1,3] & Ken E. Giller[2]
[1]*International Centre for Research in Agroforestry, P.O. Box 30677, Nairobi, Kenya*
[2]*University of Zimbabwe, Box MP 167, Mt. Pleasant, Harare, Zimbabwe.*
[3]*Corresponding author**

Key words: agroforestry, nitrogen, organic matter management, phosphorus, rock phosphates

Abstract

Tropical upland regions have both the world's highest population growth rates and a preponderance of soils less suitable for agriculture. Nitrogen deficiency is almost universal, while acid infertility and the related problem of P deficiency affect more than 40% of tropical soils. Nutrient depletion, as opposed to inherent infertility, affects large areas due to continuous cropping with few inputs. Nitrogen management requires a continual supply of N that can be achieved through fertilization, green manuring, legume rotations or leguminous tree-shrub fallows. Rotations and fallows usually require foregoing one or more staple crop harvests, which is not feasible for smallholder farmers in land-limited areas. In such areas there are few options for farmers except mineral N fertilizers. Phosphorus availability can be built up in soils, but to achieve this, external inputs of inorganic P are essential. Use of P-efficient crops, or mobilisation of soil organic P by various means, are temporary solutions since they involve only offtake with no addition. Phosphorus additions may be either as soluble processed fertilizers or indigenous phosphate rocks (PRs). Most tropical PR deposits are unreactive and require processing before use. Some lower-cost options include partial acidulation, blending with soluble P sources or microbial solubilization. The relevance of microbial solubilization in particular requires testing at larger scale. There are no magic solutions to soil nutrient deficiencies or toxicities; to maintain productivity mineral fertilizers are necessary. They should be used in judicious amounts and coupled with improved organic matter management.

Introduction

Increasing world population requires that agricultural production increase so as to keep up with basic food needs. This is especially true in tropical regions, where population growth rates are the highest in the world. At the same time, soils which are less suitable for agricultural production are relatively more abundant in tropical regions compared to the temperate zone, while the chemical inputs needed for high productivity are often in scarce supply in remote areas and beyond the economic grasp of smallholder farmers.

Soils of low inherent fertility, particularly Oxisols and Ultisols, cover large areas of the tropics worldwide. Soils with one or more chemical constraints to agricultural development (primarily acidity and related problems) occupy about 47% of South America and 59% of Southeast Asia, compared with 22% of North America and 33% of Europe (Dent, 1980). While Dent (1980) estimates only 18% of African soils to suffer from chemical constraints, this figure rises to 32% when only soils not affected by drought or other non-chemical constraints are considered (Woomer and Muchena, 1996).

The situation in sub-Saharan Africa can be contrasted with much of the tropics, as Africa has large areas of soils which were once moderately fertile, but in which continuous cropping with few external inputs has led to fertility depletion. Recent studies give an estimate of the magnitude of the problem. Smaling et al. (1997) have made estimates of nutrient depletion at farm, national and continental scales in sub-Saharan Africa; estimated annual per-hectare losses in Africa are 22 kg N, 2.5 kg P and 15 kg K. Loss estimates vary widely, but are negative throughout sub-Saharan

* FAX No.: +254-2-524001; Tel.: +254-2-524000; E-mail: p.smithson@cgiar.org

This paper previously appeared in *Plant and Soil* **245**, 2002, pp. 169–180.

Africa, excluding South Africa. Symptomatic of the problem is that per capita food production has continued to decline over the past two decades in sub-Saharan Africa, while it has increased modestly in Asia and Latin America (World Bank, 1996a,b).

In this paper we focus on two of the most widespread constraints to tropical crop production: N deficiency and P deficiency, including high P fixation. The use of organic inputs is highlighted as a low cost, if partial, solution to the nitrogen deficiency which is pervasive in most low input cropping systems. Phosphorus management will require inorganic inputs, and some lower-cost options for P fertilisation are considered.

Table 1. Estimates of the N_2-fixation potential of various grain legumes in the tropics. Source: Giller (2001)

Species	N_2 fixation potential (kg N ha^{-1} $year^{-1}$)
Arachis hypogaea	21–206
Cajanus cajan	0–166
Cicer arietinum	0–124
Glycine max	15–188
Lens culinaris	4–83
Phaseolus vulgaris	0–125
Vigna radiata	0–107
Vigna unguiculata	9–201

Management practices for problem soils

Nitrogen

Nitrogen is the plant nutrient taken up in the greatest quantities by crops, and is almost universally deficient in all but the most extensive agricultural systems. It is very difficult to build large reserves of soil N, unless in recalcitrant, poorly available forms, so a nearly continual supply of N is necessary (Giller et al., 1997). Mineral N fertilizer can supply the N required, and in many cases this is the logical choice for N management (indeed some would argue the only choice—see Quiñones et al., 1997). Commercial fertilizers are expensive and beyond the means of many smallholders, so organic approaches to N management are important, particularly those based on N_2-fixing legumes.

Organic N additions can be achieved through green manuring, legume rotations or fallowing with leguminous shrubs or trees. Even under the best circumstances, however, the amounts of N_2 fixation needed to maintain productivity may be difficult to achieve, particularly where P is also deficient. Giller and Cadisch (1995) estimated an *average* of 30 kg N ha^{-1} $year^{-1}$ must be fixed in order to offset the estimated N losses in sub-Saharan Africa, but legumes in P-deficient soils may fix as little as an average of 10 kg N ha^{-1} (Giller et al., 1997).

Potential of biological nitrogen fixation

The proportion of N in a legume crop which derives from biological nitrogen fixation (BNF) may vary widely, but rates up to 5 kg N ha^{-1} day^{-1} have been recorded (Giller, 2001). The proportion of N from N_2-fixation in crops ranges from zero, usually where environmental stresses are severe and prevent nodulation, to 98% in crops growing in ideal conditions. Amounts of N_2 fixed up to 450 kg ha^{-1} $crop^{-1}$ have been recorded in the tropics. Legume rotations or intercrops are common in smallholder agriculture, but since most involve a harvest offtake in grain or fodder, net N additions are small or even negative (Figure 1; Giller and Cadisch, 1995). Partial exceptions which give both edible yield and useful additions of N are pigeonpea and cowpea (*Vigna unguiculata*) (Giller et al., 1997). Common bean (*Phaseolus vulgaris*), grown by smallholders throughout the tropical highlands, is unfortunately rather poor at fixing N_2 due to its susceptibility to environmental stresses (Table 1). Peoples and Craswell (1992) noted the N balance from several studies on several grain legume crops ranged from −74 to +136 kg N ha^{-1}. Soybean

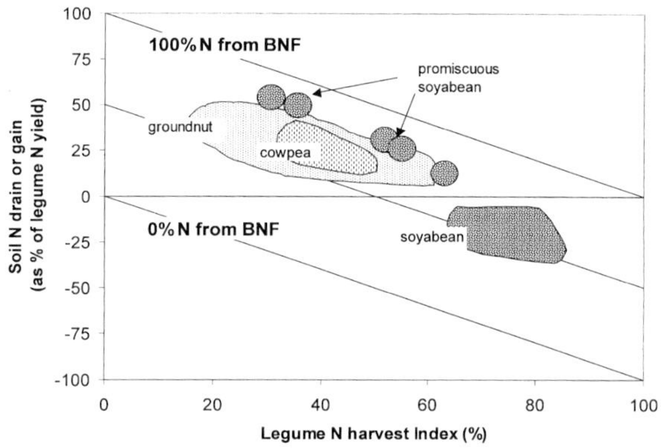

Figure 1. Gain or loss of soil N as a function of the fraction of legume N removed in legume harvest (the N harvest index) and the percentage of legume N derived from N_2 fixation. Source: modified from Giller and Cadisch (1995).

(*Glycine max*) tends to have negative balances, while cowpea and pigeonpea tend toward positive balances. Exceptions to this rule are promiscuously nodulating soybean varieties used in southern Africa which have relatively low harvest indices and produce abundant biomass (Mpepereki et al., 2000). Promiscuous soybean varieties give reasonable yields under smallholder conditions, and the stover contributes up to 60 kg N ha^{-1} for succeeding crops (Figure 1).

Regardless of the amount of N fixed, the efficiency with which the organic N is taken up by a following cereal crop is generally low, around 10–20% of the N contained in the legume biomass (Palm, 1995; Giller and Cadisch, 1995). Cereal yield increases after legumes relative to continuous cereal crop production range from 10 to 353%, equivalent to N fertilization of 7–68 kg ha^{-1} (Peoples and Craswell, 1992). The main conclusion is that N benefits to subsequent crops from N_2-fixation are highly variable, and qualitatively at least, tend to be inversely proportional to the edible legume grain harvest (MacColl, 1989). As pulses are a major source of protein in the diets of the poor, this must be viewed as a largely beneficial tradeoff.

Green manures, rotations and intercrops

Green manure crops or longer-term fallows can supply substantial N, but involve the loss of one or more season's marketable harvest, and are therefore best suited for areas where population density is low. A promising green manure crop with a long history of human use is *Mucuna pruriens* (mucuna or velvetbean). Mucuna produces high quality forage, and produces upwards of 10 Mg ha^{-1} dry matter in Honduras (Buckles and Triomphe, 1999). Mucuna may be relay established in maize, but timing is critical, with early undersowing causing substantial grain yield reductions (Buckles and Triomphe, 1999; Fischler and Wortmann, 1999). Mucuna is best suited to a humid bimodal rainfall regime (Buckles, 1995), and requires available land which can be taken out of grain production; in Honduras, for example, minimum farm size for uptake of the mucuna–maize rotation is about 1.6 ha (Buckles, 1995).

The requirement that there be excess land to make green manure rotations feasible is not absolute, however. *Tephrosia vogelii* and *Crotalaria grahamiana* are among the species being tested in densely populated western Kenya, where landholdings average 0.2–1.6 ha (ICRAF, 1998). The legumes are relay established in maize, and allowed to grow through the 'short rains' season, when farmers often do not crop maize due to unreliable rains and pest and disease problems. Maize yields after a single-season *Crotalaria grahamiana* fallow in Siaya District, Western Kenya averaged 4.6 Mg ha^{-1}, compared to 2.0 Mg ha^{-1} for continuous maize and 2.8 Mg ha^{-1} for maize following the weedy fallow practiced by farmers (Bashir Jama, personal communication).

In cases where high population density forces continuous food cropping, grain legume intercrops or rotations may be the best option. Grain legumes in rotation with cereal crops can result in a net N benefit to the cereal crop; for example, Kumar Rao et al. (1983) found maize (*Zea mays*) after sole crop pigeonpea gave yields about equal to maize with 45 kg added N ha^{-1}. In smallholder systems, however, intercrops are more common than rotations, and the benefit to a following crop may not be as great. In the same study, Kumar Rao et al. found that maize after a pigeonpea/sorghum intercrop resulted in little benefit to the maize, though sorghum benefited from the intercropped pigeonpea. It is likely to be generally true that residual benefits will be greater following a sole legume crop rather than a legume–cereal intercrop.

Agboola and Fayemi (1972) in Nigeria found that either rotation or interplanting with legumes increased maize yields relative to unfertilized controls. *Calopogonium* was the best of several species tested, giving yield increases approximately equivalent to gains from 55 kg inorganic N ha^{-1}. Control yield was 1.5 Mg ha^{-1}, while *Calopogonium* treatments yielded 2.2 Mg ha^{-1}.

MacColl (1989), studying legume–maize rotations in Malawi found total residual N (over 3 years of maize cropping) from a previous 2-year pigeonpea crop of 24–110 kg N ha^{-1}. Annual legumes in the same study gave 0 to 52 kg ha^{-1} residual N, with soybean tending to leave little or no residual N. Calculations from MacColl's data indicate that total maize yield from 5 years continuous maize cropping was similar to the total from 3 years of maize following the 2-year pigeonpea, so the pigeonpea yield could be considered as the 'profit' from the system. The relevance of a system which requires a 2-year loss of staple grain production may be questioned, however. In southern Malawi, pigeonpea is commonly grown in intercrops with maize where the maize population is held at the same density as in the sole crop. The slow initial development of the pigeonpea results in little penalty in maize yield and up to 80 kg N ha^{-1} can be added to the soil from fallen pigeonpea leaves (Sakala, 1998). The slow decomposition of these sen-

Table 2. Estimates of biomass and N production by selected woody legumes. Source: selected from compilations of Szott et al. (1991), Kang and Mulongoy (1992) and Danso et al. (1992)

Species	Biomass production (Mg ha^{-1} year^{-1})	N in biomass (kg N ha^{-1} year^{-1})
Gliricidia sepium	11	171–205
	1.4	na
	na	150
	5.5	169
Erythrina spp.	8.1	198
	1.9	34
Inga edulis	5.6	136
	12.5	na
Calliandra calothyrsus	6.1	218
Leucaena leucocephala	7.4	247
	na	700–1050
	na	468
Sesbania sesban	na	239–927
Sesbania rostrata	na	1120–1660

na, data not available.

esced leaves means that there is little benefit of N to the companion crop (Sakala et al., 2000), but a substantial addition for subsequent crops of maize.

Agroforestry options

Alley cropping and variants The abundance of leguminous trees and shrubs in the tropics, and numerous descriptions of agroforestry practice in indigenous agriculture, caused optimism in the 1970s and 1980s that tropical N deficiencies could be eliminated through agroforestry technologies, with the added benefit of sorely needed wood products. Tree legumes fix similar quantities of N as herbaceous legumes (Table 2), so the prerequisite of substantial N inputs from BNF is met. Much of the variability in amounts of N_2-fixation can be attributed to the wide range in biomass production, often due to soil infertility (Szott et al., 1991).

Agroforestry was largely a descriptive science until the late 1980s (Young, 1997), and many claims about deep nutrient capture and other benefits of tree–crop combinations were untested. As more rigorous studies have been conducted, and flaws in earlier designs discovered, the early hopes for quick agroforestry-based solutions for N deficiencies have not been realized. In fact, hedgerow intercropping or alley cropping has been shown to be unsuccessful in many cases, with most successful examples being on fertile, well-watered soils (Szott et al., 1991).

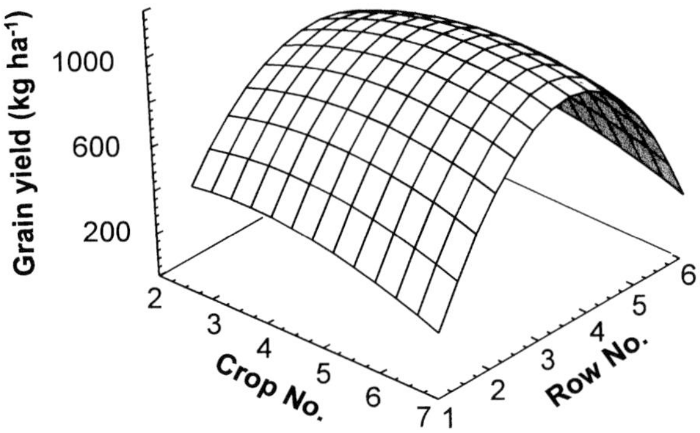

Figure 2. Rice yields in an alley cropping system in Peru as a function of time and distance from hedgerow (rows 1 and 6 are adjacent to hedgerow). Source: Fernandes et al. (1993).

In subhumid or drier climates, Mathuva et al. (1998) and others have shown that competition for water by the alley-cropped trees usually outweighs the positive fertility effects. Even in humid areas, presence of a strongly acidic subsoil will restrict rooting of both crops and trees to the surface soil, thus increasing competition for water and nutrients, especially near the hedgerow (Figure 2; Fernandes et al., 1993). In both of these studies, maize or rice (*Oryza sativa*) yields in alley-cropped treatments were usually poorer than in unfertilized controls.

Alley cropping's major benefit is addition of N, and if N is not the most limiting nutrient at a site the benefits are usually small (particularly if P is severely limiting; Young, 1997). On an infertile site in Zambia, Matthews et al. (1992) also found no benefits of alley cropping for several crops relative to unfertilized controls over a 4-year study due to poor biomass production by the trees. One variant of alley cropping, however, has seen some success on-station in southern Malawi. In this system, gliricidia (*Gliricidia sepium*) is interplanted with maize in every furrow rather than in widely spaced alleys. The trees are pruned severely, to about 30 cm height, several times per year and at least twice during the cropping season, to maximise biomass input and minimize competition (ICRAF, 1998; Ikerra et al., 1999). In on-station tests, gliricidia increased topsoil inorganic N and maize yields relative to unfertilized controls (Ikerra et al., 1999). Yields with gliricidia were greater than those with 48 kg N ha^{-1} inorganic fertilizer, but combination of gliricidia and inorganic N gave less than additive yield increases (ICRAF, 1998). Nitrogen inputs in the combination treatments approached 240 kg N ha^{-1} in some years, 2.5 times local recommended N rates, suggesting that yields may have been within the plateau region of the response curve.

It is noteworthy that the soil at the initial test site was able to provide sufficient P and K for plant growth, with little alteration of soil fertility status with depth to >1 m (Ikerra et al., 1999). Relatively deep and fertile soils appear to be a prerequisite for success of simultaneous agroforestry systems (Szott et al., 1991). Subsoil moisture in the long dry season of southern Africa is an additional factor which will influence the applicability of this technology.

Fallow systems The problems often encountered with simultaneous systems such as alley cropping have caused attention to shift towards sequential or relay systems such as improved fallows. In eastern Zambia, traditional fallows of up to several decades have decreased to 1–5 years due to population increase. Unfertilized maize yields average around 1.5 Mg ha^{-1} in the area, but maize responds strongly to N, 112 kg N ha^{-1} giving 4.9 Mg ha^{-1} maize grain in one study (Kwesiga and Coe, 1994).

Several species have been tested as improvements on the traditional bush fallow, including *Sesbania sesban*, *Tephrosia vogelii* and pigeonpea among others. Two-year sesbania fallows were found to be the most successful, maintaining better yields for at least two seasons after the fallow period. Economic analysis showed that the 2-year fallow became more profitable by the fourth year than continuous unfertilized maize, despite the labour requirements for fallow establishment and loss of 2 years' yield (Kwesiga et al., 1999). Another potential effect of fallows is the capture of N leached beyond the rooting zone of maize, and brought to the surface by the deeper-rooted trees during the fallow. In Oxisols and oxic Alfisols of western Kenya, subsoil N accumulation may be as high as 70–315 kg N ha^{-1}, and 18-month sesbania fallows significantly reduced the subsoil nitrate (Hartemink et al., 1996).

While improved fallows have shown some success, nitrogen accumulation in the fallows and yield increases of subsequent crops are related to the duration of the fallow (Kwesiga and Coe, 1994). Longer fallows require a surplus of land, which is not available in many areas. In densely populated southern Malawi, a relay fallow with *Sesbania sesban* has been tested, a system which does not require loss of any cropping seasons. In this system, sesbania is interplanted with maize about 1 week after maize planting. The growth of sesbania is suppressed during the cropping season, but grows vigorously during the 7-month dry season, producing biomass which is incorporated during fallow clearance and land preparation for the next maize crop. As for other similar technologies, however, loss of the usual pigeonpea crop is a drawback.

On-farm trials of a similar system in southern Malawi gave maize yield increases after sesbania biomass incorporation of <0.1–0.4 Mg ha^{-1}, from unfertilised yields of 0.7–1.3 Mg ha^{-1} (Phiri et al., 1999a). A yield increase of 0.6 Mg ha^{-1} in sesbania plots during the first season of sesbania establishment was difficult to explain, but was postulated to be due to improved water relations. Soil nitrate levels were increased by the sesbania additions, and the N accumulation was correlated to sesbania biomass production (Phiri et al., 1999b).

Besides N inputs, much of the appeal of woody fallows is due to the wood production, important in fuel-scarce areas. The offtake of nutrients in the wood harvest must not be ignored. Buresh et al. (1997) estimated P removal from different land uses during four seasons (2 years) in western Kenya. In three seasons of sesbania fallow followed by one season of maize, 11.9 kg P ha^{-1} were removed in wood plus 1.3 Mg ha^{-1} maize, compared to 14.9 kg P ha^{-1} removed in four seasons of continuous maize (total 6.8 Mg ha^{-1} grain yield).

Manure management

The potential value of livestock manure varies widely among locations and farming systems. In densely populated areas far from markets, many households own no livestock and therefore have no access to manure (Shepherd and Soule, 1998). However, where markets exist, for example the Nairobi market near the Central Highlands of Kenya, small-scale dairying is almost ubiquitous, even on very small farms averaging <1 ha. In this case, small land holdings act to force improved manure management, as almost all cattle are raised in confinement and are stall-fed.

Lekasi et al. (1998) surveyed Murang'a and Kiambu districts in Kenya's Central Highlands, with average farm size of 0.45 ha for their 'small farm' category, 1.1 ha for 'medium' and 2.8 ha for 'large' farms. Number of cattle owned was not strongly related to farm size, so small farms produced more manure per unit area, averaging 8.2 Mg dry weight ha^{-1} $year^{-1}$ of manure, or 114 kg N ha^{-1} $year^{-1}$. Smallholders manage manure by various means, including addition of bedding and feed refusals (primarily crop wastes and other low quality materials) to the manure heap. This practice is that recommended by Myers et al. (1994) to reduce manure N losses. The integration of livestock into smallholder farming systems, and management of the manure resource provided by livestock, deserves increased attention. However, in these areas the poorer farmers tend to have no livestock and yet they grow napier grass (*Pennisetum purpureum*) for sale to livestock owners, resulting in net transfers of nutrients from poor to more wealthy households.

Manure or other organic additions serve to maintain soil organic matter content, though usually at lower levels than under native vegetation or perennial crops. Swift et al. (1994) reviewed a 16-year trial at Kabete, Kenya, on soils similar to those of the agriculturally important Central Highlands. With no inputs, soil organic C declined from $>2\%$ to about 1.1%. Even with the addition of 10 Mg dry weight ha^{-1} $year^{-1}$ manure, soil C decreased to about 1.8%. As Giller et al. (1997) noted, building large stocks of soil organic matter in continuously cultivated systems is difficult. Despite the decrease in soil organic matter, yields were maintained at near original levels by various combinations of organic and inorganic fertilization, though 5 Mg dry weight ha^{-1} of manure alone was insufficient to maintain yields.

Kapkiyai et al. (1999) studied soil parameters in the same experiment at Kabete. The treatment with the greatest inputs, 10 Mg dry weight ha^{-1} manure with residue retention and inorganic NP fertilization, still resulted in a loss of about 0.43 Mg C ha^{-1} $year^{-1}$. Various measures of soil quality, including N mineralization potential, particulate organic matter and others, were strongly related to manure additions in particular. The same general trends in organic quality parameters were detected by Murage et al. (2000) in a comparison of 12 farmers' fields identified as productive or non-productive by the farmers. Total C, particulate C and 'labile' ($KMnO_4$-reactive) C were all significantly higher in productive fields. These findings show the importance of organic matter amount and quality in smallholder systems. They also serve to show the rather large additions required to maintain yields and even partially maintain soil carbon.

Phosphorus

After nitrogen, phosphorus is usually the most limiting nutrient for crop production. Phosphorus must be added at much greater rates than is removed by crops due to the conversion of soil P to plant-unavailable forms, or P fixation. High P fixation presents management problems of its own, but is much less widespread than simple P deficiency, which will be the major thrust of this section. Phosphorus deficiency and/or severe P fixation are likely in acidic soils dominated by 1:1 clay minerals, particularly those with clayey or loamy texture, and with substantial Fe and Al oxide contents. High P fixation due to allophane is primarily limited to soils of volcanic origin, occurring in only about 1% of the developing world, and will not be discussed further here.

There are species and varietal differences in tolerance to low soil P, and in the ability to utilize poorly soluble P sources. A classic example is tea (*Camellia sinensis*), an Al accumulator. Jayman and Sivasubramaniam (1975) showed that roots of tea plants produced malic acid, which can solubilize Al-P compounds. Among food crops, pigeonpea (*Cajanus cajan*) solubilizes soil P by exudation of organic acids (Ae et al., 1990); and groundnut (*Arachis hypogaea*) acquires Fe-bound P by a yet unexplained mechanism at the cell wall (Ae and Otani, 2000). While crop selection and plant breeding has promise for increasing crop P efficiency (Johansen et al., 1995), continued crop offtake must at some point be countered by increased soil P supply. In the short term this may be through increased availability of native soil P or by external additions. It is a given that in the longer term, external P additions will be required. Examples will emphasize the East African Highlands, where P problems and opportunities are similar to those in much of the developing world.

The East African highlands comprises portions of Ethiopia, Kenya, Uganda, Tanzania, Rwanda and Burundi, and an estimated 26% of the arable land in the region suffers from P deficiency (KARI, 1994). Within the densely populated Lake Victoria region of western Kenya, data from several hundred farmers' fields show that more than 80% are seriously P-deficient, indicated by <5 mg kg^{-1} bicarbonate-extractable P (Jama et al., unpublished data; Niang et al., unpublished data). Yields of the staple crop maize average <1 Mg ha^{-1} under usual smallholder conditions in this area (Sanchez et al., 1997). Responses to P are significant even at rates as low as 10 kg P ha^{-1} (Jama et al., 1997).

Native soil P

Plant-available P has generally been considered to be inorganic orthophosphate (P_i), but soil organic P (P_o) may also be a potential source of P by mineralization of P_o (Adepetu and Corey, 1977). This is particularly true in highly weathered tropical soils, which often have a greater percentage of total P as P_o (Sharpley et al., 1987). In unfertilized systems, bicarbonate- and NaOH-extractable P_o has been correlated with soybean (*Glycine max*) yields in Hawaii (Linquist et al., 1997), while NaOH P_o best explained crop yields in an unfertilized Peruvian Ultisol (Beck and Sanchez, 1994). In Kenya, Maroko et al. (1999) found that microbial biomass P and P in particulate soil organic matter were significantly increased in a *Sesbania sesban* fallow compared to continuous maize. Unfertilized maize yields after the fallow were best related to the same P fractions, while soil test P_i (e.g. Olsen P) explained little of the variability in yields.

Soil heating can also increase soil P availability, presumably by partial oxidation of soil P_o. The effect may be incidental as in slash burning after land clearance, or intentional as in various indigenous practices involving mounding or trenching of soil in combination with organic materials, which are then burned. The effects can persist for more than a decade in some cases (Strømgaard, 1992). Laboratory heating to 200 °C for 48 h, simulating a soil-plus-organics burning system known as *guie* in Ethiopia, caused increased available P (Sertsu and Sanchez, 1978). A similar system, studied under field conditions in Bhutan, produced highly variable increases in available P (Roder et al., 1993). On the other hand, increased P sorption capacity has been noted after soil heating (Kwari and Batey, 1991), indicating caution in attempting wide scale adoption of such practices. In any case, recycling of the limited soil P_o resource by whatever means cannot solve the fertility depletion problem, and some external sources of P are essential for increasing soil P stocks.

Organic P sources

Both organic and inorganic sources of P should be considered, but unlike plant N, plant P content is too low to provide enough P for annual crops, though biomass additions may be important as supplements. Livestock manures may be an exception to this generalisation, with P content often around 0.5% or higher (Palm et al., 1997). In western Kenya, Jama et al. (1997) obtained maize yield increases from 10 kg P ha^{-1} added as cattle manure on a P-deficient Oxisol. Manure is the most common soil amendment in the area, but a majority of farmers own no livestock and therefore have no source of manure (Shepherd and Soule, 1998). Low P content of most organics, low availability and competing uses will generally preclude exclusive use of organics for P fertilization requirements. As for N management, a possible exception is on dairy farms in the Central Highlands of Kenya. Smaller farms in this region produce an average of about 50 kg manure P ha^{-1} $year^{-1}$ (Lekasi et al., 1998). Where nearby markets are available, integrated livestock-cropping operations may offer a viable option for P fertilisation.

Inorganic P sources

Almost all soluble P fertilizer used in the East Africa Highlands is imported, high priced and often unavailable when needed (Sanchez et al., 1997). Soluble P fertilizers are probably essential for the needed growth in African agriculture (Quiñones et al., 1997), but better use of indigenous phosphate rock (PR) deposits is also necessary. Two sources in the region typify the range of materials available. The Minjingu PR deposit near Lake Manyara in Tanzania is moderately reactive, containing 13% total P and 3% neutral ammonium citrate (NAC)-soluble P (Smithson, unpublished data). The Minjingu deposit is small and apparently unique, consisting of Pliocene deposits of bird and fish bones (P. van Straaten, personal communication). More typical of the region's deposits is the igneous and relatively unreactive Busumbu deposit in eastern Uganda, which contains from 4.8 to 14% total P and 0.3 to 1.0% NAC-soluble P (Smithson, unpublished data).

Direct use of phosphate rocks

Minjingu PR is suitable for direct application to row crops in acidic soils (Buresh et al., 1997; Sanchez et al., 1997). Soil pH less than about 5.5 is required for effective dissolution (Sanchez, 1976; Rajan et al., 1996), a criterion met by about 47% of nearly 1800 soils sampled from the western Kenya study areas (Smithson, unpublished data). ICRAF began work with Minjingu PR in western Kenya in 1996, comparing PR with TSP in terms of immediate and residual effects, using annual applications of 50 kg P ha^{-1} or a single application of 250 kg P ha^{-1}. Early results were encouraging, with Minjingu PR averaging around 75–85% relative agronomic effectiveness [RAE, defined as 100× (yield increase with PR/yield increase with TSP)] depending upon P rate (Mutuo et al., 1999).

Similar experiments have been installed at a number of other sites in the area, and in 28 P-responsive sites (of a total of 34 sites), Minjingu PR at 100 kg P ha^{-1} gave first-season RAEs averaging about 70% (Lijzenga, 1998, ICRAF, 1999). Overall performance of PRs must be judged over several years, because in general the first-season RAE of PR is less than that of TSP, but due to slow dissolution may surpass the RAE of TSP in later years, as the rapidly dissolved TSP is subsequently adsorbed by soil sesquioxides and becomes less available (Rajan et al., 1996). The final decision is an economic one, depending on the RAE of PRs over several seasons, and the relative cost compared with soluble sources.

Use of unreactive PRs: blending and processing options

Busumbu PR is typical of unreactive PRs and is unsuitable for direct application to annual crops. Savini (1999) tested Busumbu PR in a pot study using a P-deficient western Kenya Oxisol sufficient in N and K. He found that while TSP produced linear yield increases of maize and cowpea up to 100 kg P ha^{-1}, Busumbu PR at 50 kg ha^{-1} did not raise yields significantly above zero-P treatments.

Organic materials in combination with PRs may increase P availability by reducing P sorption, presumably by competition for P adsorption sites by organic anions. Organic additions may also directly enhance PR solubility, due both to lowered pH and to decomposition-related production of metal-complexing organic acids. Reduction in P sorption has been shown with incorporated green manures, crop residues or animal manures (Syers and Lu, 1990; Nziguheba et al., 1998).

Certain groups of bacteria and fungi can solubilize soil P or PRs (Goldstein, 1986). Despite laboratory demonstrations of P solubilization, many early

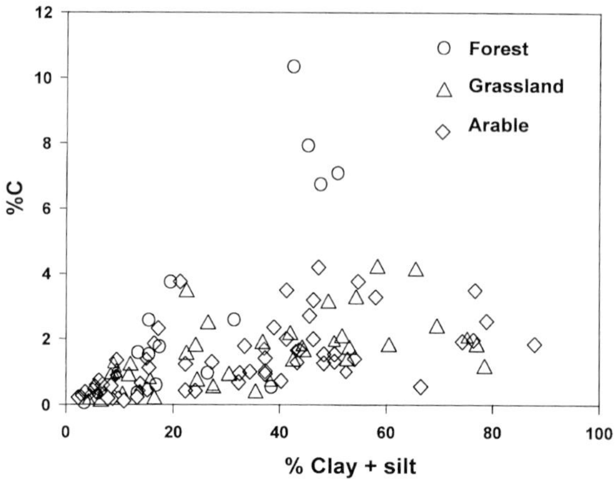

Figure 3. Total carbon in African soils under different managements as a function of soil clay+silt percentage. Source: Giller et al. (1997).

field tests of microbial solubilizers were unsuccessful (Tinker, 1980). Even so, research continues, and a fungal seed inoculum based on *Penicillium bilaii* has shown some success in Canada (Leggett et al., 2000; Kucey and Leggett, 1989). Whether microbial P solubilization can be exploited successfully in smallholder tropical agriculture remains to be demonstrated.

Targeting technologies for smallholders

Given the potentials and the limitations of approaches to managing soil fertility based on mineral fertilizers, rock phosphates and legume-based or improved organic matter management systems, what can be done to target the technologies better? Selection of well-adapted germplasm and determination of the most efficient and economic rates of addition of mineral fertilizers is an obvious first step. Knowledge of the best way to manage problem soils, such as the need to apply frequently small amounts of lime to correct acidity and nutrients to correct deficiencies in sandy soils ('little and often') compared with larger, less-frequent doses in heavier textured soils with greater buffering capacities. These are basic rules well-known to all good managers of land, whether farmers or advisors, and can be equally well applied to the management of phosphorus fertilizers.

Similar rules are being developed for the use of organic amendments. The potential for 'building' soil organic matter reserves in sandy soils is very limited as the capacity of soils to store organic matter is strongly related to their clay and silt contents (Figure 3; Giller

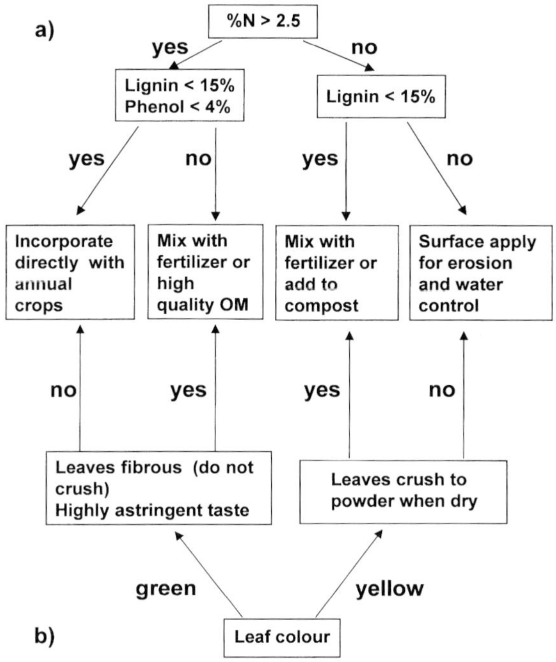

Figure 4. (a) Decision tree for organic matter management based on plant chemical parameters. (b) Tentative organic management decision tree suitable for extensionists, based on properties observable in field. Source: (a) Palm et al. (1997), Mafongoya et al. (1998); (b) Giller (2000), Palm et al. (2001).

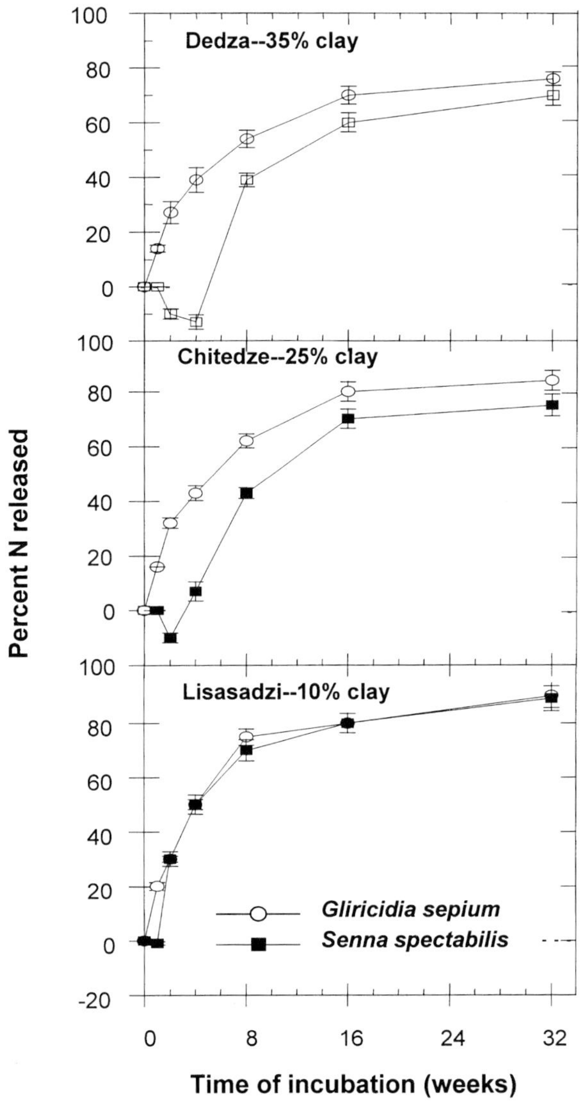

Figure 5. Decomposition of two organic materials of differing quality in Malawian soils of varying clay percentage. Source: Itimu (1997).

et al., 1997). Large additions of cattle manure or organic composts can result in leaching of organic matter in sandy soils and ideally systems which provide for small, repeated additions of organic amendments are more appropriate. By contrast there is more potential for building the SOM content in heavy-textured soils which, in a similar way to liming, can be managed with larger additions of organic matter. It therefore follows that the residual effect of leys or fallows is likely to be longer-lived in soils with a heavier texture.

The utility of organic amendments in supplying N through mineralization is known to be directly linked to the chemical composition or 'quality' of the material (Swift et al., 1979; Cadisch and Giller, 1997). Based on a detailed database, called the Organic Resource Database (http: // www.wye.ac.uk / BioSciences / soil / ord2.htm) which has detailed resource quality characteristics of more than 1500 plant residues and other organic matter resources, a simple decision tree has been constructed which can be used as a guide on the best uses of organic resources on farm. This decision tree is based on the N, lignin and polyphenol contents of the organic resources which have been shown to be the major factors regulating short-term mineralization from the majority of plant residues (Figure 4a; Palm et al., 1997; Mafongoya et al., 1998). This decision tree has been translated into 'rules of thumb' which can be applied in the field by extension workers to enable a deeper discussion of the roles of organic manures with farmers (Figure 4b; Giller, 2000; Palm et al., 2001).

Linking to the above discussion on the role of soil texture in targeting of technologies, the rates of decomposition tend to be much faster in coarse-textured, sandy soils (van Veen et al., 1985). This means that the influence of different qualities of organic residues on decomposition which may be clear in soils with more clay and silt, tends to be masked in sandy soils due to the rapid rates of decomposition (Figure 5). Again this emphasises the need for clear targeting of technologies on the basis of the characteristics of the soil as well as the needs of the farmer within the cropping system.

Conclusions

There are no magic solutions to soil nutrient deficiencies or toxicities. The examples above represent what are perhaps the most widespread problems in the tropics. In the case of P certainly, and N most probably, fertilizer additions should be seen as necessary. An organic-based system, coupled with judicious use of moderate doses of inorganic fertilizers, offers a viable path to increased agricultural productivity in the tropics.

References

Adepetu J A and Corey R B 1977 Organic phosphorus as a predictor of plant-available phosphorus in soils of southern Nigeria. Soil Sci. 122, 159–164.

Ae N and Otani T 2001 The role of root cell wall in phosphorus acquisition by groundnut. *In* Plant Nutrient Acquisition. Eds. N Ae and K Okada. Springer, Tokyo (in press).

Ae N, Arihara J, Okada K, Yoshinara T and Johansen C 1990 Phosphorus uptake by pigeon pea and its role in cropping systems in the Indian subcontinent. Science (Washington, DC) 248, 477–480.

Agboola A A and Fayemi A A 1972 Effect of soil management on corn yield and soil nutrients in the rain forest zone of western Nigeria. Agron. J. 64, 641–644.

Beck M A and Sanchez P A 1994 Soil phosphorus fraction dynamics during 18 years of cultivation on a Typic Paleudult. Soil Sci. Soc. Am. J. 58, 1424–1431.

Buckles D 1995 Velvetbean: a 'new' plant with a history. Econ. Bot. 49, 13–25.

Buckles D and Triomphe B 1999 Adoption of mucuna in the farming systems of northern Honduras. Agrofor. Syst. 47, 67–91.

Buresh R J, Smithson P C and Hellums D T 1997 Building soil phosphorus capital in Africa. *In* Replenishing Soil Fertility in Africa. Eds. R J Buresh, P A Sanchez and F Calhoun. pp. 114–149. SSSA Spec. Pub. 51, Soil Sci. Soc. America, Madison, WI, USA.

Cadisch G and Giller K E (Eds.) 1997 Driven by Nature: Plant Residue Quality and Decomposition. CAB International, Wallingford, Oxon, UK. 409 pp.

Danso S K A, Bowen G D and Sanginga N 1992 Biological nitrogen fixation in trees in agro-ecosystems. *In* Biological Nitrogen Fixation for Sustainable Agriculture. Eds. J K Ladha, T George and B B Bohlool. pp. 177–196. Kluwer Academic Publishers, Dordrecht, The Netherlands.

Dent F J 1980 Major production systems and soil-related constraints in Southeast Asia. *In* Soil-related constraints to food production in the tropics. International Rice Research Institute, Los Baños, Philippines, pp. 79–106.

Fernandes E C M, Davey C B and Nelson L A 1993 Alley cropping on an acid soil in the upper Amazon: mulch, fertilizer, and hedgerow root pruning effects. *In* Technologies for Sustainable Agriculture in the Tropics, Eds. J Ragland and R Lal. pp. 77–96. ASA Spec. Pub. 56, Am. Soc. Agron., Madison, WI, USA.

Fischler M and Wortmann C S 1999 Green manures for maize–bean systems in eastern Uganda: Agronomic performance and farmers' perceptions. Agrofor. Syst. 47, 123–138.

Giller K E 2000 Translating science into action for agricultural development in the tropics: an example from decomposition studies. Appl. Soil Ecol. 14, 1–3.

Giller K E 2001 Nitrogen Fixation in Tropical Cropping Systems, 2nd ed. CAB International, Wallingford, Oxon, UK.

Giller K E and Cadisch G 1995 Future benefits from biological nitrogen fixation: an ecological approach to agriculture. Plant Soil 174, 255–277.

Giller K E, Cadisch G, Ehaliotis C, Adams E, Sakala W D and Mafongoya P L 1997 Building soil nitrogen capital in Africa. In Replenishing Soil Fertility in Africa. Eds. R J Buresh, P A Sanchez and F Calhoun. pp. 151–192. SSSA Spec. Pub. 51, Soil Sci. Soc. America, Madison, WI, USA.

Goldstein A H 1986 Bacterial mineral phosphate solubilization: historical perspectives and future prospects. Am. J. Alt. Agric. 1, 57–65.

Hartemink A E, Buresh R J, Jama B and Janssen B H 1996 Soil nitrate and water dynamics in sesbania fallows, weed fallows, and maize. Soil Sci. Soc. Am. J. 60, 568–574.

ICRAF 1998 ICRAF Annual Report 1997. International Centre for Research in Agroforestry, Nairobi, Kenya. 204 pp.

ICRAF 1999 Technical report for the project on 'Strategic Soil Fertility Replenishment and Recapitalization Research in Western Kenya', July 1998–June 1999. International Centre for Research in Agroforestry, Nairobi, Kenya. 6 pp.

Ikerra S T, Maghembe J A, Smithson P C and Buresh R J 1999 Soil nitrogen dynamics and relationships with maize yields in a gliricidia-maize intercrop in Malawi. Plant Soil 211, 155–164.

Itimu O A 1997 Nitrogen dymanics and root distribution of *Gliricidia sepium* and *Senna spectabilis* in maize (*Zea mays*)-based alley cropping systems in Malawi. PhD thesis, Wye College, University of London, Wye, Kent, UK.

Jama B, Swinkels R A and Buresh R J 1997 Agronomic and economic evaluation of organic and inorganic phosphorus in western Kenya. Agron. J. 89, 597–604.

Jayman T C Z and Sivasubramaniam S 1975 Release of bound iron and aluminium from soils by the root exudates of tea (*Camellia sinensis*) plants. J. Sci. Food Agric. 26, 1895–1898.

Johansen C, Lee K K, Sharma K K, Subbarao G V and Kueneman E A (Eds.) 1995 Genetic manipulation of crop plants to enhance integrated nutrient management in cropping systems—1 Phosphorus: proceedings of an FAO-ICRISAT expert consultancy workshop. International Crops Research Institute for the Semi-Arid Tropics, Patancheru, Andhra Pradesh, India.

Kang B T and Mulongoy K 1992 Nitrogen contribution of woody legumes in alley cropping systems. *In* Biological Nitrogen Fixation and Sustainability of Tropical Agriculture. Eds. K Mulongoy, M Gueye and D S C Spencer. pp. 367–375. Wiley-Sayce, Chichester, UK.

Kapkiyai J J, Karanja N K, Qureshi J N, Smithson P C and Woomer P L 1999 Soil organic matter and nutrient dynamics in a Kenyan nitisol under long-term fertilizer and organic input management. Soil Biol. Biochem. 31, 1773–1782.

KARI 1994 Fertilizer use recommendations, Vol 1–22. Fertilizer Use Recommendation Project, Kenya Agricultural Research Institute, Nairobi, Kenya.

Kucey R M N and Leggett M E 1989 Increased yields and phosphorus uptake by Westar canola (*Brassica napus* L.) inoculated with a phosphate-solubilizing isolate of *Penicillium bilaji*. Can. J. Soil Sci. 69, 425–432.

Kumar Rao J V D K, Dart P J and Sastry P V S S 1983 Residual effect of pigeonpea (*Cajanus cajan*) on yield and nitrogen response of maize. Exp. Agric. 19, 131–141.

Kwari J D and Batey T 1991 Effect of heating on phosphate sorption and availability in some north-east Nigerian soils. J. Soil Sci. 42, 381–388.

Kwesiga F R and Coe R 1994 The effect of short rotation *Sesbania sesban* planted fallows on maize yields. For. Ecol. Manage. 64, 199–208.

Kwesiga F R, Franzel S, Place F, Phiri D and Simwanza C P 1999 *Sesbania sesban* improved fallows in eastern Zambia: their inception, development and farmer enthusiasm. Agrofor. Syst. 47, 49–66.

Leggett M E, Gleddie S and Holloway G 2001 Phosphorus solubilizing microorganisms and their use. *In* Plant Nutrient Acquisition. Eds. N Ae and K Okada. Springer, Tokyo (in press).

Lekasi J K, Tanner J C, Kimani S K and Harris P J C 1998 Manure management in the Kenya Highlands: practices and potential. Henry Doubleday Research Association, Emmerson Press, Kenilworth, UK. 35 pp.

Lijzenga M 1998 Maize response to NPK in relation to fertility indices in western Kenya. MSc Thesis, Wageningen Agricultural University, The Netherlands.

Linquist B A, Singleton P W and Cassman K G 1997 Inorganic and organic phosphorus dynamics during a build-up and decline of available phosphorus in an Ultisol. Soil Sci. 162, 254–264.

MacColl D 1989 Studies on maize (*Zea mays*) at Bunda, Malawi. II. Yield in short rotations with legumes. Exp. Agric. 25, 367–374.

Mafongoya P L, Giller K E and Palm C 1998 Decomposition and nitrogen release patterns of tree prunings and litter. Agrofor. Syst. 38, 77–97.

Maroko J B, Buresh R J and Smithson P C 1999 Soil phosphorus fractions in unfertilized fallow-maize systems on two tropical soils. Soil Sci. Soc. Am. J. 63, 320–326.

Mathuva M N, Rao M R, Smithson P C and Coe R 1998 Improving maize (*Zea mays*) yields in semiarid highlands of Kenya: agroforestry or inorganic fertilizers? Field Crops Res. 55, 57–72.

Matthews R B, Holden S T, Volk J and Lungu S 1992 The potential of alley cropping in improvement of cultivation systems in the high rainfall areas of Zambia I. *Chitemene* and *Fundikila*. Agrofor. Syst. 17, 219–240.

Mpepereki S, Javaheri F, Davis P and Giller K E 2000 Soyabeans and sustainable agriculture: Promiscuous soyabeans in southern Africa. Field Crops Res. 65, 137–149.

Murage E W, Karanja N K, Smithson P C and Woomer P L 2000 Diagnostic indicators of soil quality in productive and non-productive smallholders' fields of Kenya's Central Highlands Agric. Ecosyst. Environ. 79, 1–8.

Mutuo P K, Smithson P C, Buresh R J and Okalebo J R 1999 Comparison of phosphate rock and triple superphosphate on a phosphorus-deficient Kenyan soil. Commun. Soil Sci. Plant Anal. 30, 1091–1103.

Myers R J K, Palm C A, Cuevas E, Gunatilleke I U N and Brossard M 1994 The synchronisation of nutrient mineralisation and plant nutrient demand. *In* The Biological Management of Tropical Soil Fertility. Eds. P L Woomer and M J Swift. pp. 81–116. John Wiley and Sons, New York.

Nziguheba G, Palm C A, Buresh R J and Smithson P C 1998 Soil phosphorus fractions and adsorption as affected by organic and inorganic sources. Plant Soil 198, 159–168.

Palm C A 1995 Contribution of agroforestry trees to nutrient requirements of intercropped plants. Agrofor. Syst. 30, 105–124.

Palm C A, Myers R J K and Nandwa S M 1997 Combined use of organic and inorganic nutrient sources for soil fertility maintenance and replenishment. *In* Replenishing Soil Fertility in Africa. Eds. R J Buresh, P A Sanchez and F Calhoun. pp. 193–217. SSSA Spec. Pub. 51, Soil Sci. Soc. America, Madison, WI, USA.

Palm C A, Giller K E, Mafongoya P L and Swift M J 2001 Management of organic matter in the tropics: Translating theory into practice. Agric. Ecosyst. Environ. 83, 27–42.

Peoples M B and Craswell E T 1992 Biological nitrogen fixation: Investments, expectations and actual contributions to agriculture. *In* Biological Nitrogen Fixation for Sustainable Agriculture. Eds. J K Ladha, T George and B B Bohlool. pp. 13–39. Kluwer Academic Publishers, Dordrecht, The Netherlands.

Phiri A D K, Kanyama-Phiri G Y and Snapp S 1999 Maize and sesbania production in relay cropping at three landscape positions in Malawi. Agrofor. Syst. 47, 153–162.

Phiri R H, Snapp S and Kanyama-Phiri G Y 1999 Soil nitrate dynamics in relation to nitrogen source and landscape position in Malawi. Agrofor. Syst. 47, 253–262.

Quiñones M A, Borlaug N E and Dowswell C R 1997 A fertilizer-based green revolution for Africa. *In* Replenishing Soil Fertility in Africa. Eds. R J Buresh, P A Sanchez and F Calhoun. pp. 81–95. SSSA Spec. Pub. 51, Soil Sci. Soc. America, Madison, WI, USA.

Rajan S S S, Watkinson J H and Sinclair A G 1996 Phosphate rocks for direct application to soils. Adv. Agron. 57, 77–159.

Roder W, Calvert O and Dorji Y 1993 Effect of burning on selected soil parameters in a grass fallow shifting cultivation system in Bhutan. Plant Soil 149, 51–58.

Sakala W D 1998 Nitrogen dynamics in maize/pigeonpea intercropping in Malawi. PhD Thesis, Wye College, University of London.

Sakala W D, Cadisch G and Giller K E (2000) Interactions between residues of maize and pigeonpea and mineral N fertilizers during decomposition and N mineralization. Soil Biol. Biochem. 32, 679–688.

Sanchez P A 1976 Properties and management of soils in the tropics. John Wiley and Sons, New York, NY, USA.

Sanchez P A, Shepherd K D, Soule M J, Place F M, Buresh R J, Izac A-M N, Mokwunye A U, Kwesiga F R, Ndiritu C G and Woomer P L 1997 Soil fertility replenishment in Africa: an investment in natural resource capital. *In* Replenishing Soil Fertility in Africa. Eds. R J Buresh, P A Sanchez and F Calhoun. pp. 1–46. SSSA Spec. Pub. 51, Soil Sci. Soc. America, Madison, WI, USA.

Savini I 1999 The effect of organic and inorganic amendments on phosphorus release and availability from two phosphate rocks and triple superphosphate in phosphorus fixing soils. MSc Thesis, University of Nairobi, Nairobi, Kenya.

Sertsu S M and Sanchez P A 1978 Effects of heating on some changes in soil properties in relation to an Ethiopian land management practice. Soil Sci. Soc. Am. J. 42, 940–944.

Sharpley A N, Tiessen H and Cole C V 1987 Soil phosphorus forms extracted by soil tests as a function of pedogenesis. Soil Sci. Soc. Am. J. 51, 362–365.

Shepherd K D and Soule M J 1998 Assessment of the economic and ecological impacts of agroforestry and other soil management options on western Kenyan farms using a dynamic simulation model. Agric. Ecosyst. Environ. 71, 131–145.

Smaling E M A, Nandwa S M and Janssen B H 1997 Soil fertility in Africa is at stake. *In* Replenishing Soil Fertility in Africa. Ed. R J Buresh, P A Sanchez and F Calhoun. pp. 47–61. SSSA Spec. Pub. 51, Soil Sci. Soc. America, Madison, WI, USA.

Strømgaard P 1992 Immediate and long-term effects of fire and ash-fertilization on a Zambian miombo woodland soil. Agric. Ecosys. Environ. 41, 19–37.

Swift M J, Seward P D, Frost P G H, Qureshi J N and Muchena F N 1994 Long-term experiments in Africa: developing a database for sustainable land use under global change. *In* Long-term Experiments in Agricultural and Ecological Sciences. Eds. R A Leigh and A E Johnson. pp. 229–251. CAB International, Wallingford, Oxon, UK.

Swift M J, Heal O W and Anderson J M (Eds.) 1979 Decomposition in Terrestrial Ecosystems. Blackwell Scientific Publications, Oxford, UK. 372 pp.

Syers J K and Lu Ru-kun 1990 Inorganic reactions influencing phosphorus cycling in soils. *In* Phosphorus requirements for sustainable agriculture in Asia and Oceania. International Rice Research Institute, Los Baños, The Philippines, pp. 191–197.

Szott L T, Fernandes E C M and Sanchez P A 1991 Soil–plant interactions in agroforestry systems. For. Ecol. Manage. 45, 127–152.

Tinker P B 1980 Role of rhizosphere microorganisms in phosphorus uptake by plants. *In* The Role of Phosphorus in Agriculture. Eds. F E Khasawneh, E C Sample and E J Kamprath. pp. 617–654. Soil Sci. Soc. America, Madison, WI, USA.

van Veen J A, Ladd J N and Amato M 1985 Turnover of carbon and nitrogen through the microbial biomass in a sandy loam and a clay soil incubated with [^{14}C(U)]glucose and [^{15}N](NH_4)SO_4 under different moisture regimes. Soil Biol. Biochem. 17, 747–756.

Woomer P L and Muchena F N 1996 Recognising and overcoming soil constraints to crop production in tropical Africa. Afr. Crop Sci. J. 4, 503–518.

World Bank 1996a African Development Indicators. 1996 World Bank, Washington, DC.

World Bank 1996b Natural resources degradation in sub-Saharan Africa: Restoration of soil fertility. Africa Region World Bank, Washington, DC.

Young A 1997 Agroforestry for Soil Management, 2nd ed. CAB International, Wallingford, Oxon, UK.

J.J. Adu-Gyamfi (Ed.), Food security in nutrient-stressed environments: exploiting plants' genetic capabilities: 289–300, 2002.

Farmer preferences and legume intensification for low nutrient environments

S.S. Snapp[1,3] & S.N. Silim[2]
[1]Department of Horticulture, Michigan State University, East Lansing, MI 48824-1325, formerly ICRISAT, P.O. Box 1096, Lilongwe, Malawi. [2]Principal Scientist Agronomy, ICRISAT P.O. Box 39063, Nairobi, Kenya. [3]Corresponding author*

Received 10 May 2000. Accepted in revised form 14 December 2000

Abstract

Improved varieties of legumes adapted to nutrient deficiency have the potential to improve food security for the poorest farmers. Tolerant varieties could be an inexpensive and biologically smart technology that improves soils while minimizing fertilizer costs. Yet other technologies that improve productivity and appear to be biologically sound have been rejected by farmers. To translate benefits to smallholder farmers, research on low-nutrient tolerant genes and crop improvement must keep farmer preferences and belief systems in the forefront. We review farmer participatory research on legume-intensification and soil fertility management options for smallholder farmers in Africa, including recent results from our work in Malawi and Kenya. We suggest that indeterminate, long-duration legumes are the best bet for producing high quality residues, compared to short-duration and determinate genotypes. This may be due to a long period of time to biologically fix nitrogen, acquire nutrients, photosynthesize and grain fill. Also, the indeterminate nature of long-duration varieties facilitates recovery from intermittent stresses such as drought or pest pressure. However, indeterminate growth habit is also associated with late maturity, moderate yield potential and high labour demand. These traits are not necessarily compatible with smallholder criteria for acceptable varieties. Malawi women farmers, for example, prioritized early maturity and low-labour requirement, as well as yield potential. To address complex farmer requirements, we suggest the purposeful combination of species with different growth habits; e.g. deep-rooted indeterminate long-duration pigeonpea interplanted with short-duration soyabean and groudnut varieties. On-farm trials in Malawi indicate that calorie production can be increased by 30% through pigeonpea-intensified systems. Farmers consistently indicate strong interest in these systems. In Kenya, a 55% yield increase was observed for a doubled-up pigeonpea system (a double row of pigeonpea intercropped with three maize rows) compared to traditional, low density intercrops. However, the need for improved pigeonpea varieties with high intercrop suitability, including reduced early branching, was highlighted by a farmer preference study in the same area. These examples illustrate the potential for participatory research methodologies to drive biophysical research in farmer-acceptable directions.

Introduction

Varieties adapted to low fertility soils open up new possibilities for the most food insecure, resource-poor farmers. Smallholder farmers rarely have sufficient cash to invest in fertilizers, or labour to invest in making compost. In contrast to fertilizers, it is possible for even the poorest farmers to try out new varieties. The technology has the potential to reach more farmers because seeds can move from farmer to farmer, seeds are relatively low cost, and farmers can produce their own seeds. The limited investment requirement with new varieties improves adoption potential, but it is only one aspect of farmer acceptance. Understanding client perceptions and priorities is also crucial to facilitating farmer adoption (Chambers and Ghildyal, 1985; Sperling et al., 1993).

To investigate farmer perceptions, new participatory research approaches help elicit farmer selection criteria and strategies (Kitch et al., 1998). Im-

* FAX No: 517-432-2242. E-mail: snapp@msu.edu

This paper previously appeared in *Plant and Soil* **245**, 2002, pp. 181–192.

portantly, a participatory action research agenda can provide information to researchers about farmers beliefs while simultaneously facilitating farmer experimentation (Defoer et al., 1998). Socio-economic surveys are also a key tool to understanding farmer practice and belief (Ahmed et al., 1998). Farmers are not a uniform group by any means. We discuss examples of how gender can influence farmer preferences, such as demand by women farmers for varieties and systems with lower labour requirements (Kolli and Bantilan, 1997). Women farmers are also often concerned with, and may have a high degree of indigenous knowledge concerning, secondary uses of crops and quality factors, such as taste of leaves in bean and cowpea (Ferguson, 1994; Kitch et al., 1998).

In this paper, we focus on recent research evaluating legume-based technologies appropriate to smallholder farmers in Southern and Eastern Africa. Farmer preferences include complex criteria, such as: Adaptation to local conditions, yield security, cash returns, grain quality traits, tolerance to abiotic and biotic stresses, market context, production of secondary benefits (e.g. fuelwood, pesticides, soil fertility) and low labour requirements (Fischler et al., 1996; Sperling et al., 1993). Farmer criteria often incorporate a combination of traits, which may be negatively correlated or even mutually exclusive (CARE, 1998). One approach to this conundrum is for researchers to develop a wide range of varieties, and integrative systems that include both early and late maturity genotypes. Intercrop systems have been shown to reduce food security risk through compensation (Willey et al., 1997). We contend that intercrop systems can be designed to not only buffer against food shortages, but simultaneously to conserve land through longer foliage cover. Through case studies, we suggest that combining varieties with a range of growth types is required to address multiple objectives such as enhanced yield potential, reduced labour demand and soil enhancing properties.

Incorporating farmer involvement in a systematic fashion is a key factor to addressing the complex criteria held by smallholder farmers. Testing and farmer evaluation of varieties may be best done in the context of a multiple species cropping systems. Examples from the literature, and are own experiences, suggest that evaluation of variety performance in sole crops is inadequate: intercrop and rotation systems must be considered from inception (Ferguson, 1994; ICRISAT, 1998; Willey et al., 1997). This may be particularly true for varieties targeted at low soil fertility and environmentally stressed farming systems (Fischler et al., 1996). It is a challenging job to develop varieties that meet multiple farmer criteria, and tolerate low nutrient soils. Farmers require firsthand knowledge to assess intercropping and legume-intensification technologies, which often require substantial labour and skilled management. Furthermore, there may be collateral benefits, such as weed suppression, that farmers need to observe for themselves (Kanyama-Phiri et al., 2000; Snapp et al., 1998B).

Farmer perceptions of returns and constraints

Farmer preferences are determined by their perceptions of the returns possible, and the constraints they face. Researchers usually monitor returns by measuring crop yields per land area. However, farmers assess a range of complex traits, and their perceptions of returns may be different than researchers. Farmer perceptions and beliefs need to be assessed as well as monitoring performance and economic returns (Heong and Escalada, 1999). This particularly holds for crops grown to provide variety in diet and to provide secondary benefits such as building material, fuelwood, fodder, pesticide and medicinal. Cowpea is an example of a crop generally grown to meet multiple objectives, yet plant breeding efforts initially focused almost exclusively on grain yield and improving the harvest index (Kitch et al., 1998). This reduced the amount of forage produced, which conflicted with a key farmer objective, and it ignored the use of cowpea leaves for home consumption.

In 1998, a semi-formal survey was conducted in Malawi by one of the authors (S. Snapp) to elicit criteria for evaluating legume-intensified systems from 114 farmers through open-ended questions on negative and positive traits. Surveys were reviewed, and 99 farmer responses were complete enough to use for analysis. Results were summarized by categories for the most common responses, as a percentage of responses for negative and positive traits (after Kitch et al., 1998). The range of criteria used to evaluate legume ‘best bets’ are presented, for male farmers (Table 1) and for female farmers (Table 2). Overall, 50% of positive and 70% of negative traits were concerned with returns, where returns were yields and secondary products that enhanced food security, market sales or multiple minor uses. As the technologies were tested with farmers who had volunteered to test soil fertility management practices, it was expected that soil improvement would be one of the criteria

Table 1. Criteria used by male farmers in evaluating performance of soil fertility technologies in multilocational trials in 1997. The criteria were elicited by asking farmers to list positive and negative traits associated with each technology they had evaluated in on-farm trials

Positive traits	Technology[a]			
	Mz control	Maize/PP	Leg/PP	Maize/Tv
	Percentage of total responses[b]			
Less labour	1.2			
Labour less/2crops		3.1	3.8	
Less weeds		3.1	5.4	2.3
Increased land efficiency		2.7	3.4	
2 crops/area				
Few pests		1.1	0.8	
Good emergence	0.8			
Water conservation				4.5
Early Harvest	2.6	1.9	3.1	
Good yields/food security		9.6	10	4.2
Fuelwood		2.7	0.4	2.3
Increased soil fertility		7.3	6.9	5
Pesticide				0.8
Increased cash sales		5	6.1	
SUM	***4.6***	***36.5***	***39.9***	***19.1***
Negative traits				
Increased labour	0.6	0.6	1.7	5.1
Increased weeds	1.7			
Limited seed access		5.1	6.2	4.5
Expensive fertilizer	2.3			
Low yields	15.3	1.7	2.2	7.9
Slow growth/late harvest		3.9	6.7	1.1
Decreased soil fertility	1.1			
Pest problems (on crop & in storage)	1.7	3.4	1.1	1.7
Low market price (legumes)		1.1	5.1	
Market availability		1.1	2.8	
Livestock damage		6.7	4.8	
P'pea establishment		1.1	1.7	
SUM	***22.7***	***24.7***	***32.3***	***20.3***

[a]Technologies described in table 4. [b]There were 58 participating farmers in three locations, giving a total of 261 positive traits and 178 negative traits when asked to list and explain their evaluation critera.

mentioned. However, soil fertility was generally a minor component among farmer criteria, and women farmers in particular had limited interest in legume ability to improve soil fertility (Table 2). As has been observed in earlier work in Malawi, women farmers tend to prioritize cropping systems which produced the most food for consumption by children (Sahn et al., 1992).

Genotypes with reduced labour requirements is an important farmer objective, as illustrated by cowpea research in West Africa (Kitch et al., 1998) and bean research in East Africa (Sperling et al., 1993). These are some of the few documented examples of participatory research on legume improvement. Their results, taken together with our own experience, suggest that labour is the most important component of farmer criteria, after returns. Snapp's survey (as described

Table 2. Criteria used by female farmers in evaluating performance of soil fertility technologies in multilocational trials in 1997. The criteria were elicited by asking farmers to list positive and negative traits associated with each technology they had evaluated in on-farm trials

Positive traits	Technology[a]			
	Mz control	Maize/PP	Leg/PP	Maize/Tv
	Percentage of total responses[b]			
Less labour	0.9			
Labour less/2crops		4.3	4.3	
Less weeds		4.3	7.1	2.4
Increased land efficiency				
2 crops/area		2.9	3.8	
Few pests		1.4	0.9	
Good emergence	0.9			
Water conservation				1.3
Early Harvest	6.2	2.9	4.3	
Good yields/food security		10.5	11.4	1.3
Fuelwood		2.4	0.9	2.4
Increased soil fertility		5.3	6.1	2.4
Pesticide				0.8
Increased cash sales		3.2	5.2	
SUM	***8.3***	***37.2***	***44***	***10.6***
Negative traits				
Increased labour	0.3	0.6	2	8.6
Increased weeds	3.3			
Limited seed access		2.6	3.3	1.3
Expensive fertilizer	2			
Low yields	14.5	3.9	1.3	6.7
Slow growth/late harvest		4.6	6.6	3.3
Decreased soil fertility	2.6			
Pest problems (on crop & in storage)	1.3	2	1.1	1.3
Low market price (legumes)		0.5	4.6	
Market availability		0.3	2.1	
Livestock damage		5.9	4.6	4.6
P'pea establishment		1	3.1	
SUM	***24***	***21.4***	***28.7***	***25.8***

[a]Technologies described in Table 4. [b]There were 41 participating farmers in three locations, giving a total of 210 positive traits and 152 negative traits when asked to list and explain their evaluation critera.

above) indicated three different categories of farmer's concern regarding labour demands of different legume systems (Tables 1 and 2). Overall, labour was about the third most often cited criteria. Men farmers also indicated significant concern about the availability and cost of inputs. Women farmers were less concerned about soil fertility returns than men, and were particularly concerned about increased labour demands in this assessment of 'best bet' legumes to improve soil fertility (Table 2). Gendered labour constraints is supported by the findings of Kolli and Bantilan (1997), their research on groundnut production systems indicated that labour constraints is one of the biggest concerns of women farmers.

Soil improvement and genotype tolerance to low nutrient soils are a more variable concern among farmers, from strong interest to little or no interest. Farmer participatory research using local expert farmers to evaluate bean varieties found that tolerance to poor soils was the third most-often cited positive attribute of varieties chosen for home evaluation (Sperling et al.,, 1993). Kitch et al., 1998 used a very similar participatory methodology and found that farmers did not use nutrient tolerance as a selection criteria for cowpea. This may be due to species differences, as cowpea varieties generally already have a high tolerance for low nutrient soils, compared to common bean varieties that have limited tolerance for edaphic stresses (Snapp et al., 1998A; Wortmann et al., 1995). A number of researchers have suggested that only relatively well-off farmers have a strong interest in, and sufficient resources, to invest in soil ameliorating practices (Ghassali et al., 1999; Witcombe, 1999). For example, one village out of 11 in Northern Syria was found to have adequate resources and other traits – necessary but not sufficient conditions for them to rehabilitate degraded areas through seeding legumes (Ghassali et al., 1999). Only the wealthiest group of farmers in a Mali case study invested in soils through improved residue recycling (Defoer et al., 1998).

Although smallholder farmers have diverse criteria for judging genotypes and cropping systems, the point should also be made that the role of the market is increasing around the world. Kitch et al. (1998) speculate that an increasing market focus has tended to narrow farmer preferences for traits in new cowpea varieties. Farmers often capitalize on market opportunities with new varieties. Thus, a new variety may be a means to expand beyond their traditional varieties, that are already effectively addressing local nitches and tastes. This is illustrated in West Africa through a markedly uniform rating of new cowpea varieties by local expert farmers: About one-third of the selection criteria related to yield performance, one-third to quality preferences and one-third to labour-related criteria (Kitch et al., 1998). The ratings were generally constant across a wide range of locations, years and diverse socio-economic groups of farmers; the authors ascribe this uniformity to the strong market influence.

Multiple farmer criteria and agromorphological traits

Farmers often have complex criteria and are considering diverse agromorphological characteristics of varieties, and cropping systems (Fischler et al., 1996). An example of the tradeoffs farmers evaluate is presented from research conducted in partnerships with farmers in highland Central Malawi (Snapp et al., 1998A). The research evaluated the relative effectiveness of genetic tolerance and fertilizer strategies in bean systems. Soils of the region are moderately acidic (pH ∼ 5.3) and low in soil phosphorus and nitrogen status (Snapp, 1998A). A comparison was conducted on farmer fields to evaluate yield and biological nitrogen fixation of adapted bean varieties. Locally grown varieties were compared to varieties selected for performance in low nutrient soils, results of a decade long endeavor of CIAT and national scientists from five countries. Although there was some variation in tolerance to poor soils, the largest gains in bean productivity, by far, were obtained by adding 30 kg P fertilizer per ha (Snapp et al., 1998A). This indicates the long-term nature of genetic improvement work to improve variety tolerance to low nutrient soils. Combined use of targeted fertilizers and intercrop systems are important strategies, particularly for the near term: Genetic improvement may take many years to enhance variety tolerance to low nutrient soils.

To summarize findings regarding crop tolerance to poor soils, we constructed a matrix based on our on-farm trial data from Malawi. The matrix illustrates comparative performance in terms of potential soil benefit (estimated from crop residue nitrogen contribution) and yield under Malawi smallholder agriculture conditions (Figure 1). This indicated that legume crops vary in tolerance to low nutrient status. A general pattern emerged: Legume genotypes of short duration and determinate growth habit generally contributed minimally to soils, whereas long duration and indeterminate growth habit legumes were associated with soil benefits and tolerance to poor fertility (Figures 1 and 2A). Mucuna (*Mucuna pruriens* (L.) DC var. utilis) was one of the few crops which demonstrates consistent tolerance to poor soils, tremendous biomass production and high grain yields (Figure 2B). This may explain researcher observation in Benin of high farmer interest and a four-fold increase in experimentation with mucuna over 3 years (Versteeg and Koudokpon, 1993). In contrast, over the same time period, Versteeg and Koudokpon (1993) found moderate interest in a pigeonpea fallow system, and a decline of interest in hedgerow cropping systems. There are barriers to growing legumes in Malawi as well. This is indicated by the small proportion of cultivated land currently dedicated to legume production in Malawi: Less than 5% in the Southern Lakeshore area of Malawi, and 24% in the groundnut growing region of Central Malawi (S. Snapp, D. Rohrbach, H.A. Freeman and F. Simtowe, unpublished baseline survey, 1999). Yet, some smallholder farmers may be inter-

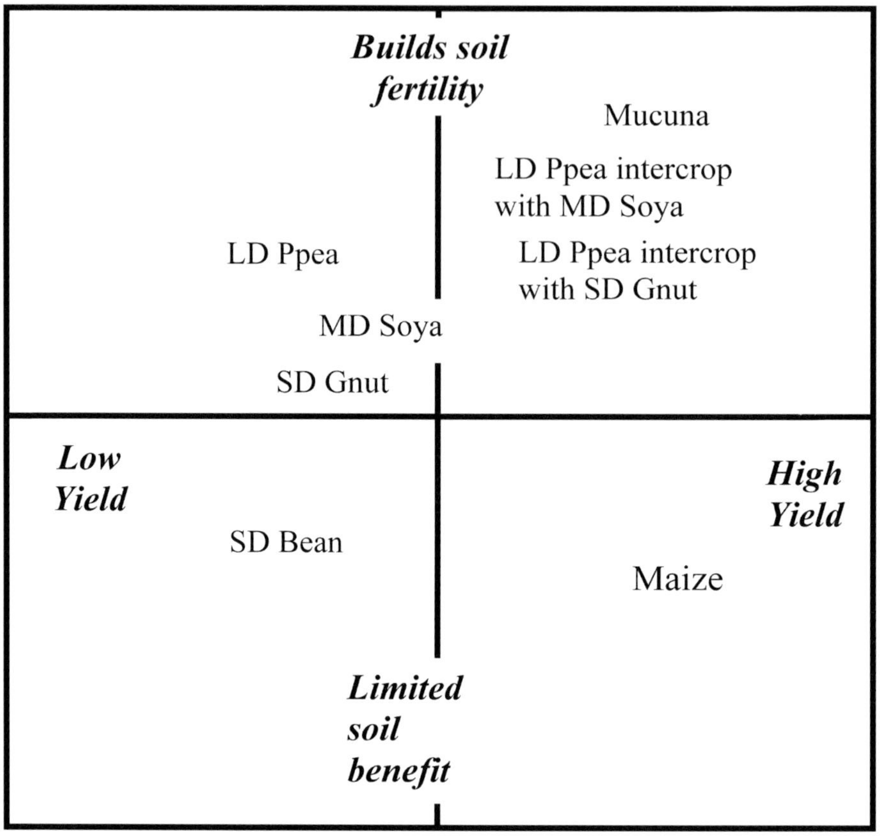

Figure 1. Schematic matrix to illustrate crop residue contribution to soil nitrogen and yield potential, from on-farm trials in Malawi. Crop varieties were categorized by growth duration. Ppea = pigeonpea, Gnut= groundnut, Soy=soyabean, bean = common bean, LD=long duration; MD=medium duration; SD=short duration.

ested in increasing legume production. The survey indicated that there was a 30% increase in experimentation with legume varieties and cropping systems over 2 years, with no increase in farmer experimentation with manures.

Contribution of crop species to improving soil is another assessment criteria, closely related to tolerance of poor soils. A clear distinction must be made between crop species that biologically fix N, and those that do not. Cereals, such as sorghum, and cassava may be very tolerant of low soil fertility, but they do not enhance soil N status. In contrast, legumes have the potential to both tolerate low soil fertility and contribute to soil rehabilitation. Legumes do not all contribute to ameliorating soil fertility, as a large amount of N is removed in grain and fodder. Recycling residues and growing legumes with high quality residues are crucial to deriving soil fertility benefits from legume intensification (Defoer et al., 1998).

Legumes with a indeterminate growth habit may be the best bet for smallholder farmers interested in ameliorating soils. Indeterminancy is related to high pest tolerance, consistent growth on low nutrient soils, and production of high quality residues. Indeterminacy improves pest resistance through a compensatory ability to regrow and thus mitigate pest damage. The soil benefits of indeterminate legumes include high nitrogen fixation rates, due to the ability to exploit favorable growth periods over a long growing season, and a protective soil cover from leaf biomass and senescent material. Our findings from on-farm trials in Malawi are similar to studies in Benin (Versteeg and Koudokpon, 1993): Aboveground, leafy biomass production levels of mucuna from 2 to 10 t/ha, and pigeonpea from 1 to 4.5 t/ha (and see Figure 2A).

Indeterminate legumes are thus recommended for ameliorating soils in low fertility, environmentally stressed conditions. However, indeterminacy is associated with two negative traits that must be considered.

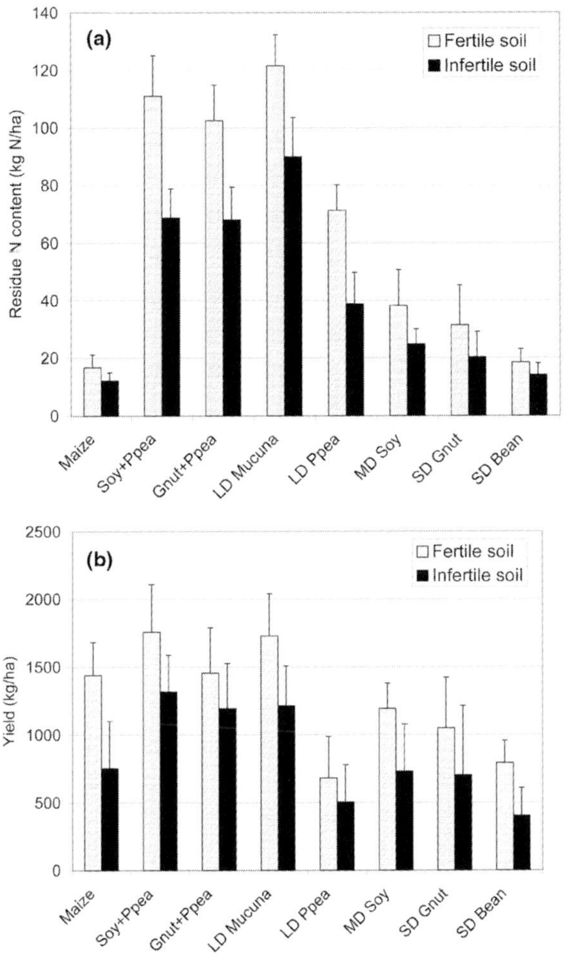

Figure 2. (A) Data from first year of a rotation and intercrop system comparison in Malawi, averaged across 1997–98 and 1998–99 growing seasons. Average and standard deviation (bars) are reported for residue nitrogen from trials conducted at high fertility sites (organic carbon above 1.4%) and low fertility sites (organic carbon below 1.4%). Trials carried out with 46 farmers, with the exception of groundnut (26 farmers) and soyabean (20 farmers). Above-ground crop residue dry weight biomass measurements were conducted at approximately 3 weeks before grain harvest. A sub-sample of residues were ground and analyzed to determine nitrogen percentage by acid digestion and nitrogen determination, to calculate total nitrogen content of residues. Trial design and preliminary results described in Kanyama-Phiri et al. (2000). Abbreviations were as in Figure 1. (B) Data from first year of a rotation and intercrop system comparison in Malawi, averaged across 1997–98 and 1998–99 growing seasons (see Figure 2A). Grain yield performance is reported as an average and standard deviation (bars) for high fertility sites (organic carbon above 1.4%) and low fertility sites (organic carbon below 1.4%). Trial description and preliminary results described in Kanyama-Phiri et al. (2000). Abbreviations were as in Figure 1.

That is: High labour demand at harvesting and a long period to wait before food is produced. An example is cowpea, where a higher labour demand is associated with a long duration, spreading growth habit (Kitch et al., 1998). It is interesting that farmers are interested in selecting for both types of cowpea, bush and spreading: These fit different niches (Kitch et al., 1998). Farmer ranking of groundnut traits suggests that bunch types are seen as less labour intensive for harvest compared to spreading types, and often are early producers (D. Boughton, pers. comm., 1997). Of particular concern for the unimodal semi-arid tropics, early maturing groundnut varieties can provide urgently needed food during the 'hungry time' before cereal harvest. In terms of tolerance to infertile soils, groundnut is somewhat unpredictable (Figure 2B). Some large seed types are associated with long duration genotypes that have limited seed fill characteristics, and poor yields under drought-stress or poor soil fertility, whereas other long-duration groundnuts are tolerant of low nutrients (Syamasonta, 1990). Providing a range of plant growth types–tailored to different marketing and household niches–may be an under-exploited approach for breeders to consider.

A wide range of pigeonpea varieties are grown in S.E. Africa, ranging from determinate to indeterminate types, and extra short duration to long duration types ratooned and grown over 2 years (ICRISAT, 1998). The potential of pigeonpea could be exploited further through developing varieties that maximize different features, such as: Intercropping performance (canopy architecture), extended soil cover over a long period, providing a bonus grain or vegetable crop, fuel wood and high quality residues. As pigeonpea is known to fix high amounts of nitrogen and has the ability to solubilize phosphorus from unavailable forms (Ae et al., 1990), emphasis should be given to selecting pigeonpea cultivars with high nutrient residues. Such improved pigeonpea could be used as a biological technology to improve soil fertility. Pigeonpea is one of the few crops with potential to enhance soils with minimal labour inputs and low seed costs, compared to other green manure and agroforestry species (Snapp et al., 1998B).

Designing intensified systems

Multiple evaluation criteria include short-term issues of food security, income generation and labour demands, and long-term issues of soil fertility regeneration and resource conservation. Designing improved varieties is one approach, where ideally low nutrient tolerant varieties can be bred which have minimal

Table 3. Pigeonpea and maize yields from on-farm trials at six sites in Semi-arid Eastern Kenya (ICRISAT, 1998). Standard error for pigeonpea yield: variety is 0.023, for agronomy (paired row versus single row) is 0.024, and variety by agronomy is 0.041. Standard error for maize yield: variety is 0.084, for agronomy is 0.078, and variety by agronomy is 0.118.

Agronomy	Improved Variety	Farmer Variety	Mean
	Pigeonpea Yield (t/ha)		
Improved: Paired Row	1.46	1.50	1.48
Farmer: Single Row	0.81	1.09	0.95
	Maize (t/ha)		
Improved: Paired Row	1.25	1.54	1.41
Farmer: Single Row	1.23	1.46	1.35

labour demands and contribute large amounts of biological fixed nitrogen through high quality residues. However, one variety cannot meet all demands, and combining genotypes in purposively designed intercropping systems is another approach (Willey et al., 1997).

Participatory on-farm research in Kenya has evaluated pigeonpea varieties (farmer and improved) grown in different intercrop systems with maize (ICRISAT, 1998). The commonly grown 'farmer pigeonpea intercropping system' involves one pigeonpea row per three rows of maize. This single row system was evaluated in comparison to a paired pigeonpea row (two rows grown close together, as one row) intercropped with three rows of maize. Results are promising as the paired row system had 59% higher pigeonpea yields than the single row system (Table 3). The intercrop systems were designed to incorporate pigeonpea as a bonus crop; where maize was the dominant food crop. Overall, this goal was met as the pigeonpea intercrop systems did not reduce maize yields, compared to sole crop maize. However, the improved pigeonpea variety, Kat81/3/3, was a spreading or 'bushy' type and it was not a good intercrop partner, with multiple branches shading maize and causing occasional yield reductions. In contrast, the farmer variety of pigeonpea was an excellent intercrop partner, not reducing maize yields (Table 3).

Researchers in Kenya have recently reviewed pigeonpea breeding program objectives, to incorporate a wider range of criteria, including: High yielding, and erect or semi-erect growth type with minimal 'bushy' characteristics (ICRISAT, 1998). Although there is a ways to go in selecting intercrop-adapted varieties some progress has been achieved. Pigeonpea varieties ICEAPs 00020,00040,00053 are now being evaluated in five countries, and have received a very favourable response from end users (Figure 2). The criteria used by farmers to evaluate long duration pigeonpea include 'intercropping ability', which was not present in improved variety 00053, the least favoured by farmers of the new pigeonpea varieties.

In Malawi, the dominant system is sole crop maize, and researchers are working with farmers to experiment with different systems that intensify the use of long duration, indeterminate legumes. Starting in 1996, an interdisciplinary group evaluated legume intensified cropping systems in over 100 on-farm trials using a satellite trial design where researcher evaluation of system performance is cross-checked with farmer evaluation (Kanyama-Phiri et al., 2000; Snapp, 1999). Promising intercrops combined long duration pigeonpea (LDP) with maize, and with a short duration groundnut variety (SDG), or with soyabean (Table 4).

The design of the SDG/LDP system was developed to combine SDG high yield potential and low labour demand (compared to spreading type groundnut varieties) with LDP soil coverage and biological nitrogen fixation. The SD groundnut met farmer criteria for early harvest, increased food security and exploitation of a market niche. The upright, bunch growth type of the short duration groundnut also requires less labour to harvest, as concentrated pods require less soil to be dug. However, the soil fertility contribution of SD groundnut can be limited (Figure 2A). Yields can also be erratic under nutrient deficient conditions, note high standard deviation (Figure 2B). Slow, early growth aboveground by long duration pigeonpea minimized competition with the short duration groundnut, and contributed a 'bonus' grain crop, as well as substantial soil fertility benefits (Figure 2A, B). Presumably, the intercropped pigeonpea fixed more nitrogen (due to a longer growth period), exploited nutrients below the more limited groundnut root system, and recycled nutrients which would have been lost after the SD groundnut harvest.

Results from this farmer participatory research indicated that doubling up SD groundnut and LD pigeonpea had consistently high land area equivalent ratios and calorie production was 28% higher than sole cropped, unfertilized maize (Kanyama-Phiri et al., 2000; Figure 2B). The MD soyabean/LD pigeonpea intercrop was successful at producing grain combined

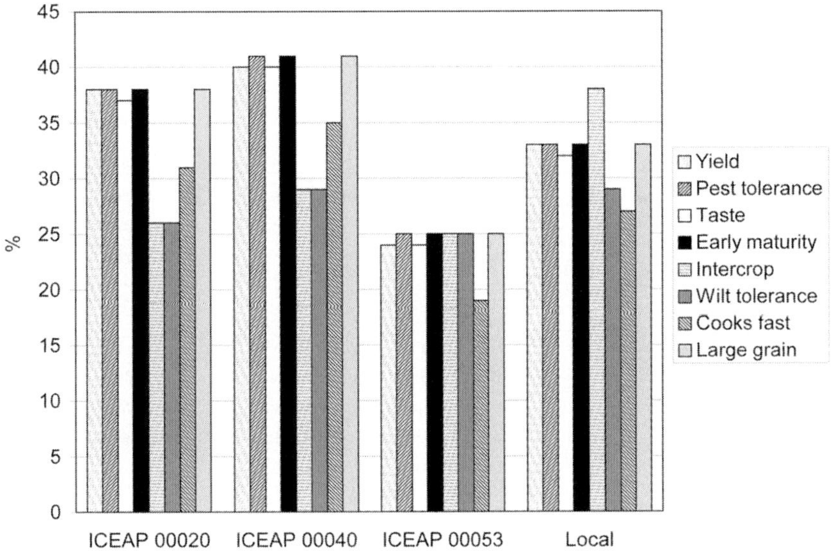

Figure 3. Results of preference rating exercise conducted with farmers in Kenya evaluating 5 pigeonpea varieties to identify desirable traits and compare improved and local pigeonpea varieties (African Development Bank Report No. 12, 1999).

with soil benefits as well (Figure 2A, B). Rating exercises conducted with farmers who had participated in the research demonstrated marked interest in the two pigeonpea intensified systems (Table 5). Women farmers in particular preferred 'doubling up legume systems' to their current cropping systems. The labour-efficiency of weeding doubled up legume systems was ranked high by farmers (Table 5). Farmer selection criteria highlighted problems as well, primarily due to concerns about high cost and unavailability of legume seeds (Tables 1 and 2). These legume systems are in a research, development and dissemination process with a focus on flexibility: they are potentially best bets among a basket of options for smallholder farmers. The emphasis is on a range of options which farmers can adapt to their needs, in contrast to blanket recommendations or bundled systems (Ahmed et al., 1997). Flexibility and options are particularly important for farmer adoption of soil improving technologies (Okali et al., 1994).

The question remains, what are the long-term environmental consequences of widespread adoption of low-nutrient adapted varieties and cropping systems? Questions have been raised about the sustainability of crop varieties that effectively exploit nutrients in a low-nutrient environment. Preliminary results from on-farm measurements of nutrient input and outputs in the intercrop systems indicate a complex picture: Intercropping maize with LD pigeonpea can enhance soil coverage, reduce erosion and significantly improve the phosphorus balance, compared to maize monoculture systems with little or no fertilizer inputs (Snapp, 1998B). It is possible that the ability of long duration, indeterminate pigeonpea varieties to solubilize phosphorus (effectively mining the soil for P), may be balanced by the enhanced soil cover and reduced erosion also associated with a pigeonpea intercrop system. The potential to biologically enhance availability of soil P is often related to ability to reduce losses, and thus increase retention of nutrients in plant-available forms (Snapp, 1998B).

Conclusions

The major challenge researchers face in designing cultivars and cropping systems that meet farmer preferences is that high performance is demanded, yet low labour inputs and edaphic stresses are the norm under smallholder environments. Given this, farmers require a wide range of crop types and systems to meet diverse demands.

Researchers should focus on identifying opportunities and designing a range of best bet options. This includes varieties that have low labour demands, as well as high yield potential in nutrient-poor soils. A range of short duration varieties, to meet needs for food and market opportunities early in the growing season, should be complemented with long-duration varieties that perform very well under edaphic stress

Table 4. Soil fertility improving, best bet legume intensification technologies, description in terms of biological and farmer considerations

Technology	Population density (×1000)	Biological characteristics	Farmer perceptions of characteristics
1. Maize control	Maize: 37	Maize hybrid MH18, three maize plants per planting stations, 0.9 m×0.9 m.	Current farmer practice throughout Malawi, productive with minimal labour
2. Maize + pigeonpea (PP) intercrop	Maize: 37 PP: 37	Temporal compatibility. PP variety ICP 9145 planted at the same time as maize, 3 plants per planting station spaced halfway between each maize station. PP grows slowly, which reduces competition with maize.	PP is a bonus crop, low density system minimizes impact on maize yields.
3. Legume + PP intercrop year 1, rotation with maize year 2 ***(Legume = groundnut or soyabean)***	G'nut: 74 Or Soyabean: 222 PP: 37	Groundnut variety JL 24 or CG 7 was grown as a single row 7 cm spacing on ridges spaced at 0.9 m spacing. To enhance residue biomass quantity and quality, Ppea is intercropped with the shortduration grain legume. Soyabean = double row of 10 cm spaced planted along each ridge. Indeterminate variety Magoye that does not require inoculum (nodulates with indigenous Rhizobium).	Legume seed density takes into account expense of g'nut seed and farmer-adoptable seeding rates. Ppea is a bonus crop. Higher density of soyabean is possible than groundnut, give the cost of the seed.
Maize + tephrosia relay intercrop	Tephrosia: 20 kg/ha Maize: 37	Temporal compatibility, enhanced by planting tephrosia at 1st weeding. Tephrosia has an initially slow growth habit. Green manure screening studies have shown the widespread adaptability of Tephrosia to Malawi agroecosystems, producing about 2 t/ha as a relay intercrop.	For a green manure system to be adopted by farmers, it must minimize labour required. Seed is broadcast along ridge and incorporated by weeding operation.

and provide secondary benefits such as enhanced soil fertility through high-quality residues, and weed suppression. Presenting farmers with the combination of short and long-duration varieties in intercrop systems should also be considered, and genetic improvement efforts need to take into account breeding for superior intercrop performance. Opportunities may exist to take advantage of occasionally extended growing opportunities, with bonus 'extra yield' potential from intercrops, combined with more consistent soil protection.

Acknowledgements

The Rockefeller Foundation is gratefully acknowledged for funding and all manner of support for the research carried out in Malawi. The authors are indebted to the many farmers, extension workers and research colleagues in Southern and Eastern Africa who have made possible this research on legume intensification cropping systems and management options for low nutrient soils.

Table 5. Farmer rating of technology traits across all sites, where scale used for rating was Very low=1, Low=2, High =3, Very high=4. Technologies were rated independently. Data from five Malawi case study sites with a total of 110 participating farmers, S. Snapp, 1999. Technologies described in Table 4

Location	Weeding & other labour Requirements	Seed availability	Contribution to food security	Contribution to cash sales	Contribution to soil fertility
Technology					
Mz	3.1	3.3	2.2	2.3	1.5
MzPP	2.5	1.9	3.4	2.9	3.1
GPP[a]	2.2	1.7	3.3	3.4	3.1
MzT	2.8	1.3	2.0	1.9	1.8
LSD	0.4	0.5	0.4	0.6	0.5

[a]GPP = groundnut/pigeonpea rotation technology for all locations except Bembeke where soyabean was the shortduration grain legume substituted for groundnut, due to the high altitude.

References

Ae N, Arihara J, Okada K, Yoshihara T and Johansen C 1990 Phosphorus uptake by pigeon pea and its role in cropping systems of the Indian subcontinent. Science 248, 477–480.

African Development Bank 1999 Improvement pigeon pea in Eastern and Southern Africa: project completion report No. 12, ICRISAT, Nairobi, Kenya.

Ahmed M M, Rohrbach D D, Gono L T, Mazhangara E P, Mugwira L, Masendeke D D and S Alibaba 1997 Soil fertility management in communal areas of Zimbabwe: Current practices, constraints and opportunities for change. Results of a diagnostic survey. Southern Eastern Africa Region Working Paper No. 6. International Crops Research Institute for the Semi-Arid Tropics (ICRISAT), Bulawayo, Zimbabwe.

CARE International 1998 Malawi Participatory Livelihood Assessment Exercise (Synthesis Report). CARE, Lilongwe, Malawi, mimeo. 75 pp.

Chambers R and Ghildyal B P 1985 Agricultural research for resource-poor farmers: The farmer-first-and-last model. Agricul. Admin. 20, 1–30.

Defoer T, De Groote H, Hilhorst T, Kante S and Budelman 1998 Participatory action research and quantitative analysis for nutrient management in southern Mali: A fruitful marriage. Agric. Ecosyst. Environ. 71, 215–228.

Ferguson A 1994 Gendered science: a critique of agricultural development. Am. Anthro. 96, 540–552.

Fischler M, David S, Farley C, Ugen M and Wortmann C 1996 Applying farmer participatory research methods to planning agricultural research: Experiences from Eastern Africa. J. Farm. Syst. Research-Extension 6, 37–54.

Ghassali F, Cocks P S, Osman A E, Gintzburger G, Christiansen S, Semaan A and Leybourne M 1999 Rehabilitation of degraded grasslands in North Syria: Use of farmer participatory research to encourage the sowing of annual pasture legumes. Expl. Agric. 35, 489–506.

Heong K L and Escalada M M 1999 Quantifying rice farmers' pest management decisions: Beliefs and subjective norms in stem borer control. Crop Protect. 18, 315–322.

ICRISAT 1998 Improvement of pigeonpea in Eastern and Southern Africa: Project completion report, December, 1998. P.O. Box39063, Nairobi, Kenya.

Katayama K, Ito O, Matsunaga R, Adu-Gyamfi J J, Rao T P, Anders M M and Lee K K 1995 Nitrogen balance and root behavior in four pigeonpea-based intercropping systems. Fertil. Res. 42, 315–319.

Kanyama-Phiri G Y, Snapp S S, Kamanga B and Wellard K 2000 Towards Integrated Soil Fertility Management in Malawi: Incorporating participatory approaches in agricultural research. Managing Africa's Soils No. 11. IIED, UK. 27 pp.

Kitch L W, Boukar O, Endondo C and Murdock L L 1998 Farmer acceptability criteria in breeding cowpea. Expl. Agric. 34, 475–486.

Kolli D R and Bantilan M C S 1997 Gender-related impacts of improved agricultural technologies: Identification of indicators from a case study. Gender, Techn. Devel. 1, 372–393.

Okali C, Sumberg J E and Reddy K C 1994 Unpacking a technical package: Flexible messages for dynamic situations. Expl. Agric. 30, 299–310.

Sahn D E, Van Frausum Y and Shively G 1992 The adverse nutrition effects of taxing export crops in Malawi. Cornell Food and Nutrition Policy Program, Working paper No. 29. Washington, D.C., USA 31 pp.

Snapp S S 1998A Soil nutrient status of smallholder farms in Malawi. Comm. Soil Sci. Plant Anal., 29, 2571–2588.

Snapp S S 1998B Phosphorus and sustainability of sub-Saharan Africa smallholder farms. *In* Phosphorus in Plant Biology: Regulatory Roles in Molecular, Cellular, Organismic and Ecosystem Processes. Eds. JP Lynch and J Deikman. pp. 59–72. American Society of Plant Physiologists, Madison, Wisconsin, USA.

Snapp S S, Aggarwal V D and Chirwa R M 1998A Note on phosphorus and genotype enhancement of biological nitrogen fixation and productivity of maize/bean intercrops in Malawi. Field Crops Res. 58, 205–212.

Snapp S S, Mafongoya P L and Waddington S 1998B Organic matter technologies to improve nutrient cycling in smallholder cropping systems of Southern Africa. Agric., Eco. Envir. 71, 187–202.

Snapp S S 1999 Mother and baby trials: A novel trial design being tried out in Malawi. Target Newsletter 17, 8.

Sperling L, Loevinsohn M E and Ntabomvura B 1993 Rethinking the farmer's role in plant breeding: Local bean experts and on-station selection in Rwanda. Expl. Agric. 29, 509–519.

Syamasonta M B 1990 A review of 'pops' research in Zambia. In Proceedings of the fourth regional groundnut workshop for

Southern Africa, 19–23 March, 1990, Arusha, Tanzania. pp. 17–21, ICRISAT.

Versteeg M N and Koudokpon V 1993 Participative farmer testing of four low external input technologies, to address soil fertility decline in Mono Province (Benin). Agric. Systems 42, 265–276.

Willey R W, Mohammed N and Morse S 1997 Factors affecting compensation in intercropping when one component fails. Trop. Agric. 74, 313–316.

Witcombe J R 1999 Do farmer-participatory methods apply more to high potential areas than to marginal ones. Outlook Agric. 28, 43–49.

Wortmann C S, Lunze L, Ochwoh V A and Lynch L 1995 Bean improvement for low fertility soils in Africa. African Crop Sci. J. 3, 469–477.

J.J. Adu-Gyamfi (Ed.), Food security in nutrient-stressed environments: exploiting plants' genetic capabilities: 301–311, 2002.

Nutrient balances: A guide to improving the sorghum- and groundnut-based dryland cropping systems in semi-arid tropical India

T. J. Rego, B. Seeling, V. Nageswara Rao, G. Pardhasaradhi, J. V. D. K. Kumar Rao, R. J. K. Myers & C. Johansen
Natural Resource Management Program, International Crops Research Institute for the Semi-Arid Tropics (ICRISAT), Patancheru 502 324, Andhra Pradesh, India

Key words: alfisols, cropping systems, crop uptake, farmers' fields, nutrient balances, nutrient mining, organic inputs

Abstract

Information on soil–plant nutrient balance in India and elsewhere is scarce and mostly generalized. Earlier nutrient balance studies and their criticism are discussed. An on-farm nutrient balance study 1995/1996 and 1996/1997, was designed to understand nutrient balances resulting from farmers' practices in semi-arid tropical regions of India. This study involved sorghum-based and groundnut-based cropping systems in low rainfall areas. Farmers were selected by systematic survey and analyses of factors affecting farmers' decision-making on nutrient inputs. Intensive plot-specific nutrient input and output measurements were made on 53 farmers' fields for sorghum-based systems and 45 farmers' fields for the groundnut-based systems. Topsoil mineral nitrogen (N) contents observed in 2 years at the beginning of the crop season at both locations were surprisingly high, and exchangeable potassium (K) contents also indicated sufficient supply in most fields. However, available phosphorus (P) in the majority of fields at both locations was just sufficient, and around threshold levels for most of the crops. The nutrient balance in sorghum-based systems indicates a moderate to high negative balance of K leading to soil mining for K supply with these systems. Nitrogen and P balances were generally positive. Although the groundnut-based system accumulates 53% of its N requirements through biological nitrogen fixation (BNF), negative balances of N and K were observed mainly due to low-input applications. Application of K together with N is suggested. Better-nodulating groundnut cultivars and efficient *Rhizobium* strains need to be introduced. The negative balance of K was estimated for sorghum-based systems. High positive balances were observed in the systems whenever a commercial crop was sown. The study also indicated farmers' preference to apply excess quantities of farmyard manure (FYM) to commercially important crops even in dryland farming systems.

Introduction

Nutrient balance studies in sub-Saharan Africa have shown that the depletion of soil nutrients is one of the most important constraints to sustainable agriculture in this region. The average nutrient loss was determined as 22 kg N, 2.5 kg P, and 15 kg K ha^{-1} year^{-1} in 1982–1984 and predicted to reach 26 kg N, 3 kg P, and 19 kg K ha^{-1} year^{-1} in the year 2000 (Stoorvogel et al., 1993). However, countrywide balances like these have limitations due to their scale. A district level study in Kenya determined a much greater depletion of 112 kg N, 3 kg P, and 70 kg K ha^{-1} year^{-1} (Smaling et al., 1993).

Information on nutrient depletion in India is surprisingly scarce. However, nutrient depletion in dryland agriculture is expected where input of purchased fertilizers and FYM are traditionally low (Jha and Sarin, 1984). Nutrient balances can be calculated on various scales, e.g., at national or regional level, on farms of a particular farming system, or in fields of a particular cropping system. Much of the published data in India is at the national and regional level and deals only with mineral fertilizer inputs and estimates of crop nutrient uptake.

Table 1. Nutrient balance in 1987–1988 in the Southern Plateau and Hills Region of India[a]

Region Nutrient	Nutrients (1000 t) Consumed	Utilizable	Removed by crops
Telengana			
N	169	68	65
P	45	7	17
K	14	9	81
Rayalaseema			
N	69	28	38
P	18	3	15
K	12	7	60
Karnataka			
N	212	85	178
P	54	8	44
K	61	37	230
Tamil Nadu			
N	144	57	73
P	27	4	19
K	57	35	95
Total			
N	594	238	354
P	144	21	95
K	144	88	466

[a]Modified from Das et al. (1991). Utilizable nutrients were calculated with nutrient use efficiencies, N=40%, P=15% and K=60%.

Large-scale fertilizer input–crop uptake balances

The gap between fertilizer inputs and nutrient removal in India has remained almost constant during the past 20 years, despite a steady increase in fertilizer consumption (Tandon and Narayan, 1990). In the years 1988/1989, crop nutrient ($N+P_2O_5+K_2O$) uptake was 20.7×10^6 t whereas fertilizer inputs amounted to 11.0×10^6 t leaving a gap of 9.8×10^6 t or 55 kg ha^{-1} (Biswas and Tewatia, 1991). There are considerable differences between the nutrient balances in different agroecological regions. The presentation of data as aggregated total nutrient (NPK) consumption and removal makes it impossible to identify the depletion of a specific nutrient. It also prevents any estimate of nutrient loss mechanisms other than nutrient removal with harvested products.

Large-scale fertilizer input–crop uptake balances corrected for nutrient use efficiency (NUE) and biological nitrogen fixation (BNF)

A different approach was taken by Das et al. (1991). In addition to listing N, P, and K separately, these authors assumed that 80% of the N uptake of chickpea and groundnut came from N_2-fixation (some crops had to be omitted from the nutrient balance because of missing data). This reduced the nutrient gap considerably. Nitrogen and phosphate showed positive balances, and only K application showed a considerable deficit (Table 1).

In order to take nutrient losses into account, Das et al. (1991) corrected fertilizer consumption data with fertilizer use efficiency (FUE) data (Table 1, Utilizable nutrients). However, this seems to be incorrect since FUE is only a measure of nutrients not taken up by plants and consists of nutrients lost and nutrients remaining in the soil. In the case of P and K, NUE of added nutrients is often quite low, but a large proportion of these nutrients remains in the soil and is eventually available for plants. Nitrogen losses reported from dryland systems were 7–27% on deep Vertisols (Moraghan et al., 1984a) and 30% on shallow Vertisols and Alfisols (Katyal, 1988; Moraghan et al., 1984b). Here too, the NUE of 40% for N is not the right correction factor as researchers cited it was measured only for the purpose of a nutrient balance. If this is taken into account then again only the K balance shows evidence of nutrient depletion.

Large-scale fertilizer input–crop uptake balances for different cropping systems

Fertilizer input is less favoured for rainfed systems than for the region as a whole. For example, although 84% of the cropped area in the Indian state of Karnataka is rainfed, only 22% of this land receives fertilizers that amount to only 23% of the total applied fertilizer-N (Table 2). The situation for P and K-fertilizers is about the same. In addition, FYM is applied preferentially to irrigated land and cash crops (Jha and Sarin, 1984).

Small-scale fertilizer input–crop uptake balances for different cropping systems

Although dryland systems seem to be more likely to have negative nutrient balances, most of the work on nutrient balances of specific cropping systems in India was done on irrigated systems, most likely because

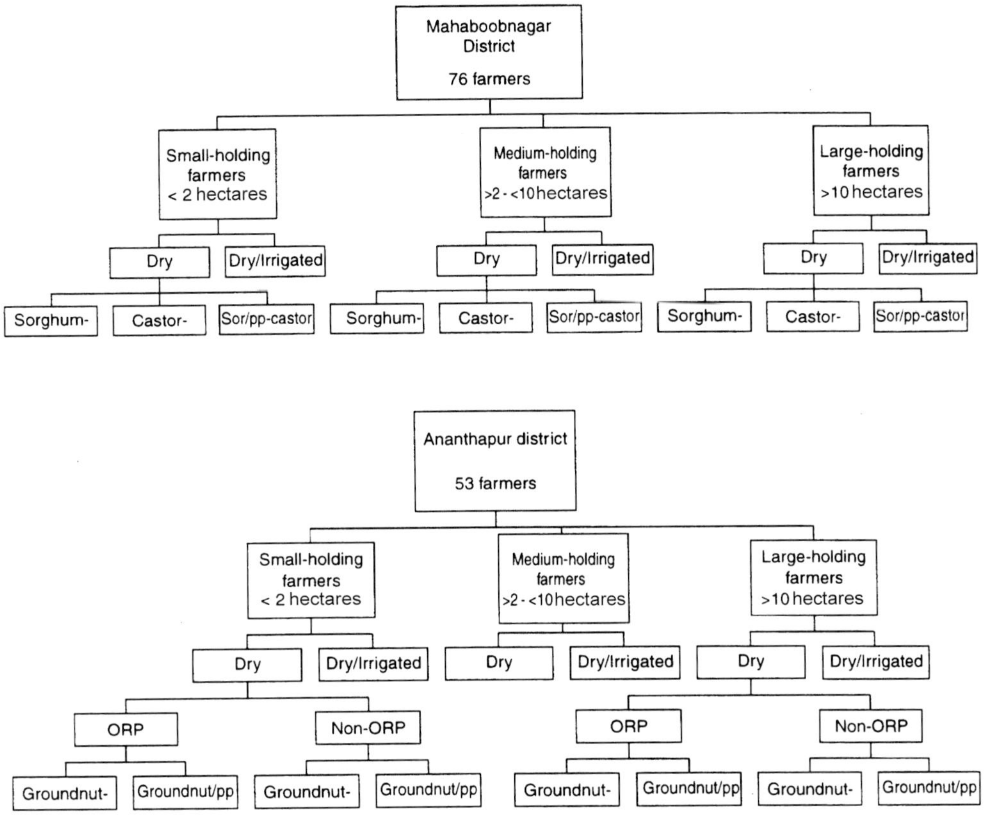

Figure 1. Systematic selection of farmers for nutrient balance studies in Mahboobnagar and Ananthapur districts of Andhra Pradesh, India during 1995–1997.

these systems are more important economically and for national food security. Another deficit of these nutrient balance studies is that most of the work in India was done on research stations with recommended fertilizer rates, and therefore is hardly relevant to the real-world situation.

Based on fertilizer input and crop uptake, Nambiar and Gosh (1987) calculated the K-balances of irrigated rotations in long-term experiments from several locations in India. Even the application of the recommended K-fertilizer dose was not sufficient to balance crop removal. Thakur et al. (1990) reported NPK balances from trials in northern India with maize–wheat, rice–wheat, soybean–wheat, and groundnut–wheat rotations fertilized at the recommended rates. The N and P balances were positive in all rotations with the exception of a slightly negative balance in the soybean–wheat rotation (-4 kg N ha^{-1} $year^{-1}$). However, the inconsistent response to K application in long-term fertilizer trials shows that most soils buffer crop demand for K for several years even in intensive cropping systems. This is further supported by the fact that the difference in extractable soil-K between fertilized and non-fertilized treatments remained fairly constant over a period of 11 years (Nambiar and Gosh, 1987).

The nutrient balances of intensive cropping systems are a good example of nutrient mining caused by the imbalanced application of nutrients. Imbalanced application of N and P increases biomass production if other nutrients are not limiting. For K, the applied amounts are often considerably less than the large quantities removed in crop residues and fodder, consequently the budget deficit is large. In rainfed cropping systems with low fertilizer inputs, yield level and nutrient mining depend more on the inherent soil fertility than on applied nutrients.

Nutrient balances in sorghum- and groundnut-based cropping systems on Alfisols in SAT India: A case study

In order to obtain reliable data on nutrient mining in dryland cropping systems of the Indian semi-arid trop-

Table 2. NPK application in rainfed agriculture of three states of the Southern Plateau and Hills Region of India[a]

State	Cropped area rainfed (%)	Rainfed area fertilized (%)	Total fertilizer nutrients applied to rainfed crops		
			N	P	K
Karnataka	84	22	23	26	24
Andhra Pradesh	54	13	10	12	14
Tamil Nadu	43	9	3	4	4

[a] Source: Tandon (1981).

ics (SAT) a nutrient balance study was conducted on farmers' fields. An analysis of the dryland cropping systems on Alfisols had shown that two cropping systems were of primary importance under semi-arid conditions, one sorghum-based and the other groundnut-based. The nutrient balance study was limited to these systems and to the major nutrients N, P, and K. Crop statistics were used to identify a typical district where these cropping systems are used.

Traditionally a rotation of sorghum and castor is grown in the rainy season on Alfisols, in some parts of SAT India. Alfisols in these regions are typically shallow or very shallow (0.30–1.00 m) with low water-holding capacities ranging from 50 to 180 mm. The prevalent cropping systems are about 110–135 days in duration, as the rainy season starts by the second week of June and recedes by the end of September. The annual rainfall varies between 540 and 705 mm.

The term 'rotation' is applied loosely in this context, in the sense that farmers *intend* to rotate sorghum with castor. The frequency of the rotation depends on the quality and the location of the land, on market prices, and on the availability of seed. Sorghum and castor are frequently grown as intercrops with pigeonpea in this system. In addition to being a subsistence food grain, the sorghum crop is the most important source of fodder for draft and milch animals in a farming system that still relies largely on draft animals for farm operations.

The center of the groundnut-growing area is Ananthapur district in the southern tip of Andhra Pradesh state. But the groundnut-based system is also followed in adjoining districts where rainfall patterns are similar. Intermittent dry periods occur after the emergence of the crop, and part of the area is covered by extremely shallow soils of low water-holding capacities, that aggravate the problems of drought. Groundnut seems to be the only dryland crop that can tolerate this unfavourable rainfall pattern. For this reason, and because groundnut is a valuable oilseed cash crop, the area is basically a groundnut monoculture.

Table 3. Initial survey questions

Family size and structure, non-farm income, hired labour
Farm size
Livestock
Soil types (area)
Dryland/irrigated land (area)
Dryland crops (area, fertilizer application, FYM application)
Irrigated crops (area, fertilizer application, FYM application)
Fallow (area)

Material and methods

Initial survey and selection of the villages

The basic cropping systems on Alfisols in SAT India were determined and delineated through the use of crop statistics and geographic information systems (GIS). The sorghum- and the groundnut-based systems were selected because they were the most important dryland cropping systems based on ICRISAT mandate crops. Subsequently an initial survey of each system was carried out in two logistically suitable locations to determine if these locations were representative of the target system. The initial survey covered demographic and agricultural information (Table 3), and was carried out with the support of local NGOs and government organizations. Based on the survey, the villages Nusikottala and Krishnamreddypally in Ananthapur district were selected for the groundnut-based system, and Amisthapur, Elkicherla, and Seripalli villages of Bhootpur mandal in Mahboobnagar district for the sorghum-based system.

In Mahaboobnagar district, a sample group of 76 farmers were surveyed and stratified (Figure 1) according to farm size as; small (<2 ha), medium (2–10 ha), and large (>10 ha). Within this sample, 53 farmers from three villages were identified for the study. Accessibility to dry and irrigated land or only

Table 4. Area and proportion of crops in the sorghum- and groundnut-based cropping systems (sample group) in 1994/1995

	Mahaboobnagar		Ananthapur	
	ha	%	ha	%
Sorghum	43.4	28		
Sorghum/pigeonpea intercrop	8.5	5		
Castor	65.9	42		
Cotton	8.7	6		
Groundnut	5.5	3	235.5	87
Groundnut/pigeonpea	–	–	29.7	11
Sunflower	4.9	3		
Maize	1.2	1		
Horse gram	0.8	1		
Finger millet	1.6	1		
Fallow/others	8.1	5	5.4	2

dryland was also considered in sub-grouping the farmers. Farmers who own irrigated land generally have more investable resources for agriculture and their preference for applying FYM to crops influenced the selection. The subgroups of farmers were further classified based on the cropping systems they grew in 1995/1996 rainy season. Farmers selected for each cropping system were directly proportional to the percentage of farmers who showed preference for the cropping system in the main sample. The average proportion of each sub-group was determined based on the initial survey (Table 4). The nutrient balances of the fields were determined in the second season regardless of the crop growing in term. In Bhootpur mandal 68% of the farmers also sow irrigated crops. The selection of farmers and fields were similar for the groundnut-based system in Ananthapur, except that there was another criterion, i.e., farmers' exposure to agricultural extension activity through operational research projects of the Agricultural Research Station, Ananthapur (Figure 1). In Ananthapur district, 45 farmers from two villages were stratified according to the above criteria.

Soil samples collection and analyses

Initial soil samples of 10 cores were taken in each field in consecutive years before the beginning of the rainy season. In the first year, the full profile depth was only sampled in selected fields. In the second year, all the fields were sampled to the depth of the 'murram' layer (semi-weathered stratum). The soil samples were analyzed for mineral N content (Keeney and Nelson, 1982), available P content (Olsen and Sommers, 1982), and exchangeable K (Thomas, 1982).

Nutrient input and output measurements

The authors attempted to establish a partial nutrient balance for N, P, and K for the two cropping systems in the 1995/1996 and 1996/1997 cropping seasons. The conceptualized five inputs (mineral fertilizers, organic manure, atmospheric deposition, BNF, sedimentation) were identified and five outputs (harvested product, residue removal, leaching, gaseous losses, water erosion) for an ideal nutrient balance study determined. Of these factors, nutrient application with mineral fertilizers, organic manure, BNF through nodulation, and the nutrient removal from harvested products and residue removal were measured. Inputs from deposition and sedimentation were not measured, nor were outputs due to factors other than harvest products and residual removal. Reliable estimates could not be made on these variables due to the lack of research data.

During both cropping seasons, at the beginning of the season and at harvest times, plot-specific input data was collected from individual farmers. This was done with specific reference to the survey number of the plot, extent of land, quantities of FYM, sheep penning, organic cakes, or mineral fertilizers applied, and an estimation of BNF by the natural abundance method (Shearer and Kohl, 1986) for groundnut in Ananthapur. BNF was not measured in intercropped pigeonpea in the sorghum-based systems. Outputs were calculated through the harvest of crop samples in terms of grain and fodder, and related to individual farmer-obtained yields. Nutrient inputs and crop yields for the preceding crop season were also obtained to understand the effect of large quantities of FYM applied into the system by several farmers. Samples of organic manure were taken and analyzed for NPK content. In the case of mineral fertilizers, the fertilizer type was determined during interviews with farmers. The NPK contents based on manufacturers' grading of these fertilizers were considered in the calculation.

The number of fields harvested is not identical to the number of fields and crops selected for the study, since farmers frequently change their plans according to weather conditions. In addition, crop failure later in the rainy season due to intermittent drought, disease or pest pressure may completely prevent planting in the current year. This may account for the relatively higher percentage of fallow mentioned in the official crop statistics compared to the percentage of planned fallow observed in the initial survey.

Crop nutrient removal was generally determined through crop cuttings. Four sub-samples per field were taken. Each sub-sample consisted of two 2-m rows. In some cases it was not possible to obtain yield data, since farmers had already harvested the field. In these cases farmers supplied the yield information and a sample of the plant material was taken for mineral nutrient analysis. Grain and vegetative parts were analyzed separately. Plant N and P contents were determined with a Technicon® auto analyzer and K contents with an atomic absorption spectrophotometer after digestion of plant samples in conc. H_2SO_4 containing 0.5% Se at 360 °C.

Results

Soil nutrient content before the beginning of the rainy season

Mineral nitrogen

The mineral N content before the beginning of the rainy season was surprisingly high in both years and locations. The most frequent topsoil N content (0–15 cm) was 10–15 mg N kg^{-1} in 1995 and 15–20 mg N kg^{-1} in 1996 (Figure 2). The distribution of the Mahboobnagar data resembled a normal distribution, but the Ananthapur data showed greater frequencies at lower mineral N contents. The conjecture that a greater proportion of legumes in the cropping system might lead to higher mineral-N contents in the topsoil of the Ananthapur fields than in the Mahboobnagar fields was not confirmed. However, the mineral-N content in the topsoil says little about total soil-N supply, for that, the mineral-N content in the soil profile provides more information (Figure 3). To obtain this information, the average mineral-N contents of each layer, and for each location were calculated from the individual values for each field. The values for the different layers are calculated from different numbers of replicates due to differences in sampling depth. But small standard errors demonstrate that the number of replicates was sufficient even at lower depths. It also demonstrates that, despite the differences in the topsoil-N content of individual fields, the large number of replicates allowed calculation of a mean concentration for the topsoil that adequately described the most frequently observed topsoil mineral-N content. Nevertheless, the average N-contents in Figure 3 and particularly the summation of the amounts of N have to be considered with care, since they depict a soil profile up to 90 cm in Mahboobnagar and up to 120 cm in Ananthapur. In many fields, the soil was considerably shallower.

Available phosphate

Available phosphate (Olsen and Sommers, 1982) in the majority of fields was below 10 mg P kg^{-1} in the topsoil (0–15 cm) For most crops this is the threshold value for sufficient P supply (Figure 4). The largest number of fields had a P content of 5–7.5 mg P kg^{-1} in Mahboobnagar and 2.5–5.0 mg P kg^{-1} in Ananthapur. However, in both years a number of fields in Mahboobnagar tested high in available P with Olsen-P contents above 15 mg P kg^{-1}.

Exchangeable potassium

Exchangeable K contents indicated that the K supply was sufficient in most fields (Figure 5). In Mahboobnagar, most fields had a medium K status (50–130 mg K kg^{-1}). Low K contents ($<$50 mg K kg^{-1}) were only found in five fields in 1996. High K contents ($>$130 mg K kg^{-1}) were found in six fields in 1995 and nine fields in 1996. In Ananthapur, most fields had a medium K status, but, six fields had a low K status in both years, and one had a high K status in 1996.

Nutrient input

Nitrogen inputs in sorghum-based 2-year rotations ranged between a low of 87 kg N ha^{-1} and a high of 218 kg N ha^{-1}. These wide variations in input application resulted from the inclusion of castor, cotton, or

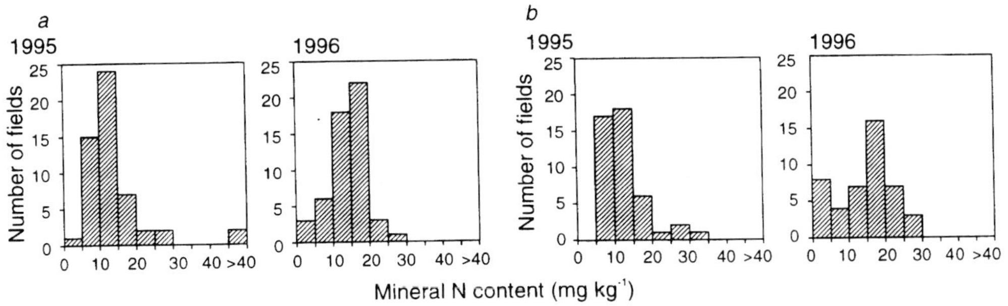

Figure 2. Mineral N content in the topsoil (0–15 cm) of fields selected for the nutrient balance study in (a) Mahboobnagar district (53 fields) and (b) Ananthapur district (45 fields).

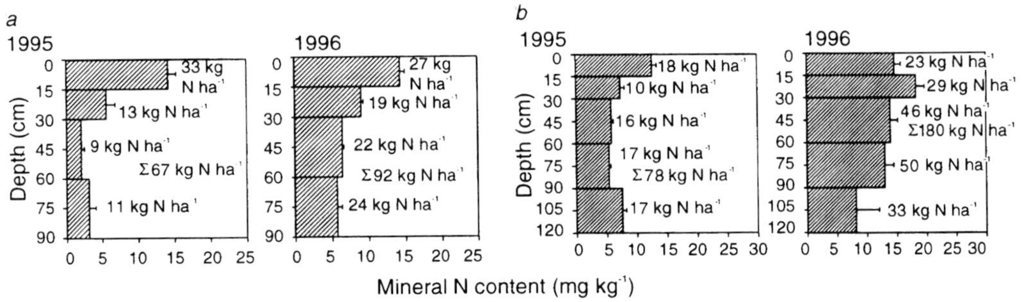

Figure 3. Average mineral N content in the soil profiles of fields selected for the nutrient balance study in (a) Mahboobnagar district (53 fields) and (b) Ananthapur district (45 fields).

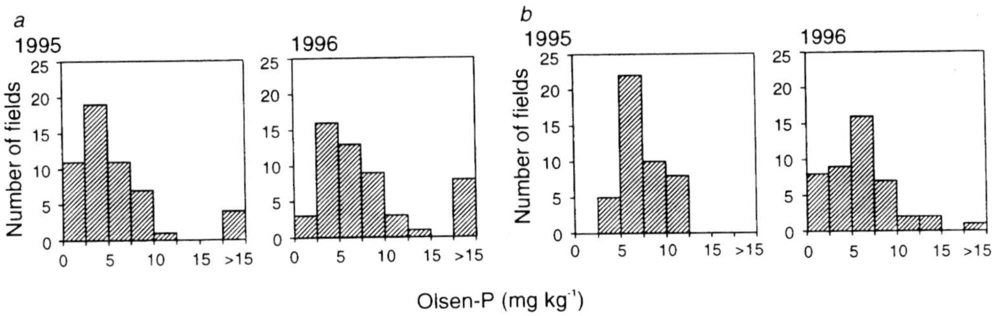

Figure 4. Available P content (Olsen-P) in the topsoil (0–15 cm) of fields selected for the nutrient balance study in (a) Mahboobnagar district (53 fields) and (b) Ananthapur district (45 fields).

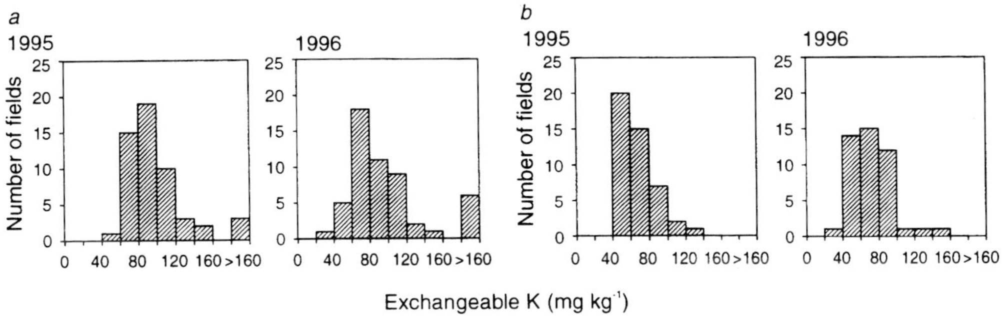

Figure 5. Exchangeable potassium content (exch. K) in the topsoil (0–15 cm) of fields selected for the nutrient balance study in (a) Mahboobnagar district (53 fields) and (b) Ananthapur district (45 fields).

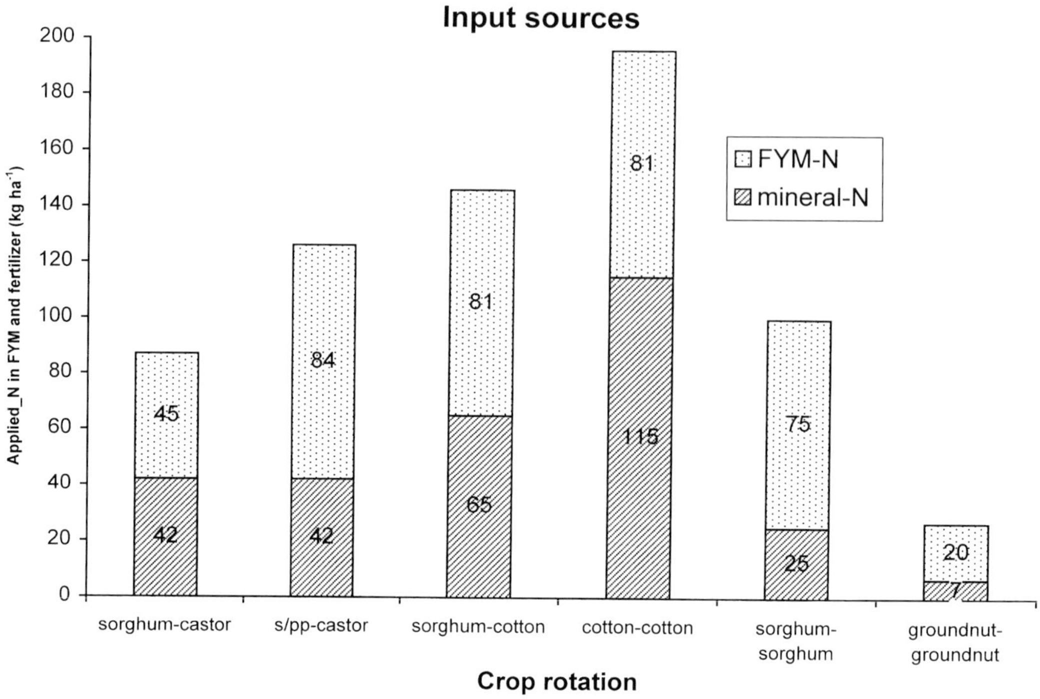

Figure 6. Nitrogen inputs in the form of fertilizer-N and farmyard manure into different rotations for 2 years.

sorghum intercropped with pigeonpea in the rotation. This is a clear indication of farmers' preference to apply high inputs to commercial and highly remunerative crops even in dryland systems. Another interesting aspect is that most of the N in this input pool came from FYM application (Figure 6). Maize crops also received farmers' attention in terms of N inputs as the market demand and price for grain and fodder were attractive. The sole sorghum rotation received very little attention in terms of N inputs. Groundnut being a leguminous oilseed crop, farmers generally apply N in the smaller quantity that comes in part as a complex fertilizer (e.g., di-ammonium phosphate, 18 N–46 P_2O_5–0 K), and the remainder in the form of FYM. Hence most of the N applied to the groundnut crop came from FYM.

Crop yields

The large rainfall and particularly rainfall distribution differences between years might also be the reason for the yield differences observed between years (Table 5). In 1996, early rains followed by intermittent drought caused failure of the first sorghum sowing in many fields. Later sorghum sowings suffered the shoot fly problems typical for this region. Other, slower-growing crops like castor, cotton, and pigeonpea suffer less from drought and late sowing because their initial water requirements are smaller, they have deeper root systems, and are not affected by the same pests.

In the groundnut-based system, the yield differences between the two sample villages and the 2 years were small. That indicates the groundnut system has greater yield stability than the sorghum-based system. However, for the large number of replicates the high standard error shows that the yield differences between individual fields were still large.

Nutrient uptake

The crop or cropping system with the highest N-uptake was the sorghum/pigeonpea intercrop. However, part of this N comes from atmospheric N fixation. Of the crops that rely only on applied or soil-N, cotton in 1995 and maize in 1996 were the crops with the highest N uptake. Phosphorus uptake was highest in the sorghum/pigeonpea intercrops followed closely by sole sorghum crops. The sorghum/pigeonpea intercrops also had the highest K-uptake, surpassed only by maize in 1996.

Table 5. Average yields (cotton, lint and seed, groundnut, pods) in the sorghum-based (Mahboobnagar district) and groundnut-based cropping system (Anantapur district)

System	1995			1996		
	Replicate	Yield (kg ha^{-1})	SE mean	Replicate	Yield (kg ha^{-1})	SE mean
Sorghum-based						
Sorghum	16	2079	329	14	618	67
Castor	20	331	65	15	459	46
Cotton	8	428	81	11	499	53
Pigeonpea	14	449	110	4	403	216
Maize	1	0	na	2	1977	494
Groundnut-based						
Groundnut	45	1427	66	44	1158	59

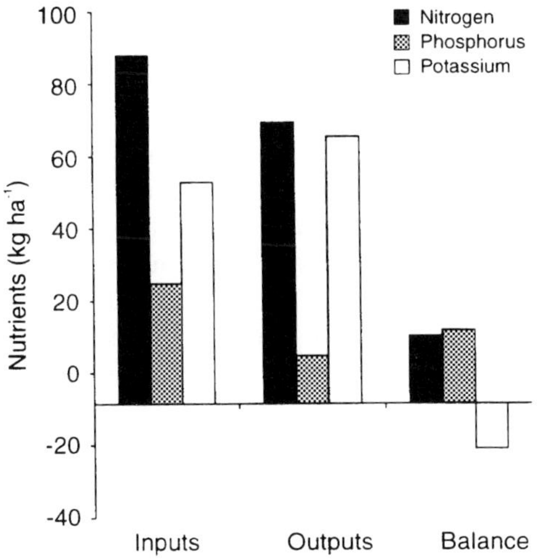

Figure 7. Soil–plant NPK balances observed in farmers' fields with 2-year sorghum–castor rotation as managed by farmers in Mahaboobnagar district of Andhra Pradesh, India during 1995/1996 and 1996/1997 cropping seasons.

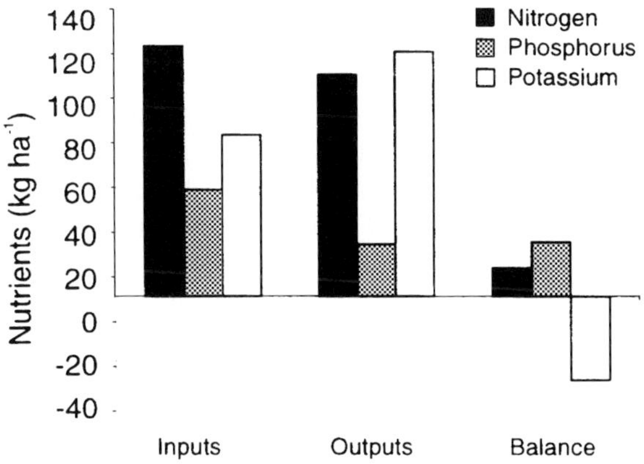

Figure 8. Soil–plant NPK balances observed in farmers' fields with groundnut monocropping as managed by farmers in Ananthapur district of Andhra Pradesh, India during 1995/1996 and 1996/1997 cropping seasons.

Nutrient balances

From the results, overall nutrient balances were estimated in a few of the cropping systems. In a sorghum–castor rotation, the N, input was 87 kg ha^{-1}, as against an output of 77 kg N ha^{-1}, indicating a net annual gain of +10 kg ha^{-1} (Figure 7). As N losses were not estimated in this study, and they may have been quite significant, the +10 kg N ha^{-1} might not indicate the correct N balance. For P and K that are less susceptible to loss than N, the estimates clearly reveal the positive P and negative K balances of the cropping system. With an improved sorghum/pigeonpea–castor system, the nutrient balances of N (+24), P (+26), and K (−16) kg ha^{-1} were estimated (Figure 8). The positive balance for N does not include an estimate for BNF of pigeonpea, which, if included, may increase the positive N balance. Nutrient balances for the mirror images of the two systems, namely castor–sorghum (+45 kg N, +35 kg P, +1 kg K ha^{-1}), and the castor–sorghum/pigeonpea (+58 kg N, +44 kg P, +32 kg K ha^{-1}) were positive. But these NPK balances seemed to be different in their object systems due to seasonal variability. Over a longer period, similar nutrient balances were observed in a long-term cropping systems experiment at ICRISAT, Patancheru, India.

Continuous cropping of sole groundnut was on average positive for P (+37 kg ha^{-1}) and negative for K (–20 kg ha^{-1}) (Figure 9). For N when an estimate of 1BNF was not included the balance was negative

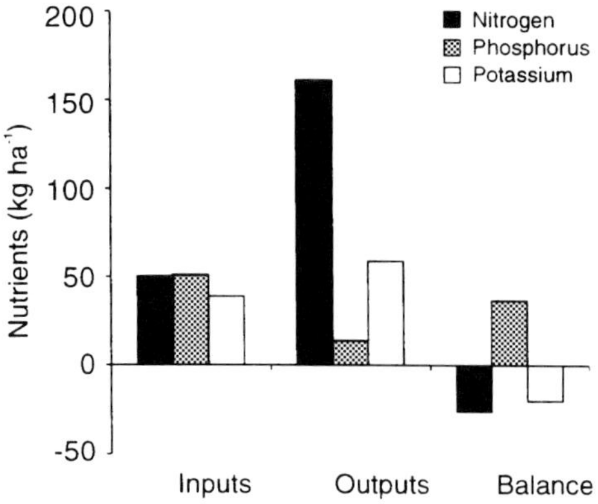

Figure 9. Soil–plant NPK balances observed in farmers' fields with sorghum/pigeonpea–castor as managed by farmers in Mahaboobnagar district of Andhra Pradesh, India during 1995/1996 and 1996/1997 cropping seasons.

(–111 kg ha^{-1}). However, observations indicate that BNF in 20 farmers' fields ranged between 41 and 65%, and averaged around 53% of the total N uptake. If the average BNF of 53% contribution to groundnut N uptake is considered, then the balance is also negative (–26 kg ha^{-1}). In order to have a positive N balance the BNF would need to be >68% of the total uptake. The soil analysis data from these farmers' fields indicate very little build up of soil N in spite of groundnut monocropping for more than 40 years. The less-efficient BNF in the farmers' fields may be one of the causes for poor build up of soil-N, while there is scope to increase BNF up to 92% (Peoples and Herridge, 1990).

A 2-year rotation of cotton, and the sorghum/pige onpea intercrop system showed reasonably high positive balances of N, P as indicated by +129 kg N ha^{-1} and +78 kg P ha^{-1}, mainly due to the application of fertilizer inputs to cotton. Farmers in this region grow medium-duration (135–150 days) cotton and castor cultivars. Farmers tend to apply more than the recommended quantities of fertilizers and FYM to either cotton or castor, that can not be effectively utilized due to the slow initial growth habit of these crops. During the later stages of crop growth, they generally suffer drought stress and the reduced growing season also affects nutrient uptake by these crops in this region. Nutrient residues thus remain in the soil for the next season, and help sequential crops in the rotation. Potassium imbalance has emerged as a common constraint in allmost all the cropping systems. The intensity and the extent of the imbalance will be pronounced in course of time and is based on the K-supplying capacity of the soil.

Discussion

There is a need to emphasize some of the salient features of this study considering the criticism evoked by the previous studies on nutrient balances. This study was mainly conducted in two districts representing the drier regions of Telangana and Rayalaseema in Andhra Pradesh. Among several cropping systems cultivated in these regions, nutrient balances were studied in both traditional and intensively cultivated improved systems. All the three major nutrients were studied individually for their balance in the soil–plant system of a crop rotation. Biological fixation of N in different farmers' fields was also measured in the groundnut-based system. The entire study was conducted on exclusively dryland cropping systems, although due consideration was given to the fact that a proportion of the farmers had access to irrigation. This study is free from calculating nutrient balances from experimental plots to which recommended levels of fertilizers were applied. Since the study comprised of a cross-section of farmers who had applied both fertilizers and FYM, based on their own perceptions and available resources, the data obtained was more realistic for consideration by policy makers. However, the variability in nutrient balances caused by seasonal variation and the resultant crop production, would be narrowed down if a longer-duration study were conducted using similar methodology.

Conclusions

This study provides factual information on the nutrient balances of different cropping systems on farmers' fields, that are typical of the region. The decreasing efficiency of BNF in groundnut monocropping over a long period suggests that more efficient BNF *Rhizobium* strains should be used to inoculate crops in that region. Measuring N losses precisely, using isotope ^{15}N would help in understanding the N dynamics in these environments. Phosphorus inputs met the crop requirements in most of the cropping systems that were studied. Potassium application should be recommended together with N input in dryland cropping systems, either through FYM or fertilizer inputs in order to reduce nutrient mining. Positive nutrient balances in maize- and cotton-based cropping systems should be attributed to farmers' preferences for fertilizer application to commercial crops.

References

Biswas B C and Tewatia R K 1991 Nutrient balance in Agro–Climatic regions of India an overview. Fertil. News 36, 13–17.

Das S K, Sharma K L, Neelam Saharan and Bhaskara Rao U M 1991 Nutrient balance and sustainable Agriculture in Southern Plateau and Hill Region of India. Fertil. News 36, 44–49.

Jha D and Sarin R 1984 Fertilizer use in Semi-arid Tropical India. ICRISAT Research Bulletin No 9, International Crops Research Institute for the Semi-Arid Tropics, Patancheru, Andhra Pradesh, India. 76 pp.

Katyal J C 1988 Nitrogen fertilizers – their use and management in the Indian semi-arid tropics. *In* Proceedings of a Colloquium Soil Fertility and Fertilizer Management in Semi-arid Tropical India. Ed. C B Christianson. pp 61–70. International Fertilizer Development Center (IFDC) Muscle Shoals, AL, USA.

Keeney D R and Nelson D W 1982 Nitrogen-inorganic forms. *In* Methods of Soil Analysis. Part II. 2nd edition. Eds A L Page, R H Miller and D R Keeney. pp 643–698. Am. Soc. Agron. Soil Sci. Soc. Am. Madison, WI, USA.

Moraghan J T, Rego T J, Buresh R J, Velk P L G, Burford J R, Singh S and Sahrawat K L 1984a Labelled nitrogen fertilizer research with urea in the semi-arid tropics. 2 Field studies on a Vertisol. Plant Soil 80, 21–23.

Moraghan J T, Rego T J and Buresh R J 1984b Labelled nitrogen fertilizer research with urea in the semi-arid tropics. 3 Field studies on an Alfisol. Plant Soil 82, 193–203.

Nambiar K K M and Ghosh A B 1987 Crop response to K and soil K balance in long-term fertilizer experiments. PR II Res. Rev. Series 4, 125–134.

Olsen S R and Sommers L E 1982 Phosphorus. *In* Methods of Soil Analysis. Part II. 2nd edition. Eds A L Page, R H Miller and D R Keeney. pp 403–430. Am. Soc. Agron. Soil Sci. Soc. Am. Madison, WI, USA.

Peoples M B and Herridge D F 1990 Nitrogen fixation by legumes in tropical and subtropical agriculture. Adv. Agron. 44, 155–223.

Shearer G and Kohl D H 1986 N_2-fixation in field settings: Estimations based on natural ^{15}N abundance. Aust. J. P. Physiol. 13, 699–756.

Smaling E M A 1993 An agroecological frame work for integrating nutrient management, with special reference to Kenya. Ph.D Thesis, Agric. Univ., Wageningen, The Netherlands.

Smaling E M A, Stoorvogel J J and Windmeijer P N 1993 Calculating soil nutrient balances in Africa at different scales: II District Scale. Fertil. Res. 35, 237–250.

Stoorvogel J J, Smailing E M A and Jansen B H 1993 Calculating soil nutrient balances at different scales. I Supra-national scale. Fertil. Res. 35, 227–235.

Tandon H L S and Narayan P 1990 Fertilizer in Indian Agriculture, Past, Present and Future (1950–2000) FDCO, New Delhi, India. 160 pp.

Thakur R, Sahay G and Sarkar A K 1990 Nutrient balance in acid sedentary soil (Haplustalf) of Ranchi as influenced by crop sequences. J. Indian Soc. 38, 233–236.

Thomas G W 1982 Exchangeable cations. *In* Methods of Soil Analysis. Part II. 2nd edition. Eds A L Page, R H Miller and D R Keeney. pp 643–698. Am. Soc. Agron. Soil Sci. Soc. Am. Madison, WI, USA.

J.J. Adu-Gyamfi (Ed.), Food security in nutrient-stressed environments: exploiting plants' genetic capabilities: 313–317, 2002.

Preplant moisture and fertility conditions as indicators of high and stable yields in rainfed cropping systems

C. F. Yamoah[1], G. E. Varvel[2] & J. J. Adu-Gyamfi[3]
[1]*Soil Research Institute, Council for Scientific and Industrial Research (CSIR), Academy Post Office, Kwadaso, Kumasi, Ghana*
[2]*Department of Agronomy, University of Nebraska, Lincoln, NE 68583-0949, USA*
[3]*International Crops Research Institute for the Semi-Arid Tropics (ICRISAT), PMB 3491, Kano, Nigeria*

Key words: preplant moisture, standard precipitation index, rotations

Abstract

Rainfed cropping systems are vulnerable to weather fluctuations. A crop rotation experiment conducted from 1984 to 1998 at the University of Nebraska USA, was analysed to determine the effect of preplant weather conditions on yields of maize (*Zea mays* L.), soybean (*Glycine max.* (L.) Merr.) and sorghum (*Sorghum bicolor* (L.) Moench). The Standardized Precipitation Index (SPI) was used as a drought classification index. Nitrogen (N) fertilizer and preplant moisture status significantly ($P<0.05$ to $P<0.008$) increased maize yields in monocropping and in rotation with soybean. As expected, N fertilizer did not increase soybean yields ($P=0.78$) but preplant moisture significantly increased yields of soybean ($P<0.02$). Preplant moisture did not have an effect on sorghum yields but yields were significantly increased ($P<0.02$) by N-fertilizer in sole-cropped sorghum. Additionally, preplant moisture significantly ($P<0.01$) predicted the subsequent season's maize yields in soybean-maize rotation ($R^2=0.71–0.85$) and soybean yields ($R^2=0.66–0.81$). Newer maize cultivars (1992–1998) outyielded the earlier cultivars (1985–1991) by 11%, despite the decreasing trend of preplant moisture status.

Introduction

Rainfed cropping systems are vulnerable to weather fluctuations. Maize or sorghum–soybean rotations are commonplace in the Corn (maize) Belt of Nebraska, USA (Clegg, 1982; Franzleubbers et al., 1994; Peterson and Varvel, 1989). Long-term studies ranging from 12 to 20 years have been conducted in Nebraska to compare rotations to continuous cropping systems. Empirical analyses of these studies showed that crops in rotations were more high-yielding, relatively stable, and profitable than those in monocultural systems (Clegg, 1982; Helmers et al., 1986).

There is sufficient understanding of the significant role played by precipitation in rainfed cropping systems (Delcourt and van Kooten, 1995; Dirks and Bolton, 1981; Teigen and Thomas, 1995; Thompson, 1962; Williams et al., 1975). The aspects of climate that are most influential in driving crop yields are precipitation and temperature. The two variables are stochastic and therefore difficult to estimate their magnitude and frequency. The Standardized Precipitation Index (SPI) is a drought classification index (McKee et al., 1993) that could be applied to crop management to lessen the extent of damage caused by inclemment climate. The SPI was previously used to evaluate how variability in pre-plant precipitation impacted sorghum yields in dryland cropping systems in Nebraska (Yamoah et al., 1998). The objective of this study is to use a measure of pre-plant moisture conditions to determine the subsequent season's crop performance.

Materials and methods

Data from a long-term crop rotation experiment conducted at the University of Nebraska USA, from 1984 to 1998, was analyzed to determine the effect of preplant moisture condition on yields of maize, sorghum,

Table 1. Crop yields as influenced by nitrogen (N) fertilizer and April 8-month Standardized Precicipitation Index (SPI), 1984–1998, Nebraska, USA

Cropping system	Nitrogen applied ($kg\ ha^{-1}$)	Yield ($kg\ ha^{-1}$)	
		Wet (+SPI)	Dry (+SPI)
Continuous maize (CM)	0	4200	2880
	90	6910	6340
	180	8140	6970
Soybean–maize rotation (S–M)	0	6800	5150
	62	8750	6880
	124	8400	7180
Continuous soybean (CS)	0	2390	2000
	34	2290	2000
	68	2270	2030
Maize–soybean (M–S)	0	2540	2170
	62	2260	2050
	124	2420	2140
Continuous sorghum (CGS)	0	3830	3330
	90	6190	6110
	180	6450	7000
Soybean–sorghum (S–GS)	0	6180	5900
	62	6600	6590
	124	6760	6700
Significance level			
CM (fertilizer)		0.008	
CM (precipitation)		0.05	
S-M (fertilizer)		0.02	
S-M (precipitation)		0.01	
CS (fertilizer)		0.78	
CS (precipitation)		0.02	
M-S (fertilizer)		0.15	
M-S (precipitation)		0.02	
CGS (fertilizer)		0.02	
CGS (precipitation)		0.96	
S-GS (fertilizer)		0.20	
S-GS precipitation)		0.52	

[a] CM, CS, CG, continuous cropping with maize, soybean, and sorghum; S–M, S–G, 2 year rotations with soybean–maize, soybean–sorghum.
CM, CG fertilizer rates: low=90 kg ha^{-1} $year^{-1}$; high=180 kg ha^{-1} $year^{-1}$. CS fertilizer rates: low=34 kg ha^{-1} $year^{-1}$; high=68 kg ha^{-1} $year^{-1}$. S–M, S–G average fertilizer rates: low=62 kg ha^{-1} $year^{-1}$; high=124 kg ha^{-1} $year^{-1}$).

and soybean. The experimental design was a split-plot design with four replications. The main cropping-system treatments were: (1) monocultures of maize, sorghum, or soybean; (2) 2-year soybean–maize or soybean–sorghum rotations; (3) 4-year soybean–sorghum–oat+clover–maize rotations; and (4) 4-year oat+clover–sorghum–soybean–maize rotations. Subplots were nitrogen (N) levels of 0, 90, or 180 kg ha^{-1} for maize and sorghum and N levels of 0, 34, or 68 kg ha^{-1} for soybean and clover.

Maize cultivars (Asgrow Rx717, 1984–8197; Pioneer 3377, 1988–1991; Pioneer 3379, 1992–1995) were used, sown at 47 000 seeds ha^{-1} in the first week of May. Soybean cultivars (Pella, 1984–1987; Pella 1986, 1988–1991; Dunbar, 1992–1995) were grown in rotation with maize sown at 370 000 seeds ha^{-1} from mid-May to June. Sorghum cultivars (NK2244, 1984–1987; Pioneer 8358, 1988–1995) were sown in the last week of May at the rate of 170 000 seeds ha^{-1}. A six-row planter was used, and each crop in the rotation appeared every year in the course of the study. Subsequent cultivation practices were reported earlier by Varvel and Peterson (1990).

The preplant moisture condition was calculated as the Standardized Precipitation Index (SPI), a measure of drought (McKee et al., 1993). The SPI was computed from precipitation values standardized according to the formula: SPI=$X - X_{\mathrm{m}}$/Std, where, X is precipitation for a given period, X_{m} is long-term precipitation mean, and Std is the standard deviation. The April 8-month SPI was used to capture the moisture status from September to April. The crops were grown under rainfed conditions therefore crop water use was dependent on precipitation. An SPI value of zero indicates average conditions, and values of greater than +2 or less than −2 generally indicate extreme wet or dry conditions associated with events that occur only 5% of the time. Furthermore, long-term climatic records in Nebraska provide evidence that there is moisture recharge and surplus from September to May, and a deficit during the summer months from June to August. Crops are sown in May and early June and usually attain physiological maturity by October. Therefore, it is hypothesized that favourable preplant moisture conditions could increase microbial activity, and result in early crop establishment and high yields. This thesis stemmed from the concept of drought elaborated by Wilhite and Glantz (1985). This paper is based only on the monoculture and 2-year rotation systems.

Results and discussion

Results showed that maize and soybean yields in years with moist April 8-month SPI were higher ($P<0.05$) than those that were dry (Table 1). Sorghum did not respond to preplant moisture conditions ($P>0.055$). Nitrogen fertilizer significantly ($P<0.02$) increased yields of monocropped sorghum, but not when it was grown in rotation with soybean (P=0.20). Maize

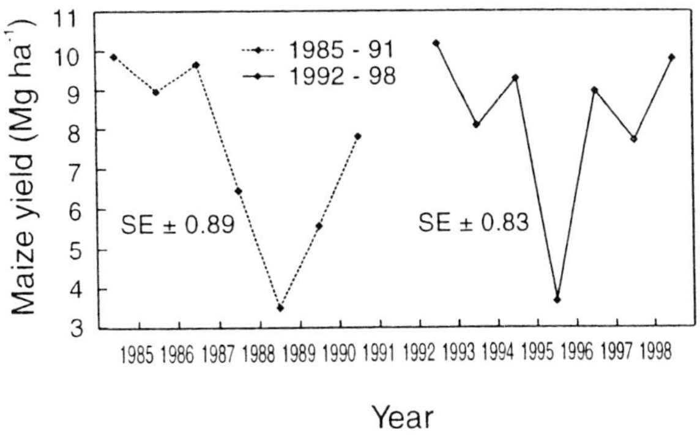

Figure 1. Average maize yields 1985–9191 versus 1992–1998.

responded to N-fertilizer both when monocropped and in rotation with soybean, but soybean did not ($P<0.008$ to $P<0.02$). Maize yields in monoculture and rotation were increased by 17% at the high N-rate (180 kg ha^{-1}) when preplant conditions were moist. Maize normally responds to fertilizer under good moisture conditions (Dirks and Bolton, 1981). But sorghum yields were reduced by 2–8% in moist preplant conditions with fertilizer application.

Regression models have been used to predict crop yields from standardized and non-standardized weather variables in north America (Teigen and Thomas, 1995; Thompson, 1962), and Canada (Delcourt and van Kooten. 1995). The main weather variables implicated in the above studies were precipitation and temperature. In the present study, second-order regression equations using preplant April 8-month SPI as independent variables and subsequent crop yields as dependent variables from 1984 to 1995, described maize and soybean yields in rotation fairly well (Table 2). Preplant April 8-month SPI accounted for up to 85% of variability in maize yields and 81% in soybean. Continuous maize yields were unaffected by preplant moisture status ($P>0.05$). Similarly, sorghum yields were indifferent to preplant moisture conditions (Table 2). The resilience of sorghum to adverse weather is attributed to peculiar anatomical features documented by Doggett (1970). These special features include a large root system, the presence of silica in the root endodermis, and rapid stomatal recovery after drought.

Correlation between preplant moisture and time (1985–1998) was −0.33 (P=0.25); correlation between preplant moisture and time (1985–1991) was −0.64 (P=0.12) and that from 1992 to 1998 was 0.37 (P=0.41). Despite the decreasing trend of preplant moisture (Figure 1), maize cultivars in maize–soybean

Table 2. Second-order regression models relating crop yields (Y) to pre-season April 8-month Standardized Precipitation (X), Mead, Nebraska, USA 1994/1995: model: $Y = a + bX - cX^2$

Cropping system	a	b	c	R^2	Significance level
Maize					
CM (0 N)[a]	3950.7	463.9	376.2	0.33	0.2060
CM (low N)	6996.3	402.5	718.1	0.23	0.3480
CM (high N)	8078.5	752.0	851.6	0.37	0.1570
S–M (0)	6398.3	932.2	654.4	0.71	0.0070
S–M (low N)	8963.8	1408.9	1503.1	0.85	0.0006
S–M (high N)	8914.7	889.2	1390.7	0.71	0.0070
Soybean					
CS (0)	2407.6	311.5	408.5	0.81	0.0010
CS (low N)	2358.9	288.2	389.9	0.81	0.0010
CS (high N)	2353.5	248.9	401.7	0.75	0.0040
M–S (0 N)	2611.3	222.1	466.6	0.67	0.0110
M–S (low N)	2418.9	157.7	495.4	0.71	0.0070
M–S (high N)	2491.3	182.1	430.0	0.66	0.0140
SG–S (0 N)	2716.1	364.7	399.2	0.71	0.0070
SG–S (low N)	2656.7	368.6	357.3	0.66	0.0130
SG–S (high N)	2835.5	364.8	415.9	0.67	0.0120
Sorghum					
CGS (0 N)	2834.7	863.7	258.5	0.32	0.2090
CGS (low N)	5752.1	488.0	135.1	0.05	0.8040
CGS (high N)	6468.6	16.6	133.5	0.01	0.9810
S–GS (0 N)	5644.8	534.7	6.4	0.11	0.6020
S–SG (low N)	6639.9	395.7	141.5	0.06	0.7690
S–SG (high N)	6479.3	291.2	75.7	0.04	0.8350

[a] CM, CS, CG, continuous cropping with maize, soybean, and sorghum; S–M, S–G, 2 year rotations with soybean–maize, soybean–sorghum.
CM, CG fertilizer rates: low=90 kg ha^{-1} year^{-1}; high=180 kg ha^{-1} year^{-1}. CS fertilizer rates: low=34 kg ha^{-1} year^{-1}; high=68 kg ha^{-1} year^{-1}. S–M, S–G average fertilizer rates: low=62 kg ha^{-1} year^{-1}; high=124 kg ha^{-1} year^{-1}).

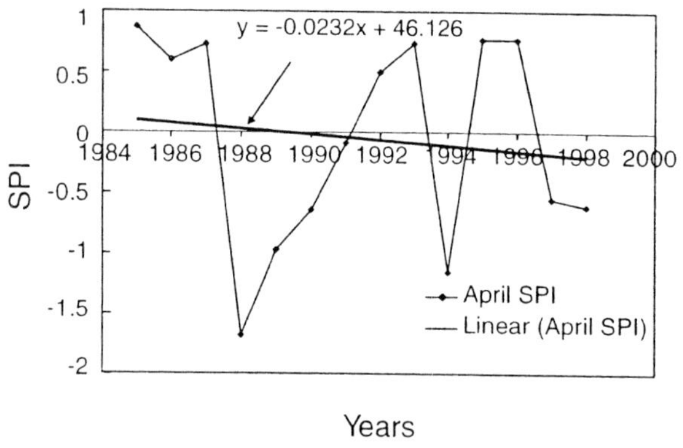

Figure 2. Trend of April 8-month SPI from 1984 to 1998.

rotation that received 90 kg fertilizer ha^{-1} from 1992 to 1998 outyielded the cultivars grown from 1985 to 1991 by 11% (Figure 2). This may be due in part the success of breeders' efforts to incorporate drought tolerance in the newer cultivars. Studies in Nebraska revealed that maize yields in maize–soybean rotation systems are high-yielding and stable across time (Varvel and Peterson, 1990).

There was a reasonable agreement ($P<0.05$) between actual maize yields and yields predicted by the regression models using the April 8-month SPI values from 1996 to 1998 (Figure 3). This underscores the importance of preseason rainfall in early crop establishment and its effect on subsequent yields. The preseason moisture level can therefore be used as an indicator to warn farmers and policymakers of impending droughts, so they can prepare strategies to mitigate their effects.

The use of information on preseason moisture levels to predict yields may be appropriate under the

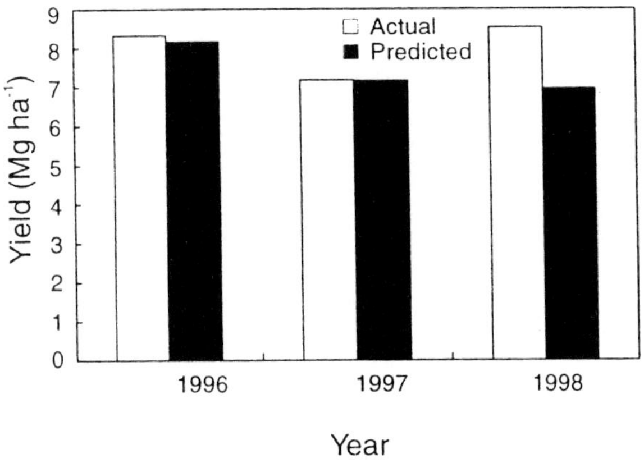

Figure 3. Actual versus predicted yields of maize–soybean rotation based on preseason April 8-month Standardized Precipitation Index, 1996–1998.

following conditions: (1) on clayey soils with high moisture-retention capabilities; (2) on deep soils rich in organic matter that can store moisture; and (3) on deep soils with vegetative cover or surface crop residue mulch materials during off-season to prevent loss of soil moisture. No-till and crop-rotation practices reduce pre-plant soil moisture loss and could therefore benefit subsequent crop yields. Knowledge of pre-plant moisture conditions is important for such management decisions as crop choices (e.g., whether to plant sorghum, a relatively drought-tolerant crop, or maize), or fertilizer application practices. These affect input costs and the overall profitability of farming systems.

References

Clegg M D 1982 Effect of soybean on yield and nitrogen response on subsequent sorghum crops in Eastern Nebraska. Field Crops Res. 5, 233–239.

Delcourt G and van Kooten G C 1995 How resilient is grain production to climatic change? Sustainable agriculture in a dryland cropping region of western Canada. J. Sust. Agric. 5, 37–57.

Dirks V A and Bolton E F 1981 Climatic factors contributing to year-to-year variation in grain yield of maize on Brookston clay. Can. J. Plant Sci. 61, 293–305.

Doggett H 1970 Sorghum. Longmans, London UK. 187 pp.

Franzleubbers A J, Francis C A and Walters W T 1994 Nitrogen fertilizer response potential of maize and sorghum in continuous and rotated crop sequences. J. Prod Agric. 7, 93–284.

Helmers G A, Langemeier M R and Atwood J 1986 An economic analysis of alternative cropping systems for east-central Nebraska. Am. J. Alternative Agric. 1, 153–158.

McKee T B, Doesken N J and Kleist J 1993 The relationship of drought frequency and duration to time scales. Preprints, 8th Conference on Applied Climatology, 17–22 January 1992, Anaheim, California, USA. pp 179–184.

Peterson T A and Varvel G E 1989 Crop yield as affected by rotation and N rate: III. Corn. Agron. J. 81, 735–738.

Teigen L D and Thomas M Jr 1995 Weather and Yield, 1950–1994: Relationships, Distributions, and Data. Commercial Agriculture Division, Economic Research Service, U S Department of Agriculture. Staff Paper No. 9527.

Thompson L M 1962 Evaluation of weather factors in the production of wheat. J. Soil Water Cons. 17, 149–156.

Varvel G E and Peterson T A 1990 Residual soil nitrogen as affected by continuous 2 year and 4 year crop rotations. Agron. J. 82, 958–962.

Wilhite D A and Glantz M H 1985 Understanding the drought phenomenon: the role of definitions. Water Int. 10, 110–120.

Williams G D V, Joynt M I and McCormick P A 1975 Regression analysis of Canadian prairie crop-district cereal yields, 1961–1972, in relation to weather, soil and trend. Can. J. Soil Sci. 55, 43–53.

Yamoah C F, Varvel G E, Francis C A and Walman W J 1998 Weather and management impact on crop yield variability in rotations. J. Prod. Agric. 11, 219–225.

J.J. Adu-Gyamfi (Ed.), Food security in nutrient-stressed environments: exploiting plants' genetic capabilities: 319–329, 2002.

Role of modelling in improving nutrient efficiency in cropping systems

P. S. Carberry[1,4], M.E. Probert[2], J.P. Dimes[3], B.A. Keating[2] & R.L. McCown[1]
[1]*CSIRO Sustainable Ecosystems, Agricultural Production Systems Research Unit (APSRU), PO Box 102, Toowoomba, Qld. 4350, Australia.* [2]*CSIRO Sustainable Ecosystems, Agricultural Production Systems Research Unit (APSRU), 306 Carmody Rd, St Lucia, Qld. 4067, Australia.* [3]*Natural Resource Management Program, ICRISAT, Matopos Research Station, PO Box 776, Bulawayo, Zimbabwe.* [4]*Corresponding author**

Received 10 May 2000. Accepted in revised form 24 November 2000

Key words: APSIM, nitrogen, phosphorus, simulation modelling

Abstract

The applicability of models in addressing resource management issues in agriculture has been widely promoted by the research community, yet examples of real impacts of such modelling efforts on current farming practices are rare. Nevertheless, simulation models can compliment traditional field experimentation in researching alternative management options. The first objective of this paper is, therefore, to provide four case study examples of where models were used to help research issues relating to improved nutrient efficiency in low-input cropping systems. The first two cases addressed strategies of augmenting traditional farming practices with small applications of chemical fertilizer (N and P). The latter two cases explicitly addressed the question of what plant genetic traits can be beneficial in low-nutrient farming systems. In each of these case studies, the APSIM (Agricultural Production Systems Simulator) systems model was used to simulate the impacts of alternative crop management systems.

The question of whether simulation models can assist the research community in contributing to purposeful change in farming practice is also addressed. Recent experiences in Australia are reported where simulation models have contributed to practice change by farmers. Finally, current initiatives aimed at testing whether models can also contribute to improving the nutrient efficiency of smallholder farmers in the SAT are discussed.

Introduction

McCown et al. (1992) reported on a 10 year research effort in utilizing experimental research and simulation modelling in attempting to identify a development path for subsistence agriculture in semi-arid Kenya. They have suggested that a strategy of augmenting traditional soil enrichment practices, based on manure and legumes, with modest amounts of fertilizer is economically feasible for many farmers and provides the best prospects for food security in this climatic zone. McCown et al. (1992) provided a strong argument that the use of the simulation model permitted the assessment of fertilizer use in a way that would not have been possible using traditional field experimentation alone. Subsequent research has confirmed both the apparent attractiveness of fertilizer use in Kenya and the applicability of models to address such research questions (Rötter and Van Keulen, 1997). However, both McCown et al. (1992) and Rötter and Van Keulen (1997) conclude that implementation of apparently advantageous changes to current farming practice remains an intractable problem.

In reflecting on this research program in Kenya, McCown and Cox (1994) suggested that there were six lines of endeavour that were worthy of further exploration in addressing soil fertility decline in Africa:

1. Find more efficient ways to capture, store and use manure on crops;
2. Test various FASE (Fertilizer-Augmented Soil Enrichment) strategies for combining applications of chemical fertilizer with manure and crop residues;
3. Extrapolate research results on legumes in systems;

* FAX No: 746881193. E-mail: peter.carberry@cse.csiro.au

This paper previously appeared in *Plant and Soil* **245**, 2002, pp. 193–203

4. Research the economics of FASE strategies for different price-cost scenarios;
5. Contribute to national policies on commodities, food security and fertilizers;
6. Keep fertilizer on the policy agenda.

While the application of simulation models were explicitly recommended in the second and third areas, one could easily imagine modelling analyses contributing to other recommended areas of future activity, particularly as inputs into economic and policy analyses.

The McCown and Cox (1994) recommendations address both research activities to broaden our understanding of the biophysical system under study and operations research that uses our current understanding to predict likely outcomes for alternative management strategies that may be imposed on the farming system. Also included is possible intervention via policy actions. However, important questions are whether intervention directly with farmers is possible and whether simulation models can contribute beyond the research phase to assist in achieving improved farming practices. While not suggesting close analogy between developed and subsistence agriculture, it is relevant in the context of this paper to refer to recent experiences in Australia that have demonstrated how simulation models can contribute to farmer exploration and acceptance of practice change (Carberry and Bange, 1998; McCown et al., 1998).

Clearly, there has been significant past research effort into utilizing modelling in exploring management strategies in nutrient-stressed environments. The objectives of this paper are firstly to provide some case study examples of how models can contribute to research on nutrient efficiency in cropping systems and, secondly, to address the question of whether these models can assist the research community in contributing to purposeful change in farming practice.

In fulfilling our stated objectives, opportunity is taken to report on the current status of the systems model APSIM (Agricultural Production Systems Simulator), particularly in regard to current capabilities in addressing the issues raised by McCown and Cox (1994). Accordingly, applications addressing the use of manure and the incorporation of legumes in cropping systems will be explored.

APSIM

The Agricultural Production Systems Simulator (McCown et al., 1996) is a software environment that enables the biophysical simulation of farming systems. The distinguishing characteristics of APSIM are its:

1. Flexibility, achieved via a library of 'plug-in/pull-out' modules representing the biological, physical, environmental, managerial and economic components of farming systems;
2. Ability to simulate the production and resource consequences of a range of crops, pastures and trees grown in response to a variety of management practices, species mixtures and rotational sequences; and
3. Emphasis on high standards in software engineering, incorporating software coding standards, version control, maintenance and development protocols, and extensive documentation.

While the simulation capability of APSIM is continually expanding with the addition of new modules of crops (groundnut, faba bean, canola, lupins, hemp), pastures (lucerne, mucuna), trees (*Pinus radiata*, *Eucalyptus globulus*) and soil processes (phosphorus, manure, acidification), most recent progress has been in expanding the testing and application of APSIM to systems issues. Examples include the assessment of cereal-legume rotations (Probert et al., 1998), ley farming systems (Carberry et al., 1996a), intercropping systems (Carberry et al., 1996b), alley farming systems (Nelson et al., 1998), tree windbreak systems (Meinke et al., 2000), crop-weed associations (Keating et al., 1999), genetic traits within cropping systems (Robertson et al., 2000b), seasonal climate forecasting systems (Hammer et al., 2000), drought policy (Keating and Meinke, 1998), impacts on deep drainage (Asseng et al., 1998) and on-farm participatory research trials (Robertson et al., 2000a).

The above-referenced examples not only demonstrate the broad applicability of APSIM to a wide range of systems issues, but also provide evidence of APSIM applications both in the commercial agricultural systems of Australia and in the smallholder farming systems of the semi-arid tropics. APSIM is well placed, therefore, to contribute to research on improving the nutrient efficiency of low-input cropping systems.

Case studies in smallholder agriculture

Four case studies are used to demonstrate how APSIM could contribute to research on nutrient efficiency in low-input cropping systems. The first two address the

Table 1. Characteristics of the two manure sources, low quality manure (LQM) and high quality manure (HQM)

	%C	%N	Relative Composition [a]	Rates of application	
				DM (t/ha)	N (kg/ha)
LQM	10	0.6	0: 0.01: 0.99	5, 10, 20	30, 60, 120
HQM	15	1.5	0: 0.20: 0.80	2.5, 5, 10	37.5, 75, 150

[a] Proportions in three pools with decreasing rates of decomposition; for comparison, the composition of crop residues is 0.20: 0.70: 0.10.

recommendation for a FASE strategy of augmenting traditional farming practices with small applications of chemical fertilizer (McCown et al., 1992; McCown and Cox, 1994). The first case study utilizes the recently developed APSIM-Manure module to compare the application of manure of different qualities with inorganic fertilizer. The second study reports on the analyses of Keating et al. (1999) who explored why on-farm agronomic efficiencies of fertilizer nitrogen (N) (kg grain per kg N fertilizer applied) are often less than on-station efficiencies, due possibly to the impacts of low levels of weed competition.

The latter two cases explicitly address the question of whether models can assist in identifying how plants' genetic diversity may be exploited in low nutrient farming systems. Accordingly, the third case study addresses the question of the impact of different morphologies of cowpea intercropped with maize under low-input production systems. For this purpose, we reproduce the analyses reported by Robertson et al. (1999). The final case study uses the newly developed APSIM-SoilP module to explore the production and resource consequences of selecting for plants with more efficient root systems for extracting soil phosphorus (P).

Case study 1: The application of manure and N fertilizer

Manure recovered from open 'bomas' (corals) by subsistence farmers of semi-arid Africa is generally of poor quality in that it contains only low concentration of nutrients (Probert et al., 1995). Working in Zimbabwe, Tanner and Mugwira (1984) reported that manure with low concentrations of nutrients was an ineffective source of nutrients even in low-input cropping systems.

Using APSIM v1.55, the following analyses explore the response of maize to inputs of manure of two different qualities (Table 1) on a Chromic Luvisol soil in Kenya. The weather record used was for Katumani (latitude 1.5° S) extending from 1957 to 1997 (80 seasons due to the bimodal rainfall pattern). The effectiveness of manure in the APSIM-Manure module is determined by concentrations of nitrogen and carbon (C) and also by susceptibility to decomposition. This latter effect is controlled by considering the organic C and N as three fractions (carbohydrate, cellulose, lignin) with differing rates of decomposition (Probert et al., 1997).

In this simulation exercise, manure was applied and incorporated each year on 15 September (to represent the local practice of moving manure to the cropping lands prior to cultivation and before the commencement of the short rains) at the rates shown in Table 1. Maize crops were simulated in both the short-rain (sowing window 16 October–20 November) and the long-rain (15 March–1 May) seasons each year using the cultivar KCB sown at 3 plants m^{-2}. For comparison, there were treatments that received 0, 20, 40, 60 and 80 kg of fertilizer-N, as nitrate, applied to each crop at sowing. A subset of the manure treatments also received an input of 20 kg N as fertilizer at sowing.

In the first set of simulations, the model was re-initialised each year on 1 September. Thus, the simulations provide a comparison of the responses to the manure and fertilizer for the 40 years of the weather file (Figure 1). The effect of manure on the crop grown in the long rains is the residual effect from manure applied the previous September.

The material denoted HQM performed similarly to that in the experiments of Ikombo (1984) using manure collected from dairy yards. He reported that the application of 8 t ha^{-1} gave high and consistent yields, close to those obtained with the standard rate of mineral fertilizer (40 kg N ha^{-1}). Good residual effects were also found in the following long-rain season. In contrast, the specified LQM was a much less effective source of N for the short-rains crop and its residual effect was also inferior for the next long-rains crop. The poor response may help explain why seemingly very high rates of manure were being applied by some farmers (Probert et al., 1995).

The response to an extra 20 kg fertilizer-N applied at planting is shown in Table 2. The effect of manure and fertilizer in these simulations is essentially additive, with the expected decline in response as one moves up the response surface. The two manures simulated here have low C:N ratios (16.7 for LQM and 10.0 for HQM, see Table 1). When such materials decompose, there is net mineralization of N. When

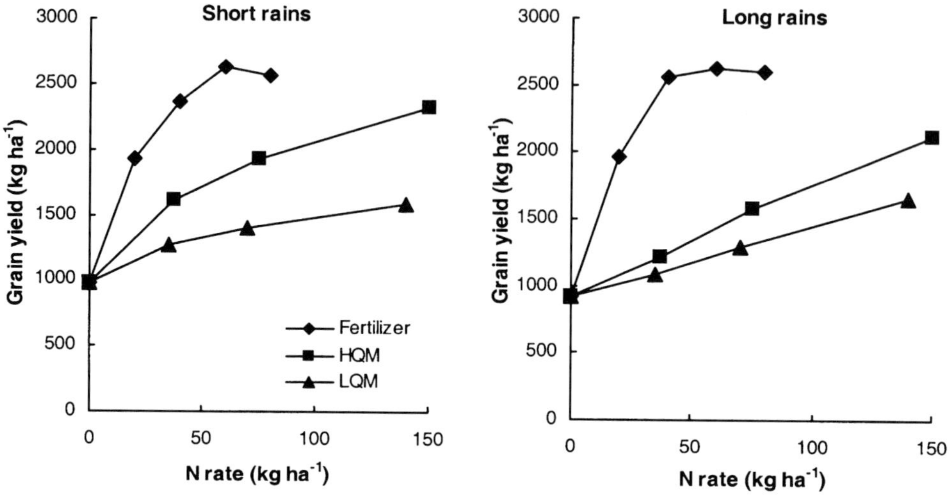

Figure 1. Simulated maize yields (averaged over 40 years) in response to high (HQM) and low quality manure (LQM) incorporated into soil in September each year. The fertilizer inputs were made to crops grown in both the short and long-rain seasons.

Table 2. Simulated response to additional 20 kg ha^{-1} of fertilizer-N applied at sowing, expressed as yields of maize (kg ha^{-1}). Data reported are averages for the 40 crops simulated

Treatment	Short rains		Long rains	
	Yield	Response to extra 20 kg N ha^{-1}	Yield	Response to extra 20 kg N ha^{-1}
Control	981	953	913	1052
Fertilizer 20 kg N ha^{-1}	1934	434	1965	589
5 t ha^{-1} LQM	1267	766	1080	982
2.5 t ha^{-1} HQM	1819	577	1215	942

materials with high C:N ratios are added to soil their decomposition causes net immobilization of N. Under these conditions, an interaction between manure and fertilizer will arise if the immobilization demand can not be met from the unfertilized soil.

Where manure is used to replenish/maintain soil fertility in farming systems, the residual effects from previous applications become cumulative. The situation above where the simulations were re-initialized each year corresponds to an experiment that aims to compare different inputs (manure, fertilizer). To simulate cumulative residual effects, a second set of simulations were undertaken whereby the model was initialized once only on 1 September 1957. The same manure treatments were applied every year prior to the short rains and compared with treatments where fixed amounts of fertilizer-N were applied to every crop.

Cumulative yields in four of the manure treatments relative to that for fertilizer-N (using 60 kg ha^{-1} of N applied to each crop as the standard) are plotted through time in Figure 2. It can be seen that the effect of manure increases during the first few years after application. In the case of the HQM, the phase where performance is improving relative to fertilizer lasts about 5 years. For the LQM, the 'catching up' phase lasts longer but the advantage of the higher quality material is maintained.

It needs to be stressed that these simulations have only examined manure as a source of N compared to fertilizer-N. There will be other situations where manure provides other benefits, such as supplying other nutrients or preventing the soil from acidifying.

Case study 2: Effects of weed competition on agronomic efficiency

Shamudzarira et al. (1999) reported that simulated agronomic efficiencies of fertilizer-N (kg grain per kg

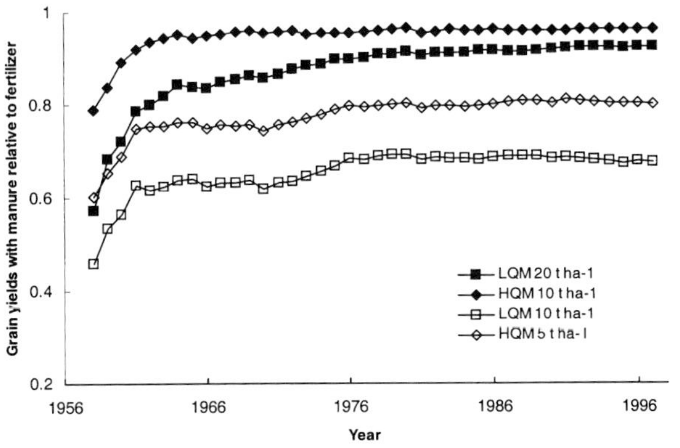

Figure 2. The effect of repeated application of different rates of high and low quality manures on the yields of maize. The cumulative yields (for crops in both the short and long-rain seasons) are expressed as a proportion of the cumulative yields obtained with 60 kg ha^{-1} of fertilizer-N applied to each crop.

N fertilizer applied) were similar to those measured from on-station trials in Zimbabwe but were generally higher than those found in farmers' fields. In exploring this issue further, Keating et al. (1999) used APSIM to explore the hypothesis that this differential may be due in part to the impact of low levels of weed competition on maize response to fertilizer. This second case study examines the analyses of Keating et al. (1999).

Keating et al. (1999) configured APSIM v1.55 to simulate inter-species competition between maize and a grass weed as per the intercropping capability described by Carberry et al. (1996b). A maize production system was simulated (cv. Makoholi) growing on an infertile light textured soil (118 mm of plant available water, seasonal mineralisation capacity of approximately 16 kg N ha^{-1}) for the climate record (1981–1991) of Bulawayo, Zimbabwe (20.1° S, 28.4° E). Maize sowing (2 plants m^{-2}) was conditional on rainfall after 23 November. The weed population was 25 seeds m^{-2} and, as seeds were assumed to germinate from depth (100 mm), the weed population emerged approximately 5 days after the emergence of the maize crop.

The model configuration provided a seasonally variable, but generally low level of weed competition. In the presence of N fertilizer, APSIM simulated that maize-weed competition reduced yields (Figure 3a). As the systems were all strongly N constrained, this competition was primarily for N and not for water or light. Competition was least in the unfertilised systems–due to there being insufficient N for the weeds to establish. The value of the fertilizer-N (agronomic efficiency) declined from 54 to 32 kg

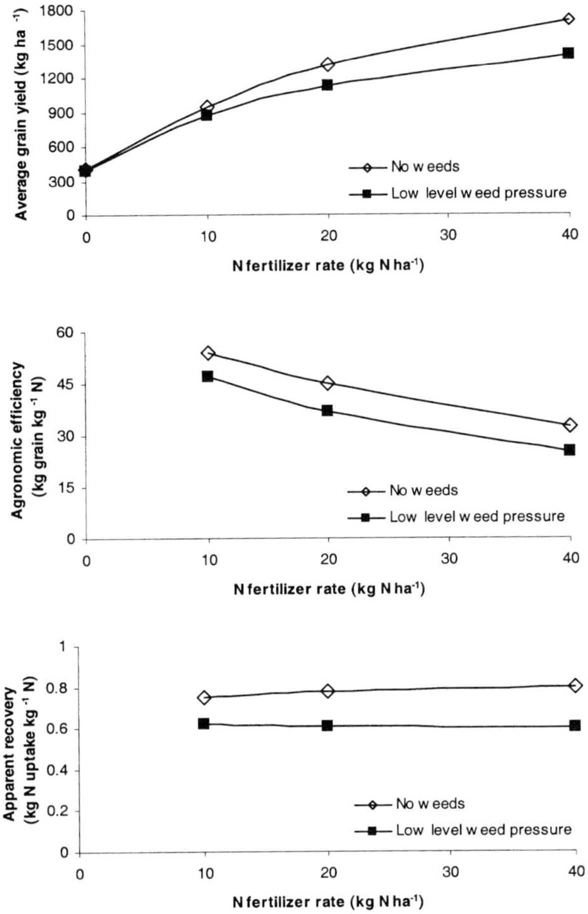

Figure 3. Effects of weed competition on simulated N response in maize, Bulawayo, Zimbabwe. (a) Average grain yield (kg ha^{-1}), (b) Agronomic efficiency (kg grain kg^{-1} N fertilizer), (c) Apparent N fertilizer recovery (kg N uptake to tops kg^{-1} N applied).

grain kg^{-1} N as fertilizer rate was increased from 10 to 40 kg N ha^{-1}. Agronomic efficiency was reduced in the presence of weed competition (e.g. 45–25 kg grain kg^{-1} N) (Figure 3b). These shifts in agronomic efficiency were associated with shifts in apparent recovery of the fertilizer-N, that was reduced from 0.8 without weeds to 0.6 with weed competition (Figure 3c).

At the low fertilizer rates relevant to the economic circumstances of Zimbabwe's smallholders, small amounts of weed competition can significantly reduce the benefits of the fertilizer inputs. On-farm experimentation on fertilizer usage in these low-input systems needs to quantify and/or manage such constraints. Experiments on weed competition tend to give highly variable results depending on rainfall patterns, soil constraints, crop agronomy and weed demography. Systems simulation models such as APSIM can help to address this variability and assess the likely impact of 'less than ideal' agronomy on investments in inputs such as N fertilizers.

Case study 3: Cowpea canopy morphology in maize / cowpea intercrops

This case study has been previously described by Robertson et al. (2000b) as an example of using models to assist in the identification of genetic traits best-adapted to particular cropping systems. It addresses the question of the impact of different morphologies of cowpea, when intercropped with maize under low-input production systems. As cereal-legume intercrops have been widely promoted as a means of improving the biological efficiency of resource use in low-input systems (Fukai, 1993; Willey 1979a,b), identifying genotypes with traits best suited to intercropping systems is a key requirement in maximizing the benefits of such systems.

Robertson et al. (2000b) configured APSIM v1.55 to simulate maize and cowpea sole and intercrops grown under rainfed conditions for 31 years (1957–1988) using climatic data at Katumani (latitude 1.5° S) in semi-arid eastern Kenya. Simulations treated each year independently; all parameters were re-initialised at sowing–soil water was initialised at the lower limit of plant available water and soil mineral N was initialised to 30 kg ha^{-1}. Maize cultivar KCB was sown during the short and long-rain seasons, with no fertilizer application, consistent with local practice. A sowing density of 4 maize and 5 cowpea plants m^{-2} was used, and all crop residues were removed at harvest.

For this exercise, cowpea morphology was modified in the APSIM-Cowpea module by changing both the maximum attainable height of the canopy (80 vs. 160 cm), to mimic bush versus climbing types, and the radiation extinction coefficient (0.47 vs. 0.80) to mimic types with erect versus prostrate leaf posture. Results were evaluated in terms of the impact on both maize and cowpea grain yields.

In narrow (50 cm) maize rows, the cowpea yield responded best to increases in maximum height, with little response to an increase in the extinction coefficient of the cowpea (Figure 4). However, there was an interaction, where the biggest response to increasing height was at the larger extinction coefficient. In contrast, in the wide maize rows (100 cm), the biggest increase in cowpea yield was obtained when the ex-

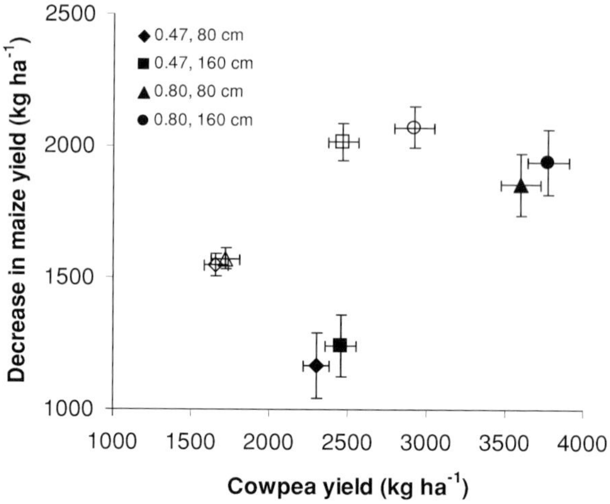

Figure 4. Relationship between the mean decrease in maize yield *versus* cowpea yield for cowpea genotypes differing in extinction coefficient (0.47, 0.80) and maximum height (80, 160 cm) grown in 50 cm maize row spacing (hollow symbols) and 100 cm row spacing (filled symbols). Error bars are ± standard error of the mean.

tinction coefficient was increased, with a much smaller response to increasing height (Figure 4).

This case study highlights the complex interactions between agronomic management (such as row spacing) and the performance of different genotypes in competing mixtures. Models such as APSIM can help analyse some of these interactions by quantifying the likely trade-off between increasing the yield of one species at the expense of the other.

Case study 4: Simulating the effect of a more phosphorus-efficient plant

Crops that are efficient at utilizing soil-P can be conceived as employing one of two strategies. Firstly, without access to a greater pool of available soil-P, some plants are able to take up more P through a more efficient root system–i.e. more roots and/or greater root absorbing power (Nye and Tinker, 1977). The second strategy would rely on gaining access to P that was unavailable to other plants; this could be achieved where root exudates result in solubilization of otherwise unavailable forms of P (Kirk, 1999).

The manner in which soil-P dynamics and P uptake by a crop are represented in APSIM should allow the capture of the consequences of a cultivar with a more efficient root system, but at present there is no ability in the model to cope with concepts such as root exudates. Simulated P uptake is limited either by the crop's demand to maintain its internal P concentration, or by

Table 3. Predicted yield advantage of the more efficient cultivar with increasing cropping history for systems without and with inputs of P fertilizer. Values shown are averages of the ratios of the yields for the two cultivars

Crop number	Without fertilizer	With 5 kg P $ha^{-1}crop^{-1}$
1–20	1.21	1.11
21–40	1.18	1.05
41–60	1.16	1.05
61–80	1.13	1.02

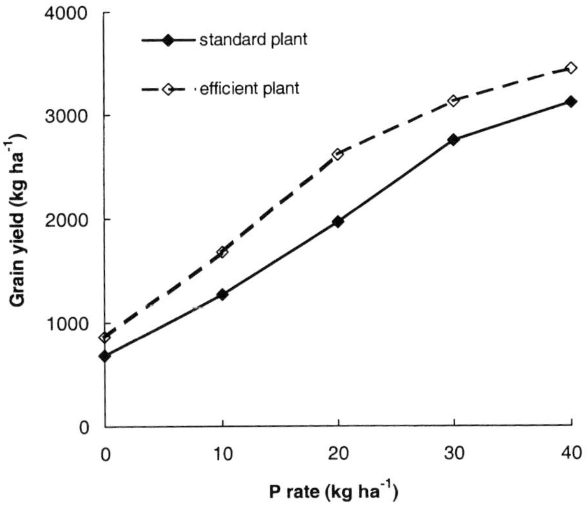

Figure 5. Simulated yields of maize cultivars differing in their efficiency in utilizing soil-P in response to increasing inputs of P fertilizer applied as a banded application. Data shown are for short rains 1958.

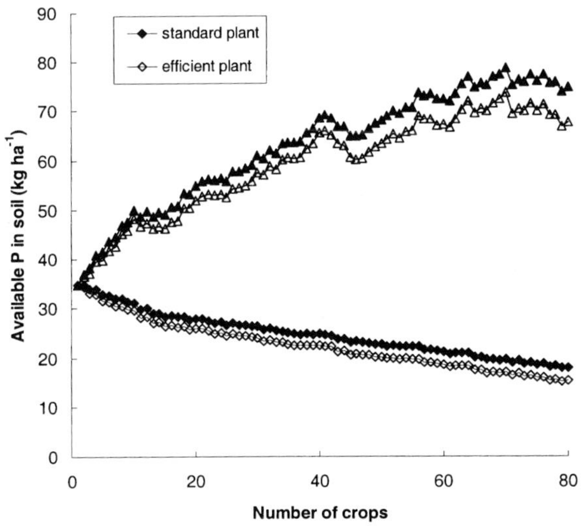

Figure 6. Prediction of soil-P status under continuous maize cropping with fertilizer-P inputs of 0 and 5 kg $ha^{-1}crop^{-1}$ for two cultivars differing in their efficiency in utilizing soil-P.

the supply of P from the soil. The latter is defined on a soil layer basis as a function of the amount of available P, the soil's P-sorption properties, the moisture content and the distribution of roots. The 'P uptake coefficient' of this function has some semblance to the root absorbing power of the plant.

The SoilP routines in APSIM v1.55 were used to examine two situations. The first was in the absence of fertilizer-P inputs when it was hypothesised that a cultivar with a greater ability to take up available P from soil will have some short-term advantage but that this advantage may decline in the longer-term as the non-renewable soil-P resource is depleted. Secondly, where modest inputs of P are used, it might be expected that residual P would accumulate in the soil and eventually the advantage of the P-efficient cultivar would be cancelled.

The simulation study involved maize (cultivar KCB) grown on a typical Chromic Luvisol in the semi-arid tropics of Kenya. Each crop was fertilized with 50 kg N ha^{-1}. The soil initially contained 4 mg kg^{-1} bicarbonate-extractable P in the surface soil, declining with depth, and had moderate capacity to sorb P. The weather file used was that for Katumani (1957–1997). The parameterization for the 'standard' maize plant is that which has been found to give satisfactory prediction of the experiments on Mutua Farm described by Probert and Okalebo (1992), with the P-uptake coefficient calibrated to a value of 4 (unitless). For a more P-efficient cultivar, this coefficient was increased by 25% to a value of 5.

When grown on a soil with a much higher P status, there was no difference between the two cultivars (non-limited yield was 3889 kg ha^{-1} for the season shown) (Figure 5). The longer-term effects are shown in Figures 6 and 7, and summarized in Table 3. In the absence of fertilizer-P, the run down in the soil-P status through time was slightly quicker for the more efficient cultivar (Figure 6). The relative yields (the ratio of crop yield to non-limited yield) for both cultivars decreased over time (Figure 7) and many crops produced low yields (below 500 kg ha^{-1}) irrespective of cultivar. There was some tendency for the advantage of the more efficient cultivar to diminish as it progressively depleted the soil-P resource (Table 3).

Repeated applications of fertilizer-P to each crop resulted in increasing soil P status and an increase in the relative yields of the two cultivars (Figures 6 and

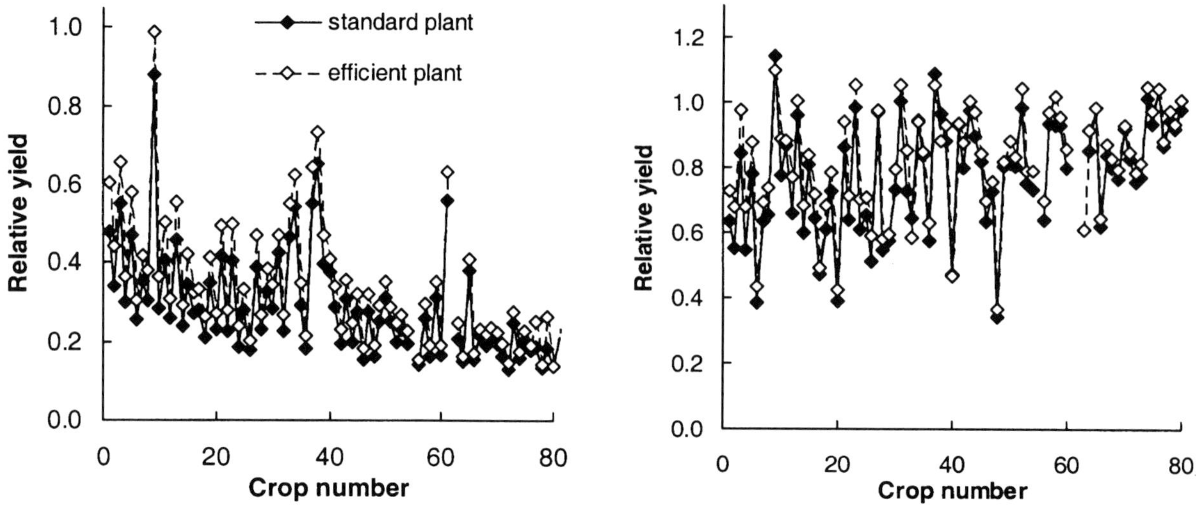

Figure 7. Relative yields of maize for two cultivars differing in efficiency in utilizing soil-P. The left hand frame shows the predictions for a cropping system without inputs of fertilizer-P; the right hand frame is for a cropping system that has 5 kg ha^{-1} of fertilizer-P applied to every maize crop.

7). Simulations were run with inputs of 3, 5 and 10 kg ha^{-1} as banded P to each crop but results are shown only for the 5 kg ha^{-1} rate. As the residual effects of past applications accumulated, the difference between the two cultivars became smaller (Table 3).

These simulations showed that the P routines in APSIM can represent the expected response of soil-P and crop growth–with the exception of dealing with the impacts of root exudates. The simulated results supported the hypothesis that a crop with a greater ability to take up P from soil has a yield advantage over a crop with less ability to source soil-P. However, the simulations indicated that this advantage persisted for at least the 40 years of this case study even though soil P was specified as a fixed, non-renewable resource.

Can models assist in improving farming practices?

The Australian experience

FARMSCAPE (Farmers, Advisers, Researchers, Monitoring, Simulation, Communication And Performance Evaluation) is an acronym employed to represent a participatory R&D approach that explicitly addresses the question of relevance of systems models to commercial farming in Australia (Carberry and Bange, 1998; McCown et al., 1998). It involves research to explore whether farmers and their advisers could gain benefit from tools such as soil characterisation and sampling, seasonal climate forecasts and, in particular, simulation modelling and, if so, how such tools could be delivered cost-effectively to industry.

FARMSCAPE has been based on the key elements identified in its name:

1. Close collaboration of farmers, their advisers and researchers in discovering together how best to explore management options;
2. Implementation of research on commercial farms, especially incorporating improved soil monitoring to gain better knowledge of actual soil water and nitrogen resources in individual fields;
3. Application of APSIM to exploring management options of interest to participants in interactive, co-learning 'What-if?' analysis and discussion sessions (WifADs), with the pre-requisite requirement that simulations be credible against participants' real-world experiences;
4. Broader communication of project outcomes not only through public extension activities but particularly through agribusiness client services, and
5. Continual assessment of project activities and impacts via formal evaluation processes.

As part of FARMSCAPE a number of evaluation activities have been undertaken. The following is an extract from an evaluation activity undertaken by Coutts et al. (1998):

> "*The evaluation process provided strong evidence that (FARMSCAPE) was having a positive impact on: learning within each participant group, attitudes, decision-making and practice. The eval-*

uation highlighted the complexities in the management of dryland crops and the limitations of simulation-aided decision making. However, the evaluation has shown that simulation, adequately contextualized, was valued by participating farmers and advisers in (a) gaining insights into production system function and (b) augmenting their farming experience in making judgements required in tactical responses and the evolution of improved production strategies."

FARMSCAPE has helped demonstrate that a key to farm managers in Australia valuing simulation is the positioning of these simulations in the context of their own farming situation. A simulator enables information to be specified to an individual field, its results can be tested against one's own crop performance and a simulator such as APSIM can be used to explore a range of issues of interest to participants. In Australia, a market now exists for timely and high quality interactions based on soil monitoring and simulation amongst a growing sector of the farming community.

Relevance of models to smallholder agriculture

The productivity of dryland crop production systems throughout the semi-arid tropics is of clear concern (Shapiro et al., 1999). One explanation for the apparent low incentive of farmers to invest scarce capital in soil fertility may be due to their perception of the benefits and risks associated with investments such as inorganic fertilizer. Simulation models are appropriate tools to help explore such investment options under climatic uncertainty, not only in terms of returns and risks but also future productivity. However, are such tools relevant to farmers in the semi-arid tropics?

In addressing this question, both the International Crops Research Institute for the Semi-Arid Tropics (ICRISAT) and Centro Internacional de Mejoramiento de Maïz y Trigo (CIMMYT) are currently collaborating with APSRU on two projects targeted at testing the relevance of APSIM to smallholder agriculture in the semi-arid tropics–the initial focus is in both India and Zimbabwe. Both projects are aimed at using APSIM with farmers, national agricultural research systems (NARS) professionals and non-government organizations (NGOs) in the evaluation of farm management options. Accordingly, a number of on-farm case studies are being implemented as part of these projects, whereby participatory on-farm research trials are negotiated and conducted with farmers on issues identified via simulation analyses relevant to the participating farmers' own livelihood strategies.

Whether the model has a role in direct intervention with smallholder farmers is viewed differently among the researchers involved in these projects and, consequently, this question forms the basis of a researchable issue in itself. The consensus view is that APSIM may identify management options worth testing in on-farm experimentation, with the expectation that the process of identification and development of new technologies could be expedited. Some consider that it is also possible to directly engage smallholder farmers with the model in 'What if?' analysis and discussions sessions on issues of interest to them. Whether this is the case is being explored using a Participatory Action Research (PAR) methodology, where each researcher-farmer interaction involving APSIM follows an iterative cycle of *planning* the interaction, *acting* on the plans, *observing* and documenting what happens and *reflecting* on the interaction's outcomes in order to repeat the process a next time.

Conclusions

Strong argument that simulation models can have a significant role in contributing to research on nutrient efficiency in cropping systems has been presented. Four case studies were used to address the primary objective of demonstrating how models can contribute to such research. Two cases addressed issues of augmenting traditional manuring practices with fertilizer, and another two cases applied APSIM to identifying plant traits of benefit in low-input cropping systems. While these cases were innovative in addressing issues that capitalized on APSIM's emerging systems capabilities (i.e. manure, weeds, intercropping & P), the application of models in this manner is not new (see Cooper and Hammer, 1996; Muchow and Bellamy, 1991). In fact, on the issue of trait identification, some have argued that the contribution of crop physiology and modelling to plant breeding has been relatively modest to date (Jackson et al., 1996).

The second objective of questioning whether models can assist the research community in contributing to purposeful change in farming practice is really the key question. The FARMSCAPE experience with commercial farmers in Australia provides strong evidence that this can indeed be the case. The current initiatives of ICRISAT and CIMMYT will test whether the models can also contribute to improving the nutrient efficiency of smallholder farmers in the SAT.

Acknowledgements

The work on smallholder agriculture referred to in this paper is being undertaken in collaboration with researchers at ICRISAT and CIMMYT and has received funding support from the Australian Centre for International Agricultural Research (ACIAR).

References

Asseng S, Fillery I R P, Anderson G C, Dolling P J, Dunin F X and Keating B A 1998 Use of the APSIM wheat model to predict yield, drainage and NO3 leaching in a deep sand. Aust. J. Agric. Res, 49, 363–377.

Carberry P S and Bange M P 1998 Using systems models in farm management. *In* Proc. 9th Aust. Cotton Conf. 10–14 August, Gold Coast Australia. The Australian Cotton Growers Research Organisation, 153–160.

Carberry P S, McCown R L, Muchow R C, Dimes J P, Probert M E, Poulton P L and Dalgliesh N P 1996a Simulation of a legume ley farming system in northern Australia using the Agricultural Production Systems Simulator. Aust. J. Exp. Agric. 36, 1037–1048.

Carberry P S, Adiku S G K, McCown R L and Keating B A 1996b Application of the APSIM cropping systems model to intercropping systems. *In* Dynamics of Roots and Nitrogen in Cropping Systems of the Semi-Arid Tropics. Eds. O Ito, C Johansen, J J Adu-Gyamfi, K Katayama, J V D K Kumar Rao and T J Rego, pp. 637–648. Japan International Research Center for Agricultural Sciences, International Agricultural Series No 3.

Cooper M and Hammer G L (Eds) 1996 Plant Adaptation and Crop Improvement. CAB International, ICRISAT and IRRI, Wallingford, UK. 636 pp.

Coutts J A, Hochman Z, Foale M A, McCown R L and Carberry P S 1998 Evaluation of participative approaches to RD&E: A case study of FARMSCAPE. Proceedings of the 9th Australian Society of Agronomy Conference: pp. 681–682 Australian Society of Agronomy: Wagga Wagga, NSW

Fukai S (Ed.) 1993 Intercropping – bases of productivity. (Special issue) Field Crops Res 34, 239–467.

Hammer G L, Carberry P S and Stone R 2000 Comparing the value of seasonal climate forecasting systems in managing cropping systems. *In* Applications of Seasonal Climate Forecasting in Agricultural and Natural Ecosystems – The Australian Experience. Eds. Hammer GL, Nicholls N and Mitchell C, pp. 183–195. Kluwer Academic Publishers, Dordrecht, The Netherlands.

Ikombo B M, 1984 Effects of farmyard manure and fertilizers on maize in semi-arid areas of Eastern Kenya. East. Afr. Agric. For. J. 44, 266–274.

Jackson P, Robertson M J, Cooper M and Hammer GL 1996 The role of physiological understanding in plant breeding: From a breeding perspective. Field Crops Res. 49, 11–37

Keating B A and Meinke H 1998 Assessing exceptional drought with a cropping systems simulator: A case study for grain production in north-east Australia. Agric. Syst. 57, 315–332.

Keating B A, Carberry P S and Robertson M J, 1999 Simulating N fertilizer response in low-input farming systems 2. Effects of weed competition. ESA Symposium on "Modelling Cropping Systems", Spain, June 1999.

Kirk G J D 1999 Modelling root-induced solubilization of nutrients. *In* Food Security in Nutrient-Stressed Environments: Exploiting Plants' Genetic Capabilities, Summary and Recommendations of an International Workshop, 27–30 Sep 1999, ICRISAT, Patancheru India. Ed. JJ Adu-Gyamfi, pp. 62–63. ICRISAT and JIRCAS, Patancheru.

Meinke H, Carberry P S, Cleugh H, Poulton P L and Hargreaves J N G H 2001 Modelling crop growth and yield under the environmental changes induced by windbreaks 1. Model development and validation. Agroforestry Syst (in press).

McCown R L, Keating B A, Probert M E and Jones R K, 1992. Strategies for sustainable crop production in semi-arid Africa. Outlook Agric. 21, 21–31.

McCown R L and Cox P G 1994 Agric Syst research in Africa and Australia: Some recent developments in methodology. *In* Soil Fertility and Climatic Constraints in Dryland Agriculture. Eds. E T Craswell and J Simpson, pp. 125–133. Australian Centre for International Agricultural Research, Canberra.

McCown R L, Hammer G L, Hargreaves J N G, Holzworth D P and Freebairn D M, 1996 APSIM: A novel software system for model development, model testing, and simulation in agricultural research. Agric. Syst. 50, 255–271.

McCown R L, Carberry P S, Foale M A, Hochman Z, Coutts J A and Dalgliesh N P 1998 The FARMSCAPE approach to farming systems research. Proceedings of the 9th Australian Society of Agronomy Conference. Eds. D L Michalk and J E Pratley, pp. 633–636. Australian Society of Agronomy: Wagga Wagga, NSW.

Muchow and Bellamy J A (Eds) 1991 Climatic risk in crop production: Models and management in the semiarid tropics and subtropics. CAB International, Wallingford, UK. 548 pp.

Nelson R A, Dimes J P, Paningbatan E P and Silburn D M, 1998 Erosion/productivity modelling of maize farming in the Philippine uplands. Part I: Parameterising the Agricultural Production Systems Simulator. Agric. Syst. 58, 129–146.

Nye P H and Tinker P B 1977 Solute Movement in the Soil-Root System. Blackwell Scientific Publications: Oxford.

Probert M E and Okalebo J R 1992 Effects of phosphorus on the growth and development of maize. *In* A Search for Strategies for Sustainable Dryland Cropping in Semi-arid Eastern Kenya. Ed. M E Probert, pp. 55–62. ACIAR Proceedings No. 41. Australian Centre for International Agricultural Research, Canberra.

Probert M E, Okalebo J R and Jones R K 1995 The use of manure on smallholders' farms in semi-arid eastern Kenya. Exp. Agric. 31, 371–381.

Probert M E, Dimes J P, Keating B A, Dalal R C and Strong W M 1997 APSIM's water and nitrogen modules and simulation of the dynamics of water and nitrogen in fallow systems. Agric. Syst. 56, 1–26.

Probert M E, Carberry P S, McCown R L and Turpin J E 1998 Simulation of legume-cereal systems using APSIM. Aust. J. Agric. Res. 49, 317–327.

Robertson M J, Carberry P S and Lucy M 2000a Evaluation of a new cropping option using a participatory approach with on-farm monitoring and simulation: A case study of spring-sown mungbeans. Aust. J. Agric. Res. 51, 1–12.

Robertson M J, Carberry P S, Wright G C and Singh D P 2000b Using models to assess the value of traits of food legumes from a cropping systems perspective. *In* International Food Legumes

Conference September 1997. Ed. R L Knight, pp. 265–278. Adelaide, Australia.

Rötter R and Van Keulen H 1997 Variations in yield response to fertilizer application in the tropics: II. Risks and opportunities for smallholders cultivating maize on Kenya's arable land. Agric. Syst. 53, 69–95.

Shamudzarira Z, Robertson M J, Mushayi P T, Keating B A, Waddington S, Chiduza C and Grace P 1999 Simulating N fertilizer response in low-input farming systems 1. Fertilizer recovery and crop response. ESA Symposium on "Modelling Cropping Systems", Spain, June 1999.

Shapiro B I, Osborne D and Ahmed M, 1999 Ex-ante impact assessment and economic analysis of breeding for nutrient efficiency and alternate strategies, a conceptual model and research issues: The socio-economists' view. *In* Food Security in Nutrient-Stressed Environments: Exploiting Plants' Genetic Capabilities, Summary and Recommendations of an International Workshop, 27–30 Sep 1999, ICRISAT, Patancheru India. Ed. JJ Adu-Gyamfi. pp. 24–26. ICRISAT and JIRCAS, Patancheru

Tanner P D and Mugwira L 1984 Effectiveness of communal area manures as sources of nutrients for young maize plants. Zimb Agric J 81, 31–35.

Willey R W 1979a Intercropping – its importance and research needs. Part 1. Competition and yield advantages. Field Crop Abst. 32, 1–10.

Willey R W 1979b Intercropping – its importance and research needs. Part 2. Agronomy and research approaches. Field Crop Abst. 32, 73–85.

J.J. Adu-Gyamfi (Ed.), *Food security in nutrient-stressed environments: exploiting plants' genetic capabilities*: 331–335, 2002.

Combining genetic improvement with natural resource management

R. M. Welch[1] & C. Johansen[2]
[1] *United States Department of Agriculture-Agricultural Research Service (USDA-ARS), U.S. Plant, Soil, and Nutrition Laboratory, Cornell University, Ithaca, New York, NY 14853-2901, USA*
[2] *Natural Resource Management Program, International Crops Research Institute for the Semi-Arid Tropics (ICRISAT), Patancheru 502 324, Andhra Pradesh, India*

Introduction

Ultimate success in exploiting the genetic capabilities of plants to grow in nutrient-stressed environments of the semi-arid tropics (SAT) requires a holistic view of food systems to ensure that genetic selections for improved yields on nutrient-poor soils will actually be adopted by farmers. This session focused on using existing knowledge of soil fertility and plant nutrition to develop farming technologies that could increase crop productivity on the nutrient-depleted soils of small-farm holders in the SAT. Knowledge gaps were identified for future research opportunities in low-nutrient input farming systems. Additionally, the important role that simulation models can play in synthesizing existing knowledge, determining future research and development needs, and in providing information applicable to policy and economic decision-making was discussed. Brief overviews of the papers presented during this session follow.

Appropriate farm management practices for low-nutrient soils

Because of poorly developed transportation and market infrastructures, limited access to credit facilities and the small size of land holdings, frequent and adequate use of mineral fertilizers in the tropics is not an option for most farmers. Thus, lower cost options supplying essential nutrients need to be developed if smallholder farmers are to achieve increased crop productivity on infertile soils. Some potential options for improved nitrogen (N) and phosphorus (P) availability and the control of aluminium (Al) toxicity in acid soils in the region were presented.

Nitrogen

Various cropping techniques have been employed to increase available N in soils such as green manures, legume rotations, or incorporating leguminous shrubs or trees in fallow land. Because of the harvest off-take of grain or fodder legumes, most of these rotations do not significantly increase net soil-N. Most effective of the grain legumes tested were pigeonpea and cowpea. The use of green manure crops contributes significant quantities of N to subsequent crops but causes farmers to forego one or more marketable crops. Tree legumes, herbaceous legumes, and agroforestry options can provide net increases in soil-N under some circumstances. However, simultaneous agroforestry systems, including alley cropping techniques, may result in excessive competition with adjacent seasonal crops especially in nutrient-poor soils under low moisture conditions. More attention should now be given to sequential cropping systems (e.g., relay cropping, 1–3-year leguminous fallows). Fallow duration is an important factor, but requires more land so is not an option for typical smallholders in regions of increasing human population. Therefore, N fertilizers are necessary in many cases. Possibly, marginal lands not currently cropped (if available) should be planted with legumes and the biomass produced used as fodder or as organic matter for soil enrichment.

Phosphorus

Infertile Oxisols and Ultisols do not contain enough soil-available P to allow maximum production of available crop varieties in the SAT. Thus, additions of inorganic sources of P are essential to good soil fertility. Crops such as pigeonpea and groundnut are capable of accessing poorly available P in the soils of these regions. There is also substantial genetic diversity in P use efficiency among most cultivated crops grown in these regions that has not been exploited fully. However, selecting for P-efficient crops will not sustain crop productivity indefinitely. Mining of the soil-P by P-efficient crops will ultimately deplete the available soil-P stores and exacerbate the nutrition for less P-efficient ones. Burning forests can increase available soil-P levels for several years, but this technique is not sustainable because successive crops will eventually deplete native P supplies in the soil. Phosphorus-rich manures from off-farm sources can also contribute to soil-P fertility but this is often impracticable for smallholder farmers and causes a transfer of P from surrounding, perhaps more fertile, areas.

Ultimately, sources of inorganic P must be applied to maintain or increase soil-P fertility. Choosing to use either processed P fertilizers or indigenous phosphate rocks is an economic decision based on both relative agronomic effectiveness and the cost of phosphate sources. If processed P fertilizers are not an option, then moderately soluble indigenous phosphate rock sources can be used if applied to acidic soils. However, poorly soluble rock sources must either be processed or used exclusively in perennial crop or permanent pasture systems where slow P release is desirable. Options for processing P-rock sources include local processing of superphosphate, partial acidulation, and microbial solubilization. Production of superphosphate requires some industrial capacity and relatively high-grade P-ores that are not normally available in many regions. Low-grade P-ores can be partially acidulated but this also requires some industrial capacity. Microbial processing has yet to be demonstrated as effective in field experiments using community-scale operations. Better recycling of crop residues from P-efficient annual crops could be an option worth pursuing. Possibly, the use of microbial-P solubilizers as seed inoculants could be developed for tropical agricultural systems because modest success using this technique in temperate agricultural regions has been reported.

Aluminium toxicity

Liming soils is seen as the ultimate solution to Al toxicity in acid soils. The high cost of transporting lime to farmers' fields from lime deposits is seen as the major constraint in treating acid soils in the tropics. Increasing soil organic matter status, permitting chelation of free Al, is an option worthy of further investigation in the tropics. Breeding for Al tolerance has not been a major goal in plant breeding efforts in this region. Recent evidence suggests that it may be possible to genetically transform crop plants such that they exude greater quantities of the organic acids that neutralize the effects of Al toxicity. Such approaches would indeed better adapt crop plants to acid soil conditions, but would not necessarily improve the soil condition itself (except perhaps by allowing greater biomass production and consequent return of organic matter to the soil). While genetic enhancement of crop productivity constraints offer prospects for the medium term, the long-term goal of improving the soil condition, by lime or organic matter addition, must remain.

The problems associated with low-nutrient status and Al toxicity in tropical regions will not be solved easily. Ultimately, locally available and affordable sources of both P and N fertilizer must be developed. The use of organic-based agricultural systems together with efficient use of moderate levels of inorganic fertilizers offer the most promising means of dealing with soil fertility constraints in the tropics.

Farmer preferences for crop varieties in low-nutrient environments

Both low soil fertility and inadequate resources are primary limitations to improving crop productivity in the developing world. Integrating farmer preferences with cropping system approaches were viewed as effective means of attaining sustainable and acceptable production levels on nutrient-poor soils. Understanding farmer preferences and multiple selection criteria are seen as very important factors in achieving farmer acceptance of new technologies. Farmer decision-making processes should be studied along with genetic improvements in crops adapted to nutrient-poor environments to maximize impact. To meet multiple farmer objectives a purposeful combination of nutrient-efficient crops and improved cropping system designs (including long-duration and short-duration genotypes) were employed in a case study in Malawi.

Important interactions were shown to exist between crop yield, labour requirements, and crop tolerance to infertile soils in legume-intensification systems. Using a matrix of grain yields versus adaptation to nutrient-deficient conditions, it was shown that farmers could accept lower yields if labour demands were low enough and if secondary benefits were high enough (e.g., significant improvements in soil nutrient status and availability of fuel wood). There was a close association between high yield potential and labour demand.

Mucuna spp. can suppress weeds and produce high biomass and seed yields when grown on relatively nutrient-poor soils; its propagation requires relatively low labour demands. However, because mucuna grain is toxic it is only useful as a green manure or in cover-crop rotations. Intercropping systems were designed to meet farmer criteria for consistent yields, low labour demands, and tolerance to nutrient-poor soils. Combined with genetic improvements, this approach appears to have many advantages because purposeful combination of species and growth duration in intercropping system maximizes resource exploitation and reduces the risk of crop failure.

Other studies in Malawi in on-farm trials were carried out using a satellite trial design where system performance (evaluated by researchers) was compared to farmer assessment. Intercrops showing promise included long-duration pigeonpea genotypes with either short-duration groundnut genotypes or short-duration soybean genotypes. Farmers showed great interest in short-duration groundnut/long-duration pigeonpea intercrops. Limitations to farmer acceptance included concerns for the high cost and short supply of legume seeds.

The soil 'mining' issue associated with nutrient-efficient genotypes was addressed. Intercropping with long-duration pigeonpea can reduce soil erosion and significantly improve nutrient balance with little or no fertilizer additions. Both P and N budgets obtained from on-farm measurements suggest that soil 'mining' of nutrients by crops can be effectively balanced by reducing soil erosion and improving retention of available nutrients. It was concluded that intensification of cropping systems with low-nutrient adapted, long-duration leguminous species will improve the natural resource base.

Nutrient balances: A guide to improving the management of sorghum- and groundnut-based dryland cropping systems in semi-arid tropical India

Nutrient balances (N, P and K) for farmers fields in two dryland cropping systems (sorghum-based and groundnut-based) on Alfisols were carried out in an Indian SAT region using geographical information system (GIS) analysis and crop statistics. Determinations were made for mineral fertilizer and organic manure inputs from farmer interviews. In addition, nutrient removal was also determined from harvested products and residue removal. Soil samples from farmers fields were analyzed for mineral N, available P, and exchangeable K content.

Overall nutrient balances were calculated for a few cropping systems. In a sorghum–castor bean rotation, N balance was positive with a net annual gain of 10 kg ha^{-1}. However, N losses were not determined in this study and they may have been significant. Balances of P and K were also estimated as showing a positive P and a negative K balance for the cropping system. In an improved sorghum/pigeonpea–castor system, N and P were in positive balance, while K was estimated to be in negative balance. The positive N balance did not include biological N fixation that could have contributed significant amounts of N. Other cropping systems were also investigated.

Similar nutrient balances were seen in a long-term cropping systems experiment at ICRISAT. In a continuously cropped groundnut system, positive P and negative K balances were noted. If biologically fixed N was not included in the balance calculations, negative N balances resulted. If only 50% of the groundnut-N uptake came from fixed N, the N balance was also negative. In farmers' field trials (20 field sites) biologically fixed N ranged from 41 to 65%, averaging 53%. Positive N balance would require that >68% of the N come from biological N fixation. After 40 years of continuous groundnut monocropping, no significant N accumulation in the soils was found. Low soil biological N-fixation rates may be one reason contributing to the poor accumulation of soil-N in the region. Thus, more efficient N-fixing *Rhizobium* strains need to be used for legume root inoculation in this region.

The use of the stable isotope, ^{15}N, should be expanded in research to determine N losses precisely in different cropping systems in farmers' fields. Doing so would provide valuable data needed to understand the dynamics of N balance in these cropping systems. The use of P fertilizer inputs apparently meets the crop requirements in most cropping systems studied, but K and N inputs need to be increased in dryland cropping systems to off-set losses from nutrient mining. In maize and cotton systems, positive nutrient balances were attributed to farmers' preferences for fertilizer application to these high-value crops.

The reported studies gave only a broad estimate of micronutrient balance. The study pinpointed the various knowledge gaps in sources of gain or loss of nutrients, that will require further component research to quantity. Further, such balances provide only a static picture of the situation. It is desirable to understand the long-term dynamics of nutrient flows in the target systems. This is now becoming more feasible to attempt with the advent of more user-friendly crop and nutrient models.

Preplant moisture and fertility conditions as indicators of high and stable yields in rain-fed cropping systems

Cropping systems dependent on rain are vulnerable to weather fluctuations. Data was collected from experiments on long-term maize, sorghum, and soybean rotations (monoculture and 2-year rotations) in the Corn (maize) Belt of Nebraska in the USA.

Standard Precipitation Indices (SPI), indicative of preplant moisture, were calculated and used to quantify yearly drought conditions. Moisture status was determined by using 8-month (September to April) SPI values. A SPI value of zero indicated average conditions while positive values greater than or less than 2 indicated extreme conditions associated with weather events that occurred only 5% of the time. Hypothetically, favorable preplant moisture conditions could be responsible for increased soil microbial activity resulting in early crop establishment and higher yields.

Maize and soybean yields were sensitive to preplant soil water content with moist soil producing higher yields. However, sorghum did not respond to preplant moisture conditions. Nitrogen fertilizer effects varied with crop and with preplant moisture conditions. Yields of sorghum in monoculture (and not in rotation with soybean) increased in response to N fertilizer additions while maize responded to N fertilizers in both monoculture and in rotation with soybean. Soybean did not respond to N fertilizers. Maize yield normally increases in response to N application under moist preplant conditions. However, sorghum yields usually were reduced by 2–8% when soil moisture was high before sowing.

Regression models were constructed to predict crop yields from weather data. The most important weather data used included precipitation and temperature. A second-order regression equation that included preplant SPIs as independent variables and subsequent crop yields as dependent variables, described maize and soybean yields in rotation satisfactorily. Sorghum yields were not affected by preplant moisture conditions. The resilience of sorghum to adverse weather conditions may be the result of certain peculiar anatomical characteristics of the plant. The regression model provided reasonable predictions between actual maize yields and predicted yields. Possibly, the utility of the model constructed from this study could be applied to provide warnings to farmers and policy makers of impending droughts so that strategies could be implemented to mitigate their adverse effects.

Role of modeling in improving nutrient efficiency in cropping systems

Simulation modeling in conjunction with experimental research was used to identify a development path for subsistence agriculture in the semi-arid region of Kenya. Manure- and legume-based strategies to augment traditional soil enrichment practices along with modest amounts of fertilizer inputs was judged economically feasible for many farmers. Possibly, this approach provides the best prospects for food security in this region. Simulation modeling allowed for the assessment of fertilizer use in a way that was not possible using only traditional field experiments. Research confirmed the value of fertilizer use in Kenya and the applicability of the models to address such research questions. Unfortunately, using these findings to actually change current farming practices has not yet been possible. Considerable follow-up with extension-oriented projects is required. Future research areas identified for focus by the models included:

- Develop more efficient means of utilizing manure on crops (including capturing, storing, and using).
- Determine which Fertilizer-Augmented Soil Enrichment (FASE) strategies should be used to improve the efficiency of use of chemical fertilizers combined with manure and crop residue applications.
- Transfer results obtained through research to legume systems.
- Carry out research to determine the most economical applications of FASE strategies for different price-cost scenarios.
- Contribute input into national policies on commodities, food security, and fertilizers.
- Make sure that fertilizer investments remain on the policy agenda.

The models developed could be used not only in the areas of identifying future research activities but also as a source of inputs into economic and policy analyses. Both research activities to understand the biophysical system and operations research (using current understanding) to predict likely outcomes for alternative management farming system strategies can be addressed using the models.

Interventions via policy action could also be developed by using simulation modeling. Whether or not direct interventions with farmers are possible, or if simulation models can contribute beyond the research phase to improve farming practices are unknown. In Australia, simulation models have contributed to farmer exploration and to their acceptance of practice change. The challenge is to determine if this can be applied to resource-poor smallholder farmers.

The current systems model, Agricultural Production Systems Simulator (APSIM) was described particularly in regard to current capabilities in addressing important issues. Plans are being made to explore the worth of APSIM in addressing the use of manure and legumes in cropping systems. Some attention will also be given to using the model to identify how plants' genetic potential might be exploited to improve yields in low-nutrient input farming systems.

Conclusion

In the short term, the use of legume crops in rotations with staple food crops, improved utilization of manure and crop residues, and genetic improvement in crop nutrient efficiencies can all help improve N and P balances for crop production in the SAT. However, in the long term these technologies will not provide the needed P and other nutrient inputs required to sustain production for the distant future. Thus, there is a great need to develop more affordable, locally available sources of N, P, and K fertilizer for farmers in the SAT.

Clearly, more emphasis should be given to transferring technology obtained from basic research and applied research findings to farmers if sustainable ways to improve food security in this region are to be found. This will require closer cooperation between agricultural researchers and other experts studying food system functionality. Getting farmers to adopt new technologies directed at improving nutrient-use efficiency in crops and more sustainable low-input agricultural practices must consider the food system in its entirety. Without farmer participation and technology adoption, research findings will go untried. More holistic food system approaches should be encouraged whereby agricultural scientists interact with other specialists (such as sociologists, economists, and policy makers) to ensure that their recommendations are practical and adopted by farmers. These new practices need to have a sustainable impact on improving food security for these regions of the world with minimum adverse impact on the environment.

J.J. Adu-Gyamfi (Ed.), Food security in nutrient-stressed environments: exploiting plants' genetic capabilities: 337–338, 2002.

Closing remarks

T. Hamazaki
Director, Environmental Resources Division, Japan International Research Center for Agricultural Sciences (JIRCAS), 1–2 Ohwashi, Tsukuba, Ibaraki 305-8686, Japan

On behalf of the Director General of the Japan International Research Center for Agricultural Sciences (JIRCAS), and the Ministry of Agriculture, Forestry and Fisheries (MAFF), I would like to thank all the participants of the workshop for their contributions. I am also delighted that Dr L. D. Swindale, who helped initiate the Project in 1984 and who today is at ICRISAT as the Director General, was able to be with us to open the meeting.

As you are all aware, this workshop was held to culminate the activities of the Government of Japan Special Project at ICRISAT. Many of the scientists who were involved with the Project are here today and join me in thanking the Management of ICRISAT for their unflinching support to them and their families whilst they were here in India.

During the 15 years of the Project its scientists conducted strategic and basic research aimed at improving crop nutrient and water uptake and utilization efficiencies through field management practices, and/or by exploiting morphological and physio-genetic systems to improve the adaptation of ICRISAT mandate crops to the low-nutrient and low-moisture environments of the semi-arid tropics (SAT). The Project also trained Research Fellows, Visiting Scientists, and Scholars and improved the creativity and job competence of its support staff.

The valuable research achievements of the Project from 1984 to 1999 have been highlighted in their presentations by the Project scientists. Among these achievements, the discovery that piscidic acid is exuded by pigeonpea and that this compound enhances pigeonpea's ability to take up more phosphorus (P) from iron phosphates in the soil is highly significant. As a follow-up to Phase I, further studies to examine genotypic variability in the solubilizing activity of root exudates were conducted here at ICRISAT and at the National Institute of Agro-Environmental Sciences (NIEAS) in Japan. These led to Dr Ae proposing the new concept of 'contact reaction' which postulates the role of root cell walls in nutrient acquisition by groundnuts and pigeonpea.

I am delighted that during Phase III, there was an effort to identify physiological and agronomic traits to help improve the efficiency of nutrient uptake and use by ICRISAT mandate crops. Because crop productivity in the SAT is often limited by the lack of available nutrients (particularly P), enhancing the efficiency of crops' ability to exploit sparingly available nutrients from the soil will eventually accelerate the adoption of improved cultivars by farmers, thus improving productivity in these harsh environments where agriculture is such an unpredictable enterprise.

Being a pedologist, I had little previous knowledge of nutrient physiology, or water management, but I have learned a lot since I took over as the Coordinator of GoJ Special Projects at ICRISAT. I am glad to know that ICRISAT has started to work on cross-programme projects that aim to exploit the joint research application of genetically based, statistical, crop-physiological models to overcome the current limitations to breeding for low-nutrient environments.

The current workshop solicited a range of scientists' views on whether it is socio-economically viable, and environmentally sustainable to breed for enhanced crop productivity for low-nutrient environments. Questions have been raised about the sustainability of crop varieties that effectively exploit nutrients in low-nutrient environments. Will these varieties mine the soils and further exacerbate land degradation? This question has been well addressed during the meeting.

Major breakthroughs have been achieved in breeding for resistances to pests, diseases, and drought. It is now time to seriously consider breeding for nutrient acquisition and use efficiency. Recent studies by Drs Wissuwa, Yano, and Ae reported in Theoretical and Applied Genetics, 1998 on mapping quantitative trait loci (QTL) for P-deficiency tolerance in rice provide evidence that could have application in helping to enhance P-use efficiency in ICRISAT mandate crops. This potential is encouraging in the light of the use of modern molecular biological tools to enhance P-use efficiency as demonstrated by other workshop participants.

Apart from improving nutrient use efficiencies through genetic manipulation, there is also an urgent need to fine-tune the technologies available for improving field management practices to enhance crop yields. Soil fertility management and enhancement are vital to food security in western and southern Africa. Pigeonpea was identified as a unique crop capable of increasing the pool of available P in the cropping systems in which it is grown. Besides, it is documented that pigeonpea leaf fall contributes to 4–7 kg P ha^{-1} during the cropping season. There is a need to quantify the advantages of pigeonpea over indigenous and such introduced leguminous crops as cowpea, groundnut, and mungbean in relation to soil fertility replenishment. The deep-rooting ability of pigeonpea allows the crop to exploit water from lower soil layers. But, the problem of ensuring that efficient crops reach smallholder farmers who need them is a key bottleneck that remains to be addressed.

Clearly, despite the advances made by this Project during the past 15 years, there is a need for more research into the decomposition of piscidic acid in Alfisols and the biochemical and molecular mechanisms of P acquisition. Though Phase III identified physiological traits responsible for P uptake and use efficiency, there is still a need to develop mapping populations, or to identify previously existing mapping populations, based on pairs of adapted parents of each crop that differ greatly in P uptake ability and/or use efficiency. Such pairs could be used to map the QTL that contribute to components of these traits.

An important feature of the GoJ Project work was its interdisciplinary approach. During all three Phases the team members were chosen for their appropriate blend of disciplines.

In conclusion, the GoJ Project based at Patancheru helped to enhance the productivity of upland crops in the SAT through research, and enhanced our knowledge base of mechanisms of P and N acquisition by crops. The Project also helped to strengthen interdisciplinary efforts among scientists from Japan and their counterparts in ICRISAT. Once again, on behalf of MAFF, and the Director General of JIRCAS, Dr Maeno, I would like to thank ICRISAT's Management and Staff for their cooperation and support to the Project scientists. I thank all the Project Scientists, Drs Arihara, Ae, Okada, Ito, Matsunaga, Katayama, Tobita, Nakamura, and Nakano for their dedication and hard work. I thank the Organizing Committee for their efforts to ensure high standards of presentation and discussions. I wish to express my sincere thanks to all the workshop participants for the kindness with which they shared their knowledge and experience with us. Thanks are also due to Drs Adu-Gyamfi and Ishikawa, who coordinated this culmination of 15 years' successful work. My special thanks to those who assisted the Japanese teams in the field and laboratories; without their help it would have been impossible to attain the results presented in the three Project workshops. At last, but not least many thanks to Dr Johansen, for the important role in ensuring the excellent research results throughout the Project. I trust that these achievements highlighted in the commemorative publication, will not be an end, but a means to an end.

J.J. Adu-Gyamfi (Ed.), Food security in nutrient-stressed environments: exploiting plants' genetic capabilities: 339–340, 2002.

Future research needs, priorities, and strategies

J.J. Adu-Gyamfi
International Crops Research Institute for the Semi-Arid Tropics (ICRISAT), PMB 3491, Kano, Nigeria

During the final workshop discussion sessions it was agreed that optimizing phosphorus (P) nutrition of sorghum, pigeonpea, groundnut, and cowpea is a high priority area that requires immediate attention. Whereas considerable progress has been made in breeding for pest and disease resistant cultivars of crops, little has been achieved in breeding for nutrient uptake and use by crops. In the low-P, adverse rainfall regions of the semi-arid tropics (SAT), no improved cultivar has a reasonable chance of achieving substantial and sustainable yield levels in farmers' fields unless the critical constraint of P deficiency is addressed by adapting crops to the existing soil conditions.

Consequently, the following research needs were identified by participants as priorities. These research needs build on achievements of the 15 years' research activities of the Government of Japan Project at the International Crops Research Institute for the Semi-Arid Tropics (ICRISAT) Patancheru, India.

Researchable issues

- The need to exploit mechanisms by which wild species of some of ICRISAT's mandate crops are adapted to low-P conditions. Screening for adaptation of crops to low-P soils has been done on-station rather than in the target environments where P bioavailability is low. It was recommended that future projects should screen and select from large collections of genotypes (including wild relatives) of pigeonpea, groundnut, cowpea and sorghum for their capacity to produce high yields in both P-sufficient and P-deficient field conditions in the Sahel and Guinea Savanna zones.
- Carboxylic anions from root exudates play a critical role in enabling dessert grasses or wild relative species to survive in low-P environments. The current methodology used to screen large number of germplasm accessions for their P-solubilizing ability is laborious and requires sophisticated and expensive equipment. It was therefore recommended that future projects focus on developing rapid, simple and easy screening systems to quantify the P-solubilizing activity of carboxylic anions from the root exudates of a large number of genotypes. Such prospects should examine genotypic variability in the mechanisms by which these crops solubilize P from rock phosphates. The degradation of carboxylic and piscidic acids in soils needs to be studied in detail. Their physical properties and metal stability constants, genetic variability in their synthesis or excretion, and its quantification in the rhizosphere all need to be established.
- With the evolution of new molecular biotechnological tools the need to create transgenic plants that over-express enzymes of the biosynthetic pathway for organic acid anions, and to test their exudation pattern was identified. Particular genes can then be mapped to a genome of the crops in question as future molecular markers.
- Pigeonpea has been shown to increase the pool of available P in the cropping systems in which it is grown. It is documented that the leaf fall contributes to 4–6 kg P ha^{-1} during the cropping season. Quantifying the advantages of pigeonpea over indigenous and introduced leguminous crops (e.g., cowpea, groundnut) in relation to soil fertility replenishment was identified as a researchable area.
- The advantage of pigeonpea over other crop species in its ability to utilize iron-bound P and/or locally available P sources of rock phosphate in cropping systems under field conditions needs to be quantitatively evaluated. This would involve quantifying and documenting the nutrient content of fallen leaves, and measuring the rate of decomposition of the fallen litter, and the amount of nutrients (N and P) released into the soil.
- The development of a simple and easy screening system for target traits that would improve the N-fixation ability of leguminous crops, and permit

quantification of their N-fixing abilities in cropping systems in on-farm conditions was identified as another area for future research.

- On-farm experimentation using a farmer participatory approach to the intensification of pigeonpea from the view point of biologically cycled P in cereal/legume cropping systems in order to improve the available P in such systems was highlighted.
- Socio-economic and risk analysis of pigeonpea-based technologies and identification of the critical factors influencing farmers' preference of pigeonpea varieties were identified as worthy of attention.

The target regions for the proposed research are western and southern Africa. Delegates endorsed the suggestion that a project proposal that builds on the achievements of the Government of Japan Project, and on the intensive discussions during the workshop, be developed and submitted for funding to a range of donors.

J.J. Adu-Gyamfi (Ed.), *Food security in nutrient-stressed environments: exploiting plants' genetic capabilities:* 341–344, 2002.

Participants

N. Ae
Division of Soil Sciences, National Institute of Agro-Environmental Sciences (NIAES), 3-1-1 Kannondai, Tsukuba, Ibaraki 305-8604 Japan
Tel 81 298 38 8271
Fax 81 298 38 8271
E-mail noriae@niaes.affrc.go.jp

J.J. Adu-Gyamfi
International Crops Research for the Semi-Arid Tropics (ICRISAT), Sabo Bakin Zuwo Road, PMB 3491, Kano, Nigeria
Tel 234 64 662050
Fax 234 64645352
E-mail j.j.adugyamfi@cgiar.org

M.Y. Ali
Bangladesh Agricultural Research Institute (BARI), On-Farm Research Division, Barind Station, Rajshshahi 6000, Bangladesh

J. Arihara
Department of Crop Physiology and Quality, National Agricultural Research Center (NARC), 3-1-1 Kannondai Tsukuba, Ibaraki 305-8666 Japan
Tel 81 298 38 8392
Fax 81 298 38 8392
E-mail arihara@narc.affrc.go.jp

A. Bationo
TSBF-CIAT, c/o World Agroforestry Center (ICRAF), United Nations Avenue, Gigiri, Box 30677, Nairobi, Kenya
Tel 254 252 4754
Fax 254 252 4763
E-mail: a.bationo@cgiar.org

P. Bramel
International Crops Research Institute for the Semi-Arid Tropics (ICRISAT), Patancheru, 502 324, Andhra Pradesh, India
Tel 91 40 2329 6161
Fax 91 40 2324 1239
E-mail p.bramel-cox@cgiar.org

P.S. Carberry
Commonwealth Scientific and Industrial Research Organisation (CSIRO), Tropical Agriculture Agricultural Production Systems Research Unit (APSRU), P O Box 102, Toowoomba, Queensland 4350 Australia
Tel 61 7 4688 1377
Fax 61 7 4688 1193
E-mail peter.carberry@tag.csiro.au

F.D. Dakora
Research Development, Cape Technikon, Room 2.8 Admin Bldg., P.O. Box 652, Cape Town 8000, South Africa
Tel 21460 3878
Fax 21460 3887
E-mail: dakora@ctech.ac.za

J. Dunlop
AgResearch Grasslands Research Centre, Private Bag 11008, Palmerston, North New Zealand
Fax 64 6 351 8032
E-mail dunlopj@agresearch.cri.nz

G.R. Findenegg
Wageningen Agricultural University, Broekemalaan 18, 6703 GM Wageningen, The Netherlands
Tel 31 317 412034
Fax 31 317 425675
E-mail g.findenegg@diva.nl

K. Fujita
Faculty of Applied Biological Science, Hiroshima University, 1-4-4 Kagamiyama, Higashi-Hiroshima 739-8528, Japan
Fax 81 824 24 0791.
E-mail fujiko@ipc.hiroshima-u.ac.jp

P. Gyaneshwar
Soil and Water Sciences Division, International Rice Research Institute (IRRI), MCPO Box 3127, 1271 Makati City, Philippines
Tel 63 2 845 0563
Fax 63 2 891 1292
E-mail g.prasad@cgiar.org

S.D. Hainsworth
Editorial and Publishing Services Plot 200, Road 14 Jubilee Hills, Hyderabad 500 033, Andhra Pradesh, India
Tel 91 40 235 44367
E-mail suehains@hd2.vsnl.net.in

T. Hamazaki
Environmental Resources Division, Japan International Research Center for Agricultural Sciences (JIRCAS), 1-2 Ohwashi Tsukuba, Ibaraki 305-0851, Japan
Tel 81 298 38 6306
Fax 81 298 38 6651
E-mail hamazaki@jircas.affrc.go.jp

C.T. Hash Jr
International Crops Research Institute for the Semi-Arid Tropics (ICRISAT), Patancheru,
502 324, Andhra Pradesh, India
Tel 91 40 2329 6161
Fax 91 40 2324 1239
E-mail c.t.hash@cgiar.org
cth8@cornell.edu

S. Ishikawa
National Institute for Agro-Environmental Sciences, 3-1-3, Kannondai, Tsukuba, Ibaraki, 305-8604, Japan
Tel 81 235 38 8315
Fax 81 235 38 8315
E-mail isatoru@niaes.affrc.go.jp

O. Ito
Japan International Research Center for Agricultural Sciences (JIRCAS), 1-2 Ohwashi Tsukuba, Ibaraki 305-0851, Japan
Tel 81 298 38 6306
Fax 81 298 38 6651
E-mail osamuito@jircas.affrc.go.jp

M.L. Izaguirre-Mayoral
Instituto Venezolano de Investigaciones Cientificas (IVIC), Laboratorio de Biotecnología y Virología Vegetal, Centro de Microbiología y Biología Celular Apartado Postal 21827, Caracas 1020-A Venezuela
Tel 58 2 5041189
Fax 58 2 5041382
E-mail mizaguir@pasteur.ivic.ve

H.H. Jensen
Royal Veterinary and Agricultural University KVL, Hojbakkegaard Alle 13, DK-2630 Taastrup, Denmark
Tel 45 35283391
Fax 45 35282175
E-mail hhj@kvl.dk

C. Johansen
Apartment 2B Paldale, Plot 6 Road 140, Gulshan-2, Dhaka, Bangladesh
E-mail ljohanse@citechco.net

G.J.D. Kirk
Soil and Water Sciences Division, International Rice Research Institute (IRRI), MCPO Box 3127, 1271 Makati City, Philippines
Tel 63 2 845 0563
Fax 63 2 845 0606
E-mail g.kirk@cgiar.org

L. Krishnamurthy
International Crops Research Institute for the Semi-Arid Tropics (ICRISAT), Patancheru,
502 324, Andhra Pradesh, India
Tel 91 40 2329 6161
Fax 91 40 2324 1239
E-mail l.krishnamurthy@cgiar.org

Jagadish Kumar
International Crops Research Institute for the Semi-Arid Tropics (ICRISAT), Patancheru,
502 324, Andhra Pradesh, India
Tel 91 40 2329 6161
Fax 91 40 2324 1239
E-mail j.kumar@cgiar.org

Sailaja Koti
International Crops Research Institute for the Semi-Arid Tropics (ICRISAT) Patancheru,
502 324 Andhra Pradesh, India
Tel 91 40 2329 6161
Fax 91 40 2324 1239
E-mail sailakoti@hotmail.com

D.R. Lukiwati
Faculty of Animal Agriculture, Diponegoro University, Campus Tembalang, Semarang, Indonesia, 50275
E-mail: dwiretno_lukiwati@yahoo.com

S. Madhavan
Department of Biochemistry, PO Box 880 664, University of Nebraska, Lincoln, NE 68588-0817, USA
Tel 1 402 472 2932
Fax 1 402 472 7842
E-mail msoundar@unlnotes.unl.edu

T. Nakamura
Japan International Research Center for Agricultural Sciences (JIRCAS), 1-2 Ohwashi Tsukuba, Ibaraki 305-8686, Japan
Tel 0298 38 6313
Fax 0298 38 6316
E-mail takuwan@jircas.affrc.go.jp

H. Nakano
Hokkaido National Agriculture Experiment Station (HNAES), Shinsei, Memuro, Kasaigun Hokkaido 082-0071 Japan
Tel 81 155 62 9297
Fax 81 155 62 2926
E-mail nakanohi@mmr.affrc.go.jp

R. Ortiz
International Institute of Tropical Agriculture (I ITA), Oyo Road, PMB 5320, Ibadan, Nigeria
Tel 234 2 241 2626
Fax 234 2 241 2221
E-mail r.ortiz@cgiar.org

J.M. Peacock
International Center for Agricultural Research in the Dry Areas (ICARDA), Arabian Peninsula Regional Program (APRP), PO Box 13979, Dubai United Arab Emirates
Tel 971 4 2957338
Fax 971 4 2958216
E-mail j.peacock-t@cgiar.org

T.J. Rego
International Crops Research Institute for the Semi-Arid Tropics (ICRISAT) Patancheru, 502 324 Andhra Pradesh, India
Tel 91 40 2329 6161
Fax 91 40 2324 1239
E-mail t.rego@cgiar.org

Z. Rengel
Soil Science and Plant Nutrition, Faculty of Agriculture, The University of Western Australia, Nedlands WA 6907 Australia
Tel 61 8 9380 2557
Fax 61 8 9380 1050
E-mail zrengel@agric.uwa.edu.au

O.P. Rupela
International Crops Research Institute for the Semi-Arid Tropics (ICRISAT), Patancheru, 502 324, Andhra Pradesh, India
Tel 91 40 2329 6161
Fax 91 40 2324 1239
E-mail o.rupela@cgiar.org

N.P. Saxena
International Crops Research Institute for the Semi-Arid Tropics (ICRISAT), Patancheru, 502 324, Andhra Pradesh, India
Tel 91 40 2329 6161
Fax 91 40 2324 1239
E-mail n.saxena@cgiar.org

B.I. Shapiro
International Crops Research Institute for the Semi-Arid Tropics (ICRISAT), Patancheru, 502 324, Andhra Pradesh, India
Tel 91 40 2329 6161
Fax 91 40 2324 1239
E-mail b.shapiro@cgiar.org

T.R. Sinclair
United States Department of Agriculture – Agricultural Research Service (USDA-ARS), Agronomy, Physiology, and Genetics Laboratory, University of Florida, PO Box 11095, Gainesville, FL 32611-0965 USA
Tel 352-392-6180
Fax 352-392-6139
E-mail trsincl@gnv.ifas.ufl.edu

F.W. Smith
Commonwealth Scientific and Industrial Research Organization (CSIRO), Tropical Agriculture Cunningham Laboratory, 306 Carmody Road St Lucia, Queensland 4067, Australia
Tel 07 3214 2288
Fax 07 3378 3214
E-mail frank.smith@tag.csiro.au

P.C. Smithson
International Centre for Research in Agroforestry (ICRAF), PO Box 30677, Nairobi, Kenya

Tel 254 2 521450
Fax 254 2 521001
E-mail p.smithson@cgiar.org

S.S. Snapp
Department of Horticulture, Michigan State University, East Lansing, MI 48824-1325, USA
Tel 1 517 355 5187
Fax 1 517 353 0890
E-mail snapp@msu.edu

S.P. Srivastava
Agricultural Research Station, Rampur Chitwan, Nepal
Tel 97 756 29301
Fax 97 715 21197

C. Sudha Rani
Regional Agricultural Research Station, Nandyal, Kurnool 518 501, Andhra Pradesh, India
E-mail sudhac@hotmail.com

L.D. Swindale
International Crops Research Institute for the Semi-Arid Tropics (ICRISAT), Patancheru, 502 324 Andhra Pradesh, India
Tel 91 40 23296161
Fax 91 40 2241239
E-mail l.swindale@cgiar.org

T. Tadano
Faculty of Agriculture, Hokkaido University, Kita-9, Nishi-9, Kita-ku, Sapporo 060-8589 Japan
Tel 81 11 706 3846
Fax 81 11 706 3846
E-mail tadano@chem.agr.hokudai.ac.jp

M. D. Winslow
International Crops Research Institute for the Semi-Arid Tropics (ICRISAT), Patancheru, 502 324, Andhra Pradesh, India
Tel 91 40 2329 6161
Fax 91 40 2324 1239
E-mail m.winslow@cgiar.org

R.M. Welch
United States Department of Agriculture – Agricultural Research Service (USDA-ARS), US Plant, Soil, and Nutrition Laboratory, Cornell University Ithaca, New York, NY 14853-2901 USA
Tel 1 607 255 5434
Fax 1 607 255 1132
E-mail rmw1@cornell.edu

A. Yamamoto
Experimental Farm of Tohoku University, Tohoku University, Miyagi 989-67, Japan
Tel 81 229 84 7311
Fax 81 22 251 1341
E-mail akiko@comax.net

C.F. Yamoah
Soil Research Institute, Council for Scientific and Industrial Research (CSIR), Academy Post Office, Kwadaso, Kumasi Ghana
Tel 233 51 50353
Fax 233 51 50308
E-mail soils@africaonline.com.gh

T. Yoneyama
National Agricultural Research Center (NARC), 3-1-1 Kannondai, Tsukuba, Ibaraki 305-8666, Japan
Tel 81 298 38 8392
Fax 81 298 38 8392
E-mail yoneyama@narc.affrc.go.jp

T. Yoshihara
Laboratory of Bioscience and Chemistry, Hokkaido University, Graduate School of Agriculture, Kita -9, Nishi- 9, Kita-ku Sapporo 060-8589 Japan
Tel 81 11 706 2505
Fax 81 11 706 2505
E-mail yosihara@chem.agr.hokudai.ac.jp

Y.G. Zhu
Department of Soil and Water, Collaborative Research Centre for Molecular Plant Breeding, Centre for Plant Symbioses, Waite Campus, The University of Adelaide, Glen Osmond, SA 5064 Australia
Tel 61 8 83037436
Fax 61 8 83036511
E-mail yongguan.zhu@adelaide.edu.au